Undergraduate Texts in Mathematics

Undergraduate Texts in Mathematics

Undergraduate Texts in Mathematics are generally aimed at third- and fourth-year undergraduate mathematics students at North American universities. These texts strive to provide students and teachers with new perspectives and novel approaches. The books include motivation that guides the reader to an appreciation of interrelations among different aspects of the subject. They feature examples that illustrate key concepts as well as exercises that strengthen understanding.

More information about this series at http://www.springer.com/series/666

Calvin Jongsma

Introduction to Discrete Mathematics via Logic and Proof

Springer

Calvin Jongsma
Dordt University
Sioux Center, IA, USA

ISSN 0172-6056   ISSN 2197-5604   (electronic)
Undergraduate Texts in Mathematics
ISBN 978-3-030-25360-8   ISBN 978-3-030-25358-5   (eBook)
https://doi.org/10.1007/978-3-030-25358-5

Mathematics Subject Classification (2010): 03-01, 05-01, 06-01, 03B05, 03B10, 03E20, 05C45, 05C10, 05C15, 06E30

This Springer imprint is published by the registered company Springer Nature Switzerland AG
The registered company address is: Gewerbestrasse 11, 6330 Cham, Switzerland

Preface

Discrete Mathematics is both very old and very new. Counting and arithmetic go back to prehistoric times, logic to the ancient Greek period. But *Mathematical Logic, Set Theory, Boolean Algebra,* and *Graph Theory,* among other areas of *Discrete Mathematics,* emerged only during the past two centuries. The field has surged during the current Information Age both because of its importance for digital technology and because high-speed computers make it possible to go beyond classical topics and methods. Modern telecommunication, transportation, and commerce all depend heavily on its results. Since the early 1980s, a number of mathematicians, computer scientists, and professional organizations have promoted *Discrete Mathematics* as the mathematics needed for our time. **Introduction to Discrete Mathematics via Logic and Proof** contributes to this trend from a basic logical perspective.

Topics Selected

Discrete Mathematics investigates matters like integers and networks instead of real numbers and continuous functions—things that involve *discrete* rather than *continuous* entities. This means that it encompasses a large collection of somewhat disjoint subjects, including algebraic structures, algorithms, combinations and permutations, discrete probability, finite state machines, formal languages, graphs and networks, induction and recursion, logic, relations, sets, and more. A *Discrete Mathematics* textbook can't hope to cover all of these topics without becoming either a superficial collage or an encyclopedic handbook. My response to this profusion of riches is to weave a semester's worth of core topics into a *less expansive but more integrated whole.*

We start by studying *Mathematical Logic* (Chapters 1–2) because it is so vital to understanding the structure and nature of mathematical proofs, one of the key goals of this text. Doing so also lays the formal groundwork for understanding mathematical developments in computer engineering. We then explore the proof techniques of mathematical and structural induction and definition by recursion (Chapter 3), topics central both to mathematics and computer science. We round out the first unit by looking at some matters connected to natural numbers and integers (*Peano Arithmetic,* divisibility).

The second part of the text focuses on fundamental topics in *Set Theory* and *Combinatorics* (Chapters 4–5). Beginning with elementary set-theoretic operations and relations, we then use them to study some significant counting techniques—also topics important both for mathematics and computer science. We next discuss numerosity/cardinality and venture a bit into the realm of infinite sets. Although this topic is further removed from the practical needs of computer scientists, its relevance to theoretical computability can be seen in its connection to the Halting Problem.

Assuming the framework of *Set Theory*, we then move into an area that is more algebraic in nature (Chapters 6–7). After looking at functions and relations in general, we investigate some algebraic structures that involve them. Equivalence relations give rise to partitions, which provide a theoretical foundation for treating both the integers and modular arithmetic using quotient structures. The theory of partial-order relations and lattices gives us the background for introducing *Boolean Algebra*, a topic that nicely ties together the earlier ones of *Propositional Logic* and *Set Theory*. While this material is more abstract and theoretically demanding, it also provides a satisfying conceptual basis for treating the more concrete topics of electronic circuits, Boolean functions, and K-maps.

The final unit looks at *Graph Theory* (Chapter 8). Like earlier topics, this one could be expanded into a course on its own, but we restrict our attention to a few key themes initially arising out of recreational interests— traversing edges and vertices, drawing planar graphs, and coloring maps. This concluding chapter also shows that *Discrete Mathematics* is a dynamic field of interest to practitioners and theoreticians alike.

Intended Audiences

Introduction to Discrete Mathematics via Logic and Proof is an outgrowth of my having regularly taught two mathematics courses on the intermediate undergraduate level—a *transition course* to prepare mathematics majors for proof-based upper-level courses, and a *discrete mathematics* course for computer science and engineering majors. At one point, given decreasing enrollments and some overlap of material, these courses were merged into one designed to meet the needs of both groups. Not finding a suitable textbook, I wrote up a set of class notes and exercises, which eventually grew to become this text. In one form or another, I've taught this content for about three decades with good success.

My mathematics students' ability to read and write proofs improved greatly from studying the foundational material on logic, and being exposed to *Set Theory*, relations, and functions gave them a strong foundation for understanding and constructing arguments in their upper-level courses. Those who went on to pursue graduate-level work in mathematics were well prepared for investigating more abstract mathematics. Computer science and engineering students found that the material related to their areas gave context and grounding to topics in their respective fields.

Over the past few years, an early edition of individual chapters of this text has been available online. In this form, tens of thousands of PDFs have been downloaded in over 150 countries, especially Chapters 6 and 7 by people interested in digital engineering and computer science. It is my hope, therefore, that this text will assist those interested in computer-related fields as well as those continuing in mathematics.

Goals and Approach

My goals for students in writing this text, and the approach I've taken to achieve them, can be summarized as follows:

1. *To Learn How to Read and Write Proofs*

   A principal goal of this text is to help students learn the critical skill of reading and writing proofs. There are widely divergent opinions about how best to accomplish this. Earlier, many of us learned to do proofs by the *sink-or-swim method*. We were thrown into an upper-level course where the content was abstract and the methodology was proof-based, and we were expected to pick up the appropriate mathematical habits and techniques by osmosis. Watching a sage-on-the-stage perform clever deductive maneuvers with complex concepts, we then tried it on our own in the homework and on the tests. Those of us who were able to wend our way through the haze and finally decipher the essentials of the logical process were deemed worthy of continuing in the field; those who didn't survive the ordeal were judged not to have had what it takes to succeed.

   Today's student apprentices rightly expect a more welcoming experience. Learning how to evaluate and construct proofs can still be challenging, but reflecting on the deductive process itself in the context of familiar and elementary content makes this less threatening. This is what I aim to achieve in the somewhat leisurely but systematic study of logic in the first unit of the text. It's likely that students will have already seen proofs and made some of their own, but to prepare them for creating increasingly formal and rigorous arguments, it is important to spend time learning how this is done, figuring out what makes deductions work.

   Because the logical approach taken in this text is somewhat unique, I'd like to say a few more words about its character and value. Mathematicians are mostly familiar with logic as the *foundational system* developed by Bertrand Russell around 1900, in which form it appears to be a *specialized foundational study of tautologies and axiom systems*. As such, it offers relatively little of interest to most practicing mathematicians. But there is a more fundamental way to conceptualize logic, better known to contemporary philosophers, that goes back to Aristotle and that was developed in a modern form by mid-twentieth-century Polish logicians. This approach takes logic as the *systematic study of conclusive argumentation*. Here tautologies are barely considered, the focus instead being on *rules of inference* that validate proofs of conclusions from premises. This makes a study of logic essential to *mathematical methodology*, providing students with a look under the deductive hood of mathematics, so to speak.

   This *natural deduction approach to logic* equips students to construct deductions the way mathematicians and others argue all the time. The text uses proof diagrams at the outset to help students become familiar with the various proof strategies connected to logic, but this scaffolding is gradually abandoned and proofs are written in a more standard informal manner. Coupled with lots of practice, students internalize the techniques they're learning and begin to master the basics of proof construction and better comprehend the deductive organization that mathematics is known for. And, as they see how logic undergirds various proof techniques, students begin to enjoy doing proofs instead of being paralyzed by their fear of them. Knowing some logic gives students insight into and power over the inferential code, just as knowing proper grammar can improve writing.

To meet this goal of understanding proofs, **Introduction to Discrete Mathematics via Logic and Proof** provides a more substantial treatment of logic and proof than most other texts in *Discrete Mathematics*, but students have found that this yields ample dividends, both for later in the course and for other courses they take. An additional benefit of studying *Mathematical Logic* in the first part of the book is that it supplies a solid theoretical background for discussing *Boolean Algebra*, combinational logic circuits, and Boolean functions in Chapter 7.

2. *To Master Core Concepts and Techniques*

   Besides logic, the text takes up a number of other core topics in *Discrete Mathematics*—mathematical induction and recursion, elementary *Set Theory, Combinatorics*, functions and relations, and *Graph Theory*—exploring ideas and procedures that are foundational for many areas of mathematics and computer science.

3. *To Make Connections*

   A guiding principle behind the topics I've chosen to include in this text is that they be both *central and interconnected*, so that students will experience coherence in what they're studying. Covering common proof techniques in the first part of the text helps accomplish this with respect to methodology, but there are important subject-matter connections as well. The natural deduction rules used for creating proofs have formal counterparts in the main laws governing set-theoretic operations, and these eventually get abstracted as the properties postulated or proved in *Boolean Algebra* and physically realized in electronic circuits. And since *Set Theory* forms a unifying field for mathematics, its ideas and notation come up repeatedly throughout the text. In its analysis of infinity, it also makes contact with philosophy.

4. *To Become Adept with Abstract Thinking and Formal Notation*

   In addition to overcoming the hurdle of learning how to do proofs, students must become familiar with an abstract mode of thinking and formal symbolism in order to progress further in their fields of study. This is important for developing computer programs as well as for learning advanced mathematics. These ideas are introduced in Chapter 2's discussion of *First-Order Logic's* syntax and semantics, but they also occur in several other places—in Section 3.3 on *Structural Induction*, Section 3.4 on *Peano Arithmetic*, Section 6.4 on *Integers and Modular Arithmetic*,

and Section 7.3 on *Boolean Algebra*. Encountering an abstract formal
approach in different contexts interspersed with more concrete related
material helps students gradually become proficient with these contemporary practices.

5. *To Value Historical Context and Foundational Issues*
   Paying attention to historical and foundational issues can give peda-
   gogical insight into mathematics as well as make broader connections to
   other areas. I've therefore included brief discussions of such things when
   they're germane to the matter being considered. Besides providing a
   historical setting for topics like *Graph Theory*, I touch on some foundational issues connected with *Propositional Logic, First-Order Logic, Set
   Theory*, and *Peano Arithmetic*.

Prerequisites and Course Emphases

The material in this text should be accessible to students regardless of major, because the content prerequisites for studying each topic are covered
as needed. Nothing beyond a solid secondary mathematics preparation and
a willingness and ability to think more abstractly are required to use this
textbook. A prerequisites graph for chapter content is provided below.

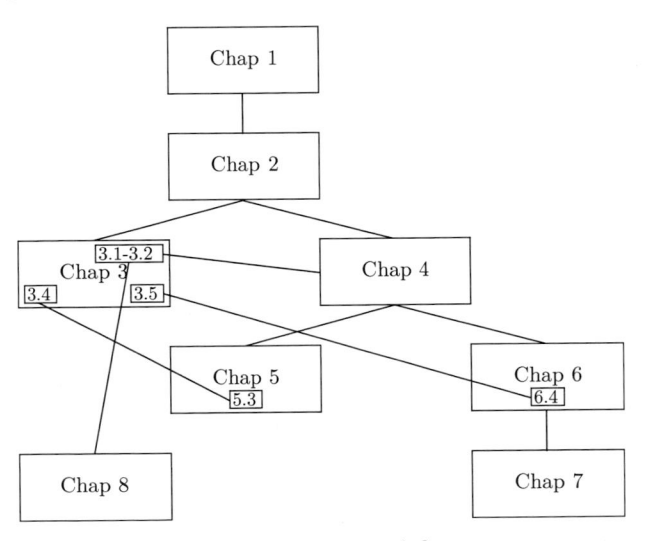

This text offers more than enough material for a one-semester course in
Discrete Mathematics on the intermediate college level, so an instructor can
choose to emphasize slightly different aspects of the material, depending on
interest and clientele. Each section is intended to be used for a full 50-minute
class period, though a few sections could benefit from additional time. Students may find Section 7.3, for instance, which introduces *Boolean Algebra*
as an abstract axiomatic system, more accessible if it is covered over two periods. And Section 3.3 on recurrence relations and structural induction can

be split into two parts and just one done for a more relaxed period. The
Exercise Sets provide a wealth of interesting problems (there are about 2300
of them) of varying levels of difficulty, and in some cases these can be used
to expand a section. Those who would like to work further with divisibility
or Euler's phi function, for example, can draw from the Exercise Sets in
Sections 3.1, 3.5, 4.3, and 4.5. Those wanting to develop *Peano Arithmetic*
in the context of *Set Theory* can use a sequence of exercises from Exercise
Set 5.3. This gives instructors some flexibility to tailor the text as they see fit.

   As noted above, this text has been used for classes containing several
majors—mathematics and secondary education mathematics majors as well
as computer science and engineering majors. It blends together topics of
importance and interest to a wide range of students. Instructors may weight
the material toward one or another of these groups, though, depending on
the composition of their classes. The following guideline chart may help in
this respect.

Common Core Sections	Computer-Science-Focused Sections	Mathematics-Focused Sections
1.1–1.9		
2.1, 2.3–2.4		2.2
3.1–3.2	3.3(.2–.3), 3.5	3.3(.1), 3.4, 3.5
4.1–4.5		
5.1–5.2	5.3(.1–.2, .13)	5.3
6.1–6.3		6.4
7.1–7.3	7.4–7.6	
	8.1–8.4	8.1–8.4

   Those who wish to spend less time on logic may decide to combine some
sections in Chapter 1 (say, covering the main ideas in Sections 1.1–1.3 in
two periods), and they may choose to omit Section 2.2 and cover the key
proof strategies of Sections 2.3–2.4 in one period. Classes aimed at mathematics majors might omit the material on structural induction and strings
in Section 3.3 and some of the final sections in Chapter 7 on *Boolean Functions*. Classes aimed at computer science and engineering students might
want to include those sections but omit Section 3.4 on *Peano Arithmetic* or
Section 6.4 on *The Integers and Modular Arithmetic* or not delve too deeply
into the transfinite *Set Theory* of Chapter 5. Of course, it's always valuable
for students to learn about cognate ideas that lie just outside their major.

For Students: Reading a Mathematics Text

I'll end with a few words directed primarily to students.

   Logic helps us achieve deductive rigor and provides us with concise notation for formulating results, but I won't be using a Spartan definition-

theorem-proof style of writing. I first try to contextualize and motivate results before formulating and proving them. Proofs will be more detailed at first, but as you become familiar with the deductive process, they become more informal, though still using correct terminology and valid argumentation. A growing fluency with logical procedures and symbolism should enable you to follow the text's arguments and discussion without much difficulty.

A mathematics text is not a novel, though, so you can't read it like a story. You should first figure out the overall drift of a section by reading the introductory and concluding material and by looking at the main headings. Then go back through the text more slowly and make sure you understand the central points. Minor details are occasionally left to be filled in; use a pencil and paper to flesh out the discussion where necessary. Examples are important for understanding ideas and procedures, but don't expect them to be templates for working all of the exercises. The more deeply you comprehend the mathematics, the better you will be able to apply it to a wide variety of situations.

The text does require a degree of mathematical maturity on your part—being able to follow moderately paced deductive arguments and to think somewhat abstractly—but no exceptional mathematical ability or specific college-level mathematical knowledge is presupposed, not calculus or linear algebra or any course building upon them. Even gifted first-year college students should be able to master the material, although the text is pitched at a slightly higher level than this. The material in the first unit on logic and proof should help you read (and make) proofs later in the text. Students have told me over the years that they found this material both interesting and invaluable for being able to navigate complex material in later courses.

Like any activity worth mastering, mathematics requires genuine effort. As you've probably heard before, mathematics is not a spectator sport. You can only become competent in a mathematical topic by working a fair number of problems. Each section of the text has plenty of them for you to try. Some are rather routine, asking you to demonstrate familiarity with the basic ideas, results, and methods. Others are more challenging, requiring you to make connections or expand on what is present in the text. As you work the problems, don't get discouraged if you don't immediately see how to solve them—setbacks can be as instructive as success if you persist until you've made some progress. No one grasps everything the first time through, but with perseverance, understanding should follow. As in most areas of life, success is more a matter of persistence and learning from your mistakes than innate genius.

I hope you come to enjoy this material as much as I do and as my students have. If you have corrections or suggestions that you'd like to share for making the textbook better, I'll be happy to hear from you. You can contact me at calvin.jongsma@dordt.edu.

Acknowledgements

I'd like to start by thanking John Corcoran, whose writings on the history and uses of logic first convinced me that logic is more than a specialized branch of mathematics and that it could pay large dividends for learning how to do proofs.

I also want to express my appreciation to Dordt University (then Dordt College) for granting me release time to develop an early edition of this text and to the mathematics, computer science, and engineering students over the years who have given me feedback and encouragement as this text was being developed. In particular, I'd like to thank Kyle Fey, a student assistant at the time who helped me one summer to develop a number of the book's exercises. Kyle is now a coauthor with me of the **Solution Manual** for this text, available to instructors who adopt the text.

My thanks go out as well to Jim Bos, Dordt College's computer-savvy registrar, who in the dark ages of Plain TEX helped me puzzle out some of the mysteries behind constructing proof diagrams and truth tables, and to then-colleague Rick Faber, who transcribed my hand-drawn diagrams for functions into a form I could easily incorporate into the text.

I'm indebted to the following course alumni and colleagues for reviewing portions of the text and offering suggestions for improvements, most of which I've happily accepted: James Boer, Jim Bradley, Gary De Young, Craig Disselkoen, Duane Einfeld, Rick Faber, Kyle Fey, Mike Janssen, Bonnie Jonkman, Richard Stout, Kevin Vander Meulen, Timothy Vis, Chase Viss, and Chris Wyenberg.

Finally, I'd like to thank my wife, Sally, not only for her patience as I worked through numerous drafts of this text, but also for her professional editorial assistance in making the final product more readable.

Sioux Center, IA, USA   Calvin Jongsma
June 2019

List of Notations

Logical Acronyms

Add: Introduction rule *Addition* for ∨

Assoc: Replacement rule *Association* for ∧ or ∨

BE: Elimination rule *Biconditional Elimination* for ↔

BI: Introduction rule *Biconditional Introduction* for ↔

Bicndnl: Replacement rule *Biconditional* for ↔

BICon: Introduction rule *BI, Contrapositive Form* for ↔

BiTrans: Introduction/elimination rule *Biconditional Transitivity* for ↔

Cases: Elimination rule *Proof by Cases* for ∨

Cndnl: Replacement rule *Conditional* for →

Comm: Replacement rule *Commutation* for ∧ or ∨

Conj: Introduction rule *Conjunction* for ∧

Conpsn: Replacement rule *Contraposition* for →

CP: Introduction rule *Conditional Proof* for →

CycBI: Introduction rule *Cyclic BI* for ↔

DeM: Replacement rule *De Morgan's* for ¬∧ or ¬∨

Dist: Replacement rule *Distributivity* for ∧ and ∨

DN: Replacement rule *Double Negation* for ¬

DS: Elimination rule *Disjunctive Syllogism* for ∨

EG: Introduction rule *Existential Generalization* for ∃

EI: Elimination rule *Existential Instantiation* for ∃

EN: Replacement rule *Existential Negation* for ¬∃

EO: Introduction rule *Either-Or* for ∨

Exp: Replacement rule *Exportation* for →

F: *False*

FOL: *First-Order Logic*

HS: Introduction/elimination rule *Hypothetical Syllogism* for →

Idem: Replacement rule *Idempotence* for ∧ or ∨

Iden: Introduction rule *Law of Identity* for =

iff: if-and-only-if

LEM: *Law of Excluded Middle*

LNC: *Law of Non-Contradiction*

Mod PMI: *Modified Proof by Mathematical Induction*

MP: Elimination rule *Modus Ponens* for →

MT: Elimination rule *Modus Tollens* for →

NBE: Elimination rule *Negative Biconditional Elimination* for ↔

NE: Elimination rule *Negation Elimination* for ¬

Neg Cndnl: Replacement rule *Negative Conditional* for ¬ →

NI: Introduction rule *Negation Introduction* for ¬

PL: *Propositional Logic*
PMI: *Proof by Mathematical Induction*
Prem: Introduction rule *Premises* for premises
Reit: Introduction rule *Reiteration* for repeating a proposition
Simp: Elimination rule *Simplification* for $\wedge$
Spsn: *Supposition*, used as a premise in a subproof
Sub: Elimination rule *Substitution of Equals* for $=$
Sym: Introduction/elimination rule *Symmetry of Equals* for $=$
Trans: Introduction/elimination rule *Transitivity of Equals* for $=$
T: *True*
UG: Introduction rule *Universal Generalization* for $\forall$
UI: Elimination rule *Universal Instantiation* for $\forall$
UN: Replacement rule *Universal Negation* for $\neg\forall$
Uniq Exis: Replacement rule *Unique Existence* for $\exists!$
wff, wffs: *Well-formed formula(s)*

Logical Symbols

$\neg P$ *not-P*
$P \wedge Q$ *P and Q*
$P \vee Q$ *P or Q*
$P \veebar Q$ *P exclusive-or Q*
$P \rightarrow Q$ *if P then Q*
$P \leftrightarrow Q$ *P if-and-only-if Q*
$P \triangle Q$ *P* NAND *Q*
$P \triangledown Q$ *P* NOR *Q*
$\mathcal{P} \vDash Q$ *P logically implies Q*
$\mathcal{P} \vDash_{\mathcal{A}} Q$ *P logically implies Q relative to a theory with axioms $\mathcal{A}$*
$P \Vdash Q$ *P is logically equivalent to Q*
$P \Vdash_{\mathcal{A}} Q$ *P is logically equivalent to Q relative to a theory with axioms $\mathcal{A}$*
$\mathcal{P} \vdash Q$ *P proves Q*
$P \dashv\vdash Q$ *P is interderivable with Q*
$P :: Q$ *P can be validly substituted for Q and conversely*
$\forall x$ *for all x*
$\exists x$ *there exists an x*
$\exists! x$ *there exists a unique x*
ε *the empty string*
$s \cdot t$ *concatenation of string s with string t*

Mathematical Symbols

$\mathbb{A}$ *set of algebraic numbers*
$\mathbb{C}$ *set of complex numbers*
$\mathbb{N}$ *set of natural numbers $\{0, 1, 2, \ldots\}$*
$\mathbb{N}^{+}$ *set of positive natural numbers $\{1, 2, \ldots\}$*
$\mathbb{Q}$ *set of rational numbers*

$\mathbb{R}$ *set of real numbers*
$\mathbb{Z}$ *set of integers*
∎ *end-of-proof symbol*
✓ *end-of-subproof/verification-of-assertion symbol*
ϕ *the golden ratio* $\frac{1+\sqrt{5}}{2}$
$a \mid b$ *a divides b*
$\varphi(n)$ *Euler's φ function*
$\gcd(a, b)$ *the greatest common divisor of a and b*
$\operatorname{lcm}(a, b)$ *the least common multiple of a and b*
$x \in S$ *x is a member of set S*
$\{x \in U : P(x)\}$ *the set of all x in U satisfying condition P(x)*
$S \subseteq T$ *S is a subset of T*
$S \subset T$ *S is a proper subset of T*
$S \supseteq T$ *S is a superset of T*
$S \supset T$ *S is a proper superset of T*
$\emptyset$ *the empty set*
$S \cap T$ *the intersection of sets S and T*
$S \cup T$ *the union of sets S and T*
$S - T$ *the difference of sets S and T*
$S \oplus T$ *the symmetric difference of sets S and T*
$\overline{S}$ *the complement of set S*
$\mathcal{P}(S)$ *the power set of S*
$S \times T$ *the Cartesian product of sets S and T*
S^n *the n-fold Cartesian product of set S with itself*
$|S|$ *the cardinality of set S*
$_nP_k = P(n, k)$ *the number of permutations of n things k at a time*
$_nC_k = C(n, k) = \binom{n}{k}$ *the number of combinations of n things k at a time*
$S \sim T$ *S is equinumerous to T*
$S \preceq T$ *S is less numerous than or equinumerous to T*
$S \prec T$ *S is less numerous than T*
$\aleph_0$ *aleph nought, the cardinality of countably infinite sets*
$2^{\aleph_0}$ *2 to the aleph nought, the cardinality of $\mathbb{R}$*
$f : D \longrightarrow C$ *a function f from domain D to codomain C*
$f[S]$ *the set of images of set S under function f*
$f^*[V]$ *the set of pre-images of set V under function f*
$g \circ f$ *the function composed of f followed by g*
f^{-1} *the inverse function of f*
xRy *x is related to y by relation R*
$\operatorname{Dom}(R)$ *the domain of the relation R*
$\operatorname{Rng}(R)$ *the range of the relation R*
$\widehat{R}$ *the converse of relation R*
$R_1 \circ R_2$ *the relation composed of relations R_1 and R_2*
$[a]$ *the equivalence class containing a*

$\leq$ *a partial order on some set S*
$<$ *a strict order on some set S*
$x \wedge y$ *the meet of x and y in a lattice $\mathcal{A}$*
$x \vee y$ *the join of x and y in a lattice $\mathcal{A}$*
$\bar{x}$ *the complement of x in a complemented lattice $\mathcal{A}$*
$0, 1$ *the extreme elements in a bounded lattice $\mathcal{A}$*
$\mathcal{B}$ *the two-element Boolean algebra $\{0, 1\}$*
$\deg(v)$ *the degree of vertex v*
K_n *the complete graph on n vertices*
$K_{m,n}$ *the complete bipartite graph on $m + n$ vertices*
$\Delta(G)$ *the maximum degree of a graph G*
$\chi(G)$ *the chromatic number of graph G*

Contents

Chapter 1
Propositional Logic

We all draw conclusions every day—it's what humans do, usually without thinking about the process. We hear a sound and conclude what caused it. We adjust a recipe for four people down to one for two or up to one for six. We troubleshoot problems with our computer, eliminating possible causes in order to determine what's making it act in an unexpected way.

Some activities, though, are more logical than others. All fields of study use reasoning to develop and organize their results, but mathematics does this more systematically than almost any other area. In fact, deductive reasoning is so pervasive in mathematics, particularly in advanced courses, that it's worth learning about the variety of reasoning strategies it employs. It's also important to know which arguments are valid and which ones are not. This is especially valuable as you begin taking proof-based courses.

The following two examples, which don't require much mathematical background, illustrate how conclusions in geometry and real-number arithmetic are established by a logical argument based upon simpler truths.

✤ Example 1.0.1
Show that however many diagonals are drawn between the vertices of a convex polygon, at least three vertices or two pairs of vertices have the same number of edges (their *degrees*) connecting them to other vertices.

Solution
· Consider a polygon with $n \geq 3$ sides. Each vertex is connected to 2 adjacent vertices and at most $n - 3$ others, i.e., its *degree* lies between 2 and $n - 1$.
· There are thus at most $n - 2$ degree values for the n vertices, so at least two vertices must have the same degree as another, either with a common vertex or with two different vertices. ■

✤ Example 1.0.2
Must a^b be irrational when a and b are irrational?

Solution
· Most values of a^b are irrational, something we'll be able to prove in Section 5.2, but—surprisingly—there are also instances when it's rational.
· To see this, consider $\sqrt{2}^{\sqrt{2}}$.
· Either this value is rational (it's not, but ignore this fact—it's too involved to prove here), in which case we have our instance, or else it's irrational.
· If the latter is the case, take $a = \sqrt{2}^{\sqrt{2}}$ and let $b = \sqrt{2}$.
  Now a^b is rational because $\left(\sqrt{2}^{\sqrt{2}}\right)^{\sqrt{2}} = \sqrt{2}^2 = 2$. ■

© Springer Nature Switzerland AG 2019
C. Jongsma, *Introduction to Discrete Mathematics via Logic and Proof*,
Undergraduate Texts in Mathematics,
https://doi.org/10.1007/978-3-030-25358-5_1

1.1 A Gentle Introduction to Logic and Proof

This chapter and the next explore the logic that underlies deductive proofs, providing us with a basis for making and evaluating arguments. We'll begin with a general overview of the deduction process—something that will become clearer to you as we proceed. First, though, we'll look briefly at how logic and mathematics have interacted with one another over the centuries.

1.1.1 *Mathematics and Logic in Historical Perspective*

Mathematics and logic have been closely allied since ancient times, and computer science has relied on logic since it began in earnest in the 1930s. In fact, over the last century, portions of logic have become almost indistinguishable from mathematics, and parts of computer science seem like applied logic.

Mathematics is the very model of a systematic science. Results build upon one another in an orderly step-by-step fashion. Parts are interconnected by a consistent network of ideas and procedures. Statements are precisely worded, and careful reasoning proves them true. This tidy picture of mathematics with its deductive methodology is a legacy from the ancient Greeks, who first organized mathematics into an axiomatic system about 2500 years ago.

Rational debate was all-important to the Greeks. It supported their democratic institutions, and it gave rise to philosophy and mathematics. Unlike others, the Greeks saw geometry as a theoretical science, not as a set of procedures for calculating measurement results. And they prized number theory, not arithmetic, because it gave universal truths about quantities, such as that *every number greater than one has a prime divisor*. Numerical computations belonged to the workaday world of trade and government bureaucracy.

From early times, then, deductive reasoning has been associated with theoretical mathematics, not computation. Mathematical theories began with basic principles, called axioms or postulates, and with definitions that specified the meaning of terms. Theorems were strictly argued on the basis of previously accepted propositions.

The first system of logic was created around 325 B.C. by the philosopher Aristotle (Figure 1.1). His analysis of deductive argumentation was nearly the final word on the subject for over 2000 years. Aristotle's view of scientific knowledge was modeled on mathematics: you begin with self-evident truths and proceed to more complex results via deductive reasoning. Together, logic and mathematics showed how knowledge should be pursued and organized.

Fig. 1.1 Aristotle

Important advances were made in computational arithmetic and algebra in the late medieval and early modern periods. Experimentation and inductive reasoning became central features of natural science. These developments led to a decreased emphasis on logic and deductive reasoning, even within mathematics, and mathematicians began to research topics and use methods that lacked a sound theoretical basis. Powerful computational procedures in calculus allowed mathematicians to discover new results in both mathematics and physics, but they also outstripped mathematicians' ability to justify them deductively.

In the nineteenth century, mathematicians sought to solidify the theoretical foundations of calculus. They also began investigating the logical basis of number systems, calculation procedures, and algebra.

Around the middle of the century, George Boole (Figure 1.2) and Augustus De Morgan contributed to the revival of deductive logic, proposing radically new approaches. De Morgan expanded logic to treat relations as well as properties. Boole developed an unusual version of algebraic logic. Using an area of mathematics (computational algebra) far removed from deductive rigor as the conceptual vehicle for treating the science of deductive inference (logic) seemed wrong-headed to many. But Boole's work connected mathematics, logic, and computation in

Fig. 1.2 George Boole

ways that eventually gave birth to computer science in the twentieth century.

Logic was also being transformed by another group of mathematicians. Gottlob Frege, and later Bertrand Russell, thought logic should be the foundation for mathematics instead of the other way around, as Boole had advocated. Ultimately, they failed to reduce even arithmetic to elementary logic, but in the process of trying, logic was modernized in a way that connected it closely to mathematics.

Today variants of this system exist. The primary aim of the earliest logical systems was to encapsulate the *results* of mathematics *within* logic, not capture its *method of reasoning*. By the mid-1930s, however, some mathematicians began exploring a *natural deduction system*, developing logic to handle how people actually deduce conclusions from premises. This approach, which we'll use here, is better suited for analyzing and constructing mathematical proofs. It derives from the work of Stanisław Jaśkowski, a Polish mathematician, and was slightly modified by the American logician Frederic Fitch.

Over the past two centuries, then, logic has become increasingly tied to mathematics. Due to Boole's work, logic became associated with an algebra of 0 and 1. In the 1930s, Claude Shannon (Figure 1.3) designed switching circuits to physically represent Boolean expressions and used *Boolean Algebra* to simplify circuits. Soon people were building logic-gate circuits and using them for electronic calculations. Boole's union of logic and algebra thus led to the development of digital computers and the rise of computer science in the twentieth century.

Fig. 1.3 Claude Shannon

Our interest in logic here is primarily twofold. We'll mainly focus on logic's role in proof construction, the topic of the first two chapters. But elementary logic is also foundational for computer science and engineering. This latter application will take center stage in Chapter 7, where we'll unpack the various connections between algebra, logic, and circuits.

1.1.2 The Role of Proof in Mathematics

We use logical processes whenever we draw an informed conclusion from something we already know. Logic and proof are central features of mathematics and computer science because they involve this sort of reasoning.

A *proof* is a sequence of propositions that *deduces* a *conclusion* from a set of *premises* by showing that it *logically follows from* them. If the premises are true, the conclusion will also be true.

Proofs are used not only *to verify a conjecture* but also to *communicate* this to others. Diagrams, examples, and remarks explaining the intuition behind a result may be used to convey its meaning or significance and convince an audience of its correctness, but the main vehicle of mathematical communication and the final arbiter of the truth of a proposition is its deduction from already known results.

A proof not only establishes and communicates the truth of a conclusion; it also *shows why* it's true. Proofs convince us by *demonstrating how and why* results connect to things already known, providing linkage and meaning.

Finally, proofs have *organizational value.* They logically structure a field like mathematics into an interconnected deductive whole. Proofs help us learn and recall mathematical propositions by relating complex results to simpler ones. They can also alert us to the importance of key ideas and techniques that seem to keep coming up in our arguments.

1.1.3 Mathematical Proofs and Rules of Inference

Two things are required of a proof—that it begin with premises taken to be true, and that each deductive inference it makes is justified. The *truth* of a premise is a concern for mathematics, not logic. Logic focuses on the reasoning, on whether the argument is *valid.*

Mathematical proofs come in many shades of completeness. Some are sketchy, highlighting only the main points and leaving some intermediate conclusions (and maybe even some premises) for the reader to supply. A certain degree of detail may be appropriate for one audience but not another. In the end, though, whether a deduction is conclusive depends on whether it can be made rigorous, on whether it can be expanded into an argument in which each step is warranted by a sound rule of inference.

On the secondary-school level, arguments are sometimes put into a two-column format. The first column contains the argument's assertions, and the second column states the reasons. This method can appear somewhat arbitrary, though, since the reasons given may still be propositions (and so should really be in the first column). Other times, an inference rule may be cited. For example, in concluding one option from a pair of alternatives, you might put down something like *the other alternative isn't the case.*

A consistent proof analysis would place all mathematical statements in the left column, using the right column simply for citing the relevant logical rules of inference. This procedure would give a completely rigorous derivation.

Under ordinary circumstances, such an argument would be unnecessarily long and complex. But to become familiar with the main proof strategies in mathematics, we need to recognize the wide variety of inference types that exist. For this reason, we'll analyze and construct simple proofs in full detail for a while, using the two-column format just described.

1.1.4 *Inference Rules, Logical Implication, and Validity*

Deductions are constructed to show that a set of premises *logically implies* a conclusion, that a conclusion *logically follows from* (is a *logical consequence of* or a *valid inference from*) the premises. Each inference drawn in a deduction must itself be *valid*, certified by a *sound* rule of inference, i.e., by a rule that always produces *valid* conclusions from premises.

To avoid circularity here, we will take validity as primary. According to Aristotle, *a valid inference is one in which the conclusion necessarily follows from its premises.* In other words, it's one where you're stuck with the conclusion, logically speaking, once you accept the premises. If the conclusion contains information not contained in the premises, you can still wiggle out of admitting the conclusion. But if the conclusion is already implicit in the premises, you have no way to back out, and the argument is valid. This gives us an information content principle for judging the validity of an argument.

INFORMATION CONTENT PRINCIPLE FOR VALIDITY
> *An argument is valid if and only if the information contained in the conclusion is already contained in the premises jointly.*

✤ **Example 1.1.1**
Determine the validity of the following simple mathematical argument:
1) The square of any real number is nonnegative;
2) The square of i (where $i = \sqrt{-1}$) is -1, a negative number; therefore,
3) i is not a real number.

Solution
This is a valid inference: the conclusion necessarily follows from the premises. Premise 1 says all real numbers have nonnegative squares, and the second premise asserts that i's square is not nonnegative.
This information forces us to conclude that i isn't a real number.

An information diagram can illustrate this argument,
taking $\mathbb{R}$ to denote the set of real numbers and N the
set of numbers with nonnegative squares. Since i is
not in N, it follows that it cannot be in $\mathbb{R}$ either.

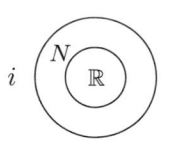

1.1.5 A Necessary Truth-Value Condition for Validity

In Example 1, the premises and the conclusion were all true. However, this
need not be the case for all valid arguments, as the next example shows.

❖ **Example 1.1.2**

Explain why the following premises logically imply the stated conclusion:
1) Every real number is either negative or positive;
2) 0 is a nonnegative real number;
   therefore,
3) 0 is a positive real number.

Solution

· This inference is valid, as is shown by an information
  diagram, letting N denote the set of negative real
  numbers and P the set of positive real numbers.

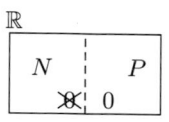

· For, if 0 is nonnegative, and *if the only options are being positive or negative*,
  then 0 must be positive. The conclusion follows from the premises. That
  the first premise and the conclusion are false is irrelevant.
· Knowing that the conclusion is false and that the argument is valid, however, we can conclude that at least one of the premises must be false.

Valid arguments can be made with either true or false premises. You might
think valid arguments with false premises are silly. Who argues from false
premises? Everyone, at one time or another. Sometimes people argue from
results they think are true but aren't. Other times they argue from results
they're sure are false in order to show that they are. This is a standard
debate strategy. It also happens in mathematics—that's how all *Proofs by
Contradiction* proceed, as we'll see later.

So, valid inferences are not limited to ones in which the premises and
conclusion are true. That being said, not every combination of truth values
for premises and conclusion are possible for a valid argument.

❖ **Example 1.1.3**

Determine the validity of the following argument:
1) All primes greater than 2 are odd;
2) 1001 is an odd number greater than 2;
   therefore,
3) 1001 is prime.

Solution

This argument is invalid. Both premises are true, but the conclusion is false:
1001 is not prime (factor it). The conclusion can't possibly be a consequence
of the premises, then, because falsehood never follows logically from truth.

If the information of a valid conclusion is already contained in its premises, and if the premises are true, then the conclusion can't possibly be false. *Truth yields truth when the inference is valid.* But this does *not* say that an argument is valid whenever its conclusion *and* premises are true. Nor does it say that if an argument is valid, then its premises *and* conclusion are true.

We can summarize these insights in the following principle.

NECESSARY TRUTH-VALUE CONDITION FOR VALID ARGUMENTS
If a valid argument has true premises, then its conclusion is true.
If an argument has true premises and a false conclusion, then it is invalid.

This principle, stated in two equivalent forms, gives a *necessary* condition for validity. It can be used to conclude that some arguments are *invalid*, but it does not yet give a *sufficient* criterion for when an argument *must* be valid.

What does all this mean for *inference rules*? Since true statements do not have false consequences, a sound inference rule won't permit us to pass from truth to falsehood. Otherwise, deductions would be incapable of grounding the truth of conclusions in the truth of their premises.

1.1.6 *Validity, Information Content, and Logical Form*

We now know that validity is somewhat independent of the truth-value configuration of premises and conclusions. But it's also independent of the *particular subject matter* of the propositions. Validity has to do with *how* the information content in the conclusion *is related to* that in the premises. What matters, therefore, is the *containment relation* between the information in the premises and the conclusion, not *what* the information actually is. This logical connection gets specified by the logical form of the propositions involved. This idea is known as the *Principle of Material Irrelevance*, but it might better be called the *Principle of Logical Form*.

PRINCIPLE OF LOGICAL FORM
The logical forms of an argument's premises and conclusion determine its validity. Arguments in the same logical form are alike valid or invalid.

It's difficult to give a precise definition of *logical form*, but we can elaborate this principle a bit further. A sentence's logical form is determined by terms that indicate general connections between information-bearing terms. They include linking words like *and, or, not, if-then,* and *is/are,* as well as quantifier words like *all, some,* and *none.* Nonlogical terms are ones that refer to particular objects, properties, and relations.

✤ **Example 1.1.4**
   Identify the logical and nonlogical terms in the following argument:
   1) All real numbers are positive, negative, or zero;
   2) $\sqrt{-1}$ is neither positive nor negative nor zero;
      therefore,
   3) $\sqrt{-1}$ is not a real number.

Solution

The nonlogical terms here are:
  1) real numbers, positive, negative, zero;
  2) $\sqrt{-1}$, positive, negative, zero;
  3) $\sqrt{-1}$, real number.
The logical terms are:
  1) all, are, or;
  2) is, neither, nor;
  3) is, not.
The word *therefore* is also a logical term but on the argument level.

To make the logical form of a proposition clear, we need a way to ignore its specific nonlogical information. To do this, we'll take our cue from algebra. To solve quadratic equations in general, we use letters to stand for the coefficients and then solve the equation in terms of those letters. We can do something similar here. We'll consider propositions abstractly, using letters in place of content words. Nonlogical material thus gets abstracted out, better revealing the remaining logical form of the sentences. Using a symbolic format, we'll still be able to argue that a conclusion follows from its premises by showing how the conclusion's information relates to that of the premises.

✤**Example 1.1.5**

Reformulate Example 4 symbolically, and establish the argument's validity.

Solution

· We'll use first letters of the terms in Example 4. This helps us ignore the meaning of the sentences and whether they're true or false, so that we can focus on the logical relations asserted to hold between the entities.

1) All R's are P's or N's or Z's;
2) I is neither a P nor an N nor a Z;
   therefore,
3) I is not an R.

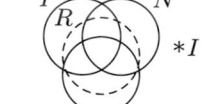

· I cannot be an R, because if it were, the first premise says it would be a P, an N, or a Z, while the second premise says it isn't any of these things. The information *connecting* R with P, N, and Z, augmented by the information about *how* I *relates* to these same three things, leads to the *information link* asserted by the conclusion and shown in the diagram.
· Thus, the premises logically imply the conclusion—the argument is valid.

The *Principle of Logical Form* allows us to strengthen our necessary condition for validity. Whenever an argument is valid, any argument in the same logical form is also valid, so *if an argument can be transformed into one with the same logical form where the premises are true and the conclusion is false, then the original argument is invalid.* This procedure, which involves reinterpreting the nonlogical terms, is called the *method of counterargument.*

GENERALIZED TRUTH-VALUE CONDITION FOR VALIDITY
  If an argument is valid, then any argument in that same logical form with
  true premises has a true conclusion.
  If an argument has true premises and a false conclusion, then any argu-
  ment in that same logical form is invalid.

✤ Example 1.1.6

Use the *method of counterargument* to show that the following is invalid.
1) $2 + 2 = 4$;
   therefore,
2) 4 is a double of 2.

Solution

Let's transform this argument by replacing *is a double of* with *is half of*,
leaving everything else the same.
   1) $2 + 2 = 4$;
      therefore,
   2) 4 is half of 2.
This argument keeps the premise true, but its conclusion is now false. Therefore, the argument is invalid. The original argument, having the same logical
form, is also invalid.

You may feel hoodwinked by the last example's solution. It may seem
illegitimate to convert *is a double of* into its opposite *is half of*, but it's not.
We've merely replaced one mathematical content phrase with another of the
same sort (both denote numerical relations) to test for validity.

Second, maybe you thought the argument was valid, even if not because
both statements are true. Given what being a double means, the conclusion
does follow. But it doesn't follow simply from the one premise given—our
counterargument shows this conclusively. A second premise is needed to spell
out the meaning of being a double. The *method of counterargument* is a
powerful logical tool for disproving things in mathematics. It's also a favorite
ploy used in philosophical debates.

The above examples show that an argument's validity is only loosely
related to the truth values of the propositions involved. An argument like
that in Example 3 with true premises and a false conclusion is invalid, but
other combinations of truth values seem compatible with being either valid
or invalid. Examples 1 and 5 presented valid arguments in which the premises
and conclusion were all true, while Example 6 had a true premise and a true
conclusion, but the argument was invalid. And although Example 2 gave a
valid argument with both true and false premises and a false conclusion, not
all arguments like this will be valid. Other combinations of truth values for
premises and conclusions are likewise compatible with both valid and invalid
arguments (see Exercise 17).

The *Generalized Truth-Value Condition for Validity* claims that if an argument can be transformed into one of the same form with true premises
and a false conclusion, it is invalid. But are there enough counterarguments

to demonstrate invalidity like this in all cases? If so, we'll have a fool-proof method for assessing validity. Naturally, it may take some ingenuity to locate a successful counterargument to show that an argument is invalid, but that's an issue of human ability, not a feature of the universe.

1.1.7 Tarski's Validity Thesis

Surprisingly, modern logic takes the approach just outlined. The necessary *Generalized Truth-Value Condition for Validity* is also taken to be sufficient. An inference is considered valid if no counterargument exists, i.e., if there is no argument in the same logical form having true premises and a false conclusion. This valid-by-default approach was first formulated by the Polish logician Alfred Tarski (Figure 1.4) in a seminal article on logical consequence (1936), so we'll call this *Tarski's Validity Thesis.*

Fig. 1.4 Alfred Tarski

TARSKI'S VALIDITY THESIS
If an argument's conclusion is true whenever its premises are, under all possible interpretations of the nonlogical terms, then the argument is valid.

Tarski's Validity Thesis combined with the *Generalized Truth-Value Condition for Validity* gives us a complete truth-value criterion for assessing an argument's validity. While the particular truth values of premises and conclusions don't determine an argument's validity, they do when all arguments of that logical form are considered.

TRUTH-VALUE CRITERION FOR VALIDITY
An argument is valid if and only if all arguments in the same form having true premises also have a true conclusion.

This way of establishing validity is still a bit nebulous. We'll have more definitive ways to show validity once we begin studying a particular system of logic. But the basic idea will be the same. Using abstract symbolism to express the logical form of an argument, we'll demonstrate its validity by showing that whenever the premises are true, the conclusion is also.

At the heart of all this lie the notions of *true proposition/sentence* and *logical form*. Before we can study valid arguments and proofs, therefore, we must investigate the logical forms propositions can possess and learn when they're true. We'll then be able to identify valid arguments and choose a system of sound inference rules. Using such a deduction system, we'll finally be able to construct conclusive deductions. We'll begin this journey in the next section with the logical system known as *Propositional Logic*.

The key ideas introduced so far are summarized in the following definitions.

DEFINITIONS
1) **Logical Implication**: A set of premises *logically implies* a conclusion if and only if every interpretation of the nonlogical terms making the premises true also makes the conclusion true.
2) **Logical Consequence**: A conclusion is a *logical consequence* of a set of premises if and only if it is logically implied by them.
3) **Valid Argument**: An argument is *valid* if and only if the premises logically imply the conclusion.
4) **Sound Rule of Inference**: a rule of inference is *sound* if and only if the arguments it warrants are valid.

EXERCISE SET 1.1

Exercises 1–4: Premises and Conclusions of Valid Arguments
Identify the premises and conclusions in the following arguments, assumed to be valid. Make explicit any statements that are assumed but not stated. If an argument contains an intermediate conclusion, identify it as such.

1.1.1. The square of a real number is nonnegative because it is either zero or it is positive.

1.1.2. All differentiable functions are continuous; $f(x) = \sin x$ is differentiable; therefore $f(x) = \sin x$ is continuous.

1.1.3. Corresponding parts of congruent triangles are congruent; if two sides and an included angle of a triangle are congruent, respectively, to two sides and an included angle of another, the triangles are congruent; $\triangle ABC$ has sides AB and BC congruent to sides DE and EF of $\triangle DEF$, and $\angle B$ congruent to $\angle E$; thus, $\triangle ABC \cong \triangle DEF$, and so sides AC and DF are also congruent.

1.1.4. The number 846 is divisible by 6 because if a number is divisible by both 2 and 3 it is divisible by 6. Moreover, 846 is divisible by 2 because 6 is; and it is divisible by 3 because $8 + 4 + 6$ is divisible by 3, and if the sum of a number's digits is divisible by 3, then the number is divisible by 3.

Exercises 5–7: Brief Explanations
Explain the following in your own words.

1.1.5. What are the various functions of a mathematical proof? Which ones seem most important? Why? How does logic contribute to this enterprise?

1.1.6. What is the difference between a proposition being a logical consequence of given premises and being a conclusion deduced from them?

1.1.7. How are sound rules of inference related to valid arguments? Which concept relates to logical implication? to making deductions?

Exercises 8–13: True or False?
Are the following statements true or false? Explain your answer.

1.1.8. Valid arguments with true premises have a true conclusion.

1.1.9. Valid arguments with a false premise have a false conclusion.

1.1.10. Valid arguments with a false conclusion have a false premise.

1.1.11. Arguments with false premises and a true conclusion are invalid.

1.1.12. Arguments with false premises and a false conclusion are invalid.

1.1.13. Arguments yielding a valid argument by reinterpretation are valid.

Exercises 14–20: Argument Analysis
Analyze the following arguments.

1.1.14. *Valid Arguments*
Tell why the following arguments are valid. Use a diagram for the premises to help you argue your case (see Examples 1, 2, and 5).
  a. *All rational numbers are algebraic; algebraic numbers are complex numbers; therefore, rational numbers are complex numbers.*
  b. *All isosceles triangles are right triangles; no scalene triangles are right triangles; therefore, no scalene triangles are isosceles triangles.*
  c. *Some prime numbers are even; all prime numbers are square-free; therefore, some square-free numbers are even.*

1.1.15. *Drawing Valid Conclusions*
What valid conclusion can be drawn from the premises of the following arguments, paraphrased from logician Lewis Carroll? Explain your answer.
  a. *Every sane person can do logic; no insane person can serve on a jury; none of your friends can do logic; therefore, ⋯ .*
  b. *Braggarts think too much of themselves; no really well-informed people are bad company; people who think too much of themselves are not good company; therefore, ⋯ .*

1.1.16. *Invalid Arguments*
Show that each of the following arguments is invalid in two ways: by showing with a diagram and an argument that the conclusion's information goes beyond the premises, and by constructing a counterargument.
  a. *All items of value are made of metal; gold rings are made of metal; therefore, gold rings are items of value.*
  b. *No penguins can fly; some pigs can fly; therefore, no pigs are penguins.*
  c. *Some civil servants do not accept bribes; some politicians do accept bribes; therefore, some civil servants are not politicians.*

1.1.17. *Truth Values and Validity*
  a. Draw up a table to indicate all possible truth-value combinations for an argument having two premises and a conclusion. Use separate columns for each premise and conclusion and fill in the entries with either T or F. How many different truth-value assignments (rows) are possible?
  b. For *each* truth-value assignment in part *a*, give two arguments of that type, if possible: one that is valid and one that is invalid. You may use arguments either from everyday life or from mathematics. Which truth-value assignments, if any, are associated only with invalid arguments? Which ones, if any, are associated only with valid arguments. Explain.

1.1.18. *True Statements about Falsehoods*
Statement n in a list of 100 statements says: *Exactly n statements in this list
are false.* Which of these 100 statements are true? Explain.

1.1.19. *Who's Telling the Truth? Who's Not?*
On a remote island, politicians never tell the truth, but everyone else does.

 a. A stranger meets three islanders and asks the first if she is a politician.
    She mumbles an answer too soft to be heard. The second islander says
    the first one denies being a politician. The third islander smiles and says
    the first one is, nevertheless, a politician. How many politicians are there
    in this trio of islanders? Explain your reasoning.

 b. A stranger meets three islanders and asks how many of them are politi-
    cians. The first one says they're all politicians. The second one disagrees,
    claiming only two of them are. The third one walks away without answer-
    ing. How many politicians are there? Explain your reasoning.

 c. A stranger comes to a fork in the road where an islander is sitting. What
    one question can the stranger ask to determine which path leads to the
    home of the island's chief politician? What further question could he ask
    to discover whether the islander is a politician?

1.1.20. *Identifying Rules of Inference*

 a. Example 1.0.2 argued that some irrational raised to an irrational power is
    rational. Identify any logical rules of inference (ones governing the use of
    terms like *and, or, not, if-then, all, some*) that were used in that argument.

 b. Read through the following proof and identify as many rules of inference
    as you can. First write the argument out in more detail and put it into a
    two-column format, with the content statements in the left-hand column
    and the rules of inference in the right-hand column.

    *Theorem: If the hypotenuse and a leg of one right triangle are congruent,
    respectively, to the hypotenuse and a leg of another right triangle, then
    the two triangles are congruent.*

    Proof:
    Let $\triangle ABC$ and $\triangle A'B'C'$ denote the two right triangles with right angles
    at A and A', $AC \cong A'C'$, and $BC \cong B'C'$.
    Extend segment AB from A to a point D opposite B so that $AD \cong A'B'$.
    Connect C and D to form $\triangle ACD$.

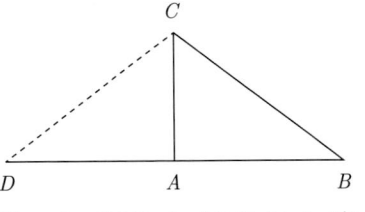 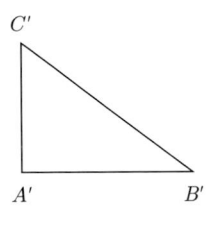

    Then by SAS, $\triangle ADC \cong \triangle A'B'C'$.
    Thus, $DC \cong B'C'$, and so $DC \cong BC$, too.
    $\triangle BCD$ is therefore an isosceles triangle, so $\angle B \cong \angle D$.
    By AAS, $\triangle ABC \cong \triangle ADC$, and so $\triangle ABC \cong \triangle A'B'C'$.   ∎

1.2 Conjunction, Disjunction, and Negation

In this section we'll begin investigating how propositions can be combined using truth-functional connectives. This part of logic is known as *Propositional Logic* (PL). We'll see how to symbolize such propositions (syntax), and we'll explain when they are true (semantics). This background will give us a sound basis for discussing valid argument forms, logical inference rules, and deductive arguments for *Propositional Logic*.

1.2.1 *Sentences, Propositions, and Truth Values*

A sentence is a complete statement, formed according to the rules of grammar and communicating a meaningful thought. We use various kinds of sentences in everyday conversation: questions, commands, assertions, and so on. PL only considers *declarative statements*, which *affirm* or *deny* something.

We can distinguish sentences (symbolic formulations) from statements (utterances) and propositions (thought contents), but given their close connections, we'll often use these terms interchangeably.

Declarative sentences are *true* just in case what they say is actually so and are *false* otherwise. This correspondence view of propositional truth is central to mathematics and goes back to Plato and Aristotle. It also underlies the modern approach to logic elaborated by Russell, Tarski, and others.

Because a given state of affairs either is or is not the case and cannot be both, *a proposition is either true or false but not both*. A PL sentence has a unique *truth value*, which is either *True* (abbreviated by T) or *False* (abbreviated by F). *Propositional Logic* is thus a *two-valued* system of logic.

1.2.2 *Truth-Functional Connectives*

Ordinary language provides many kinds of connectives for combining sentences. The only ones entertained in *Propositional Logic*, however, are *truth-functional connectives*, ones in which *the truth value of the compound sentence is uniquely determined by the truth values of the component sentences*.

The main connectives we'll consider are *and, or, not, if-then,* and *if-and-only-if*. A sentence with one or more of these connectives is a *compound* sentence; otherwise it is an *atomic* sentence. A fairly simple sentence may still be compound, while a long complex sentence may be atomic.

✤ **Example 1.2.1**

Determine which of the following mathematical sentences are compound.
  a) $1 + 2 + 3 + \cdots + 99 + 100 = \frac{100 \cdot 101}{2}$.   b) $|\pi - \frac{22}{7}| \geq 0$.
  c) If p is a prime number greater than 2, then p is odd.

Solution

  a) This first sentence is atomic; it contains no sub-sentences.
  b) The second sentence is shorter, but it is compound.

Its atomic components are: $|\pi - \frac{22}{7}| > 0$, and $|\pi - \frac{22}{7}| = 0$.

The only logical connective here is *or*, embedded in the $\geq$ notation.

c) This sentence is more compound than it looks. An obvious connective is *if-then*, but the first clause is actually a conjunction. The atomic components are: *p is a prime number*, *p is greater than 2*, and *p is odd*. The connectives here are *if-then* and *and*. Fully spelled out, this sentence is: *If p is a prime number and p is greater than 2, then p is odd.*

We'll use capital letters like P, Q, and R to name specific sentences (not parts of sentences, as in Section 1.1). Letters standing for sentences in general (sentence variables) will be put in boldface font: **P**, **Q**, and **R**.

Letters may represent either atomic or compound sentences. A single letter only indicates that a complete sentence is being denoted, not that it is atomic. In addition to letters, possibly subscripted, we'll use special symbols to stand for logical connectives. We'll use right and left parentheses (and occasionally brackets or braces) as punctuation, both to guard against ambiguity and to make sentences more readable.

1.2.3 *Syntax and Semantics of Conjunction*

The *and* connective is the simplest and least controversial of all logical connectives. It joins sentences P and Q to form the conjunction P-and-Q, which we'll symbolize by $P \wedge Q$. Think of $\wedge$ as the outline of the first letter of And. Some logic texts use the ampersand & for *and*, while others use the · of ordinary multiplication, a notation we'll adopt when we connect *Propositional Logic* to *Boolean Algebra* in Chapter 7.

Definition 1.2.1: *Truth-Value Assignment for Conjunction*
*A **conjunction** **P** $\wedge$ **Q** is true if and only if **P** is true and **Q** is true.*

We can summarize this truth-value assignment for conjunction by means of a *truth table*, devised by the late nineteenth-century American mathematical logician and philosopher C. S. Peirce. A truth table exhibits all possible combinations of truth values for the sentence variables **P** and **Q** and gives the truth value of the conjunction **P** $\wedge$ **Q** for each assignment.

We can do this in a full table, using headings for **P**, **Q**, and **P**$\wedge$**Q**, placing truth values below them. Or, following the twentieth-century American logician Quine, we can use a more compact form, putting the conjuncts' truth values below their letters and the truth value of the full conjunction below the $\wedge$. The column containing the overall truth value for the sentence can be underlined one or more times to make the compact table more readable.

P	**Q**	**P**$\wedge$**Q**
T	T	T
T	F	F
F	T	F
F	F	F

P	$\wedge$	**Q**
T	T	T
T	F	F
F	F	T
F	F	F

To make comparison easier, we'll assign truth values to **P** and **Q** in a uniform way. The final letter's column alternates $T\ F\ T\ F$, while the first letter's column has all T first and then all F. This is the most popular way to list all four truth-value assignments, though some texts use 0 for F and 1 for T and then reverse the order so it coincides with numerical order.

The word *and* does not always indicate a logical connective. It may just join a list of items, possibly in sequential order. The sentence *the sum of* 2 *and* 3 *is* 5 uses *and* in this way. We cannot expand this into *the sum of* 2 *is* 5 *and the sum of* 3 *is* 5; both 2 and 3 are needed to give the sum 5.

On the other hand, conjunction may be present when *and* is missing. Ordinary language often piles up adjectives as an abbreviated form of conjunction. The mathematical sentence $\triangle ABC$ *is a right isosceles triangle* can be expanded into the conjunction $\triangle ABC$ *is a right triangle and* $\triangle ABC$ *is an isosceles triangle*. *The number* 2 *is prime but not odd* asserts in shortened form that *the number* 2 *is prime and the number* 2 *is not odd*. The word *but* often indicates a conjunction where some sort of contrast is being stressed. To transcribe such sentences, however, we use the *and* connective.

In some mathematical formulas *and* is camouflaged by the notation. This occurs with $a \leq b \leq c$, which means $(a \leq b) \wedge (b \leq c)$. It's important to realize that such formulas are conjunctions if they're to be used properly in arguments.

1.2.4 *Syntax and Semantics of Disjunction*

The truth-functional connective *or* is our second connective. Joining sentences P and Q by *or* to obtain *P-or-Q* gives the *disjunction* of P and Q. We'll symbolize *or* by $\vee$; *P-or-Q* is written as $P \vee Q$. In some mathematical notation, though, *or* remains implicit. For example, $a \leq b$ is short for $(a < b) \vee (a = b)$; $x = \pm 1$ abbreviates $(x = +1) \vee (x = -1)$.

The most appropriate meaning of *or* is less fixed than that of *and*. Sometimes *or* is meant in an exclusive sense. This is often the case when two mutually exclusive alternatives are present. The sentence

$$\sqrt{2}^{\sqrt{2}} \ is\ rational\ or\ \sqrt{2}^{\sqrt{2}}\ is\ irrational$$

seems to use *or* in this narrow sense. In cases like this, however, where the sentences are logical opposites, truth values can't be assigned independently. Contradictories cannot be true simultaneously (row one in a truth table), which is the only truth-value assignment that distinguishes an *exclusive or* from a *nonexclusive or*.

There are cases in everyday life where *exclusive or* is intended. The *or* on a restaurant menu saying that a meal comes with either a salad or a vegetable is an *exclusive or*. Here there is no logical reason why both disjuncts cannot be true, only a business reason. If a waiter brings both a salad and a vegetable without charging extra, the menu's intent will have been violated. When both alternatives are true, an *exclusive-or* statement is false.

On the other hand, *or* can also be used in a nonexclusive sense. This is the type of *or* used in Boolean searches on the internet. The search-phrase *Aristotle or Boole* will call up items on Aristotle, Boole, and both.

Which connective to take as primary is mostly a matter of preference. *Propositional Logic* has been developed in both ways. Following an earlier tradition, George Boole used *exclusive or* in his system of logic. The modern approach, though, follows Boole's successors Jevons and Peirce in adopting the inclusive meaning for *or*, since this lends itself more readily to algebraic treatment. *Nonexclusive or* is also the simpler of the two—*exclusive or* can be defined more easily in terms of *nonexclusive or* than the other way around (see Exercise Set 1.3). Finally, *nonexclusive or* better matches how union is defined in *Set Theory*, as we'll see later. For all these reasons, we'll assume that *nonexclusive or* is intended in a mathematical sentence and formulate it with $\vee$.[1] We'll use $\veebar$ or XOR to represent *exclusive or* when it is wanted.

Definition 1.2.2: *Truth-Value Assignment for Disjunction*
A **disjunction** $\mathbf{P} \vee \mathbf{Q}$ *is true if and only if either* $\mathbf{P}$ *is true or* $\mathbf{Q}$ *is true.*
The truth table for $\mathbf{P} \vee \mathbf{Q}$ (in both formats) is the following:

P	**Q**	**P $\vee$ Q**
T	T	T
T	F	T
F	T	T
F	F	F

P	$\vee$	**Q**
T	T	T
T	T	F
F	T	T
F	F	F

1.2.5 *Syntax and Semantics of Negation*

The logical connective *not* differs from *and* and *or*. In fact, you might question whether it's even a connective, since it doesn't do any connecting. It's still a propositional operator, however—a *unary*, rather than a binary, operator. The *not*-operator applied to a sentence P turns it into its logical opposite, *not-P*. If P is the sentence $\sqrt{2}^{\sqrt{2}}$ *is rational*, its negation *not-P* is the sentence $\sqrt{2}^{\sqrt{2}}$ *is not rational*. *Not* is usually placed somewhere in the middle of an ordinary sentence, seemingly negating the verb or predicate clause, but it actually negates the entire statement—*it is not the case that* $\sqrt{2}^{\sqrt{2}}$ *is rational*.

We will use the hooked minus sign $\neg$ for logical negation. Some textbooks use a tilde ($\sim$), but since this symbol is also used in mathematics as a relation symbol while $\neg$ isn't, we'll adopt the hook. The negation of P, then, is $\neg P$, read as *not-P*. A more elaborate reading is *it is not the case that P*.

In certain mathematical contexts we slash through a sign to indicate negation. This is done, for instance, in cases like the following:

$$0 \neq 1; \qquad \triangle ABC \not\cong \triangle A'B'C'; \qquad x \notin S.$$

Wherever this works, we'll use the slash rather than pre-fixing a $\neg$, which often makes a sentence less easy to read.

---

[1] The two Latin words for *exclusive* and *nonexclusive or* are *aut* and *vel*, respectively. Might this be why $\vee$ (standing for *vel*) was chosen to denote *nonexclusive or*?

Definition 1.2.3: *Truth-Value Assignment for Negation*
*A **negation** ¬P is true if and only if* **P** *is false.*

The truth table for ¬**P** is the following:

P	¬**P**
T	F
F	T

Since *not* is a propositional operator, it can be applied to any sentence, even one that is already a negation. The meaning of a simple negation is quite clear, but multiple negation is often abused in everyday conversation. A double negative, for instance, might indicate a strongly felt negation. In two-valued logic, however, our feelings about negation don't count. Double negation is the logical opposite of negation. It asserts the original statement in an equivalent form, giving ¬¬**P** and **P** the same truth tables.[2]

1.2.6 *Complex Compound Sentences*

Using ∧, ∨, and ¬, we can make complex propositions out of simple ones. The semantics of such compound sentences is straightforward, provided the syntax is clear. Our notation should show how a full sentence is syntactically constructed from its constituent sub-sentences by truth-functional connectives. Adopting a priority convention can simplify this process.

For example, consider the following compound sentence, which could arise when solving a quadratic equation for a positive root:

$$(*)\qquad\qquad x \geq 0 \ \wedge \ x - 1 = 0 \ \vee \ x + 2 = 0.$$

Does this yield -2 as a solution? It all depends on the order in which the logical connectives operate. If ∧ is the last connective applied, then $x = 1$, and $x = -2$ must be ruled out. On the other hand, if the final connective is ∨, $x = -2$ is a solution. As it stands, this sentence is ambiguous.

We can make such sentences unambiguous in several ways. A parenthesis-free notation, devised by twentieth-century Polish logicians, uses pre-fix notation for the operators. This seems unnatural to someone accustomed to the algebraic practice of putting operation symbols between the letters (in-fix notation), but it makes sentences completely unambiguous. The above sentences would be written, using *Polish Notation* (but our connective symbols), as follows: *P and Q-or-R* as ∧*P* ∨ *QR*, and *P-and-Q or R* as ∨ ∧ *PQR*.

A more common approach uses parentheses, like in algebra. Here *P and Q-or-R* would be written as $P \wedge (Q \vee R)$, and *P-and-Q or R* as $(P \wedge Q) \vee R$.

We can use fewer parentheses by stipulating that some operations have *higher priority than* (are applied before) others. We'll give ¬ highest priority and apply it before both ∧ and ∨. Of ∧ and ∨, ∧ has priority over ∨ (like multiplication over addition in algebraic formulas).

---

[2] Intuitionist logic, however, takes double negation as weaker than the original proposition. We'll note some differences this makes as we proceed.

Priority	Symbol
1	¬
2	∧
3	∨

With this convention, we would write P *and* Q-*or*-R as $P \wedge (Q \vee R)$ but P-*and*-Q *or* R as $P \wedge Q \vee R$. When parentheses are dropped according to our conventions, formulas remain unambiguous. At times, though, to avoid any potential misunderstanding, we may still include parentheses, even if they can be omitted.

✤ Example 1.2.2

Compare the two sentence forms $\neg(\mathbf{P} \vee \mathbf{Q})$ and $\neg\mathbf{P} \vee \neg\mathbf{Q}$.

Solution

The truth tables for these sentences are as follows:

P	**Q**	**P ∨ Q**	**¬(P ∨ Q)**
T	T	T	F
T	F	T	F
F	T	T	F
F	F	F	T

P	**Q**	**¬P**	**¬Q**	**¬P ∨ ¬Q**
T	T	F	F	F
T	F	F	T	T
F	T	T	F	T
F	F	T	T	T

These tables make it clear that the two sentences make distinct claims. Their truth values disagree in the second and third rows.

The syntactic structure of a sentence reflects how it's constructed from its constituent sentences. A *production graph* nicely illustrates this, starting with sentence letters on the bottom and proceeding up the graph to more compound components, ending with the full sentence at the top. For instance, the production graphs for the two sentences in Example 2 are as follows:

The last connective joining the parts of a sentence together is the *main connective*. Recognizing a sentence's main connective is important, both for creating abbreviated truth tables and for being able to construct deductions. The main connective determines a compound sentence's overall logical structure. In the two sentences just given, this is ¬ and ∨, respectively.

✤ Example 1.2.3

Determine the main connective of the sentence form $\neg\mathbf{P} \wedge (\mathbf{Q} \vee \mathbf{R})$, which is one way to transcribe sentence (∗) above, taking $x \geq 0$ to mean $x \not< 0$. Then write down its truth table in both formats.

Solution

The main connective for $\neg \mathbf{P} \wedge (\mathbf{Q} \vee \mathbf{R})$ is $\wedge$. The full sentence's truth-value assignment in the compact table can be read off below this connective (above the double underline). Single underlines indicate main connectives of inner-level component sentences whose values must be calculated first.

P	Q	R	¬P	Q∨R	¬P∧(Q∨R)	¬	P	∧	(Q	∨	R)
T	T	T	F	T	F	F	T	F	T	T	T
T	T	F	F	T	F	F	T	F	T	T	F
T	F	T	F	T	F	F	T	F	F	T	T
T	F	F	F	F	F	F	T	F	F	F	F
F	T	T	T	T	T	T	F	T	T	T	T
F	T	F	T	T	T	T	F	T	T	T	F
F	F	T	T	T	T	T	F	T	F	T	T
F	F	F	T	F	F	T	F	F	F	F	F

Note that since this sentence has three letters, the truth table has eight rows to include all possible truth-value assignments. The last letter in the formula ($\mathbf{R}$) again has truth values that alternate $TFTF$, while truth values for the letters to its left ($\mathbf{Q}$, $\mathbf{P}$) alternate in blocks of two or four.

1.2.7 Logical Truths and Falsehoods

There are three distinct types of sentence forms with respect to truth values.

Definition 1.2.4: Logical Status of Sentences
  a) A *logically indeterminate* sentence form is one that can be either true or false under different truth-value assignments.
  b) A *logically true* sentence form is one that is true under all truth-value assignments. **Logical truths** are also called **tautologies**.
  c) A *logically false* sentence form is one that is false under all truth-value assignments. **Logical falsehoods** are also called **contradictions**.

The sentence forms in Examples 2 and 3 are logically indeterminate, because their truth values depend on more than logic. Logical truths and falsehoods, on the other hand, are true or false solely due to their logical form.

Some logical truths involve only the connectives *and*, *or*, and *not*. Two of these are associated with principles some consider *Basic Laws of Logic*. The *Law of Non-Contradiction*, abbreviated *LNC*, captures the idea that nothing can both be the case and not be the case. Thus, $\neg(\mathbf{P} \wedge \neg\mathbf{P})$ is a logical truth for any proposition $\mathbf{P}$, as a short truth table will show (see Exercise 35).

The *Law of Excluded Middle* (*LEM*) states that something either is or is not the case—a third or middle alternative is excluded. Either a sentence is true or its negation is.[3] Sentences of the form $\mathbf{P} \vee \neg\mathbf{P}$ can likewise be shown to be logical truths by a simple truth table (see Exercise 36).

---

[3] Intuitionist mathematicians dispute this law due to their view of negation.

EXERCISE SET 1.2

Exercises 1–3: Ordinary Sentence Symbolization
Determine all atomic sentences within the given statements. Symbolize each sentence with a capital letter, provide an interpretation key, and then rewrite the entire sentence using logical symbolism.

1.2.1. Either this is an easy exercise, or there's more to it than I know.

1.2.2. Little Bo Peep has lost her sheep and doesn't know where to find them.

1.2.3. Either this is more complex, or I'm mistaken; but I'm not mistaken.

Exercises 4–6: Mathematical Sentence Symbolization
Determine all atomic sentences within the given mathematical statements. Symbolize each sentence with a capital letter, provide an interpretation key, and then rewrite the entire sentence using logical symbolism.

1.2.4. $\triangle ABC$ is equilateral or isosceles or scalene.

1.2.5. n is divisible by 6 or n is not divisible by 2.

1.2.6. Either $r > 0$ and $|r| = r$, or $r \not> 0$ and $|r| = -r$.

Exercises 7–10: True or False
Are the following statements true or false? Explain your answer.

1.2.7. $P \vee Q$ is true only if exactly one of P or Q is true.

1.2.8. $\neg\neg \mathbf{P}$ has the same truth table as $\mathbf{P}$.

1.2.9. With the priority conventions for $\neg$, $\wedge$, and $\vee$, parentheses are never needed for formulating sentences involving these connectives.

1.2.10. Logic alone can never determine the truth value of a sentence.

Exercises 11–14: Defining Terms
Define each of the following terms in your own words.

1.2.11. *Truth-functional connective*   **1.2.13.** *Tautology*
1.2.12. *Main connective*   **1.2.14.** *Contradiction*

Exercises 15–17: Problematic Formulations
What is wrong with the following statements? What should be said instead?

1.2.15. $P \wedge Q$ means P is true and Q is true.

1.2.16. $P \vee Q$ says P is true or Q is true or both.

1.2.17. $\neg P$ is an abbreviation for P is false.

Exercises 18–21: Sentence Construction and Main Connectives
Give a production graph for each of the following compound sentences and identify the main connective. Then write out its truth table.

1.2.18. $P \vee \neg Q$ **1.2.20.** $(\neg P \vee Q) \wedge R$
1.2.19. $\neg(P \wedge \neg Q)$ **1.2.21.** $P \vee (\neg Q \wedge \neg R)$

Exercises 22–25: Polish Notation
Write each of the following sentences in Polish pre-fix notation, using our
connective symbols. Where does the main connective show up in the formula?

1.2.22. $P \vee \neg Q$ **1.2.24.** $(\neg P \vee Q) \wedge R$

1.2.23. $\neg(P \wedge \neg Q)$ **1.2.25.** $P \vee (\neg Q \wedge \neg R)$

1.2.26. Is the sentence form $\mathbf{P} \wedge \mathbf{Q} \wedge \mathbf{R}$ ambiguous or not? Explain.

Exercises 27–30: Sentences with Given Truth Values
Design compound sentences in the letters given to satisfy the stated condition.

1.2.27. A sentence that is true if and only if neither $\mathbf{P}$ nor $\mathbf{Q}$ is true.

1.2.28. A sentence that is true if and only if either $\mathbf{P}$ or $\mathbf{Q}$ are false.

1.2.29. A sentence that is true if and only if $\mathbf{P}$, $\mathbf{Q}$ have different truth values.

1.2.30. A sentence that is true if and only if exactly one of $\mathbf{P}$, $\mathbf{Q}$, $\mathbf{R}$ is true.

Exercises 31–34: Logical Status of Sentences
Are the following logically true, logically false, or logically indeterminate?

1.2.31. $P \vee \neg(P \wedge Q)$ **1.2.33.** $(P \wedge \neg Q) \wedge (\neg P \vee Q)$

1.2.32. $(P \vee \neg Q) \wedge (\neg P \vee Q)$ **1.2.34.** $(\neg P \wedge \neg Q) \vee (P \vee Q)$

Exercises 35–36: Tautologies
Show that the following laws are tautologies.

1.2.35. The *Law of Non-Contradiction*: $\neg(\mathbf{P} \wedge \neg\mathbf{P})$

1.2.36. The *Law of Excluded Middle*: $\mathbf{P} \vee \neg\mathbf{P}$

1.2.37. Explain why the negation of a tautology is a contradiction and the negation of a contradiction is a tautology.

1.2.38. A truth table for a formula with one sentence variable has only two rows. A truth table for a formula with two sentence variables has four rows.
   a. Explain why a formula with three sentence variables has an eight-row truth table.
   b. How many rows will a truth table have for a formula with four sentence variables? Explain.
   c. How many rows will a truth table have for a formula with n sentence variables? Support your answer as best you can.

1.2.39. Find the solution set of all ordered pairs (x, y) such that $x(1 - y^2) = 0$ and $(x + 2)y = 0$. Factoring each equation, work your solution step by step and point out where the logical connectives *or* and *and* enter the process.

1.3 Argument Semantics for Propositional Logic

We'll now focus on valid argument forms and logical implication relative to *Propositional Logic*. We'll begin by looking at the slightly simpler relation of logical equivalence. At the end of this section, we'll also introduce the metalogical concepts of consistency, independence, and completeness.

1.3.1 *Logical Equivalence in Propositional Logic*

Compound sentences $\mathbf{P}$ and $\mathbf{Q}$ are logically equivalent when they say essentially the same thing, possibly using different logical formulations. When this is the case, we write $\mathbf{P} \vDash\dashv \mathbf{Q}$[4] and read $\mathbf{P}$ *is logically equivalent to* $\mathbf{Q}$.

Definition 1.3.1: *Logical Equivalence*
  $\mathbf{P} \vDash\dashv \mathbf{Q}$ *if and only if* $\mathbf{P}$ *and* $\mathbf{Q}$ *have identical truth tables.*

It's easy to find pairs of logically equivalent sentences. It's intuitively clear, for instance, that conjunction and disjunction are commutative operators: $\mathbf{P} \wedge \mathbf{Q} \vDash\dashv \mathbf{Q} \wedge \mathbf{P}$ and $\mathbf{P} \vee \mathbf{Q} \vDash\dashv \mathbf{Q} \vee \mathbf{P}$ (see Exercise 3).

There are more interesting examples than these—for instance, the *Law of Double Negation* (*DN*) asserts that negation is its own inverse: $\neg\neg\mathbf{P} \vDash\dashv \mathbf{P}$ (see Exercise 2). Two other important equivalences are *De Morgan's Laws*, named in honor of Augustus De Morgan, though known earlier to the medieval philosopher William of Occam. They expand negated conjunctions and disjunctions: $\neg(\mathbf{P} \wedge \mathbf{Q}) \vDash\dashv \neg\mathbf{P} \vee \neg\mathbf{Q}$ and $\neg(\mathbf{P} \vee \mathbf{Q}) \vDash\dashv \neg\mathbf{P} \wedge \neg\mathbf{Q}$. We'll show the first of these and leave the other as an exercise (see Exercise 9a).

✦**Example 1.3.1**
  Show *De Morgan's Law* for negating a conjunction: $\neg(\mathbf{P} \wedge \mathbf{Q}) \vDash\dashv \neg\mathbf{P} \vee \neg\mathbf{Q}$.

Solution
  The following double truth table establishes the equivalence. Compare truth values below the two main connectives (above the double underlines).

$\neg\ (\mathbf{P} \wedge \mathbf{Q})$	$\neg\ \mathbf{P} \vee \neg\ \mathbf{Q}$
$F\ T\ T\ T$	$F\ T\ F\ F\ T$
$T\ T\ F\ F$	$F\ T\ T\ T\ F$
$T\ F\ F\ T$	$T\ F\ T\ F\ T$
$T\ F\ F\ F$	$T\ F\ T\ T\ F$

1.3.2 *Logical Implication in Propositional Logic*

In Section 1.1 we said that a set of sentences $\mathcal{P}$ logically implies a sentence Q if and only if every interpretation of the nonlogical terms making the premises $\mathcal{P}$ true also makes the conclusion Q true. For *Propositional Logic*, we can reformulate this criterion in terms of the sentences' truth values. By assigning all possible truth-value combinations to the constituent sentences, we can be certain that all cases have been taken into consideration.

Definition 1.3.2: *Logical Implication, Logical Consequence*
  a) *A set of sentences* $\mathcal{P}$ ***logically implies*** *a sentence* $\mathbf{Q}$, *written* $\mathcal{P} \vDash \mathbf{Q}$, *if and only if every truth-value assignment making* $\mathcal{P}$ *true makes* $\mathbf{Q}$ *true.*
  b) $\mathbf{Q}$ *is a **logical consequence** of* $\mathcal{P}$ *if and only if* $\mathcal{P} \vDash \mathbf{Q}$.

---

[4] This notation ought to be standard, given the one used for implication (see below), but I don't know of any other text that adopts it.

As this notation suggests, logical implication is half of logical equivalence for pairs of sentences. If $\mathbf{P} \vDash \mathbf{Q}$ and $\mathbf{Q} \vDash \mathbf{P}$, then $\mathbf{P} \Vdash \mathbf{Q}$: for $\mathbf{Q}$ is true whenever $\mathbf{P}$ is, and $\mathbf{P}$ is true whenever $\mathbf{Q}$ is.

Both $\mathbf{P}$ and $\mathbf{Q}$ may be thought of as *conditions* with respect to logical implication. For suppose the sentence $\mathbf{P} \vDash \mathbf{Q}$ is true. $\mathbf{P}$ is then a *sufficient condition* for $\mathbf{Q}$, i.e., affirming $\mathbf{P}$ is *sufficient* for concluding $\mathbf{Q}$. Similarly, $\mathbf{Q}$ is a *necessary condition* for $\mathbf{P}$; for if $\mathbf{P}$ is the case, $\mathbf{Q}$ is *necessarily* so as well. Whether or not you think $\mathbf{Q}$ is a *condition* in the strict sense, using this term for the consequent as well as the antecedent gives us a common name for both components of an implication.

Truth tables provide us with an *effective decision procedure* for testing logical implication in PL: you make an *extended truth table* with columns for each formula involved, separate the premises from the conclusion by a double vertical line, and check whether the conclusion is true whenever the premises are. The next example illustrates this process.

❖**Example 1.3.2**

Show that *conjunction implies disjunction*, but not conversely, i.e., show $\mathbf{P} \wedge \mathbf{Q} \vDash \mathbf{P} \vee \mathbf{Q}$, but $\mathbf{P} \vee \mathbf{Q} \nvDash \mathbf{P} \wedge \mathbf{Q}$. Thus, $\mathbf{P} \wedge \mathbf{Q} \nVdash \mathbf{P} \vee \mathbf{Q}$.

Solution

This is shown by the following double truth table.

$\mathbf{P} \wedge \mathbf{Q}$	$\mathbf{P} \vee \mathbf{Q}$
$T \, ⓉT$	$Ⓣ$
$T \, F \, F$	T
$F \, F \, T$	T
$F \, F \, F$	F

· Comparing the truth values of the conclusion with those of the premise, we see that whenever a conjunction is true (row one), so is the disjunction.
· The converse fails, however. The disjunction can be true while the conjunction is false (the second and third rows). Thus, disjunction does not imply conjunction, and so the two sentence forms are not logically equivalent, either. Conjunction is *logically stronger* than disjunction—more is asserted by a conjunction than by a disjunction.

Arguments with more than one premise are treated similarly. We'll list the individual sentences in the premise-set $\mathcal{P}$ left of the double turnstile $\vDash$, separated by commas, and put the conclusion $\mathbf{Q}$ to the right. Each premise receives a separate column preceding the double line and the conclusion in an extended truth table demonstrating the implication.

❖**Example 1.3.3**

Show that ruling out one alternative in a disjunction implies the other disjunct: $\mathbf{P} \vee \mathbf{Q}, \neg\mathbf{P} \vDash \mathbf{Q}$. This result underlies the strategy for playing Sudoku and for drawing detective-like conclusions from evidence.

Solution

In the following extended truth table, the third row establishes the implication: when both premises are true, the conclusion is also true.

P ∨ Q	**¬P**	**Q**
T T T	*F*	*T*
T T F	*F*	*F*
F ⓉT	*Ⓣ*	*Ⓣ*
F F̱ F̱	*Ṯ*	*F̱*

Complete truth tables are useful but aren't really necessary to show that $\mathcal{P} \models \mathbf{Q}$. You merely determine what the truth of the premises $\mathcal{P}$ dictates about the truth values of its constituent sentences and then show that **Q** must also be true under those assignments. Looking again at the last example, if ¬**P** is true, **P** must be false; but then **Q** must be true to make **P** ∨ **Q** true. So, whenever the premises are both true, the conclusion **Q** is, too.

Showing $\mathcal{P} \not\models \mathbf{Q}$ is done in the opposite way: you find an assignment that makes **Q** false while keeping the premises $\mathcal{P}$ true. Alternatively, you can construct a *counterargument*: find a concrete argument in the same form in which the premises are true but the conclusion is false.

Reading sentences such as $P \wedge Q \models P \vee Q$ shouldn't be confusing, because $\models$ and $\dashv\models$ are relation symbols, not logical operators. $P \wedge Q \models P \vee Q$ is not a compound PL sentence, but a claim *about* how sentences $P \wedge Q$ and $P \vee Q$ are logically related. It belongs to the *meta-theory* of *Propositional Logic*. Such claims are analogous to ones like $3 \leq 3 + 1$, which make order-relation claims *about* numbers and are not themselves numbers like $3 + 1$.

Logical implication notation can be used to exhibit a sentence as a tautology. Placing it after a double turnstile indicates it is true without assuming any premises: $\models \neg(P \wedge \neg P)$ and $\models P \vee \neg P$ say that these instances of the *Law of Non-Contradiction* and the *Law of Excluded Middle* are logical truths.

1.3.3 *Implication/Equivalence Relative to Mathematics*

Mathematicians are usually not concerned with pure logical implication or equivalence but with how these concepts relate to a mathematical theory. For example, in claiming that a result is logically equivalent to *Euclid's Parallel Postulate*, mathematicians mean this relative to (the rest of) *Euclidean Geometry*, not in an absolute sense.

We can define these relativized notions as follows, taking $\mathcal{A}$ as a set of axioms and the deductive theory of $\mathcal{A}$ as the set of consequences of $\mathcal{A}$.

Definition 1.3.3: *Implication and Equivalence Relative to a Theory*
   a) *A set of sentences $\mathcal{P}$ **logically implies** a sentence **Q** **relative to** $\mathcal{A}$, symbolized by $\mathcal{P} \models_{\mathcal{A}} \mathbf{Q}$, if and only if $\mathcal{A}, \mathcal{P} \models \mathbf{Q}$.*
   b) *A sentence **P** is **logically equivalent** to a sentence **Q** **relative to** $\mathcal{A}$, symbolized by $\mathbf{P} \dashv\models_{\mathcal{A}} \mathbf{Q}$, if and only if $\mathbf{P} \models_{\mathcal{A}} \mathbf{Q}$ and $\mathbf{Q} \models_{\mathcal{A}} \mathbf{P}$.*

We'll find these ideas helpful in Section 1.4 when we consider how certain misperceptions about logical implication arise in mathematical circles.

1.3.4 *Consistency, Independence, and Completeness*

When mathematicians axiomatize a theory, they have several objectives in mind. First of all, they want their axioms to be relatively simple while still providing a solid deductive foundation for the whole field. This requires trained mathematical intuition and experience, not logical expertise.

However, mathematicians also have some logical concerns. They want axioms that are consistent with one another; they want axioms that are somewhat independent (to avoid unnecessary overlap); and they want axioms that will prove all known results. We'll look at each of these logical notions in turn: *consistency*, *independence*, and *completeness*.

A set of sentences is *logically consistent* if and only if the sentences do not contradict one another, i.e., if and only if some of them being true don't force others to be false.

Definition 1.3.4: *Consistent and Inconsistent Sentences*
   a) *A set of sentences $\mathcal{P}$ is **consistent** if and only if some truth-value assignment makes all the sentences in $\mathcal{P}$ true.*
   b) *A set of sentences $\mathcal{P}$ is **inconsistent** if and only if it is not consistent.*

Consistency can be tested for PL sentences by an extended truth table. For example, $P \vee Q$, $\neg Q \vee R$, and $\neg P \wedge R$ are consistent, while $P \vee Q$, $\neg Q \vee R$, and $\neg P \wedge \neg R$ are inconsistent (see Exercise 25a).

Mathematicians and logicians occasionally discover that sentences they believe to be consistent are not so after all. From a logical point of view, this devastates the entire deductive system and calls into question the truth of the system's conclusions. As we'll see later, inconsistency makes it possible to prove *any proposition* formulated in the language of that theory. Consistency is thus an absolute prerequisite for developing an axiomatic theory. Demonstrating consistency has been a primary goal of mathematical logicians since foundational issues became a central concern a little over a century ago.

Another goal for an axiomatic theory is that its axioms be *logically independent*. This means that the sentences have no logical connection.

Definition 1.3.5: *Independent Sentences*
   *A sentence $\mathbf{Q}$ is **independent** of a set of sentences $\mathcal{P}$ if and only if neither $\mathbf{Q}$ nor $\neg\mathbf{Q}$ is a logical consequence of $\mathcal{P}$, i.e., $\mathcal{P} \not\models \mathbf{Q}$ and $\mathcal{P} \not\models \neg\mathbf{Q}$.*

This definition says that $\mathbf{Q}$ is independent of $\mathcal{P}$ if and only if one truth-value assignment makes all sentences of $\mathcal{P}$ and $\mathbf{Q}$ true and another makes the sentences of $\mathcal{P}$ true but $\mathbf{Q}$ false. In other words, $\mathbf{Q}$ is independent of $\mathcal{P}$ if and only if both $\{\mathcal{P}, \mathbf{Q}\}$ and $\{\mathcal{P}, \neg\mathbf{Q}\}$ are logically consistent (see Exercise 25c).

In many cases, we don't want to accept as an axiom something that can be proved. This strategy can be counterproductive, though. A set of completely independent axioms may make some deductions long and difficult. That being said, there are times when independence is important. If neither a result nor its negation is implied by the axioms of a theory, mathematicians must decide which alternative to adopt as a new axiom in an extension of the original theory. In the nineteenth century, for example, after more than 2000 years of unsuccessful attempts to prove *Euclid's Parallel Postulate*, mathematicians discovered that it was independent of the other axioms of geometry. This gave them logical sanction to take *Non-Euclidean Geometry* seriously. Some of the most important results in *Set Theory* and *Logic* over the past century have been independence results.

Finally, closely related to consistency and independence is the concept of *theory completeness*. A theory is *complete* if and only if no sentence in the language of the theory is independent of its axioms. Any genuine extension of a complete theory, therefore, would be inconsistent.

Definition 1.3.6: *Theory Completeness*
*A theory axiomatized by $\mathcal{A}$ is **complete** if and only if for any sentence $\mathbf{Q}$ in the language of $\mathcal{A}$ either $\mathcal{A} \models \mathbf{Q}$ or $\mathcal{A} \models \neg\mathbf{Q}$.*

Sometimes mathematicians want a complete theory, but this is not always their aim. For instance, it would be good to have a complete theory of *Arithmetic* or *Euclidean Geometry*, since these are intended to be about particular structures (the natural numbers; Euclidean space). But the theories of vector spaces and groups and topological spaces are meant to be incomplete because they axiomatize common abstract features of mathematical structures known to have some very different properties.

You may wonder whether complete consistent mathematical theories exist. They do, but well-known incompleteness results by Kurt Gödel and others apply even to relatively simple mathematical theories, like *Arithmetic*. This goes beyond *Propositional Logic*, however, and lies outside the scope of this book.

EXERCISE SET 1.3

Exercises 1–7: Basic Logical Equivalences
Show that the following equivalences hold.

1.3.1. *Idempotence Laws*
a. $\mathbf{P} \wedge \mathbf{P} \equiv\!\models \mathbf{P}$ b. $\mathbf{P} \vee \mathbf{P} \equiv\!\models \mathbf{P}$

1.3.2. *Law of Double Negation*
a. $\neg\neg\mathbf{P} \equiv\!\models \mathbf{P}$ b. $\neg\neg\neg\mathbf{P} \equiv\!\models \neg\mathbf{P}$

1.3.3. *Commutative Laws*
a. $\mathbf{P} \wedge \mathbf{Q} \equiv\!\models \mathbf{Q} \wedge \mathbf{P}$ b. $\mathbf{P} \vee \mathbf{Q} \equiv\!\models \mathbf{Q} \vee \mathbf{P}$

1.3.4. *Associative Laws*
a. $(\mathbf{P} \wedge \mathbf{Q}) \wedge \mathbf{R} \equiv\!\models \mathbf{P} \wedge (\mathbf{Q} \wedge \mathbf{R})$ b. $(\mathbf{P} \vee \mathbf{Q}) \vee \mathbf{R} \equiv\!\models \mathbf{P} \vee (\mathbf{Q} \vee \mathbf{R})$

1.3.5. *Absorption Laws*
  a. $\mathbf{P} \wedge (\mathbf{P} \vee \mathbf{Q}) \dashv\vdash \mathbf{P}$ c. $\mathbf{P} \wedge (\mathbf{Q} \vee \neg\mathbf{Q}) \dashv\vdash \mathbf{P}$
  b. $\mathbf{P} \vee (\mathbf{P} \wedge \mathbf{Q}) \dashv\vdash \mathbf{P}$ d. $\mathbf{P} \vee (\mathbf{Q} \wedge \neg\mathbf{Q}) \dashv\vdash \mathbf{P}$

1.3.6. *Distributive Laws*
Show that *conjunction distributes over disjunction*, much like multiplication
distributes over addition in arithmetic and algebra.
  a. $\mathbf{P} \wedge (\mathbf{Q} \vee \mathbf{R}) \dashv\vdash (\mathbf{P} \wedge \mathbf{Q}) \vee (\mathbf{P} \wedge \mathbf{R})$ b. $(\mathbf{P} \vee \mathbf{Q}) \wedge \mathbf{R} \dashv\vdash (\mathbf{P} \wedge \mathbf{R}) \vee (\mathbf{Q} \wedge \mathbf{R})$

1.3.7. *More Distributive Laws*
Addition does not distribute over multiplication: $3 + 5 \times 2 \neq (3 + 5) \times (3 + 2)$.
Does *disjunction distribute over conjunction*? Determine whether the following hold. Support your answer.
  a. $\mathbf{P} \vee (\mathbf{Q} \wedge \mathbf{R}) \dashv\vdash (\mathbf{P} \vee \mathbf{Q}) \wedge (\mathbf{P} \vee \mathbf{R})$ b. $(\mathbf{P} \wedge \mathbf{Q}) \vee \mathbf{R} \dashv\vdash (\mathbf{P} \vee \mathbf{R}) \wedge (\mathbf{Q} \vee \mathbf{R})$

1.3.8. Find a simpler logical equivalent to the following computer program
instruction: *while* $((x < 40$ *AND* $y > 90)$ *OR* $(x < 40$ *AND* $(y > 90$ *OR*
$z > 10)))$, *do* .... Symbolize the atomic sentences with letters to help you
determine an equivalent. Check your answer with a truth table.

1.3.9. *De Morgan's Law for Negating a Disjunction*
  a. Show the dual law to Example 1: $\neg(\mathbf{P} \vee \mathbf{Q}) \dashv\vdash \neg\mathbf{P} \wedge \neg\mathbf{Q}$.
  b. Determine an equivalent to $\neg(\mathbf{P} \wedge \neg\mathbf{P})$. Using the *Law of Double Negation*,
     simplify your answer. What law is the final sentence an instance of?
  c. Carefully interpret the meaning of the sentence $x \neq \pm 1$ by putting it
     into expanded form, using logical symbolism for the connectives involved.
     Then state an equivalent sentence using *De Morgan's Law*.

1.3.10. Show that logical equivalence is an *equivalence relation*. That is, tell
why $\dashv\vdash$ satisfies the following *reflexive, symmetric*, and *transitive* properties:
  a. *Reflexive*: $\mathbf{P} \dashv\vdash \mathbf{P}$.   b. *Symmetric*: if $\mathbf{P} \dashv\vdash \mathbf{Q}$, then $\mathbf{Q} \dashv\vdash \mathbf{P}$.
  c. *Transitive*: if $\mathbf{P} \dashv\vdash \mathbf{Q}$ and $\mathbf{Q} \dashv\vdash \mathbf{R}$, then $\mathbf{P} \dashv\vdash \mathbf{R}$.

1.3.11. Is the relation of logical implication an equivalence relation? That is,
does $\vDash$ satisfy all of the following properties? If not, which ones hold?
  a. *Reflexive*: $\mathbf{P} \vDash \mathbf{P}$.   b. *Symmetric*: if $\mathbf{P} \vDash \mathbf{Q}$, then $\mathbf{Q} \vDash \mathbf{P}$.
  c. *Transitive*: if $\mathbf{P} \vDash \mathbf{Q}$ and $\mathbf{Q} \vDash \mathbf{R}$, then $\mathbf{P} \vDash \mathbf{R}$.

Exercises 12–15: Logical Implications
Use truth tables to evaluate the following implications.

1.3.12. $\neg(\mathbf{P} \wedge \mathbf{Q}), \mathbf{P} \vDash \neg\mathbf{Q}$ **1.3.14.** $\mathbf{P} \vee \mathbf{Q}, \neg\mathbf{P} \wedge \neg\mathbf{Q} \vDash \mathbf{R}$
1.3.13. $(\mathbf{P} \vee \mathbf{Q}) \wedge \mathbf{R}, \neg\mathbf{P} \vdash \mathbf{Q} \wedge \mathbf{R}$

1.3.15. Do the following expansion and contraction laws hold? Explain.
  a. $\mathbf{P} \wedge \mathbf{Q} \vDash (\mathbf{P} \vee \mathbf{R}) \wedge (\mathbf{Q} \vee \mathbf{S})$ b. $(\mathbf{P} \vee \mathbf{R}) \wedge (\mathbf{Q} \vee \mathbf{S}) \vDash \mathbf{P} \wedge \mathbf{Q}$

Exercises 16–21: True or False
Are the following statements true or false? Explain your answer.

1.3.16. The implication symbol $\vDash$ represents a truth-functional connective.

1.3.17. $P \Dashv\vDash Q$ if and only if $P \vDash Q$ and $Q \vDash P$.

1.3.18. Mathematicians require their axiom systems to be consistent.

1.3.19. A mathematical result is independent of other statements if and only if it cannot be proved from them.

1.3.20. Many axiom systems in mathematics are incomplete by design.

1.3.21. Maximal consistent theories are complete.

1.3.22. *Defining Terms*
Give the meaning of the following terms in your own words.
  a. *Logical equivalence*   b. *Logical consequence*

Exercises 23–26: Consistency, Implication, and Independence
The following problems explore consistency, implication, and independence.

1.3.23. *Logical Falsehoods, Inconsistent Sentences, and Logical Consequences*
  a. Explain why a logical falsehood implies any sentence whatsoever.
  b. Sometimes mathematicians make a stronger claim than *a*, namely, that a *false* sentence implies any sentence. Is this correct? Explain carefully.
  c. Does an inconsistent set of sentences imply any sentence whatsoever? Why or why not?

1.3.24. *Contradictory Sentences, Inconsistent Sentences*
Contradictory sentences are ones that have opposite truth values for all truth-value assignments.
  a. Show that $\mathbf{P} \wedge \neg\mathbf{Q}$ and $\neg\mathbf{P} \vee \mathbf{Q}$ are contradictory sentences.
  b. Show that if $\mathbf{P}$ and $\mathbf{Q}$ are contradictories, then $\{\mathbf{P}, \mathbf{Q}\}$ is inconsistent.
  c. Is the converse to part *b* also true? That is, if $\{\mathbf{P}, \mathbf{Q}\}$ is inconsistent, must $\mathbf{P}$ and $\mathbf{Q}$ be contradictories? Prove it or give a counterexample.

1.3.25. *Consistency and Implication*
  a. If $\mathbf{Q}$ logically follows from a set of consistent sentences $\mathcal{P}$, must $\mathbf{Q}$ be consistent with $\mathcal{P}$? Prove it or give a counterexample.
  b. If $\mathbf{Q}$ is consistent with a set of sentences $\mathcal{P}$, must $\mathbf{Q}$ logically follow from $\mathcal{P}$? Prove it or give a counterexample.

1.3.26. *Independence*
  a. Show that $\{\mathbf{P} \vee \mathbf{Q}, \neg\mathbf{Q} \vee \mathbf{R}, \neg\mathbf{P} \wedge \mathbf{R}\}$ is a consistent set of sentences, but $\{\mathbf{P} \vee \mathbf{Q}, \neg\mathbf{Q} \vee \mathbf{R}, \neg\mathbf{P} \wedge \neg\mathbf{R}\}$ is inconsistent.
  b. Show that $\neg\mathbf{P} \wedge \mathbf{R}$ is independent of the sentences $\mathbf{P} \vee \mathbf{Q}$ and $\neg\mathbf{Q} \vee \mathbf{R}$. Are either $\mathbf{P} \vee \mathbf{Q}$ or $\neg\mathbf{Q} \vee \mathbf{R}$ independent of the other two sentences?
  c. Show from the definition that $\mathbf{Q}$ is independent of a set of sentences $\mathcal{P}$ if and only if each of the two sets $\{\mathcal{P}, \mathbf{Q}\}$ and $\{\mathcal{P}, \neg\mathbf{Q}\}$ is logically consistent.

Exercises 27–29: Other Conjunctive and Disjunctive Connectives
The following problems explore additional connectives related to $\wedge$ and $\vee$.

1.3.27. *Exclusive and Nonexclusive Disjunction*
  a. Write out $\mathbf{P} \veebar \mathbf{Q}$'s truth table, where $\veebar$ denotes exclusive disjunction.
  b. Show that $\veebar$ can be defined in terms of $\vee$, $\wedge$, and $\neg$. That is, find a sentence using these connectives that is logically equivalent to $\mathbf{P} \veebar \mathbf{Q}$.

   c. Show that $\vee$ can be defined in terms of $\veebar$ and $\wedge$. That is, find a sentence
     using these connectives that is logically equivalent to $\mathbf{P} \vee \mathbf{Q}$.

   d. True or false: $\neg(\mathbf{P} \veebar \mathbf{Q}) \models \neg\mathbf{P} \wedge \neg\mathbf{Q}$.

   e. Exploration: what laws hold for the XOR connective $\veebar$?

1.3.28. *The* NAND *Connective*

$\mathbf{P} \triangle \mathbf{Q}$ (read: $\mathbf{P}$ NAND $\mathbf{Q}$) is an abbreviation for negated conjunction (not-and), i.e., for $\neg(\mathbf{P} \wedge \mathbf{Q})$. (Think of the bottom line of $\triangle$ as negating the $\wedge$ sign above it.) This connective is often symbolized by | and is called the Sheffer stroke after H. M. Sheffer, who investigated its properties in 1913.

   a. Write a truth table for $\mathbf{P} \triangle \mathbf{Q}$.

   b. Show that $\neg\mathbf{P} \vDash\dashv \mathbf{P} \triangle \mathbf{P}$.

   c. Show that $\mathbf{P} \wedge \mathbf{Q} \vDash\dashv (\mathbf{P} \triangle \mathbf{Q}) \triangle (\mathbf{P} \triangle \mathbf{Q})$.

   d. Determine a logical equivalent involving $\triangle$ for $\mathbf{P} \vee \mathbf{Q}$.

   e. Exploration: what laws hold for the NAND connective $\triangle$?

1.3.29. *The* NOR *Connective*

Let $\mathbf{P} \triangledown \mathbf{Q}$[5] (read: $\mathbf{P}$ NOR $\mathbf{Q}$) be an abbreviation for neither-nor, i.e., for $\neg(\mathbf{P} \vee \mathbf{Q})$. (Think of the top line of $\triangledown$ as negating the $\vee$ sign below it.) This connective is often symbolized by the downward arrow $\downarrow$, which can be thought of as a slashed or negated $\vee$ sign.

   a. Write a truth table for $\mathbf{P} \triangledown \mathbf{Q}$.

   b. Show that $\neg\mathbf{P} \vDash\dashv \mathbf{P} \triangledown \mathbf{P}$.

   c. Show that $\mathbf{P} \vee \mathbf{Q} \vDash\dashv (\mathbf{P} \triangledown \mathbf{Q}) \triangledown (\mathbf{P} \triangledown \mathbf{Q})$.

   d. Determine a logical equivalent involving $\triangledown$ for $\mathbf{P} \wedge \mathbf{Q}$.

   e. Exploration: what laws hold for the NOR connective $\triangledown$?

1.3.30. *Logically Equivalent Sentences and Logical Consequences*

   a. Show that $\mathbf{P} \vDash\dashv \mathbf{Q}$ if and only if their sets of consequences agree, i.e.,
     $\mathbf{P} \vDash\dashv \mathbf{Q}$ if and only if $\mathbf{P} \models \mathbf{R}$ whenever $\mathbf{Q} \models \mathbf{R}$, and conversely.

   b. Call two sets of sentences $\mathcal{P}$ and $\mathcal{Q}$ logically equivalent when each sentence
     of $\mathcal{P}$ is logically implied by $\mathcal{Q}$ and conversely. Using this definition, gener-
     alize the result given in part *a* to cover the case when sets of sentences are
     involved. What does this result say about axiomatizing a mathematical
     theory by means of a set of sentences equivalent to a given set of axioms?

   c. Show that the extended notion of logical equivalence defined in part *b* is
     an equivalence relation (see Exercise 10).

1.4 Conditional and Biconditional Sentences

In this section we'll look at the two conditional truth-functional connectives *if-then* and *if-and-only-if*. As the latter connective is composed of *if-then* and *and*, we'll treat *if-then* first.

    We'll symbolize *if* $\mathbf{P}$ *then* $\mathbf{Q}$ by $\mathbf{P} \rightarrow \mathbf{Q}$. $\mathbf{P}$ is the sentence's *antecedent*, $\mathbf{Q}$ is its *consequent*. Other notations used by logicians and mathematicians for *if-then* are $\supset$ and $\Rightarrow$.

---

[5] Unfortunately, Tarski used $\triangle$ in some writings to denote neither-nor.

1.4.1 *Simple Motivation for If-Then's Truth Table*

We'll motivate the truth table for the conditional sentence **P** → **Q** by thinking of it as a promise. *If you do all the homework, then you'll pass the course* says you can expect a passing grade so long as you do the assigned work. Of course, you might still pass without doing the homework if your test scores are good, but doing homework guarantees that you'll pass. A promise **P** → **Q** is violated only when the antecedent **P** is true but the consequent **Q** is false. This is the logical version of *innocent until proven guilty*.

Definition 1.4.1: *Truth Values for Conditional Sentences*
*The **conditional** $P \rightarrow Q$ is false if P is T and Q is F; else it is true. In positive terms: $P \rightarrow Q$ is true if and only if P is F or Q is T.*

This yields the following truth table:

P	Q	$P \rightarrow Q$
T	T	T
T	F	F
F	T	T
F	F	T

If **P** → **Q** is true, it *cannot happen* that **P** is true and **Q** is false. Thus, it seems **P** → **Q** should be taken as equivalent to ¬(**P** ∧ ¬**Q**). Alternatively, affirming **P** → **Q** means either **P** is not the case or else **Q** is. This suggests the equivalent ¬**P** ∨ **Q**. These two alternatives are themselves equivalent and have the above truth table (see Exercises 44–45), so we have further evidence of a sort for the correctness of **P** → **Q**'s truth-value assignment.

1.4.2 *The Truth Table for If-Then: A Closer Look*

Now it's time for some skepticism. Let's look at the truth table for **P** → **Q** more closely, line by line. The second row is unproblematic: if **P** is true while **Q** is false, **P** → **Q** can't possibly be true. Truth-value assignments for the other three rows, however, don't fare as well.

Suppose $P \rightarrow Q$ denotes the sentence *If ABCD is a square, then ABCD is a rectangle* and consider the case where $ABCD$ is a square. Then both P and Q are true. Row one's truth-value assignment T seems correct because squares are rectangles with equal sides. But if → is a truth-functional connective—its truth value determined by the component's truth values—row one's truth value must also be true even when no logical connection holds between the two parts. The compound sentence *If 2 is even, then 7 is prime*, for instance, would then also be true. Should this be the case?

The truth values in the third and fourth lines of $P \rightarrow Q$'s truth table are also T. But once again, we're forced to accept some strange sentences as true under those assignments. The sentences *If $0 = 1$, then 2 is even* and *If $0 = 1$, then 2 is odd* are both true according to the standard truth-value assignment, even though in neither case does the first clause imply the last one.

The strangeness of all of these cases arises because the component sentences joined by *if-then* have no logical relation. Yet we must be able to connect *any* two sentences by → and *uniformly assign* a truth value to the result based solely on the truth values of the constituents.

It should be clear by now that we filled out the truth table for conditional sentences using a fairly specialized type of example and that the values chosen are in some sense conventional. The conditionals fully satisfying the truth table seem to be ones in which the first sentence logically implies the second; the problem conditionals are ones where this does not occur.

1.4.3 *Conditional Sentences vs. Logical Implication*

To gain a deeper understanding of the *if-then* connective and its truth table, we'll have to wander into the weeds a bit and consider how → is related to ⊨ .

Many seem to think that the connective → indicates the relation of logical implication, that we can read $\mathbf{P} \to \mathbf{Q}$ as $\mathbf{P}$ *implies* $\mathbf{Q}$. However, identifying $\mathbf{P} \to \mathbf{Q}$ with $\mathbf{P} \models \mathbf{Q}$ is wrong, on several counts.

First of all, → is a *truth-functional connective*, but ⊨ is not. The truth or falsehood of the assertion $\mathbf{P} \models \mathbf{Q}$ is largely independent of the truth values of the component sentences, as we learned in Section 1.1. Except for when $\mathbf{P}$ is T and $\mathbf{Q}$ is F, which makes $\mathbf{P} \models \mathbf{Q}$ false, $\mathbf{P} \models \mathbf{Q}$ can be either true or false—it all depends on how $\mathbf{P}$ and $\mathbf{Q}$ are logically related. Since ⊨ is not truth-functionally determined, it cannot be the same as →, which is.

Second, identifying → and ⊨ confuses different levels of language. $\mathbf{P} \to \mathbf{Q}$ is a compound sentence in the *object language*. It's a conditional assertion about the objects, such as squares and rectangles, mentioned by $\mathbf{P}$ and $\mathbf{Q}$. $\mathbf{P} \models \mathbf{Q}$ is very different. It's not about *mathematical objects* but about the *logical relation* holding between *sentences* $\mathbf{P}$ and $\mathbf{Q}$. It thus belongs to the *meta-language*, not the object language. $\mathbf{P} \to \mathbf{Q}$ and $\mathbf{P} \models \mathbf{Q}$ reside in different domains of discourse and shouldn't be equated.

To clarify this further, think of PL sentences as forming an algebraic structure. The *objects* of this system are the sentences of *Propositional Logic*. Logical connectives like ¬, ∧, ∨, and → *operate* upon these objects to yield compound formulas. Truth-value assignments are *functions* defined on the formulas, taking on the (truth) values T and F. PL also has logical *properties* and *relations*. Some sentences have the *property* of being logically true or false, while others are logically indeterminate. Logical implication and logical equivalence are *relations* holding between PL's objects.

By equating → with ⊨ , a *propositional operator* is being identified with a *relation*, something that strictly makes no sense. One would never identify a computational *operation*, such as − , with an arithmetic *relation*, such as < .[6] Subtraction *operates* on numbers and yields *numbers*. The *relation* < generates *statements* about numbers rather than numbers.

---

[6] Well, not never. In number theory, beginning students sometimes confuse the *relation* of *divides* with the *operation* of *division*.

In logic, however, conflating the operator $\to$ with the relation $\models$ is easier to do because both generate sentences, though on different language levels. Unfortunately, *if-then* is often used to express both the conditional connective and the relation of logical implication, which encourages the confusion.

1.4.4 *Conditional Sentences and Logical Implication*

Let's focus more closely on what links logical implication to the conditional connective. We'll continue the algebraic analogy from the last subsection.

To characterize a relation, we often use a property of an operation result. For example, $<$ can be defined in terms of a subtraction property: $a < b$ *if and only if* $b - a$ *is positive*. This is a familiar mathematical maneuver. Here's another example: a *divides* b *if and only if* b *divided by* a *leaves a remainder of* 0. *Divides* is defined in terms of a property of division.

Can we do something similar here? Can we say, for instance, that $\mathbf{P} \models \mathbf{Q}$ *if and only if* $\mathbf{P} \to \mathbf{Q}$ *is true*? This isn't quite right. $\mathbf{P} \models \mathbf{Q}$ holds on account of the sentences' logical forms, regardless of their particular truth values. To invoke the same degree of generality for $\mathbf{P} \to \mathbf{Q}$ as for $\mathbf{P} \models \mathbf{Q}$, we should perhaps say that $\mathbf{P} \to \mathbf{Q}$ is true irrespective of the truth values of $\mathbf{P}$ and $\mathbf{Q}$.

This connection turns out to be correct and is important enough to call a theorem. We'll state it in two forms, first for pure logical implication and then for implication relative to an axiomatized theory (see Section 1.3).

Theorem 1.4.1: *Implication Theorem*[7]
   $\mathbf{P} \models \mathbf{Q}$ *if and only if* $\models \mathbf{P} \to \mathbf{Q}$, *i.e., if and only if* $\mathbf{P} \to \mathbf{Q}$ *is logically true.*

Proof:
   For easy reference, we'll repeat the truth table for $\mathbf{P} \to \mathbf{Q}$.

$\mathbf{P}$	$\mathbf{Q}$	$\mathbf{P} \to \mathbf{Q}$
T	T	T
T	F	F
F	T	T
F	F	T

· First suppose that $\mathbf{P} \models \mathbf{Q}$. We must show that $\mathbf{P} \to \mathbf{Q}$ is true under all truth-value assignments *compatible with this given*.
   Since $\mathbf{P} \models \mathbf{Q}$, it is impossible for $\mathbf{P}$ to be true and $\mathbf{Q}$ to be false: *row two can't happen* for such a $\mathbf{P}$ and $\mathbf{Q}$.
   But this is the only case in which $\mathbf{P} \to \mathbf{Q}$ is false.
   Thus, $\mathbf{P} \to \mathbf{Q}$ is always true, i.e., $\models \mathbf{P} \to \mathbf{Q}$. ✓

· Conversely, suppose $\mathbf{P} \to \mathbf{Q}$ is always true.
   Then $\mathbf{P}$ cannot be true while $\mathbf{Q}$ is false; we must again rule out row two of $\mathbf{P} \to \mathbf{Q}$'s truth table.

---

[7] A more complex form of this theorem was proved by Tarski in 1921; Herbrand came to it independently in 1928. In its original context, this result played the role of *Conditional Proof*, which we'll discuss in Section 1.7.

Line one of the truth table is now all that's relevant for assessing whether **P** implies **Q**. Thus, whenever **P** is true, **Q** is also true.
Hence, **P** ⊨ **Q**. ∎

❖ Example 1.4.1

Show $(\mathbf{P} \vee \mathbf{Q}) \wedge (\mathbf{P} \vee \neg\mathbf{Q}) \models \mathbf{P}$. Thus, $\models [(\mathbf{P} \vee \mathbf{Q}) \wedge (\mathbf{P} \vee \neg\mathbf{Q})] \rightarrow \mathbf{P}$.

Solution

The following double truth table shows that logical implication holds; in fact, the two formulas are logically equivalent.

(**P**	∨	**Q**)	∧	(**P**	∨	¬	**Q**)	‖	**P**
T	*T*	*T*	ⓣ	*T*	*T*	*F*	*T*		ⓣ
T	*T*	*F*	ⓣ	*T*	*T*	*T*	*F*		ⓣ
F	*T*	*T*	*F*	*F*	*F*	*F*	*T*		*F*
F	*F*	*F*	*F*	*F*	*T*	*T*	*F*		*F*
−			=	−					=

By the *Implication Theorem*, the associated conditional is a tautology.

The next example highlights the fact that the *Implication Theorem* has a limited application and should not be pushed beyond what it says.

❖ Example 1.4.2

Show that $(\mathbf{P} \rightarrow \mathbf{Q}) \vee (\mathbf{Q} \rightarrow \mathbf{P})$ is a logical truth but that neither conditional implies the other.

Solution

The truth table for $(\mathbf{P} \rightarrow \mathbf{Q}) \vee (\mathbf{Q} \rightarrow \mathbf{P})$ is the following:

(**P**	→	**Q**)	∨	(**Q**	→	**P**)
T	*T*	*T*	*T*	*T*	*T*	*T*
T	*F*	*F*	*T*	*F*	*T*	*T*
F	*T*	*T*	*T*	*T*	*F*	*F*
F	*T*	*F*	*T*	*F*	*T*	*F*
−		=		−		

While $(\mathbf{P} \rightarrow \mathbf{Q}) \vee (\mathbf{Q} \rightarrow \mathbf{P})$ is logically true, it is not a conditional sentence, and the corresponding meta-language statement, **P** ⊨ **Q** or **Q** ⊨ **P**, is obviously false. **P** → **Q** and **Q** → **P** are logically indeterminate, so by the *Implication Theorem* neither of the associated implications holds.
The moral of this is that *you cannot read* **P** → **Q** *as* **P** ⊨ **Q**.

An argument similar to that given for the *Implication Theorem* shows that a *Relativized Implication Theorem* holds for axiomatic theories.

Theorem 1.4.2: *Relativized Implication Theorem*

P ⊨*ₐ* **Q** *if and only if* ⊨*ₐ* **P** → **Q**, *i.e., if and only if* 𝒜 ⊨ **P** → **Q**.

Proof:
See Exercise 61.

Having stressed the difference between $\to$ and $\vDash$, the *Relativized Implication Theorem* tells us we can relax this slightly for mathematics. If **Q** follows from **P** *relative to an axiomatic theory*, then **P** $\to$ **Q** *follows from the theory's axioms*. Conversely, if **P** $\to$ **Q** is true relative to a theory, then **P** $\vDash$ **Q** holds there as well. Thus, while **P** $\to$ **Q** doesn't say **P** *implies* **Q** (and shouldn't be read that way), its status as a theorem allows one to conclude that **P** *implies* **Q** *relative to that theory*. This is as close as we can come to justifying the mathematical custom of reading **P** $\to$ **Q** as an implication.

1.4.5 *Reading Conditional Sentences*

It's difficult to overestimate the importance of the *if-then* connective. Most mathematical theorems involve a conditional connective, sometimes as the main connective. How should these be read when formulated symbolically?

The preferred reading for **P** $\to$ **Q** is *if* **P** *then* **Q**, but the split phrasing for the connective is a bit awkward. Ideally, we'd like a connective-word xyz to substitute for $\to$ so that **P** $\to$ **Q** could be read as **P** xyz **Q**. The graphic reading **P** *arrow* **Q** sounds artificial; **P** *then* **Q** is probably a better option.

Alternatively, we can read **P** $\to$ **Q** as **P** *only-if* **Q**, which is equivalent to *if* **P** *then* **Q**. The trouble with reading **P** $\to$ **Q** this way is that many confuse *only-if* with *if*, and these two connectives indicate logically opposite directions.

Unfortunately, reading $\to$ as *implies* works all too well, reinforcing the misconception we just exposed. As there is no universally accepted alternative to this besides *if* **P** *then* **Q**, you may need to resort to that split reading.

Conditional sentences in mathematics texts are sometimes disguised by alternative formulations. **Q**, *provided* **P** means the same as **P** $\to$ **Q**. Conditional sentences are also implicit in universal statements of the form *all* **X**s *are* **Y**s. These can be reformulated as *for all* **a**, *if* **a** *is an* **X**, *then* **a** *is a* **Y**.

Mathematicians sometimes formulate a conditional result by saying **P** *is a sufficient condition for* **Q** or **Q** *is a necessary condition for* **P**, both of which mean **P** *implies* **Q**. By the *Relativized Implication Theorem*, these can be reformulated as the proposition **P** $\to$ **Q** in the object language.

1.4.6 *Priority Level of Conditionals*

The next example shows that piling up conditions in a compound conditional like **P** $\to$ (**Q** $\to$ **R**) amounts to conjoining them. It also illustrates the importance of using parentheses. We need to use parentheses to indicate which $\to$ to perform first, because $\to$ is not associative (see Exercise 31). Complex nested conditionals can occur in mathematics, but mathematicians tend to formulate them at the outset by conjoining the conditions to give (**P** $\wedge$ **Q**) $\to$ **R**, since that seems simpler to comprehend.

✤**Example 1.4.3**
   Show that **P** $\to$ (**Q** $\to$ **R**) $\models\!\models$ (**P** $\wedge$ **Q**) $\to$ **R**.

Solution

The following eight-line truth table shows the logical equivalence.

P → **(Q → R)**	**(P** ∧ **Q)** → **R**
T Ⓣ T T T	T Ⓣ
T F T F F	T F
T Ⓣ F T T	F Ⓣ
T Ⓣ F T F	F Ⓣ
F Ⓣ T T T	F Ⓣ
F Ⓣ T F F	F Ⓣ
F Ⓣ F T T	F Ⓣ
F Ⓣ F T F	F Ⓣ

Parentheses in compound sentences involving → can sometimes be omitted using the following priority convention for →.

Priority	*Symbol*
1	¬
2	∧
3	∨
4	→

With this convention, we would write the sentences in Examples 1–3 as

1) $(\mathbf{P} \lor \mathbf{Q}) \land (\mathbf{P} \lor \lnot \mathbf{Q}) \to \mathbf{P}$ (parentheses are needed, but brackets aren't);
2) $(\mathbf{P} \to \mathbf{Q}) \lor (\mathbf{Q} \to \mathbf{P})$ (all parentheses are still needed); and
3) $\mathbf{P} \to (\mathbf{Q} \to \mathbf{R})$ and $\mathbf{P} \land \mathbf{Q} \to \mathbf{R}$.

1.4.7 *Sentences Related to Conditional Sentences*

We noted above that $\mathbf{P} \to \mathbf{Q}$ is logically equivalent to two other forms, $\lnot(\mathbf{P} \land \lnot \mathbf{Q})$ and $\lnot \mathbf{P} \lor \mathbf{Q}$. The latter equivalent is simpler, but the first equivalent helps simplify negated conditionals via *Double Negation* and plays a role in some proofs by contradiction (see Section 1.8).

A third equivalent is $\lnot \mathbf{Q} \to \lnot \mathbf{P}$, the *contrapositive* of $\mathbf{P} \to \mathbf{Q}$ (see Exercise 46). A sentence and its contrapositive assert the same thing, one in a positive way, the other in a negative way.

The *converse* of $\mathbf{P} \to \mathbf{Q}$ is $\mathbf{Q} \to \mathbf{P}$, obtained by interchanging (converting) the sentence's components. A sentence and its converse assert different things; neither implies the other. A conditional statement and its converse may both be true, but this need not be the case (see Exercise 47). For example, the converse of *if ABCD is a rectangle, then its diagonals bisect one another* is false. Confusing a conditional sentence with its converse may indicate a fuzzy understanding of *if-then*.

Mathematical theorems are often more complex than a simple conditional $\mathbf{P} \to \mathbf{Q}$. For instance, it may be of the form $\mathbf{P} \land \mathbf{Q} \to \mathbf{R}$. Its (full) converse is $\mathbf{R} \to \mathbf{P} \land \mathbf{Q}$. The sentences $\mathbf{R} \land \mathbf{P} \to \mathbf{Q}$ and $\mathbf{R} \land \mathbf{Q} \to \mathbf{P}$ can be considered *partial converses.* Sometimes a partial converse can be proved when the full converse cannot be. That's often what's wanted, for many times one of the conditions (**P** or **Q**) is to be held constant. Note, though, that these partial converses are also not equivalent to the original sentence (see Exercise 48).

We can form *partial contrapositives* in a similar way. The partial contrapositives of $\mathbf{P} \wedge \mathbf{Q} \to \mathbf{R}$ are $\mathbf{P} \wedge \neg \mathbf{R} \to \neg \mathbf{Q}$ and $\mathbf{Q} \wedge \neg \mathbf{R} \to \neg \mathbf{P}$. Interestingly, both of these contrapositives are logically equivalent to the original conditional (see Exercise 49).

1.4.8 *Syntax and Semantics of Biconditional Sentences*

Sentences of the form **P** *if and only if* **Q** are *biconditional* sentences. We could introduce *if-and-only-if* similarly to how we did *if-then*. We would then also argue that the biconditional connective is intended as a truth-functional approximation to logical equivalence. Having discussed conditional sentences in detail, we can be brief here.

As with conditional sentences, alternative notations are used to denote the biconditional connective *if-and-only-if*. Some logic texts use $\equiv$; mathematicians often use $\Leftrightarrow$. Since these symbols are sometimes used to indicate logical equivalence, which is a logical relation, not a logical operator, we'll instead use $\leftrightarrow$, a counterpart to $\to$. On the other hand, since some texts use $\equiv$ as a biconditional connective, we did not choose $\equiv$ to stand for logical equivalence either, but adopted the suggestive symbol $\boxminus$ instead.

Definition 1.4.2: *Truth-Value Assignment for Biconditionals*
   *The **biconditional** $\mathbf{P} \leftrightarrow \mathbf{Q}$ is true if and only if* **P** *and* **Q** *have the same truth values.*

The truth table for $\mathbf{P} \leftrightarrow \mathbf{Q}$ is as follows:

P	**Q**	$\mathbf{P} \leftrightarrow \mathbf{Q}$
T	T	T
T	F	F
F	T	F
F	F	T

Our earlier remarks about the strangeness of the truth-value assignment for $\mathbf{P} \to \mathbf{Q}$ apply here as well. Some biconditionals will be true even when no logical relation connects their two conditions.

On the basis of the truth-value assignment for $\mathbf{P} \leftrightarrow \mathbf{Q}$, we can prove an *Equivalence Theorem* and a *Relativized Equivalence Theorem* that show how $\leftrightarrow$ is related to $\boxminus$. We will leave this for the exercises (see Exercises 63–64).

Biconditional sentences occur frequently in mathematics. They are used to state definitions (though mathematicians tend to write *if* for this when they really mean *if and only if*), but they also occur in characterizing a given concept. Theorems given in *if-and-only-if* form (here mathematical practice is fussier) provide alternative definitions for a concept. For example, if we define *an integer is odd if and only if it is not even* and then prove *an integer is odd if and only if it leaves remainder 1 when divided by* 2, this last biconditional could be taken as the definition of being odd, making the original definition a theorem needing proof. Such situations happen repeatedly in mathematics. When they do, mathematicians choose the definition that seems most fundamental or that gives the best deductive basis for what follows.

Mathematicians use more than one style for expressing biconditional sentences. Biconditional sentences are usually formulated using *if and only if*, but at times you will read that **P** *is a necessary and sufficient condition for* **Q** or that **P** *is equivalent to* **Q**. These are meta-language statements, but they can be taken as a stand-in for the associated object language biconditionals.

In compound sentences where one or more $\leftrightarrow$ occur along with other connectives, parentheses are generally needed to make the meaning clear. We'll take $\leftrightarrow$ as a connective of lowest priority but on the same level as $\rightarrow$. If both $\rightarrow$ and $\leftrightarrow$ occur in a single sentence, parentheses may be required.

Our final list of priorities for truth-functional connectives, therefore, is given in the following table:

Priority	Symbol
0	( )
1	$\neg$
2	$\wedge$
3	$\vee$
4	$\rightarrow$, $\leftrightarrow$

1.4.9 *Equivalents of Biconditional Sentences*

The double arrow in **P** $\leftrightarrow$ **Q** suggests it is made up of **P** $\rightarrow$ **Q** and **P** $\leftarrow$ **Q** (i.e., **Q** $\rightarrow$ **P**). This is also indicated by the phrase *if-and-only-if*. **P** *only-if* **Q** is symbolized by **P** $\rightarrow$ **Q**, while **P** *if* **Q** is symbolized by **Q** $\rightarrow$ **P**. The *and* between *if* and *only-if* suggests these two sentences should be conjoined. We could thus *define* **P** $\leftrightarrow$ **Q** by (**P** $\rightarrow$ **Q**) $\wedge$ (**Q** $\rightarrow$ **P**). We've instead taken $\leftrightarrow$ as a primitive connective. Constructing truth tables for the two sentences involved, the following equivalence is easily shown (see Exercise 50):

$$\mathbf{P} \leftrightarrow \mathbf{Q} \vDash (\mathbf{P} \rightarrow \mathbf{Q}) \wedge (\mathbf{Q} \rightarrow \mathbf{P}).$$

This gives the principal equivalent of **P** $\leftrightarrow$ **Q**, but three others are related to it and are sometimes useful:

$$\mathbf{P} \leftrightarrow \mathbf{Q} \vDash \neg\mathbf{Q} \leftrightarrow \neg\mathbf{P}$$
$$\mathbf{P} \leftrightarrow \mathbf{Q} \vDash (\mathbf{P} \rightarrow \mathbf{Q}) \wedge (\neg\mathbf{P} \rightarrow \neg\mathbf{Q})$$
$$\mathbf{P} \leftrightarrow \mathbf{Q} \vDash (\mathbf{P} \wedge \mathbf{Q}) \vee (\neg\mathbf{P} \wedge \neg\mathbf{Q})$$

These can be shown by truth tables or by stringing equivalences together (see Exercises 51–53).

EXERCISE SET 1.4

Exercises 1–6: Symbolizing Conditional/Biconditional Sentences

Formulate the following sentences using the connectives $\rightarrow$ and $\leftrightarrow$. You may use sentence letters or any appropriate mathematical and logical symbolism.

1.4.1. $\displaystyle\sum_{n=0}^{\infty} ar^n = \frac{a}{1-r}$ if $|r| < 1$.

1.4.2. x is a real number only if x is a rational number or an irrational number.

1.4.3. $|a| = a$, provided $a \geq 0$.

1.4.4. $a - b \neq b - a$, unless $a = b$.

1.4.5. A necessary condition for two lines to be parallel is that they be everywhere equidistant.

1.4.6. The condition $a_n \rightarrow 0$ is necessary but not sufficient for series $\sum a_n$ to converge.

Exercises 7–10: Formulating Definitions

Write definitions for the following, using $\leftrightarrow$ and any appropriate logical and mathematical symbolism. Look these up in a textbook or online, if necessary.

1.4.7. $\triangle ABC$ is an isosceles triangle.

1.4.8. l and m are perpendicular lines.

1.4.9. a divides b (symbolized by $a \mid b$).

1.4.10. c is a zero (root) of a function f.

Exercises 11–12: Truth Values of Statements

Determine the truth value for the following statements if P and R are true and Q and S are false.

1.4.11. $\neg P \wedge (Q \rightarrow S) \vee R$ **1.4.12.** $P \leftrightarrow (\neg Q \vee R) \wedge (R \rightarrow S)$

Exercises 13–16: True or False

Are the following statements true or false? Explain your answer.

1.4.13. The symbol $\models$ denotes a truth-functional connective.

1.4.14. $P \leftrightarrow Q \models\mid (P \wedge \neg Q) \vee (\neg P \wedge Q)$

1.4.15. $P \rightarrow Q$ is logically false if and only if $P \not\models Q$.

1.4.16. If $\neg P \rightarrow Q$ is true, then $P \rightarrow Q$ must be false.

Exercises 17–18: Explanations

Explain the following, using your own words.

1.4.17. Compare and contrast $\rightarrow$ with $\models$

1.4.18. State the *Relativized Implication Theorem*

Exercises 19–20: Problematic Formulations

What is wrong with the following statements, which are sometimes found in texts on logic and proof? What should be said instead?

1.4.19. P → Q means either **P** is false or else **Q** is true.

1.4.20. P ↔ Q means **P** and **Q** have the same truth values.

Exercises 21–35: Tautologies and Implication

For the following problems (some of which are classic tautologies),
  a. *determine whether the sentence is a tautology; and*
  b. *state what your answer means, according to the Implication Theorem.*

1.4.21. $P \to P \vee Q$

1.4.22. $P \vee Q \to P$

1.4.23. $P \wedge Q \to P$

1.4.24. $P \wedge (P \to Q) \to Q$

1.4.25. $(P \vee Q) \wedge \neg Q \to P$

1.4.26. $(\neg P \to P) \to P$

1.4.27. $P \to (\neg P \to Q)$

1.4.28. $[(P \to Q) \to P] \to P$

1.4.29. $(P \to Q) \vee (P \to \neg Q)$

1.4.30. $\neg(P \leftrightarrow \neg P)$

1.4.31. $P \to (Q \to R) \to (P \to Q) \to R$

1.4.32. $[P \to (Q \to R)] \to [(P \to Q) \to (P \to R)]$

1.4.33. $[P \to Q] \to [(P \to (Q \to R)) \to (P \to R)]$

1.4.34. $(P \to Q) \to [(Q \to R) \to (P \to R)]$

1.4.35. $(P \to Q) \to [(P \to \neg Q) \to \neg P]$

Exercises 36–41: Implication and Tautologies

For the following problems,
  a. *determine whether the claim of logical implication is true or false; and*
  b. *state what your answer means, according to the Implication Theorem.*
  [See Exercise 62 for multiple premises.]

1.4.36. $P \to Q,\ \neg P \vDash \neg Q$

1.4.37. $P \to Q,\ \neg Q \vDash \neg P$

1.4.38. $P \wedge Q \vDash P \to Q$

1.4.39. $P \to Q \vDash P \wedge Q$

1.4.40. $(P \to Q) \to Q \vDash P \to Q$

1.4.41. $P \vee Q,\ P \to R,\ Q \to R \vDash R$

Exercises 42–43: True Conditionals and Implication

Show that the following implications fail, but that the associated conditional sentences are true. Does this contradict the Implication Theorem? Explain.

1.4.42. $0 = 1 \nvDash 2$ is even.

1.4.43. $0 = 1 \nvDash 2$ is odd.

Exercises 44–53: Logical Equivalences

Show that the following are equivalent or not equivalent, as asked.

1.4.44. P → Q ⊨⊨ ¬P ∨ Q   (Show this with a truth table.)

1.4.45. P → Q ⊨⊨ ¬(P ∧ ¬Q)   (Show this with a truth table or by using *De Morgan's Laws* and *Double Negation*.)

1.4.46. P → Q ⊨⊨ ¬Q → ¬P   (Show this with a truth table.)

1.4.47. Show both by means of a double truth table and by a mathematical counterexample that a conditional sentence **P → Q** and its converse **Q → P** are *not* logically equivalent and that neither implies the other.

1.4.48. Show that a compound conditional **P ∧ Q → R** is *not* logically equivalent to either of its partial converses, **R ∧ P → Q** or **R ∧ Q → P**.

1.4.49. P ∧ Q → R ⊨⊨ P ∧ ¬R → ¬Q

1.4.50. $\mathbf{P} \leftrightarrow \mathbf{Q} \vDash\dashv (\mathbf{P} \rightarrow \mathbf{Q}) \wedge (\mathbf{Q} \rightarrow \mathbf{P})$

1.4.51. $\mathbf{P} \leftrightarrow \mathbf{Q} \vDash\dashv \neg \mathbf{P} \leftrightarrow \neg \mathbf{Q}$

1.4.52. $\mathbf{P} \leftrightarrow \mathbf{Q} \vDash\dashv (\mathbf{P} \rightarrow \mathbf{Q}) \wedge (\neg \mathbf{P} \rightarrow \neg \mathbf{Q})$

1.4.53. $\mathbf{P} \leftrightarrow \mathbf{Q} \vDash\dashv (\mathbf{P} \wedge \mathbf{Q}) \vee (\neg \mathbf{P} \wedge \neg \mathbf{Q})$

1.4.54. The following proposition formulates the *Basic Comparison Test*:
If $0 \le a_n \le b_n$ *for all natural numbers n, then if the series* $\sum b_n$ *converges, the series* $\sum a_n$ *converges.*
   a. Formulate this sentence as a single compound conditional in two ways, and tell why they're equivalent.
   b. Beginning with the form from part *a* that has a conjoined antecedent, formulate the partial contrapositive that retains the same *given* as the *original statement*. If the first statement is true, is the second one, too?

1.4.55. *Liouville's Theorem* in complex analysis states that *a bounded, everywhere differentiable function is constant.* (This is false for real analysis.)
   a. Write this proposition using logical notation. Use P, Q, and R for the three atomic sentences, and make all logical connectives explicit.
   b. Give two equivalent formulations for your sentence in part *a* by writing out its partial contrapositives. Then put your reformulations back into mathematical English.

Exercises 56–60: Conditional and Biconditional Equivalents
Show the following. See Exercises 1.3.27–1.3.29 for the connectives involved.

1.4.56. $P \veebar Q \vDash\dashv \neg(P \leftrightarrow Q)$

1.4.57. $P \rightarrow Q \vDash\dashv [(P \triangledown P) \triangledown Q] \triangledown [(P \triangledown P) \triangledown Q)]$

1.4.58. Using Exercise 1.3.29 and the last exercise, determine a logical equivalent for $P \leftrightarrow Q$ using only $\triangledown$.

1.4.59. Using Exercise 1.3.28, determine a logical equivalent for $P \rightarrow Q$ using only $\triangle$.

1.4.60. Using Exercise 1.3.28 and the last exercise, determine a logical equivalent for $P \leftrightarrow Q$ using only $\triangle$.

Exercises 61–64: Proofs of Theorems
Prove the following theorems.

1.4.61. *Relativized Implication Theorem:* $\mathbf{P} \vDash_{\mathcal{A}} \mathbf{Q}$ *if and only if* $\mathcal{A} \vDash \mathbf{P} \rightarrow \mathbf{Q}$.

1.4.62. *Implication Theorem, Conjunctive Form:* $\mathbf{P}_1, \mathbf{P}_2, \ldots, \mathbf{P}_n \vDash \mathbf{Q}$ *if and only if* $\vDash \mathbf{P}_1 \wedge \mathbf{P}_2 \wedge \cdots \wedge \mathbf{P}_n \rightarrow \mathbf{Q}$.

1.4.63. *Equivalence Theorem:* $\mathbf{P} \vDash\dashv \mathbf{Q}$ *if and only if* $\vDash \mathbf{P} \leftrightarrow \mathbf{Q}$.

1.4.64. *Relativized Equivalence Theorem:* $\mathbf{P} \vDash\dashv_{\mathcal{A}} \mathbf{Q}$ *if and only if* $\mathcal{A} \vDash \mathbf{P} \leftrightarrow \mathbf{Q}$.

1.5 Introduction to Deduction; Rules for AND

Truth tables give us a way to test whether a PL conclusion follows from
a set of premises. However, this gets cumbersome when more than a few
sentence variables are involved—the number of rows in a truth table grows
exponentially with the number of sentence variables (see Exercise 1.2.38).

The sheer size of such tables pushes us to find another method to demonstrate that premises imply a conclusion. Deducing a conclusion is such a
procedure. But there is a deeper and more important reason for learning to
do deductions: that's the way mathematical theories are developed.

Thus far we've ignored this deductive aspect of logic. We've discussed
when an inference from premises to a conclusion is valid, but that's different.
A sequence of premises followed by a logical consequence is not yet a *deduction* of that conclusion. Postulating Euclid's axioms, for example, and then
asserting the *Pythagorean Theorem* isn't a proof. A deduction *demonstrates*
that a conclusion follows from a set of premises by using reasoning that logically connects them via a chain of valid inferences.

1.5.1 *Setting Up a Natural Deduction System for PL*

As we noted earlier, a *deduction* (*derivation, proof*) of a sentence $\mathbf{Q}$ from
a set of sentences $\mathcal{P}$ is a sequence of sentences having $\mathcal{P}$ as premises and
concluding $\mathbf{Q}$ from $\mathcal{P}$ using logical rules of inference. Each sentence in the
derivation must be an assumption or a conclusion legitimately drawn from
earlier lines in the argument. To indicate that $\mathcal{P}$ *proves* $\mathbf{Q}$, we'll write $\mathcal{P} \vdash \mathbf{Q}$.

A system of inference rules for creating deductions is called a *deduction
system*. The first and most essential prerequisite for a deduction system is
that each rule be *sound*. *A rule is sound if and only if whenever it warrants
concluding* $\mathbf{Q}$ *from* $\mathcal{P}$, $\mathbf{Q}$ *logically follows from* $\mathcal{P}$. Extended truth tables will
help us show that PL's inference rules are individually sound. In the end,
we'll know for the full deduction system that if $\mathcal{P} \vdash \mathbf{Q}$, then $\mathcal{P} \vDash \mathbf{Q}$.

We'd also like our deduction system as a whole to be *complete*, i.e., *when-
ever* $\mathbf{Q}$ *logically follows from* $\mathcal{P}$, $\mathcal{P}$ *also proves* $\mathbf{Q}$. We have no *a priori* guarantee of this. Being provable depends upon whether a deduction system has
enough rules of the right sorts. Showing that a deduction system is complete
is more complex than showing its soundness, so we won't be able to argue
for it here. However, the deduction system we'll adopt for *Propositional Logic*
will be complete. Thus, the converse of soundness will also hold: *if* $\mathcal{P} \vDash \mathbf{Q}$,
then $\mathcal{P} \vdash \mathbf{Q}$. Putting these together gives $\mathcal{P} \vdash \mathbf{Q}$ *if and only if* $\mathcal{P} \vDash \mathbf{Q}$.

A third prerequisite is that rules of inference should be *natural* and *simple*—they should codify the valid ways we ordinarily infer conclusions. *Mir-
roring actual practice* is a rather loose requirement, but this will become more
precise as we flesh out PL's *Natural Deduction System*.[8]

---

[8] The rules in our system are modified versions of rules that go back to Stanisław
Jaśkowski, Gerhard Gentzen, and others from the mid-1930s and later.

Rules of inference permit us to state an assumption or infer a conclusion on the basis of preceding lines. As laws governing valid argumentation, inference rules depend solely upon the logical form of the sentences involved. To guarantee *simplicity*, these forms shouldn't be too complicated. It should be easy to see that a conclusion logically follows from a set of premises.

✤ **Example 1.5.1**

Analyze the simplicity of the following argument forms:

a) $\mathbf{P} \vee \mathbf{Q}, \neg \mathbf{P} \vDash \mathbf{Q}$ b) $\mathbf{P} \leftrightarrow (\mathbf{Q} \vee \mathbf{R}), \mathbf{R} \rightarrow \neg \mathbf{P} \vDash \neg \mathbf{R}$

Solution

a) Most would say that this first implication is intuitively valid: given two alternatives $\mathbf{P}$ and $\mathbf{Q}$, if $\mathbf{P}$ is not the case, then $\mathbf{Q}$ must be. We've looked at this inference before (see Example 1.3.3).

b) The second implication is far from obvious—you may wonder if it holds (it does). Such an inference is too complicated to adopt as a rule.

For our rules to be simple and natural, they should involve at most two or three sentence variables, and they should operate on a single connective, either <u>introducing</u> or <u>eliminating</u> it. Such rules are known as *Int-Elim Rules*. We'll adopt rules for *concluding* a conjunction, a disjunction, a negation, a conditional, and a biconditional, as well as rules that tell what can be *concluded from* such forms. These rules will justify taking small logical steps in a deduction, jointly enabling us to construct complex chains of reasoning.

We'll also include a number of *Replacement Rules* based on simple logical equivalents. For example, since $\mathbf{P} \vee \mathbf{Q} \vDash\dashv \mathbf{Q} \vee \mathbf{P}$, either disjunction can be substituted for the other anywhere in an argument.

Finally, there is a feature of deduction systems that moves in the opposite direction of completeness. As we add more rules to cover new deductive situations, we may be admitting redundant inference rules, ones we could omit without weakening the system's deductive power. For the sake of economy, rules whose conclusions follow using other rules may be omitted. In the extreme, this would give us a *completely independent set* of inference rules.

However, the more we whittle down our Natural Deduction System, the more difficult it will be to deduce conclusions with the remaining rules. So, without multiplying rules excessively, we'll accept any sound rule of inference that's fairly simple and intuitively obvious, even if it adds some redundancy. This will give us a moderately large system of rules, but natural ones. More rules make proof construction easier.

As we become acquainted with PL's Natural Deduction System, we'll learn proof techniques used all the time in mathematical arguments. However, because most mathematical proofs also use non-truth-functional rules of inference (see Chapter 2), we won't yet be able to give complete proofs of many mathematical results using only PL's inference rules. To compensate, we'll include some arguments from everyday life, and we'll use abstract arguments formulated using the symbolic language of *Propositional Logic*.

1.5.2 *Simple Inference Rules: Premises*

Our first inference rule is one that's easy to overlook: the *Rule of Premises* (*Prem*). Given a list of premises for an argument, you may assert any one of them anywhere in your deduction, citing *Prem* as your reason.

To record a deduction using a *proof diagram*, we'll put the deduction's sentences on the left side and cite the relevant rules of inference on the right.[9]

A deduction based upon premises **P**, **Q**, and **R** and concluding **S** would look as indicated below. The sideline indicates the argument's progression, the numbers to the left label the steps, and the single underline separates the premises from the conclusions. The final conclusion appears as the last line. The reasons given for each step cite the relevant rule of inference, followed by the line numbers to which the rule has been applied.

$$
\begin{array}{l|l l}
i & \mathbf{P} & \text{Prem} \\
j & \mathbf{Q} & \text{Prem} \\
k & \mathbf{R} & \text{Prem} \\
  & \cdots & \\
n & \cdots & \text{Inf Rule X} \quad j, m \\
  & \cdots & \\
v & \mathbf{S} & \text{Inf Rule Y} \quad m, n
\end{array}
$$

You won't use such diagrams to construct proofs in mathematics, but for a while they will help us spell out the logical detail of an argument and grasp the overall structure of a proof.

A formal argument lists all the premises at the outset, but this does *not* mean you should use them all right away. When a sentence is needed, you merely refer back to it by line number. In an informal mathematical proof, on the other hand, premises aren't usually stated until they're needed.

A closely related rule, which we'll call the *Rule of a Previously Proved Proposition* (*Prev Propn*), allows you to cite a result you've already proved. We won't have much occasion to use this rule here, but it would be used repeatedly in developing a deductive theory, where the theory's axioms and definitions are the underlying premises of the entire system, and theorems are proved by means of anything previously assumed or proved.

1.5.3 *Introduction and Elimination Rules for* **A**ND

The *Simplification Elimination Rule* for conjunction is schematized in the following way, both forms going under the name *Simp*[10]:

$$
\textit{Simp} \qquad
\begin{array}{|l}
\mathbf{P} \wedge \mathbf{Q} \\ \hline
\mathbf{P}
\end{array}
\qquad\qquad
\begin{array}{|l}
\mathbf{P} \wedge \mathbf{Q} \\ \hline
\mathbf{Q}
\end{array}
$$

---

[9] This *Fitch-style natural deduction schema* is due to F. B. Fitch (1952), though a version of it was proposed a couple of decades earlier by S. Jaśkowski.

[10] This and other rules of inference are located for easy reference in Appendix A.

In an inference rule schema, if a formula like one above the double underline occurs in a proof, you may infer the formula below the line. Double underlines are used for presenting *inference rules* (note the large $\vDash$); *deductions* use single underlines (like in $\vdash$) to separate premises from conclusions. Also, no sideline numbers appear in an inference schema—rules are not deductions but are used for making them.

Simplification is used subconsciously all the time in everyday reasoning. Given two propositions joined by *and*, we can legitimately conclude either conjunct. Such an inference rule is obviously sound (see Exercise 1).

The *Introduction Rule* for $\wedge$ goes in the other direction. If **P** and **Q** each occur in a deduction prior to a given line (in either order), you may certainly conclude **P** $\wedge$ **Q** on that line, citing the *Rule of Conjunction (Conj)*. This rule is also sound (see Exercise 2).

The schema for the conjunction inference rule is:

$$\left|\begin{array}{l} \mathbf{P} \\ \mathbf{Q} \\ \hline\hline \mathbf{P} \wedge \mathbf{Q} \end{array}\right. \qquad\qquad\qquad\qquad \textit{Conj}$$

The following examples show how the *Int-Elim Rules* for $\wedge$ are used to construct a deduction. Each of these inferences will be covered by a *Replacement Rule* (discussed next), but they're derived here using *Simp* and *Conj* to illustrate a simple deduction.

✤**Example 1.5.2**
  Show that **P** $\wedge$ **Q** $\vdash$ **Q** $\wedge$ **P**   (*commutativity* of conjunction).

Solution
  The following proof diagram establishes the claim.

$$\begin{array}{ll} 1 \; \left|\; \mathbf{P} \wedge \mathbf{Q} \right. & \text{Prem} \\ \\ 2 \; \left|\; \mathbf{P} \right. & \text{Simp 1} \\ 3 \; \left|\; \mathbf{Q} \right. & \text{Simp 1} \\ 4 \; \left|\; \mathbf{Q} \wedge \mathbf{P} \right. & \text{Conj 3, 2} \end{array}$$

This example can be reversed: **Q** $\wedge$ **P** $\vdash$ **P** $\wedge$ **Q**. In fact, the example's deduction covers this case as well, since it validates interchanging the two conjuncts. This gives us an *interderivability* result: **P** $\wedge$ **Q** $\dashv\vdash$ **Q** $\wedge$ **P**, where the two-directional $\dashv\vdash$ indicates that the formula on each side of the symbol can be derived from the one on the other side.

The next example is also half of an interderivability result, though this time the second direction requires a separate proof (see Exercise 5).

✤**Example 1.5.3**
  Show that $(\mathbf{P} \wedge \mathbf{Q}) \wedge \mathbf{R} \vdash \mathbf{P} \wedge (\mathbf{Q} \wedge \mathbf{R})$ (*associativity* of conjunction).

Solution

The following proof diagram establishes this claim.

1	$(\mathbf{P} \wedge \mathbf{Q}) \wedge \mathbf{R}$	Prem
2	$\mathbf{P} \wedge \mathbf{Q}$	Simp 1
3	$\mathbf{P}$	Simp 2
4	$\mathbf{Q}$	Simp 2
5	$\mathbf{R}$	Simp 1
6	$\mathbf{Q} \wedge \mathbf{R}$	Conj 4, 5
7	$\mathbf{P} \wedge (\mathbf{Q} \wedge \mathbf{R})$	Conj 3, 6

1.5.4 Replacement Rules Involving AND

Replacement Rules permit us to substitute equivalent sentences for one another. They will be given in the form $\mathbf{P} :: \mathbf{Q}$,[11] (read $\mathbf{P}$ *is inter-replaceable with* $\mathbf{Q}$). If $\mathbf{F}(\mathbf{P})$ is a formula containing occurrences of $\mathbf{P}$, and if $\mathbf{P} \vDash \mathbf{Q}$, we may infer $\mathbf{F}(\mathbf{Q})$, in which $\mathbf{Q}$ has been substituted *one or more times* for $\mathbf{P}$.

There are three *Replacement Rules* for $\wedge$. We mentioned the first two above—*Commutation* (*Comm*) and *Association* (*Assoc*). These rules rarely if ever appear in informal proofs. We mentally rearrange conjuncts in these and more complex ways without ever thinking about it. In formal proofs, though, we'll use and cite these rules as needed.

These rules are schematized as follows:

Comm ($\wedge$) $\mathbf{P} \wedge \mathbf{Q} :: \mathbf{Q} \wedge \mathbf{P}$

Assoc ($\wedge$) $\mathbf{P} \wedge (\mathbf{Q} \wedge \mathbf{R}) :: (\mathbf{P} \wedge \mathbf{Q}) \wedge \mathbf{R}$

The *Idempotence Replacement Rule* (*Idem*) is seldom used either, except in formal proofs to expand or contract a sentence.

This rule is schematized as follows:

Idem ($\wedge$) $\mathbf{P} \wedge \mathbf{P} :: \mathbf{P}$

1.5.5 Proof Strategy: the Backward-Forward Method

At this point, our Natural Deduction System is far from complete. We can't prove very much from the few rules we have so far (see Exercise 7). But before we move on to consider another connective's *Int-Elim Rules*, let's reflect on overall proof strategy for informal as well as formal proofs.

The first prerequisite for constructing a good argument sounds trite but bears stating: *keep in mind what the premises are and what the conclusion is.* Beginning proof-makers sometimes get confused about what they're trying to prove and what they're allowed to assume. You may finish a proof, only to discover that you proved something different than what was asked. Or you may get disoriented and end up constructing a circular argument, assuming

---

[11] The symbol :: symbolizes sameness of ratios in mathematics. Bergmann, Moor, and Nelson in *The Logic Book* (1980) use it as we do here to indicate *replaceable equivalents*.

something very similar to the final conclusion. This happens to nearly every-one at one time or another, especially in an informal argument. When you're constructing complicated mathematical arguments or axiomatically develop-ing the theory of a field you're already familiar with, it's easy to get mixed up. Before you know it, you've constructed an argument that uses as given what you've never deduced and have no right to use. The best way to avoid this is to continually remind yourself, using verbal indicators like *have* and *want*—what you already *know* and what you still need to *show*.

Assuming you have a firm grasp of the premises and conclusion, there are three basic approaches you can follow in constructing a proof. You can

- *work forward from the premises to their consequences*;
- *work backward from the conclusion to sentences from which it follows*; or
- *work using a mixture of both forward and backward reasoning*.

When working forward from the premises, you should ask yourself: *what fol-lows from this? what does this imply?* Working backward from the conclusion, you should ask: *why is this? how can this be?* Your goal is to connect the two strands of reasoning together by a logical chain of results proceeding from the premises to the conclusion.

Think of a proof as a journey through a maze. You must map out a con-nected route from start (the premises) to finish (the conclusion). You can plot a path from the starting point and move forward, but you must keep your final destination in view to end up where you want to be. Or you can plot a route from the end and go backward, but you need to know where the starting point is to link up with it. You can even map out a course from both ends and meet somewhere in the middle. Regardless, both the starting point and the destination need to be kept in mind.

We'll call this combined proof strategy the *Backward-Forward Method of Proof Analysis*.[12] In constructing a proof, it is generally best to *begin by backtracking from the conclusion*. Ask yourself what sort of sentence you need to prove and how you can get it from sentences in the forms you're given. The advantages of this approach include the following:

1. This proof strategy is *goal-oriented*, concentrating on what needs to be done. If you begin with the premises, it may be less clear which one(s) you should work with first. This is especially true in developing an axiomatic theory, where all the axioms and definitions are poten-tial premises, though you may not know which ones are relevant for a particular theorem. Focusing on the conclusion helps you avoid drawing conclusions that are true but irrelevant. Of course, in proceeding back-ward, you still have to keep the initial premises in mind, and you have to remember that wanting something is different than having it.

2. The attitude engendered by the *Backward-Forward Method of Proof Anal-ysis* is helpful for doing mathematical research. When mathematicians

---

[12] Daniel Solow in ***How to Read and Do Proofs*** (6th edition, 2013) uses a similar term. The ideas behind this method, though, have long been used for constructing proofs.

suspect a proposition to be true, they begin with preconceived notions of why that's the case. As they continue their investigation, they may modify their original hypotheses, weakening or strengthening conditions as they discover what premises the conclusion follows from. Or they may modify their conjecture. A joint *Backward-Forward* approach helps ferret these things out. Mathematicians rarely adopt a set of premises and try to see what they entail, using a purely *Forward* mode of proof. They work with conjectures and try to link them back to what they already know.

3. Finally, in proving a tautology (something we do in logic), the *Backward Method* is the only strategy initially available, since a tautology is true on the basis of no premises. There is thus nothing to work forward from.

Beginning proof-makers often don't think to use the *Backward Method*. Since premises are available at the outset, it's tempting to start with them. But this can be the wrong tack to take. You should first ask what's needed to prove the conclusion and what known results (premises, definitions, and earlier propositions) apply to this situation.

In the end, your principal *logical* guide should be the *form* of the sentences involved.[13] The main connective determines what *sort* of sentence a given proposition is, which in turn determines what can be done with it. This is particularly important for working with the abstract sentences of PL, but it also helps in constructing mathematical proofs. To draw conclusions from a premise set, you must use rules of inference that pertain to those sentence forms. You may intuitively draw conclusions from premises without knowing any systematic logic, but if you run stuck and don't know what to do, or if the field is abstract and unfamiliar, analyzing the logical forms involved may be just what you need to get started.

Let's make this advice concrete in a simple case by revisiting Example 3.

✤ Example 1.5.4

Use the *Backward-Forward Method* to show $(\mathbf{P} \wedge \mathbf{Q}) \wedge \mathbf{R} \vdash \mathbf{P} \wedge (\mathbf{Q} \wedge \mathbf{R})$.

Solution
Proof Analysis
$(\mathbf{P} \wedge \mathbf{Q}) \wedge \mathbf{R}$ is our premise; we want to conclude $\mathbf{P} \wedge (\mathbf{Q} \wedge \mathbf{R})$.

· The conclusion conjoins $\mathbf{P}$ with $\mathbf{Q} \wedge \mathbf{R}$. If we have these two conjuncts separately, we can apply *Conj*. At this point, though, neither one follows immediately from the premise, so we continue with our backward analysis.

· To get $\mathbf{Q} \wedge \mathbf{R}$, it suffices to get $\mathbf{Q}$ and $\mathbf{R}$ separately.

This concludes our backward analysis: we need $\mathbf{P}$, $\mathbf{Q}$, and $\mathbf{R}$ individually.

Proof Construction   (This verbalizes the proof diagram of Example 3.)

· Working in a forward direction, we begin with the premise $(\mathbf{P} \wedge \mathbf{Q}) \wedge \mathbf{R}$.

$\mathbf{R}$ can be gotten from this by *Simp*. This yields $\mathbf{P} \wedge \mathbf{Q}$ as well.

We can then further simplify $\mathbf{P} \wedge \mathbf{Q}$ to obtain both $\mathbf{P}$ and $\mathbf{Q}$.

---

[13] Naturally, more than logic is needed to create mathematical proofs. Intuitions about how the mathematical ideas are related are also crucial. We'll say more about this later.

· Now we're ready to combine sentences.
We first conjoin $\mathbf{Q}$ and $\mathbf{R}$ to get $\mathbf{Q} \wedge \mathbf{R}$.
Conjoining this with $\mathbf{P}$, we obtain the conclusion $\mathbf{P} \wedge (\mathbf{Q} \wedge \mathbf{R})$.   ∎

Matters will get more complex as we proceed, but the general proof construction procedure we're advocating remains the same and is at bottom simple common sense: *determine where you want to end up, assess what you need to get there, and then see what you have to get started along that path.*

EXERCISE SET 1.5

Exercises 1–2: Soundness of Inference Rules
Show that the following Int-Elim and Replacement Rules are sound.

1.5.1. *Simplification:* $\mathbf{P} \wedge \mathbf{Q} \vDash \mathbf{P}$; $\mathbf{P} \wedge \mathbf{Q} \vDash \mathbf{Q}$.

1.5.2. *Conjunction:* $\mathbf{P}, \mathbf{Q} \vDash \mathbf{P} \wedge \mathbf{Q}$.

Exercises 3–5: Proof Analysis
Give a proof analysis and then construct a formal proof to show the following.

1.5.3. $\mathbf{P} \wedge (\mathbf{Q} \wedge \mathbf{P}) \vdash \mathbf{P} \wedge \mathbf{Q}$
Prove this both with and without using *Replacement Rules*.

1.5.4. $\mathbf{P} \wedge (\mathbf{Q} \wedge \mathbf{R}) \vdash \mathbf{P} \wedge \mathbf{R}$

1.5.5. *The Associative Law for Conjunction*
   a. $\mathbf{P} \wedge (\mathbf{Q} \wedge \mathbf{R}) \vdash (\mathbf{P} \wedge \mathbf{Q}) \wedge \mathbf{R}$. Work this using only *Int-Elim Rules*.
   b. Repeat part *a*, using any inference rules for conjunction.

Exercises 6–7: Logical Implication and Deductions
The following explore connections between implication and derivability.

1.5.6. *Derivability and Validity*
   a. Tell why $\mathbf{P} \wedge \mathbf{Q}$ cannot be derived from $\mathbf{P} \wedge (\mathbf{Q} \wedge \mathbf{R})$ via *Simp*.
   b. Show that $\mathbf{P} \wedge (\mathbf{Q} \wedge \mathbf{R}) \vdash \mathbf{P} \wedge \mathbf{Q}$.
   c. By analyzing part *b*'s derivation, explain why $\mathbf{P} \wedge (\mathbf{Q} \wedge \mathbf{R}) \vDash \mathbf{P} \wedge \mathbf{Q}$.

1.5.7. *Completeness So Far*
At this point in developing PL's Natural Deduction System, which of the following claims hold? Explain your answer.
   a. $\mathbf{P}, \mathbf{Q} \vDash \mathbf{P} \wedge \mathbf{Q}$; $\mathbf{P}, \mathbf{Q} \vdash \mathbf{P} \wedge \mathbf{Q}$.
   b. $\mathbf{P} \wedge \mathbf{Q} \vDash \mathbf{P} \vee \mathbf{Q}$; $\mathbf{P} \wedge \mathbf{Q} \vdash \mathbf{P} \vee \mathbf{Q}$.
   c. $\mathbf{P} \vee \mathbf{Q} \vDash \mathbf{P} \wedge \mathbf{Q}$; $\mathbf{P} \vee \mathbf{Q} \vdash \mathbf{P} \wedge \mathbf{Q}$.
   d. $(\mathbf{P} \vee \mathbf{Q}) \wedge \neg \mathbf{Q} \vDash \mathbf{P}$; $(\mathbf{P} \vee \mathbf{Q}) \wedge \neg \mathbf{Q} \vdash \mathbf{P}$.

Exercises 8–10: True or False
Are the following statements true or false? Explain your answer.

1.5.8. *Sound* inference rules are only applied to true premises.

1.5.9. *Replacement Rules* are special types of *Int-Elim Rules*.

1.5.10. *Premises* are listed at the start of a formal deduction, but they're often not used until later in the proof.

Exercises 11–12: Explanations
Explain the following, using your own words.

1.5.11. The *Backward-Forward Method of Proof Analysis.*

1.5.12. The difference between a sentence **Q** being a logical consequence of a set of sentences $\mathcal{P}$ and being deduced from $\mathcal{P}$.

Exercises 13–16: Drawing Conclusions from Premises
What conclusions can be drawn from the following premises via the rules of inference available so far? List or describe all such sentences.

1.5.13. P, Q **1.5.15.** $P \wedge Q \to R$

1.5.14. $P \wedge Q$ **1.5.16.** $P, P \to Q$

Exercises 17–18: Finding Premises for Conclusions
Find two premise sets that will yield the following conclusions via the rules of inference given so far. Do not list any premises that would not figure in a deduction or that contain additional sentence variables.

1.5.17. $(P \vee Q) \wedge R$ **1.5.18.** $(P \wedge Q) \wedge \neg R$

Exercises 19–24: Hofstadter's MIU System
*The following MIU System is due to Douglas Hofstadter in **Gödel, Escher, Bach: an Eternal Golden Braid**, pp. 33–43 and 259–61.*

The alphabet for this system are the letters M, I, and U. Words, such as MUMI, are formed by concatenation; new words can be made from old words using the following rules (X and Y stand for any words):

1. *A word ending in I can be extended by a U: $XI \mapsto XIU$.*
   For example: MUMI generates MUMIU.
2. *A word beginning in M can be extended by duplicating what follows: $MX \mapsto MXX$.*
   For example: MUMI generates MUMIUMI.
3. *III may be replaced by U: $XIIIY \mapsto XUY$.*
   For example: UMIIIMU generates UMUMU.
4. *UU may be deleted: $XUUY \mapsto XY$.*
   For example: MUUI generates MI.

1.5.19. Derive MUIIU from MIUUI. Indicate which rules you're applying.

1.5.20. Show that $MUIIU$ and MUI can each be derived from the other.

1.5.21. Derive the word $MUIIU$ from the word MI using the above rules.

1.5.22. Derive $MUUIU$ from MI using the above rules.

1.5.23. Derive $MIUI$ from MI using the above rules.

1.5.24. *Word Derivation from MI*

If MI is the starting word for a derivation:

a. What words are formed after one application of the rules? two? three?

b. What lengths are the words formed after one application of the rules? two? three?

c. If I is valued at 1, U at 3 and M at 0, what are the values of the words produced from MI after one application of the rules? two? three?

d. *Hofstadter's MU puzzle*: derive MU from MI using the above rules, or show it cannot be done.

Exercises 25–28: Mathematical Exercises Involving Conjunction

The following problems illustrate how conjunction may be involved implicitly in mathematical statements.

1.5.25. A number k is *composite* if and only if there are numbers m and n with $m \neq \pm 1$, $n \neq \pm 1$, and $mn = k$.

a. What separate results must be shown for k to satisfy this definition? What rule of inference must then be used to conclude that k is composite?

b. If a number k satisfies the definition of being composite, what individual sentences can you conclude? Which rule of inference is used for this?

1.5.26. Rewrite the sentence *the hypotenuse of a right triangle is longer than either leg* to exhibit its logical connectives. What conclusions can be drawn from this statement? What rule of inference justifies these conclusions?

1.5.27. If you are given the two premises $-1 < x < 1$ and $z < x$, what new double inequality can you conclude? Write down an argument in proof diagram format establishing your result, citing the appropriate rules of inference. Recall that $a < b < c$ abbreviates a conjunction.

1.5.28. *Inference Rules in Mathematical Proofs*
Analyze the following set-theoretic proof, pointing out where the inference rules governing conjunctions enter in.

Theorem: *Subset Property of Intersections*
Intersection $S \cap T$ is a subset of both S and T, i.e., $S \cap T \subseteq S \land S \cap T \subseteq T$.
Proof:

1) Suppose x is any element of $S \cap T$; in symbols, $x \in S \cap T$.
2) Then by the definition of $S \cap T$, we know that $x \in S \land x \in T$.
3) Thus, $x \in S$; and so $S \cap T \subseteq S$ by the definition of being a subset.
4) Similarly, $x \in T$; and so $S \cap T \subseteq T$, too.
5) But then $S \cap T \subseteq S \land S \cap T \subseteq T$.

1.6 Elimination Rules for CONDITIONALS

Among PL's inference rules, the most important ones for mathematics are those governing conditional and biconditional arguments. Here we'll look at the main *Elimination Rules* and a few *Replacement Rules*. *Introduction Rules* and some additional *Replacement Rules* will be treated in the next section.

1.6.1 *Eliminating* If-Then: *Modus Ponens*

Regardless of how a deduction system for *Propositional Logic* is chosen, the inference rule *Modus Ponens* is usually included because it is so essential for making deductions. This rule is applied every time we conclude that something is true of a given object because it is true for all objects of that type. Such an argument has the form *all Xs are Ys; Z is an X; therefore Z is a Y*. Putting the first premise of this argument into a conditional form, we have *if Z is an X, then Z is a Y; Z is an X; therefore Z is a Y*.

Consider a typical example from geometry. Suppose we conclude that $\triangle ABC$ has congruent base angles because the opposite sides are congruent. Why does this follow? Because of the theorem that says *if a triangle is isosceles, then the base angles opposite the congruent sides are congruent*.

This argument has the form $\mathbf{P} \to \mathbf{Q}$, $\mathbf{P}$; therefore $\mathbf{Q}$. It applies the inference rule traditionally called *Modus Ponens* (*MP*), which is schematized as follows:

$$
MP \qquad \begin{array}{|l}
\mathbf{P} \to \mathbf{Q} \\
\mathbf{P} \\
\hline\hline
\mathbf{Q}
\end{array}
$$

The Latin term *Modus Ponens* and its mate, *Modus Tollens*, discussed next, go back to medieval times. *Modus Ponens affirms the antecedent* $\mathbf{P}$ of a conditional $\mathbf{P} \to \mathbf{Q}$ in order to conclude the consequent $\mathbf{Q}$.

Modus Ponens is a sound inference rule, for if $\mathbf{P} \to \mathbf{Q}$ and $\mathbf{P}$ are both true, $\mathbf{Q}$ is as well (see Exercise 1). However, the argument form known as *affirming the consequent* ($\mathbf{P} \to \mathbf{Q}$, $\mathbf{Q}$; therefore $\mathbf{P}$) is a *fallacy* (see Exercise 9a)—it's possible for $\mathbf{P}$ to be false even if both premises are true.

The following example illustrates the use of *MP*.

✚**Example 1.6.1**

Show that $P \land Q$, $P \to R$, $Q \land R \to \neg S \vdash R \land \neg S$.

Solution

The following proof diagram establishes the claim.

$$
\begin{array}{r|ll}
1 & P \land Q & \text{Prem} \\
2 & P \to R & \text{Prem} \\
3 & Q \land R \to \neg S & \text{Prem} \\
\hline
4 & P & \text{Simp 1} \\
5 & R & \text{MP } 2,4 \\
6 & Q & \text{Simp 1} \\
7 & Q \land R & \text{Conj } 6,5 \\
8 & \neg S & \text{MP } 3,7 \\
9 & R \land \neg S & \text{Conj } 5,8
\end{array}
$$

Note that step 7 cites steps 5 and 6 in reverse order, indicating how these steps correspond to the sentence forms given in the rule schema for *Conj*. Being this precise is not necessary unless you're a computer following an algorithm. Either order is permissible in your written deductions.

It should be stressed that *Modus Ponens*, like all *Int-Elim Rules*, must be applied to *whole sentences*, not merely parts of a sentence. For example, concluding Q from P and $(P \rightarrow Q) \rightarrow R$ is not only improper but invalid (see Exercise 12)—we cannot apply *Modus Ponens* to the component sub-sentence $P \rightarrow Q$. *Replacement Rules* may be applied to component parts, but *Int-Elim Rules* may not.

1.6.2 *Eliminating IF-THEN: Modus Tollens*

A second *Elimination Rule* for the *if-then* connective is *Modus Tollens* (*MT*). This rule concludes $\neg\mathbf{P}$ from $\mathbf{P} \rightarrow \mathbf{Q}$ and $\neg\mathbf{Q}$. Additionally, it concludes $\neg\mathbf{P}$ from $\mathbf{P} \rightarrow \neg\mathbf{Q}$ and $\mathbf{Q}$, which contradicts $\neg\mathbf{Q}$. Since both forms proceed similarly, we'll consider them as two forms of *Modus Tollens*.

Schematically, we have the following:

$$\begin{array}{|l} \mathbf{P} \rightarrow \mathbf{Q} \\ \neg\mathbf{Q} \\ \hline\hline \neg\mathbf{P} \end{array} \qquad \begin{array}{|l} \mathbf{P} \rightarrow \neg\mathbf{Q} \\ \mathbf{Q} \\ \hline\hline \neg\mathbf{P} \end{array} \qquad\qquad \mathit{MT}$$

To illustrate, consider the conditional sentence *if n is even, then n^2 is even* along with the negation n^2 *is not even*. From these two propositions we can conclude by *Modus Tollens* that n *is not even*.

Modus Tollens proceeds by *denying the consequent*. If the consequent $\mathbf{Q}$ of a conditional sentence $\mathbf{P} \rightarrow \mathbf{Q}$ is *not* the case while the conditional sentence *is* the case, the antecedent $\mathbf{P}$ *cannot* be the case. This rule is sound, for if $\mathbf{P} \rightarrow \mathbf{Q}$ and $\neg\mathbf{Q}$ are true, so is $\neg\mathbf{P}$ (see Exercise 2).

Or, to argue *Modus Tollens'* soundness on the basis of *Modus Ponens*: if $\mathbf{P}$ *were* the case, then according to *MP*, $\mathbf{Q}$ would have to be the case, too, but it's not; $\neg\mathbf{Q}$ is. Thus, $\neg\mathbf{P}$ must be the case. This argument uses *Proof by Contradiction*, a strategy we'll investigate in Section 1.8.

Denying the antecedent, on the other hand, is a fallacy. If we know that both $\neg\mathbf{P}$ and $\mathbf{P} \rightarrow \mathbf{Q}$ are the case, we still *cannot* conclude that $\neg\mathbf{Q}$ is. The premises can be true and the conclusion false, so such an argument is invalid (see Exercise 9b).

We'll illustrate *MT* with the following example, which also uses *MP*.

✤**Example 1.6.2**
   Show that $P \rightarrow (Q \rightarrow R)$, P, $\neg R \vdash \neg Q$.

Solution

The following proof diagram establishes the claim.

$$
\begin{array}{ll}
1 \mid P \to (Q \to R) & \text{Prem} \\
2 \mid P & \text{Prem} \\
3 \mid \neg R & \text{Prem} \\
\hline
4 \mid Q \to R & \text{MP } 1,2 \\
5 \mid \neg Q & \text{MT } 4,3 \\
\end{array}
$$

1.6.3 *Biconditional Elimination*

Biconditional sentences occur in mathematics most frequently in definitions, so we'll couch our discussion of their *Elimination Rules* in terms of them— though these rules apply to any biconditional sentence.

Definitions state the meaning of terms: *something is a so-and-so if and only if it has such-and-such a feature.* Knowing something is a so-and-so, we can conclude it has such-and-such a feature. Conversely, if something has such-and-such a feature, then it is a so-and-so. We can thus replace a defined term by its definition (*expanding* via the definition), or we can replace a definition with the term it defines (*abbreviating* via the defined term).

The following example illustrates this from different fields of mathematics.

✤**Example 1.6.3**

Show how definitions are used for mathematical properties and relations.

Solution

· We'll give one definition from arithmetic and another from geometry.
  1) *A number n is even if and only if n is divisible by* 2.
  2) *Lines l and m are parallel if and only if they have no points in common.*
· We can conclude the defining property for objects of the right sort.
  1) *An integer $n^2 + n$ is divisible by* 2 because $n^2 + n$ *is even.*
  2) *Lines $y = 2x$ and $y = 2x+1$ have no points in common* because, having the same slope, *these lines are parallel.*
· We can also draw a conclusion about being some sort of object(s).
  1) *A number whose last digit is even is even* because *it is divisible by* 2.
  2) *Lines $y - 2x = 0$ and $y - 2x = 1$ are parallel* because *these lines (equations) have no points (solutions) in common.*

Drawing conclusions from definitions occurs so automatically that we're often not aware of making an inference. Nevertheless, this follows the rule *Biconditional Elimination (BE).*

We have the following two forms for *Biconditional Elimination*:

$$
BE \qquad
\begin{array}{|l}
\mathbf{P \leftrightarrow Q} \\
\mathbf{P} \\
\hline
\mathbf{Q}
\end{array}
\qquad\qquad
\begin{array}{|l}
\mathbf{P \leftrightarrow Q} \\
\mathbf{Q} \\
\hline
\mathbf{P}
\end{array}
$$

Biconditional Elimination is a two-directional analogue of *Modus Ponens*. It's obviously a sound rule of inference (see Exercise 3).

✤**Example 1.6.4**

Show that $(P \leftrightarrow Q) \wedge R, R \rightarrow P \vdash Q$.

Solution

The following proof diagram establishes this result.

1	$(P \leftrightarrow Q) \wedge R$	Prem
2	$R \rightarrow P$	Prem
3	$P \leftrightarrow Q$	Simp 1
4	R	Simp 1
5	P	MP $2, 4$
6	Q	BE $3, 5$

1.6.4 *Negative Biconditional Elimination*

Negative Biconditional Elimination (*NBE*) is the biconditional counterpart of *Modus Tollens*. If $\mathbf{P} \leftrightarrow \mathbf{Q}$ is the case but either $\mathbf{P}$ or $\mathbf{Q}$ is not, then the other sentence also is not the case. This is clearly a sound rule (see Exercise 4).

Schematically, we have the following:

$$\mathbf{P} \leftrightarrow \mathbf{Q} \qquad \mathbf{P} \leftrightarrow \mathbf{Q}$$
$$\neg\mathbf{P} \qquad\qquad \neg\mathbf{Q} \qquad\qquad \mathbf{NBE}$$
$$\neg\mathbf{Q} \qquad\qquad \neg\mathbf{P}$$

These are the rules needed when we know that something is *not* a so-and-so or that such-and-such a feature is *not* the case. We can then negate the other part of the biconditional definition that characterizes so-and-so's or objects with such-and-such a feature. For example, to prove n^2 *is not even*, we can show n *is not even*, for n^2 *is even if and only if* n *is even*.

Here is a PL deduction that illustrates the use of *NBE*.

✤**Example 1.6.5**

Show that $P \rightarrow (Q \leftrightarrow R), P \wedge \neg Q \vdash \neg R$.

Solution

The following proof diagram establishes this result.

1	$P \rightarrow (Q \leftrightarrow R)$	Prem
2	$P \wedge \neg Q$	Prem
3	P	Simp 2
4	$Q \leftrightarrow R$	MP $1, 3$
5	$\neg Q$	Simp 2
6	$\neg R$	NBE $4, 5$

1.6.5 *Chaining Conditionals and Biconditionals*

So far we have no rules for inferring a conditional or biconditional sentence as a conclusion. We'll introduce the main proof techniques for this in Section 1.7, but let's briefly consider the case when two conditionals or biconditionals are chained together to deduce such a conclusion.

The most basic rule of this sort concludes $\mathbf{P} \to \mathbf{R}$ from $\mathbf{P} \to \mathbf{Q}$ and $\mathbf{Q} \to \mathbf{R}$ (the transitivity property of $\to$). The traditional name for this argument form is *Hypothetical Syllogism (HS)*. It, too, is a sound rule (see Exercise 5).

The schema for *Hypothetical Syllogism* is the following:

$$HS \qquad \begin{array}{|l} \mathbf{P} \to \mathbf{Q} \\ \mathbf{Q} \to \mathbf{R} \\ \hline \hline \mathbf{P} \to \mathbf{R} \end{array}$$

♣ Example 1.6.6

Show that $P \wedge S,\ P \to \neg R,\ \neg R \to \neg Q \vdash S \wedge \neg Q$.

Solution

The following proof diagram establishes the claim.

$$\begin{array}{ll}
1\ \left| \begin{array}{l} P \wedge S \end{array} \right. & \text{Prem} \\
2\ \left| P \to \neg R \right. & \text{Prem} \\
3\ \left| \neg R \to \neg Q \right. & \text{Prem} \\[4pt]
4\ \left| P \to \neg Q \right. & \text{HS } 2,3 \\
5\ \left| P \right. & \text{Simp } 1 \\
6\ \left| \neg Q \right. & \text{MP } 4,5 \\
7\ \left| S \right. & \text{Simp } 1 \\
8\ \left| S \wedge \neg Q \right. & \text{Conj } 7,6
\end{array}$$

There is also a biconditional counterpart to *Hypothetical Syllogism*. This rule has no standard name; we'll call it *Biconditional Transitivity (BiTrans)*. *BiTrans* concludes $\mathbf{P} \leftrightarrow \mathbf{R}$ because $\mathbf{Q}$ links $\mathbf{P}$ and $\mathbf{R}$ together in both directions. It is schematized as follows:

$$BiTrans \qquad \begin{array}{|l} \mathbf{P} \leftrightarrow \mathbf{Q} \\ \mathbf{Q} \leftrightarrow \mathbf{R} \\ \hline \hline \mathbf{P} \leftrightarrow \mathbf{R.} \end{array}$$

This proof technique can be extended to longer sequences of biconditionals. For instance, to prove two sets are equal, we need to show that *an object is an element of one set if and only if it is an element of the other one.* This may be done by showing that *an object belongs to one set if and only if some first thing happens,* and so on, ending with *some final thing happens if and only if the object belongs to the other set.* Stringing biconditionals together gives what we need. *Hypothetical Syllogism* can be extended in a similar way.

1.6.6 *Replacement Rules for IF-THEN & IF-AND-ONLY-IF*

Replacement Rules usually involve more than one connective. They are less simple and less natural than *Int-Elim Rules*, and they are also redundant. Although we don't need them in PL's deduction system, they help simplify deductions.

The *Conditional Equivalence (Cndnl)* rules are often used to deduce a conclusion from a conditional sentence $\mathbf{P} \to \mathbf{Q}$ in situations where *MP* and *MT* seem inapplicable. They are based upon the equivalences considered in Section 1.4 when we introduced conditionals (see also Exercises 1.4.44–45).

The *Cndnl Replacement Rules* are schematized as follows:

$$\mathbf{P} \to \mathbf{Q} :: \neg(\mathbf{P} \wedge \neg\mathbf{Q})$$
$$\mathbf{P} \to \mathbf{Q} :: \neg\mathbf{P} \vee \mathbf{Q}$$
Cndnl

Similarly, the *Replacement Rules Biconditional Equivalence (Bicndnl)* are based upon the logical equivalences noted in discussing the *if-and-only-if* connective in Section 1.4 (see also Exercises 1.4.50–53). These rules are mostly used in going from left to right, replacing a biconditional sentence by one in a logical form that may be more easily combined with other sentences in a deduction. The right to left direction is also valid, though.

The schema for the *Bicndnl* equivalences is the following:

$$\mathbf{P} \leftrightarrow \mathbf{Q} :: (\mathbf{P} \to \mathbf{Q}) \wedge (\mathbf{Q} \to \mathbf{P})$$
$$\mathbf{P} \leftrightarrow \mathbf{Q} :: (\mathbf{P} \to \mathbf{Q}) \wedge (\neg\mathbf{P} \to \neg\mathbf{Q})$$
$$\mathbf{P} \leftrightarrow \mathbf{Q} :: (\mathbf{P} \wedge \mathbf{Q}) \vee (\neg\mathbf{P} \wedge \neg\mathbf{Q})$$
Bicndnl

Let's see how these rules can be used to duplicate the work of *BiTrans*.

❖**Example 1.6.7**

Show that $P \leftrightarrow Q$, $Q \leftrightarrow R \vdash P \leftrightarrow R$ without using *BiTrans*.

Solution

The following proof diagram establishes this claim.

1	$P \leftrightarrow Q$	Prem
2	$Q \leftrightarrow R$	Prem
3	$(P \to Q) \wedge (Q \to P)$	Bicndnl 1
4	$(Q \to R) \wedge (R \to Q)$	Bicndnl 2
5	$P \to Q$	Simp 3
6	$Q \to R$	Simp 4
7	$P \to R$	HS 5, 6
8	$Q \to P$	Simp 3
9	$R \to Q$	Simp 4
10	$R \to P$	HS 9, 8
11	$(P \to R) \wedge (R \to P)$	Conj 7, 10
12	$P \leftrightarrow R$	Bicndnl 11

There is one more standard equivalence for conditional sentences (*Contraposition*), but since it is often used in the context of deducing a conditional, we'll address it in Section 1.7.

1.6.7 *BE and Overall Proof Strategy*

While the above rules of inference are still simple, their importance for constructing mathematical proofs should not be minimized. For example, in attempting to construct a proof of some unfamiliar result, such as

All widgets that squibble are whatsits,

beginning proof-makers often run up against a mental block, not knowing where to start. The best thing to do when this happens is to look at the conclusion and ask both *What does it say?* and *How can I conclude this from what I already know?* To answer these questions, you must use both mathematics and logic.

On the *logical side*, as we noted earlier, you need to know the logical structure of the sentences involved and what proof strategies are available for deducing sentences like the conclusion from sentences like the premises.

On the *mathematical side*, our present focus, you need to be clear about the meaning of the terms involved and of any alternative characterizations they may have been given earlier. You need to know what *widgets* and *whatsits* are and what *squibbling* involves before you can show that *All widgets that squibble are whatsits.* You've undoubtedly already used this approach to study the material in this text. When you were asked to show that a set of sentences implied or proved another sentence or that they were consistent, you probably had to review the definition to see what was required.

When you expand technical terms, replacing them by an equivalent expression, you're applying *BE* (or *NBE*) to the definitions. Using *BE* as you start a proof may generate some intermediate conclusion, either as an initial step forward or as a penultimate conclusion. This can lead to a *Backward–Forward Proof Analysis*. There will still be work ahead, and you may get stumped, but you'll have overcome the psychological hurdles of getting started and acquiring some ideas for what to try next.

Anytime you run stuck in a proof, whether at the outset or somewhere in the middle, call a time out to take some distance to the problem and to view your work in more general terms. Ask yourself: what exactly are the objects, properties, and relations being considered? Get beyond the particular widget in front of you and note what type of widget it is. Then ask what properties and relations squibbling widgets have in general and whether you've taken full advantage of what you know about them. If not, can you use this knowledge in the case at hand?

Thinking both mathematically and logically becomes increasingly important as you begin to study more abstract mathematics, where you're no longer dealing with the concrete concepts of numbers and shapes and graphs that you've always thought mathematics was about. Terms' definitions will often be unfamiliar and may seem to make no more sense than the sentence *All widgets that squibble are whatsits*. So review, time and again, *what* it is you are trying to show (mathematical meaning) and *how* it might be shown (logical proof strategy). Logic without mathematical insight operates in a fog, but mathematical intuition without logical rigor remains disjointed.

If you follow this advice and refuse to be daunted by terms that are initially strange, you'll be in a position to start constructing deductions in any field of mathematics. As you proceed, this approach should become second nature, and you will have internalized the procedure mathematicians use whenever they're working on proofs.

EXERCISE SET 1.6

Exercises 1–8: Soundness of Inference Rules
Show that the following Int-Elim and Replacement Rules are sound.

1.6.1. *Modus Ponens (MP):* $\mathbf{P} \to \mathbf{Q}, \mathbf{P} \vDash \mathbf{Q}$.

1.6.2. *Modus Tollens (MT):* $\mathbf{P} \to \mathbf{Q}, \neg\mathbf{Q} \vDash \neg\mathbf{P}$.

1.6.3. *Biconditional Elimination (BE):* $\mathbf{P} \leftrightarrow \mathbf{Q}, \mathbf{P} \vDash \mathbf{Q}$.

1.6.4. *Negative Biconditional Elimination (NBE):* $\mathbf{P} \leftrightarrow \mathbf{Q}, \neg\mathbf{Q} \vDash \neg\mathbf{P}$.

1.6.5. *Hypothetical Syllogism (HS):* $\mathbf{P} \to \mathbf{Q}, \mathbf{Q} \to \mathbf{R} \vDash \mathbf{P} \to \mathbf{R}$.

1.6.6. *Biconditional Transitivity (BiTrans):* $\mathbf{P} \leftrightarrow \mathbf{Q}, \mathbf{Q} \leftrightarrow \mathbf{R} \vDash \mathbf{P} \leftrightarrow \mathbf{R}$.

1.6.7. *Conditional Equivalence (Cndnl)*
   a. $\mathbf{P} \to \mathbf{Q} \vDashH \neg(\mathbf{P} \wedge \neg\mathbf{Q})$ b. $\mathbf{P} \to \mathbf{Q} \vDashH \neg\mathbf{P} \vee \mathbf{Q}$

1.6.8. *Biconditional Equivalence (Bicndnl)*
   a. $\mathbf{P} \leftrightarrow \mathbf{Q} \vDashH (\mathbf{P} \to \mathbf{Q}) \wedge (\mathbf{Q} \to \mathbf{P})$ c. $\mathbf{P} \leftrightarrow \mathbf{Q} \vDashH (\mathbf{P} \wedge \mathbf{Q}) \vee (\neg\mathbf{P} \wedge \neg\mathbf{Q})$
   b. $\mathbf{P} \leftrightarrow \mathbf{Q} \vDashH (\mathbf{P} \to \mathbf{Q}) \wedge (\neg\mathbf{P} \to \neg\mathbf{Q})$

1.6.9. Show that the following fallacious inferences are *unsound*, in the following two ways:
   i. by assigning appropriate truth values to the sentences,
   ii. by giving a counterargument.
   a. *Affirming the Consequent:* $\mathbf{P} \to \mathbf{Q}, \mathbf{Q} \nvDash \mathbf{P}$.
   b. *Denying the Antecedent:* $\mathbf{P} \to \mathbf{Q}, \neg\mathbf{P} \nvDash \neg\mathbf{Q}$.

Exercises 10–11: Completing Deductions
Fill in the reasons for the following deductions.

1.6.10. $P \to Q, (P \to Q) \to P \vdash P \wedge Q$

1	$P \to Q$	_____
2	$(P \to Q) \to P$	_____
3	P	_____
4	Q	_____
5	$P \wedge Q$	_____

1.6.11. $P \wedge Q \leftrightarrow R,\ P \to S,\ R \vdash Q \wedge S$

1	$P \wedge Q \leftrightarrow R$	_____
2	$P \to S$	_____
3	R	_____
4	$P \wedge Q$	_____
5	P	_____
6	S	_____
7	Q	_____
8	$Q \wedge S$	_____

Exercises 12–14: Logical Implication and Conclusive Deductions

Determine whether the following **implication** *claims are true. Then determine whether the* **deductions** *given are* **conclusive**. *Carefully point out where and how a rule of inference is being misused.*

1.6.12. $P,\ (P \to Q) \to R \vDash R$

1	P	Prem
2	$(P \to Q) \to R$	Prem
3	Q	MP 2, 1
4	$P \to Q$	Conj 1, 3
5	R	MP 2, 4

1.6.13. $P \wedge Q \leftrightarrow R,\ \neg R \vDash \neg Q$

1	$P \wedge Q \leftrightarrow R$	Prem
2	$\neg R$	Prem
3	$Q \leftrightarrow R$	Simp 1
4	$\neg Q$	NBE 3, 2

1.6.14. $P \to Q,\ P \to R,\ Q \wedge R \vDash P$

1	$P \to Q$	Prem
2	$P \to R$	Prem
3	$Q \wedge R$	Prem
4	$\neg P \vee Q$	Cndnl 1
5	$\neg P \vee R$	Cndnl 2
6	$(\neg P \wedge \neg P) \vee (Q \wedge R)$	Conj 4, 5
7	$\neg P \vee (Q \wedge R)$	Idem 6
8	$P \to (Q \wedge R)$	Cndnl 7
9	P	MP 8, 3

Exercises 15–16: True or False

Are the following statements true or false? Explain your answer.

1.6.15. *Modus Ponens* concludes **P** from premises $\mathbf{P} \to \mathbf{Q}$ and **Q**.

1.6.16. *Negative Biconditional Elimination* concludes $\neg\mathbf{Q}$ from premises $\mathbf{P} \leftrightarrow \mathbf{Q}$ and $\neg\mathbf{P}$.

Exercises 17–24: Deductions
Using the inference rules available so far, deduce the following.

1.6.17. $P \to (P \to Q)$, $P \vdash P \wedge Q$

1.6.18. $P \wedge (Q \to R)$, $\neg R \vdash P \wedge \neg Q$

1.6.19. $P \leftrightarrow Q$, $\neg Q \vee R$, $\neg R \vdash \neg P$

1.6.20. $P \to Q$, $\neg P \leftrightarrow (R \wedge S)$, $\neg Q \vdash \neg P \wedge R$

1.6.21. $P \wedge Q \to \neg S$, $R \to (T \leftrightarrow S)$, P, Q, $R \vdash \neg T$

1.6.22. $P \to R$, $Q \to S$, $P \wedge \neg S \leftrightarrow T$, $T \vdash R \wedge \neg Q$

1.6.23. $P \leftrightarrow Q$, $Q \to R \wedge S \vdash \neg P \vee (R \wedge S)$

1.6.24. $P \to Q$, $Q \to R$, $R \to P \vdash P \leftrightarrow R$

Exercises 25–28: Implication and Derivation
Show that the following implication claims are true, and then explain why the associated derivation claims (replacing $\models$ with $\vdash$) can't be demonstrated yet.

1.6.25. $P \wedge Q \models P \to Q$

1.6.26. $\models P \to P$

1.6.27. $\neg(P \to Q) \models P \wedge \neg Q$

1.6.28. $(P \vee Q) \wedge \neg Q \models P$

1.6.29. *Interderivability of the Laws of Logic*
Using the rules of inference available so far, show that any one of the following laws of logic proves the other two.
   i. *Law of Identity*: $\mathbf{P} \leftrightarrow \mathbf{P}$
  ii. *Law of Non-Contradiction*: $\neg(\mathbf{P} \wedge \neg\mathbf{P})$
 iii. *Law of Excluded Middle*: $\neg\mathbf{P} \vee \mathbf{P}$

Exercises 30–33: Negating Definitions
Using the definitions from Exercises 1.4.7–10, what can you conclude from the following negative information? What rule of inference are you using? [Note: you do not need to further simplify your negations at this stage.]

1.6.30. $\triangle ABC$ is not an isosceles triangle.

1.6.31. l and m are not perpendicular lines (symbolized by $l \not\parallel m$).

1.6.32. a does not divide b (symbolized by $a \nmid b$).

1.6.33. c is not a zero (root) of a function f.

Exercises 34–39: Proof Strategy
State what overall proof strategies (logical, mathematical) you would use to begin a proof of the following theorems. How does BE enter into the process? [Note: you are not expected to know these results!]

1.6.34. The points of intersection of the adjacent trisectors of the angles of a triangle form the vertices of an equilateral triangle.

1.6.35. If $ABCD$ is a Saccheri quadrilateral, then the summit CD is longer than the base AB.

1.6.36. Every non-zero finite-dimensional inner product space has an orthonormal basis.

1.6.37. If A is an invertible matrix, then $A^{-1} = \dfrac{1}{det(A)}\, adj(A)$.

1.6.38. If G is a finite group and H is a subgroup of G, then the order of H divides the order of G.

1.6.39. The set $\mathcal{T}$ of transcendental numbers is uncountable.

1.7 Introduction Rules for Conditionals

The *Int-Elim Rules* considered so far proceed directly from premises to conclusions. *Introduction Rules* for conditional and biconditional conclusions are different. They justify certain conclusions drawn from *suppositional arguments*, i.e., deductions in which something is temporarily assumed for the sake of argument. This sort of reasoning begins by saying something like *suppose such-and-such is the case*. Suppositional arguments occur throughout mathematics—they're the lifeblood of mathematical proof, as we'll see.

1.7.1 *Suppositional Inference Rules*

As noted in Section 1.5, we're developing a Jaśkowski-Fitch-style Natural Deduction System for *Propositional Logic*. One of its characteristic features, in contrast to the older Frege-Hilbert-style deduction systems, is its use of *suppositional rules of inference*. Such rules are slightly more complex than other *Int-Elim Rules*. Once mastered, however, formal deductions become simpler, more structured, and more enjoyable. In any case, they capture the underlying logic of how mathematicians and others reason all the time.

A suppositional argument temporarily assumes an auxiliary hypothesis as a premise of a *subordinate proof* or an *inner subproof*. At the conclusion of the subproof (occasionally, two subproofs), a conclusion is exported to the main body of the proof based *not simply upon the supposition, but upon the entire subargument*. At that point the supposition no longer functions as a premise—it's incorporated into the conclusion in some way and discharged.

A suppositional proof schema will look something like the following:

$$
\begin{array}{ll}
\vdots & \\
j \quad \cdots & \text{Inf Rule X h, i} \\
k \quad \cdots & \text{Spsn for Spsnal Inf Rule Y} \\
\quad \cdots & \\
m \quad \cdots & \\
n \quad \cdots & \text{Spsnal Inf Rule Y k–m} \\
\vdots &
\end{array}
$$

Subproofs are indented in a proof diagram. We underline the supposition, since it functions as a premise of the subproof, and we specify the reason for supposing it in the right-hand column. Once the appropriate conclusion has been drawn in the subproof, we draw a *related conclusion* in the *main body* of the proof, citing the appropriate suppositional inference rule applied to the entire subordinate argument.

You may wonder where suppositions come from. From anywhere! The beauty of subproofs is that they allow you to explore *what if this were the case?* Obviously, not all suppositions will be fruitful. The proposition imported into the main body of the proof at the end of a subproof depends upon what you assumed and what followed from it.

How do you decide, then, what supposition to try? You use the *Backward Method of Proof Analysis*. Ask yourself, *What sort of sentence am I trying to prove? What rule of inference permits me to conclude such a sentence?* If you need a suppositional rule, the rule itself will dictate what to suppose.

This description is still rather abstract, but the nature and use of suppositional arguments will become clearer as you see specific inference rules applied to concrete examples and begin making deductions for yourself. By the time PL's Natural Deduction System is fully in place, you'll be very familiar with several suppositional rules.

Here we'll focus on three such rules, one for conditional sentences and two for biconditional sentences. We'll also consider a few non-suppositional rules. Some of these are *Replacement Rules*, but one is a simple bookkeeping rule. Let's begin with a brief discussion of that rule, the *Rule of Reiteration*, since it's needed in most suppositional proofs.

1.7.2 *Suppositional Proofs and Reiteration*

Subordinate proofs proceed much like the main proof. Given the supposition, you conclude sentences via sound rules of inference. You're not restricted to using only the supposition, though—you can use any result deduced up to that point. To involve earlier results, first repeat them inside the subproof. This makes each subproof self-contained.

The inference rule that warrants this repetition is the *Rule of Reiteration* (*Reit*), which says that a sentence already in an argument can be repeated later, either in a subproof or in the proof itself.

The inference schema for *Reit* is the following:

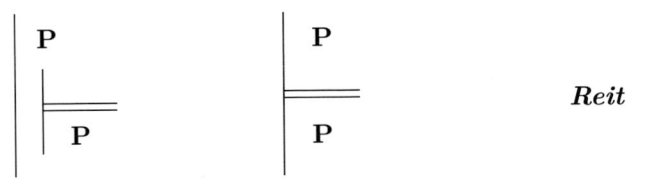

Though sentences may be moved deeper within a proof, proceeding *down and in*, sentences *within* a subproof may not be exported to the main part of the proof or into another subproof, because they depend upon a supposition, which is usually *not* validated by what occurred earlier in the overall argument. Such a maneuver would be wildly unsound, since there are no restrictions on what can be supposed.

❖**Example 1.7.1**

What is wrong with the following argument for $P \to Q \vdash Q$?

1	$P \to Q$	Prem
2	P	Spsn
3	$P \to Q$	Reit 1
4	Q	MP 3, 2
5	Q	Reit 4

Solution

The problem here is an incorrect *Reiteration* in line 5. Everything else is according to code, though no specific reason was given for the supposition.

1.7.3 *Conditional Proof*

To prove sentences of the form $\mathbf{P} \to \mathbf{Q}$, the most natural approach is to suppose $\mathbf{P}$ and deduce $\mathbf{Q}$ from it. You then conclude $\mathbf{P} \to \mathbf{Q}$.

This is the procedure regularly used in mathematics to prove conditional statements. To prove *if $\triangle ABC$ is isosceles, then its base angles are congruent*, you take an isosceles triangle and show that the base angles are congruent. In other words, you would suppose the antecedent $\triangle ABC$ *is isosceles* and then derive the consequent *its base angles are congruent*. Having demonstrated that the antecedent proves the consequent, the conditional result follows: $\mathbf{P} \to \mathbf{Q}$ is proved by showing $\mathbf{P} \vdash \mathbf{Q}$.

This sort of inference is justified by the rule *Conditional Proof (CP)*, the prototype for all suppositional rules. *Conditional Proof* was introduced independently by S. Jaśkowski and G. Gentzen in the mid-1930s.

The inference schema for this rule is as follows:

$$CP \qquad\qquad \begin{array}{|l} \mathbf{P} \\ \hline \mathbf{Q} \\ \hline\hline \mathbf{P} \to \mathbf{Q} \end{array}$$

Note that the entire subproof of $\mathbf{Q}$ from $\mathbf{P}$ is above the double underline. Only *after* you have *deduced* $\mathbf{Q}$ from $\mathbf{P}$ (along with any other permissible propositions) may you *conclude* $\mathbf{P} \to \mathbf{Q}$. This is done on the basis of a *subproof* of $\mathbf{Q}$ from $\mathbf{P}$, not, as before, on the basis of sentences of certain logical forms. The conclusion $\mathbf{P} \to \mathbf{Q}$ is a conditional statement and does not depend upon $\mathbf{P}$ actually being the case. $\mathbf{P}$ functions as a premise of the *subproof* establishing **Q**, but not of the overall argument. The supposition **P** is discharged by being incorporated into the conditional conclusion **P** → **Q**.

The next example illustrates the use of *Conditional Proof* by deducing the contrapositive of a conditional sentence from the sentence itself. We'll shortly adopt a *Replacement Rule (Contraposition)* for doing this in one step, but our deduction shows that rule is redundant. Example 3 shows that *Hypothetical Syllogism* is also redundant, but we'll also retain *HS* so we don't need to make a lengthy argument in order to chain two conditionals together.

✤ **Example 1.7.2**

   Show that **P** → **Q** ⊢ ¬**Q** → ¬**P**.

Solution

· First note that the conclusion ¬**Q** → ¬**P** is a conditional sentence. This is exactly the situation *CP* is designed for.
· So, we'll suppose ¬**Q** as a temporary premise and then prove ¬**P**.
  To use the premise **P** → **Q**, we'll reiterate it in the subproof.
  MT will then give what is wanted.
· On the basis of the subproof, we can conclude ¬**Q** → ¬**P** in the main proof.
  The proof diagram for this deduction is as follows:

1	**P** → **Q**	Prem
2	¬**Q**	Spsn for CP
3	**P** → **Q**	Reit 1
4	¬**P**	MT 3, 2
5	¬**Q** → ¬**P**	CP 2–4

✤ **Example 1.7.3**

   Show that **P** → **Q**, **Q** → **R** ⊢ **P** → **R**, without using *HS*.

Solution

The following proof diagram establishes the claim.

1	**P** → **Q**	Prem
2	**Q** → **R**	Prem
3	**P**	Spsn for CP
4	**P** → **Q**	Reit 1
5	**Q**	MP 4, 3
6	**Q** → **R**	Reit 2
7	**R**	MP 6, 5
8	**P** → **R**	CP 3–7

Let's now consider the soundness of *CP*. Why can we conclude **P** → **Q** at the end of a *CP* subargument? Because if we can prove **Q** by supposing **P**, then **Q** logically follows from **P**.[14] And if **Q** follows from **P**, then **P** → **Q** is logically true, according to the *Implication Theorem*. Thus, **P** → **Q** is true whenever **Q** is deduced from **P**, which means *CP* is sound.[15]

1.7.4 *Additional Replacement Rules for* CONDITIONALS

Section 1.6 gave the most basic *Replacement Rules* for conditional and biconditional sentences. Here we'll add two *Replacement Rules* for conditional sentences, *Contraposition* (*Conpsn*) and *Exportation* (*Exp*). These rules are applied more in one direction than the other and usually for full sentences, but like other *Replacement Rules*, they're two-directional and can be used to substitute equivalents for component parts of sentences.

Contraposition is based on the equivalence of contrapositive sentences. Mathematical proofs sometimes prove a conditional statement **P** → **Q** by proving its equivalent ¬**Q** → ¬**P**, usually via *CP*. When this happens, you may wonder why ¬**Q** is being assumed when **Q** needs to be proved: it's because *Conpsn* is being used. We'll say more about this in Section 1.8, when we compare *Conpsn* to a closely related procedure involving negations.

Contraposition is schematized as follows:

Conpsn   **P** → **Q** :: ¬**Q** → ¬**P**

Contraposition has a biconditional counterpart: **P** ↔ **Q** :: ¬**Q** ↔ ¬**P**. However, since this is not often used in proofs, we won't adopt it as a rule.

Exportation (*Exp*) is a more practical *Replacement Rule*. It tells how compound sentences involving ∧ and → can be expanded and contracted.

The two *Exportation* rules are schematized as follows:

Exp
$$\mathbf{P} \to (\mathbf{Q} \to \mathbf{R}) :: (\mathbf{P} \wedge \mathbf{Q}) \to \mathbf{R}$$
$$\mathbf{P} \to (\mathbf{Q} \wedge \mathbf{R}) :: (\mathbf{P} \to \mathbf{Q}) \wedge (\mathbf{P} \to \mathbf{R})$$

The first *Exportation* rule is rarely used because mathematicians tend to formulate nested conditionals in terms of conjunction. Compound conditionals are a bit awkward, and without parentheses their meaning is unclear (see Exercise 30). Using a conjunctive form avoids such potential confusion.

Nevertheless, some mathematical propositions do pile up conditions. The sentence *if ab = ac, then b = c, provided a ≠ 0* is an example. We can rephrase this as *if a ≠ 0, then if ab = ac, b = c*, which is of the form **P** → (**Q** → **R**). One way to prove such a proposition is by using the first *Exportation* rule to get started: *suppose a ≠ 0 and ab = ac.*

The second *Exportation* rule is most often applied to deduce a sentence in the form **P** → (**Q** ∧ **R**). In an informal proof you would merely show that

---

[14] Assuming the use of sound inference rules within the subproof, of course. This complicates the argument that *CP* is sound because subproofs may contain subproofs.

[15] Frege-Hilbert-style deduction systems, which lack *Conditional Proof* (and other suppositional rules), make up for it by proving the *Deduction Theorem*, which says **P** ⊢ **Q** if and only if ⊢ **P** → **Q**, the counterpart to our *Implication Theorem*.

both right-hand conjuncts $\mathbf{P} \to \mathbf{Q}$ and $\mathbf{P} \to \mathbf{R}$ hold. *Conj* and *Exp* validate such a procedure. The bottom *Exp* rule is a distributive law of sorts, but we'll not call it that since other distributive statements for $\to$ are *not* sound (see Exercises 31–33).

❖**Example 1.7.4**
    Show that $\neg P \vee Q, \neg P \to R \vdash \neg Q \to (R \wedge \neg P)$.

Solution
    The following proof diagram demonstrates this claim.

1	$\neg P \vee Q$	Prem
2	$\neg P \to R$	Prem
3	$P \to Q$	Cndnl 1
4	$\neg Q \to \neg P$	Conpsn 3
5	$\neg Q \to R$	HS 4, 2
6	$(\neg Q \to R) \wedge (\neg Q \to \neg P)$	Conj 5, 4
7	$\neg Q \to (R \wedge \neg P)$	Exp 6

1.7.5 *Biconditional Introduction*

It's probably clear how we can prove biconditional sentences. Since $\mathbf{P} \leftrightarrow \mathbf{Q}$ is logically equivalent to $(\mathbf{P} \to \mathbf{Q}) \wedge (\mathbf{Q} \to \mathbf{P})$, we can deduce the biconditional by showing both $\mathbf{P} \to \mathbf{Q}$ and $\mathbf{Q} \to \mathbf{P}$. But since we're not interested in concluding the conditionals, we'll just use the two subproofs as a basis for inferring the biconditional. This is a proof strategy we've already used informally—for instance, to prove the *Implication Theorem* in Section 1.4.

    This rule, called *Biconditional Introduction (BI)*, is schematized as follows:

$$\begin{array}{l} \mathbf{P} \\ \hline \mathbf{Q} \\ \mathbf{Q} \\ \hline \mathbf{P} \\ \hline\hline \mathbf{P} \leftrightarrow \mathbf{Q} \end{array} \qquad \textit{\textbf{BI}}$$

    To conclude $\mathbf{P} \leftrightarrow \mathbf{Q}$ via *BI*, we develop two subproofs, one based upon the supposition $\mathbf{P}$ and the other based upon $\mathbf{Q}$. In some cases, the second subproof will parallel the first. When this occurs, you may read something like *the other direction follows similarly*. Occasionally, instead of using the *proof* as a model, the *result* just proved may be used to make the second subproof. Example 1.5.2 (commutativity of conjunction) was such an example.

    The following two examples illustrate how to use *BI* in formal deductions. The first is fairly simple and is related to the *Bicndnl Replacement Rule*. The second illustrates that subproofs can contain subproofs. It also shows that suppositional inference rules can be used to prove results having no premises (tautologies), something you may have thought wasn't possible.

❖ Example 1.7.5

Show that $P \to Q$, $\neg P \to \neg Q \vdash P \leftrightarrow Q$ via *BI*.

Solution

The following proof diagram establishes the claim.

1	$P \to Q$	Prem
2	$\neg P \to \neg Q$	Prem
3	$\quad P$	Spsn 1 for BI
4	$\quad P \to Q$	Reit 1
5	$\quad Q$	MP 4, 3
6	$\quad Q$	Spsn 2 for BI
7	$\quad \neg P \to \neg Q$	Reit 2
8	$\quad \neg\neg P$	MT 7, 6
9	$\quad P$	DN 8
10	$P \leftrightarrow Q$	BI 3–5, 6–9

❖ Example 1.7.6

Show that $\vdash [P \to (Q \wedge R)] \leftrightarrow [(P \to Q) \wedge (P \to R)]$ via *BI*.

Solution

The formal deduction below looks long, but if you take it one piece at a time, using the following proof analysis as a guide, you'll find that it's straightforward and natural and not too difficult to follow.

Backward-Forward Proof Analysis

- To deduce our result via *BI*, we will generate two subproofs:
  $(P \to Q) \wedge (P \to R)$ from $P \to (Q \wedge R)$ (lines 1–12); and conversely,
  $P \to (Q \wedge R)$ from $(P \to Q) \wedge (P \to R)$ (lines 13–21).
- *First subproof*: to deduce $(P \to Q) \wedge (P \to R)$ (line 12), we will deduce each conjunct separately, using *CP* (lines 6, 11).
  To prove $P \to Q$ (lines 2–6), we'll suppose P (line 2) and then prove Q. Since we are given $P \to (Q \wedge R)$ as a premise and P as our supposition, *MP* allows us to conclude $Q \wedge R$ (lines 2–4).
  Using *Simp* on $Q \wedge R$ gives us Q (line 5), which is what we wanted. Having gotten Q on the assumption that P is the case, we can then conclude $P \to Q$ via *CP* in the main part of the subproof (line 6).
  The rest of the first subproof proceeds in the same way (lines 7–11).
- *The second subproof* (lines 13–21): this is easier than the first, containing only one second-level subproof. It begins with a conjunction and derives a conditional sentence by means of *CP* (lines 14–21).

- Finally, based on both subproofs, we conclude the biconditional (line 22): $[P \rightarrow (Q \wedge R)] \leftrightarrow [(P \rightarrow Q) \wedge (P \rightarrow R)]$.
- Note that our deduction would have been shorter if we had used a single subproof to deduce both Q (line 5) and R (line 10) from supposition P (line 2), because lines 7–9 simply repeat lines 2–4. While this is not the official way to apply CP, such shortcuts are often taken in informal argumentation.

Proof Diagram

1	$P \rightarrow (Q \wedge R)$	Spsn 1 for BI
2	P	Spsn for CP
3	$P \rightarrow (Q \wedge R)$	Reit 1
4	$Q \wedge R$	MP 3, 2
5	Q	Simp 4
6	$P \rightarrow Q$	CP 2–5
7	P	Spsn for CP
8	$P \rightarrow (Q \wedge R)$	Reit 1
9	$Q \wedge R$	MP 8, 7
10	R	Simp 9
11	$P \rightarrow R$	CP 7–10
12	$(P \rightarrow Q) \wedge (P \rightarrow R)$	Conj 6, 11
13	$(P \rightarrow Q) \wedge (P \rightarrow R)$	Spsn 2 for BI
14	P	Spsn for CP
15	$(P \rightarrow Q) \wedge (P \rightarrow R)$	Reit 13
16	$P \rightarrow Q$	Simp 15
17	Q	MP 16, 14
18	$P \rightarrow R$	Simp 15
19	R	MP 18, 14
20	$Q \wedge R$	Conj 17, 19
21	$P \rightarrow (Q \wedge R)$	CP 14–20
22	$[P \rightarrow (Q \wedge R)] \leftrightarrow [(P \rightarrow Q) \wedge (P \rightarrow R)]$	BI 1–12, 13–21

1.7.6 Biconditional Introduction, Contrapositive Form

Since a conditional sentence and its contrapositive are logically equivalent, we get another *Introduction Rule* for biconditional sentences from *BI*'s schema by replacing one of the subproofs with a subproof of its contrapositive. The resulting inference rule is sound since *CP* and *Conpsn* are.

We'll call this inference rule *Biconditional Introduction, Contrapositive Form (BICon)*. It's schematized as follows:

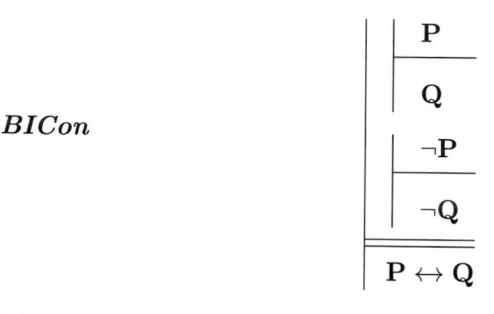

BICon

This form of *Biconditional Introduction* occurs often in mathematics. Such a proof of **P** $\leftrightarrow$ **Q** begins by supposing **P** is the case and showing **Q** is also the case. It then continues by supposing **P** is *not* the case, arguing **Q** is not the case either. The conclusion immediately follows by *BICon*.

1.7.7 *Proving Propositions Equivalent: Cyclic BI*

To show P_1 is logically equivalent to P_2 relative to some theory, we prove $P_1 \leftrightarrow P_2$. However, sometimes we want to prove that more than two propositions are equivalent. As mentioned in Section 1.4, this happens whenever a concept has important alternative characterizations. To show that $P_1, \ldots, P_n$ are logically equivalent, we could use *BI* on each pair of propositions to prove $P_i \leftrightarrow P_j$ for all i, j, but this method isn't very economical.

A widely used strategy proves the equivalences in cyclical fashion. Beginning with P_1, you prove $P_1 \to P_2$, then $P_2 \to P_3$, and so on, ending with $P_n \to P_1$. These sentences form a cycle of conditionals, so any sentence proves any other (apply *HS* repeatedly), and hence each associated biconditional also holds.

There are many variants of this procedure. Sometimes it is easier to construct several interlocking cycles rather than a single large cycle. Regardless of how it is done, such cycling is more economical than proving each biconditional separately. The rule that underlies this cyclic proof procedure is an extension of *BI* (cf. Exercise 1.6.24). We'll call this *Cyclic Biconditional Introduction* (*CycBI*).

The schema for *Cyclic Biconditional Introduction* is as follows:

CycBI

$$
\begin{array}{l}
\mathbf{P_1 \to P_2} \\
\mathbf{P_2 \to P_3} \\
\mathbf{P_3 \to P_1} \\
\hline
\mathbf{P_i \leftrightarrow P_j} \quad \text{any } i, j
\end{array}
$$

This rule is used in many areas of mathematics. One might gauge the depth and centrality of a concept by the number of other concepts it is equivalent to. Elementary *Linear Algebra*, for example, proves that the key idea of being an invertible square matrix is equivalent to many other results.

1.7.8 *Suppositional Rules and Overall Proof Strategy*

We've now introduced four suppositional *Introduction* rules—*CP*, *BI*, *BICon*, and *CycBI*—and seen how proofs are constructed with them. Suppositional rules provide modular logical structure for a proof. This is analogous to solving problems or creating structured computer programs with a top-down, stepwise-refinement procedure. Using the *Backward Method of Proof Analysis* to determine what key propositions need proving and then constructing subproofs for them, we can break up a complex proof into more manageable components. Each subproof contributes its part to organizing the whole. Proofs are no longer just a long string of propositions, one after another.

You may find working with suppositional rules difficult at first, but with practice you should be able to construct suppositional proofs with ease, regardless of length. You will not put your proofs into two-column diagrams in other contexts, but learning how to do suppositional proofs in this way is beneficial. Most mathematical propositions involve conditional and biconditional statements, so suppositional rules get used over and over again in their proofs. When you know how subproofs work, you'll realize that you can't always take an earlier result and just use it again—it all depends upon where it was proved and how it functioned earlier in the proof.

EXERCISE SET 1.7

Exercises 1–4: Soundness of Inference Rules
Show the soundness of the following Int-Elim and Replacement Rules.

1.7.1. *Contraposition:* $\mathbf{P} \to \mathbf{Q} \vDash \neg\mathbf{Q} \to \neg\mathbf{P}$

1.7.2. *Exportation*
  a. $\mathbf{P} \to (\mathbf{Q} \to \mathbf{R}) \vDash (\mathbf{P} \wedge \mathbf{Q}) \to \mathbf{R}$
  b. $\mathbf{P} \to (\mathbf{Q} \wedge \mathbf{R}) \vDash (\mathbf{P} \to \mathbf{Q}) \wedge (\mathbf{P} \to \mathbf{R})$

1.7.3. *CycBI:* $\mathbf{P_1} \to \mathbf{P_2}, \mathbf{P_2} \to \mathbf{P_3}, \mathbf{P_3} \to \mathbf{P_1} \vDash \mathbf{P_i} \leftrightarrow \mathbf{P_j}$ any i, j

1.7.4. Assuming the soundness of *CP* and various non-suppositional rules, explain why the following rules are sound.
  a. *BI*
  b. *NBI*
  c. *BICon*

Exercises 5–7: True or False
Are the following statements true or false? Explain your answer.

1.7.5. Suppositional arguments are quite common in mathematics.

1.7.6. *Reiteration* can be used whenever you want to repeat a previous line somewhere in a proof.

1.7.7. *Exportation* rules can only be used to replace whole sentences.

Exercises 8–10: Completing Deductions

Fill in the reasons for the following deductions.

1.7.8. $P \to Q \vdash P \to (R \to Q)$

1	$P \to Q$	_____
2	$\quad P$	_____
3	$\qquad R$	_____
4	$\qquad P$	_____
5	$\qquad P \to Q$	_____
6	$\qquad Q$	_____
7	$\quad R \to Q$	_____
8	$P \to (R \to Q)$	_____

1.7.9. $P \wedge Q \to R,\ P \to Q \vdash P \to R$

1	$P \wedge Q \to R$	_____
2	$P \to Q$	_____
3	$\quad P$	_____
4	$\quad P \to Q$	_____
5	$\quad Q$	_____
6	$\quad P \wedge Q$	_____
7	$\quad P \wedge Q \to R$	_____
8	$\quad R$	_____
9	$P \to R$	_____

1.7.10. $\vdash (P \to P \wedge Q) \leftrightarrow (P \to Q)$

1	$\quad P \to P \wedge Q$	_____
2	$\qquad P$	_____
3	$\qquad P \to P \wedge Q$	_____
4	$\qquad P \wedge Q$	_____
5	$\qquad Q$	_____
6	$\quad P \to Q$	_____
7	$\quad P \to Q$	_____
8	$\qquad P$	_____
9	$\qquad P \to Q$	_____
10	$\qquad Q$	_____
11	$\qquad P \wedge Q$	_____
12	$\quad P \to P \wedge Q$	_____
13	$(P \to P \wedge Q) \leftrightarrow (P \to Q)$	_____

Exercises 11–12: Logical Implication and Conclusive Deductions

Determine whether the following **implication** *claims are true. Then determine whether the* **deductions** *given are* **conclusive**. *Carefully point out where and how a rule of inference is being misused.*

1.7.11. $P \wedge Q \to R \vDash P \to R$

1	$P \wedge Q \to R$	Prem
2	Q	Spsn for CP
3	P	Spsn for CP
4	Q	Reit 2
5	$P \wedge Q$	Conj 3, 4
6	$P \wedge Q \to R$	Reit 1
7	R	MP 6, 5
8	$P \to R$	CP 2-7

1.7.12. $P \vee \neg Q, \ P \wedge Q \leftrightarrow \neg R \vDash \neg P \to R$

1	$P \vee \neg Q$	Prem
2	$P \wedge Q \leftrightarrow \neg R$	Prem
3	$P \to Q$	Cndnl 1
4	P	Spsn for CP
5	$P \to Q$	Reit 3
6	Q	MP 5, 4
7	$P \wedge Q$	Conj 4, 6
8	$P \wedge Q \leftrightarrow \neg R$	Reit 2
9	$\neg R$	MP 8, 7
10	$P \to \neg R$	CP 4–9
11	$\neg P \to R$	Conpsn 10

Exercises 13–24: Deductions

Construct formal deductions for the following.

1.7.13. $P \wedge Q \vdash P \leftrightarrow Q$

1.7.14. $P \to Q \vdash P \wedge R \to Q$

1.7.15. $P \to Q \wedge R \vdash P \to Q$

1.7.16. $P \vdash (P \to Q) \to Q$

1.7.17. $Q \to R \vdash P \wedge Q \to P \wedge R$

1.7.18. $(P \to Q) \to R \vdash Q \to R$

1.7.19. $\vdash P \to (Q \to P)$

1.7.20. $\vdash P \to [(P \to Q) \to Q]$

1.7.21. $P \to (Q \to R), \ Q \vdash P \to R$

1.7.22. $P \to R, \ Q \to \neg R \vdash P \to \neg Q$

1.7.23. $P \to Q, \ R \to S \vdash P \wedge R \to Q \wedge S$

1.7.24. $P \to Q \vdash [P \to (Q \to R)] \to (P \to R)$

1.7.25. Show that the *Law of Excluded Middle*, $\mathbf{P} \vee \neg \mathbf{P}$, can be proved via *CP* and *Cndnl*. Thus, *Cndnl* is not acceptable to intuitionists.

Exercises 26–29: Interderivability

Using only Int-Elim Rules, show the following interderivability claims.

1.7.26. $P \to (P \to Q) \vdash P \to Q$

1.7.27. $P \to (Q \to R) \vdash Q \to (P \to R)$

1.7.28. $P \to (Q \to R) \vdash (P \to Q) \to (P \to R)$

1.7.29. $P \leftrightarrow Q \vdash (P \to Q) \wedge (\neg P \to \neg Q)$

Exercises 30–33: Bogus Replacement Rules

*Show, as indicated, that the following bogus replacement rules do **not** hold. If one side implies the other, construct a deduction to show it.*

1.7.30. *Association* $(\to)$: $\mathbf{P} \to (\mathbf{Q} \to \mathbf{R})$:/: $(\mathbf{P} \to \mathbf{Q}) \to \mathbf{R}$.

1.7.31. *Distribution* $(\wedge$ *over* $\to)$: $\mathbf{P} \wedge (\mathbf{Q} \to \mathbf{R})$:/: $(\mathbf{P} \wedge \mathbf{Q}) \to (\mathbf{P} \wedge \mathbf{R})$.

1.7.32. *Distribution* $(\wedge$ *over* $\to)$: $(\mathbf{P} \to \mathbf{Q}) \wedge \mathbf{R}$:/: $(\mathbf{P} \wedge \mathbf{R}) \to (\mathbf{Q} \wedge \mathbf{R})$.

1.7.33. *Distribution* $(\to$ *over* $\wedge)$: $(\mathbf{P} \wedge \mathbf{Q}) \to \mathbf{R}$:/: $(\mathbf{P} \to \mathbf{R}) \wedge (\mathbf{Q} \to \mathbf{R})$.

Exercises 34–35: Formulating Mathematical Theorems

*Symbolize the following, using logical symbolism. Then write out its equivalent, using **Exp**. Put your final answer back into mathematical English.*

1.7.34. *Intermediate Value Theorem*

If a function f is continuous on a closed interval $[a, b]$, then if d is a number between $f(a)$ and $f(b)$, there is a number c between a and b so that $f(c) = d$.

1.7.35. *Ptolemy's Theorem*

If $ABCD$ is a quadrilateral inscribed in circle, then if AC and BD are the diagonals of $ABCD$, $AC \cdot BD = AB \cdot CD + BC \cdot AD$.

1.7.36. *Infinite Series and Convergence*

Given two positive-term series $\sum a_n$ and $\sum b_n$ with $a_n \leq b_n$ for all n, one calculus textbook proved the following two propositions independently:

(a) If $\sum b_n$ converges, then $\sum a_n$ also converges;

(b) If $\sum a_n$ diverges, then $\sum b_n$ also diverges.

What rule of inference could the author have used to shorten this work?

Exercises 37–41: Proof Strategy

Determine the underlying PL forms of the following propositions. Then identify which proof strategies might be used in deducing them.

1.7.37. If integers a and b are odd, then $a \cdot b$ is odd.

1.7.38. If $a \mid b$ and $b \mid c$, then $a \mid c$.

1.7.39. $\triangle ABC$ is a right triangle if and only if $a^2 + b^2 = c^2$.

1.7.40. If $ABCD$ is a rectangle, then diagonals AC and BD are congruent and bisect one another.

1.7.41. If S is countably infinite, then T is countably infinite if and only if T is equinumerous with S.

Exercises 42–43: Proofs

Prove the following propositions about positive integers in as much detail as you can, identifying the rules of inference you're familiar with.

1.7.42. If integers a and b are odd, then $a \cdot b$ is odd.

Definition: *a number n is odd if and only if $n = 2k + 1$ for some number k.*

1.7.43. a is even if and only if a^2 is even.

Definition: *a number n is even if and only if $n = 2k$ for some number k.*

1.8 Proof by Contradiction: Rules for NOT

In this section we'll take up the *Int-Elim Rules* for $\neg$, and we'll look at a few *Replacement Rules* that help us simplify compound negations. We'll finish by reflecting on the nature and merits of proofs that use these rules.

1.8.1 *Proof by Contradiction*

Arguments involving negation have been important for millennia and may have been responsible for the rise of logic in ancient Greece. These are often called *indirect proofs* because they draw a conclusion in a roundabout way, by assuming the opposite and showing that this leads to a patent absurdity, a contradiction. Thus, the desired result must be the case. Indirect proof is also called *Proof by Contradiction* or *Proof by Reductio ad Absurdum*.

Philosophers as far back as Zeno used *Proof by Contradiction* to argue the absurdity of their opponents' ideas. This has been a key debate strategy ever since. Aristotle discussed *Proof by Contradiction* explicitly and told how Greek mathematicians employed it to show the incommensurability of the side and diagonal of a square. The modern version of this asserts that $\sqrt{2}$ (the ratio of a square's diagonal to its side) is irrational. The classic argument for this result is as follows.

Theorem: *Existence of Irrational Numbers*
   $\sqrt{2}$ *is irrational.*

Proof:
   Suppose not, i.e., suppose $\sqrt{2}$ is rational.
   Then $\sqrt{2}$ can be expressed as a fraction m/n of integers.
   We can assume without loss of generality that m/n is in fully reduced form.
   Squaring both sides of $\sqrt{2} = m/n$, we obtain $2 = m^2/n^2$.
($*$) Clearing this of fractions gives $2n^2 = m^2$.
   Thus, m^2 is even, which implies that m itself must be even.
   So, $m = 2k$ for some integer k.
   Hence, $m^2 = (2k)^2 = 4k^2$.
   Substituting this in the equation of line ($*$) yields $2n^2 = 4k^2$.
   Thus, $n^2 = 2k^2$, which implies n^2 is even.
   Then n is even, too.
   But if both m and n are even, m/n is not in reduced form.
   This contradicts our earlier statement, so the original supposition that $\sqrt{2}$ is rational is absurd.
   Therefore, $\sqrt{2}$ is irrational.   ∎

This proof demonstrates a negative result ($\sqrt{2}$ ***is not*** *rational*) by showing that its opposite ($\sqrt{2}$ ***is*** *rational*) is absurd. You might think that the argument only shows that m/n wasn't in reduced form, but canceling the common factor of 2 and applying the same argument to the new fraction, we get the absurdity that the factor 2 can be canceled repeatedly. Put another way, it is impossible to put a fraction in reduced form, which is absurd.

This deduction uses *Negation Introduction*, since a negation is introduced as the conclusion. *Negation Elimination* proceeds similarly, concluding a positive sentence because its negation leads to a contradiction. Together these rules constitute *Proof by Contradiction* or *indirect proof*.

1.8.2 Negation Introduction and Negation Elimination

Negation Introduction (*NI*) is used to deduce a negative sentence ¬**P**. Rather than prove ¬**P** directly from the premises, we assume its opposite **P** and then, in a subproof headed by this supposition, we derive both a sentence **Q** and its contradictory opposite ¬**Q**. On the basis of this subproof, we infer ¬**P** in the main part of the proof.

Schematically we have the following:

$$
NI \qquad
\begin{array}{|l}
\quad \begin{array}{|l}
\mathbf{P} \\[4pt]
\mathbf{Q} \\[4pt]
\neg\mathbf{Q}
\end{array} \\[4pt]
\hline
\neg\mathbf{P}
\end{array}
$$

This schema indicates that **Q** and ¬**Q** are deduced somewhere in a subproof from **P** (along with any previous sentences), providing the basis for concluding ¬**P** in the main part of the proof.

Negation Elimination (*NE*) is similar to *NI* but uses a contradiction to deduce a sentence **P** that need not be a negation. Rather than proving **P** purely on the basis of the premises, we assume its negation ¬**P** as a temporary assumption and then, using it along with the premises and anything proved from them, we derive both some sentence **Q** and its contradictory ¬**Q**. On the basis of such a subproof, we then infer **P** as the conclusion.

Schematically we have the following:

$$
NE \qquad
\begin{array}{|l}
\quad \begin{array}{|l}
\neg\mathbf{P} \\[4pt]
\mathbf{Q} \\[4pt]
\neg\mathbf{Q}
\end{array} \\[4pt]
\hline
\mathbf{P}
\end{array}
$$

Negation Introduction and *Negation Elimination* are sound rules because *CP* is. In fact, one can show the logical equivalences $\mathbf{P} \rightarrow (\mathbf{Q} \wedge \neg\mathbf{Q}) \vDash\dashv \neg\mathbf{P}$ and $\neg\mathbf{P} \rightarrow (\mathbf{Q} \wedge \neg\mathbf{Q}) \vDash\dashv \mathbf{P}$ (see Exercise 5).

These two forms of *Proof by Contradiction* don't seem all that different—to deduce the conclusion, each proceeds from a supposition opposed to it and arrives at a contradictory pair of sentences. Yet there are important differences, which lead to their being accepted differently by philosophically minded segments of the mathematical and logical communities.

Negation Introduction is universally accepted. If a sentence **P** leads to an absurdity, then it cannot be the case. Logical consistency (the *Law of Non-Contradiction*) is recognized as normative for deductive reasoning.

There are some, though, who oppose the use of *Negation Elimination*. While a contradiction generated by ¬**P** indicates that it should be rejected, what follows from this? According to intuitionist mathematicians/logicians, we can only conclude what *NI* gives, namely, ¬¬**P**. They will not conclude **P**, because for them the *Law of Excluded Middle* is suspect. They take the double negative ¬¬**P** as weaker than the related positive statement **P**.

The vast majority of mathematicians and logicians hold *NE* to be a sound rule of inference. Since $\mathbf{P} \vDash\dashv \neg\neg\mathbf{P}$ in our system, we have no reason to reject *NE*. We'll include it in PL's Natural Deduction System. However, there do seem to be reasons to use it sparingly, for when no other alternative seems to work. We'll say more about this at the end of the section.

1.8.3 *Deductions Involving NI and NE*

We'll illustrate the use of *NI* and *NE* by working a few examples, beginning with a proof of the *Law of Non-Contradiction* (*LNC*). A Natural Deduction System largely avoids using logical axioms in deductions, but since tautologies like *LNC* are logical truths, we can construct arguments for them (as done before, in Example 1.7.6) that are premise-free. In fact, such deductions are a good way to show that propositions are tautologies.

✤**Example 1.8.1**
Prove the *Law of Non-Contradiction*: $\vdash \neg(\mathbf{P} \wedge \neg\mathbf{P})$.

Solution
· Let's first analyze what's needed. Since ¬ is the main connective of the conclusion ¬(**P** ∧ ¬**P**), our strategy will be to use *NI*, which requires us to suppose **P** ∧ ¬**P** and then derive two contradictory sentences from it. How to do this should be obvious: simplify **P** ∧ ¬**P** into **P** and ¬**P**.
· The proof diagram thus goes as follows:

1	**P** ∧ ¬**P**	Spsn for NI
2	**P**	Simp 1
3	¬**P**	Simp 1
4	¬(**P** ∧ ¬**P**)	NI 1-3

We can also use these rules to show **P** proves ¬¬**P**, and conversely. The former uses *NI* and so is intuitionistically valid. The more controversial direction based on *NE* is the next example.

✤**Example 1.8.2**

Prove *Double Negation Elimination*: ¬¬**P** ⊢ **P**.

Solution

The following proof diagram establishes the claim.

1	¬¬**P**	Prem
2	¬**P**	Spsn for NE
3	¬**P**	Reit 2
4	¬¬**P**	Reit 1
5	**P**	NE 2-4

As a corollary to these examples, we'll be able to prove the *Law of Excluded Middle*, **P** ∨ ¬**P**, once we have *De Morgan's Rule* available (see below).

The following examples are fairly natural valid argument forms that could be made into associated rules of inference. However, since they do not introduce or eliminate a *single* connective, we'll leave them off our official list.

✤**Example 1.8.3**

Show that $P, \neg(P \wedge Q) \vdash \neg Q$.

Solution

The following proof diagram establishes the claim.

1	P	Prem
2	$\neg(P \wedge Q)$	Prem
3	Q	Spsn for NI
4	P	Reit 1
5	$P \wedge Q$	Conj 4, 3
6	$\neg(P \wedge Q)$	Reit 2
7	$\neg Q$	NI 3-6

✤**Example 1.8.4**

Show that $\neg Q \vdash \neg(P \wedge Q)$ [roughly, a partial converse of Example 3].

Solution

The following proof diagram establishes the claim.

1	$\neg Q$	Prem
2	$P \wedge Q$	Spsn for NI
3	Q	Simp 2
4	$\neg Q$	Reit 1
5	$\neg(P \wedge Q)$	NI 2-4

1.8.4 *Replacement Rules for Simplifying Negations*

To apply *NE* when a compound sentence is being negated as a supposition, we may need to simplify the negation to proceed further with the argument. The following *Replacement Rules* help us replace compound negations with easier-to-use equivalent forms.

The rule *Double Negation* (*DN*) allows us to conclude a positive statement **P** from its double negation ¬¬**P**. This is the direction intuitionist logicians reject (see, however, Exercise 36). *Double Negation* also warrants concluding ¬¬**P** from **P**. This is a sound *Replacement Rule* (see Exercise 1).

The schema for *DN* is the following:

$$\neg\neg\mathbf{P} :: \mathbf{P} \qquad\qquad \boldsymbol{DN}$$

De Morgan's Rules (*DeM*) are based upon equivalences mentioned in Section 1.3. They can be used to expand a negated conjunction or disjunction (moving left to right), or to simplify disjunctions and conjunctions of negations (moving right to left). These formulas exhibit a duality: to negate a compound sentence, negate the individual parts and interchange ∧ with ∨.

They are schematized as follows:

$$\neg(\mathbf{P} \wedge \mathbf{Q}) :: \neg\mathbf{P} \vee \neg\mathbf{Q} \qquad\qquad \boldsymbol{DeM}$$
$$\neg(\mathbf{P} \vee \mathbf{Q}) :: \neg\mathbf{P} \wedge \neg\mathbf{Q}$$

At this point, replacing a negated conjunction with its equivalent disjunction (the first form of *DeM*) in a *Proof by Contradiction* is a dead end. To go further, we need the *Int-Elim Rules* of Section 1.9. Meanwhile, we may be able to proceed using ∨ in a roundabout way, as the following indirect proof illustrates.

✤Example 1.8.5

Show that the following *Elimination Rule* for ∨ holds: **P** ∨ **Q**, ¬**P** ⊢ **Q**.

Solution

We'll prove this by *Contradiction*, leaving a direct proof via *Cndnl* as an exercise (see Exercise 25).

1	**P** ∨ **Q**	Prem
2	¬**P**	Prem
3	¬**Q**	Spsn for NE
4	¬**P**	Reit 2
5	¬**P** ∧ ¬**Q**	Conj 4, 3
6	¬(**P** ∨ **Q**)	DeM 5
7	**P** ∨ **Q**	Reit 1
8	**Q**	NE 3–7

Simplifying a negated conditional is often needed in *NE* proofs. We'll look at this next and then compare it with a strategy it's often confused with.

1.8.5 *Simplifying Negated Conditionals*

We'll show how to simplify $\neg(\mathbf{P} \to \mathbf{Q})$ in stages:
- First replace the inside conditional with an equivalent via *Cndnl*:
  $\neg(\mathbf{P} \to \mathbf{Q}) \dashv\vdash \neg(\neg(\mathbf{P} \wedge \neg\mathbf{Q}))$.
- Then simplify further, using *DN*: $\neg(\neg(\mathbf{P} \wedge \neg\mathbf{Q})) \dashv\vdash \mathbf{P} \wedge \neg\mathbf{Q}$.
- By the transitivity of $\dashv\vdash$, conclude: $\neg(\mathbf{P} \to \mathbf{Q}) \dashv\vdash \mathbf{P} \wedge \neg\mathbf{Q}$.

Negative Conditional (*Neg Cndnl*) encapsulates this connection. While *Neg Cndnl* is mainly used in the forward direction in *NE* subproofs, we've formulated it as a bidirectional *Replacement Rule*. It is schematized as follows:

Neg Cndnl $\neg(\mathbf{P} \to \mathbf{Q}) :: \mathbf{P} \wedge \neg\mathbf{Q}$

Using *Neg Cndnl*, we can work to prove a conditional sentence $\mathbf{P} \to \mathbf{Q}$ via *NE* as shown below. We first suppose $\neg(\mathbf{P} \to \mathbf{Q})$ to start the subproof and then use *Neg Cndnl* to conclude $\mathbf{P} \wedge \neg\mathbf{Q}$. By *Simp*, this gives $\mathbf{P}$ and $\neg\mathbf{Q}$, which can be used further to try to generate a contradiction.

	$\neg(\mathbf{P} \to \mathbf{Q})$	Spsn for NE
	$\mathbf{P} \wedge \neg\mathbf{Q}$	Neg Cndnl
	$\mathbf{P}$	Simp
	$\neg\mathbf{Q}$	Simp
	$\vdots$	

This gets condensed in informal *Proof by Contradiction* arguments by saying, *suppose* $\mathbf{P}$ *is the case but* $\mathbf{Q}$ *is not*.

✤**Example 1.8.6**

Use *NE* and *Neg Cndnl* to show that $P \wedge Q \to R,\ Q \leftrightarrow \neg R \vdash P \to R$.

Solution

The following proof diagram establishes this result.

1	$P \wedge Q \to R$	Prem
2	$Q \leftrightarrow \neg R$	Prem
3	$\neg(P \to R)$	Spsn for NE
4	$P \wedge \neg R$	Neg Cndnl 3
5	P	Simp 4
6	$\neg R$	Simp 4
7	$Q \leftrightarrow \neg R$	Reit 2
8	Q	BE 7, 6
9	$P \wedge Q$	Conj 5, 8
10	$P \wedge Q \to R$	Reit 1
11	R	MP 10, 9
12	$P \to R$	NE 3-11

1.8.6 *Contraposition vs. Proof by Contradiction*

Contraposition, introduced in Section 1.7, is sometimes considered a form of *Contradiction*, so we'll briefly analyze and compare these two methods.

There are three main ways to prove $\mathbf{P} \rightarrow \mathbf{Q}$. The most direct method uses *CP*: assume $\mathbf{P}$, and prove $\mathbf{Q}$.

Proof by Contradiction (NE) is a second option. We can either assume $\mathbf{P}$ and prove $\mathbf{Q}$ by *NE*, or we can prove the entire sentence $\mathbf{P} \rightarrow \mathbf{Q}$ by *NE*, using *Neg Cndnl*. Either way, we'd use $\mathbf{P}$ and $\neg\mathbf{Q}$ to generate a contradiction.

Contraposition gives us a third alternative. Instead of deducing $\mathbf{P} \rightarrow \mathbf{Q}$ directly or using *Contraposition*, we can deduce its contrapositive, $\neg\mathbf{Q} \rightarrow \neg\mathbf{P}$. Using *CP* for *this conditional* rather than the original one, we suppose $\neg\mathbf{Q}$ and prove $\neg\mathbf{P}$.

At first glance, this looks like a *Proof by Contradiction*—aren't we given $\mathbf{P}$ and asked to prove $\mathbf{Q}$? If we suppose its opposite $\neg\mathbf{Q}$ and end up with $\neg\mathbf{P}$, which contradicts $\mathbf{P}$, aren't we concluding $\mathbf{Q}$ via *NE*?

No. *We could first suppose* $\mathbf{P}$ *and then suppose* $\neg\mathbf{Q}$, but if we prove $\neg\mathbf{P}$ from $\neg\mathbf{Q}$, *CP* yields $\neg\mathbf{Q} \rightarrow \neg\mathbf{P}$ and *Conpsn* gives us the conclusion $\mathbf{P} \rightarrow \mathbf{Q}$, *all without ever supposing* $\mathbf{P}$ *or using NE*—a more tidy proof overall. Conceptualizing *Conpsn* as *Proof by Contradiction* merely adds unnecessary lines (and another subproof layer) to the proof. In deriving $\neg\mathbf{P}$ from $\neg\mathbf{Q}$, we can't contradict $\mathbf{P}$ if we never assumed it in the first place. $\mathbf{P}$ is not automatically given; we must suppose it to use it, as we do in proving $\mathbf{P} \rightarrow \mathbf{Q}$ via *CP*.

Here's an example comparing these two types of argument.

❖**Example 1.8.7**
   Prove that *if n^2 is even, then n is even.*

Solution
   We'll give a proof by *Contradiction* followed by a proof by *Contraposition*.

Proof #1:   (Proof by *Contradiction*)
   Suppose n^2 is even but n is odd.
   Then $n \ = 2k + 1$ for some integer k.
   Thus, $n^2 = (2k + 1)^2$
$$= 4k^2 + 4k + 1$$
$$= 2(2k^2 + 2k) + 1, \text{ so } n^2 \text{ is odd.}$$
   But this contradicts the fact that n^2 is even.
   Therefore n is not odd; n is even.   ■

Proof #2:   (Proof by *Contraposition*)
   We'll prove the contrapositive equivalent: *if n is odd, then n^2 is odd.*
   Suppose n is odd.
   Then $n \ = 2k + 1$ for some integer k.
   Thus, $n^2 = (2k + 1)^2$
$$= 4k^2 + 4k + 1$$
$$= 2(2k^2 + 2k) + 1, \text{ so } n^2 \text{ is odd.}$$ ■

Note that the second proof in this example simplifies the first by deleting some unnecessary steps. You won't always be able to convert a *Proof by Contradiction* into one by *Contraposition*, but you should be alert to this possibility. *Proof by Contraposition* is generally superior to *Proof by Contradiction* in directness and simplicity, although it does require you to recognize the soundness of *Contraposition*.

Proof by Contraposition may seem like an *indirect proof* of $\mathbf{P} \to \mathbf{Q}$ (some classify it as such), but it is more directed than *Proof by Contradiction*, because to prove $\neg\mathbf{Q} \to \neg\mathbf{P}$ via *CP*, you assume $\neg\mathbf{Q}$ and seek to deduce $\neg\mathbf{P}$. You thus have the *Backward Method* to help guide the subproof. In the corresponding *Proof by Contradiction* argument, however, no *Backward Method* is available. You have no idea what the contradiction will turn out to be. You are not necessarily looking to derive $\neg\mathbf{P}$ to contradict $\mathbf{P}$.

Proof by Contraposition is also more informative than *Proof by Contradiction*, since it shows how negating the consequent leads directly to the negation of the antecedent. All that *Proof by Contradiction* tells you is that supposing the negation of the consequent while assuming the antecedent is problematic.

1.8.7 Ruminations on Indirect Proof Vs. Direct Proof

It's clear from the rules *NI* and *NE* that *Proof by Contradiction* covers conclusions of all possible sentence forms. Whether or not a sentence is a negation, you can try to prove it by showing that its opposite is absurd. Nevertheless, *Contradiction* seems especially useful for proving negations. Some propositions in mathematics, such as the above theorem asserting the irrationality of $\sqrt{2}$, naturally lend themselves to proofs by contradiction. A direct proof of a negation may be difficult to discover or may not even exist.

Proof by Contradiction is also useful when the opposite of the conclusion provides fruitful information that can be readily combined with the premises to derive other sentences. In fact, you may not see how to start a proof until you've assumed the conclusion's opposite. With *Replacement Rules* for simplifying complex negations, *Proof by Contradiction* becomes a versatile tool for constructing proofs.

Proof by Contradiction has both strengths and weaknesses. There is an advantage to assuming an additional sentence—more premises are better than fewer. However, remember that you're assuming the opposite of what you want. You don't have an additional assumption that will lead to the given conclusion. And once you've denied the conclusion, you've lost the beacon of the *Backward Method of Proof Construction*. That's a heavy loss, because you're left to derive a contradiction without knowing what the contradictory sentences will be. So while you have more sentences to work with than in a direct proof, you have no definite idea where you're heading.

Proof by Contradiction, therefore, is less directed than a direct proof. You may decide to use *NI* or *NE* on the basis of a *Backward-Forward Proof Analysis*, but once you're embarked upon the real argumentation inside the contradiction subproof, you're in uncharted territory with only the *Forward Method* at your disposal. What's more, you will often not be able to use a realistic diagram as a proof aid, because what you're assuming is (presumably) false and so may be difficult or impossible to illustrate.

There is one more drawback to *Proof by Contradiction*. In assuming the opposite of the desired conclusion, you end up showing that the supposition is absurd. This gives you a reason to believe the conclusion on negative grounds, but positive grounds are still lacking. In some hard-to-define sense, an indirect proof contains less information than a direct one, which shows more directly how the conclusion is linked to the premises. In this respect indirect proofs less clearly demonstrate their results. To put it simply, a *Proof by Contradiction* may show *that* something follows but not *why* it follows.

Nevertheless, most mathematicians accept *Proof by Contradiction* without reservation. G. H. Hardy, in ***A Mathematician's Apology*** (1940), went so far as to say that *reductio ad absurdum* "is one of a mathematician's finest weapons. It is a far finer gambit than any chess gambit: a chess player may offer the sacrifice of a pawn or even a piece, but a mathematician offers *the game*." Twentieth-century intuitionists, like L. E. J. Brouwer, however, dispute the universal validity of such inferences because of their connection to the *Law of Excluded Middle* and *Double Negation Elimination*. They reject these as general laws and permit *Proof by Contradiction* only in certain situations. David Hilbert's rejoinder was that "Forbidding a mathematician to make use of the principle of excluded middle [and hence its ally, *Proof by Contradiction*] is like forbidding an astronomer his telescope or a boxer the use of his fists" (quoted by H. Weyl in C. Reid's ***Hilbert***, 1970). Clearly, this is an issue in mathematics and logic that not everybody agrees on.

EXERCISE SET 1.8

Exercises 1–5: Soundness of Inference Rules
Show that the following Int-Elim and Replacement Rules are sound.

1.8.1. *Double Negation:* $\neg\neg\mathbf{P} \dashv\vdash \mathbf{P}$

1.8.2. *De Morgan's Rules*
  a. $\neg(\mathbf{P} \wedge \mathbf{Q}) \dashv\vdash \neg\mathbf{P} \vee \neg\mathbf{Q}$ b. $\neg(\mathbf{P} \vee \mathbf{Q}) \dashv\vdash \neg\mathbf{P} \wedge \neg\mathbf{Q}$

1.8.3. *Negative Conditional:* $\neg(\mathbf{P} \rightarrow \mathbf{Q}) \dashv\vdash \mathbf{P} \wedge \neg\mathbf{Q}$

1.8.4. *Negative Biconditional:* $\neg(\mathbf{P} \leftrightarrow \mathbf{Q}) \dashv\vdash (\mathbf{P} \wedge \neg\mathbf{Q}) \vee (\neg\mathbf{P} \wedge \mathbf{Q})$

1.8.5. *Negation Introduction* and *Negation Elimination*
Demonstrate the soundness of *NI* and *NE* in the following ways.
  a. Explain why *NI*'s derivation setup proves $\mathbf{P} \rightarrow (\mathbf{Q} \wedge \neg\mathbf{Q})$. Then show that $\mathbf{P} \rightarrow (\mathbf{Q} \wedge \neg\mathbf{Q}) \dashv\vdash \neg\mathbf{P}$. How does this establish the soundness of *NI*?
  b. Similarly, show that the proof procedure of *NE* is sound: considering its derivation setup, show that $\neg\mathbf{P} \rightarrow (\mathbf{Q} \wedge \neg\mathbf{Q}) \dashv\vdash \mathbf{P}$.

Exercises 6–8: Completing Deductions

Fill in the reasons for the following deductions.

1.8.6. $P \to (Q \to R),\ \neg R \vdash \neg P \lor \neg Q$

1	$P \to (Q \to R)$	_____
2	$\neg R$	_____
3	$\quad \neg(\neg P \lor \neg Q)$	_____
4	$\quad\quad \neg\neg P \land \neg\neg Q$	_____
5	$\quad\quad P \land Q$	_____
6	$\quad\quad P$	_____
7	$\quad\quad P \to (Q \to R)$	_____
8	$\quad\quad Q \to R$	_____
9	$\quad\quad Q$	_____
10	$\quad\quad R$	_____
11	$\quad\quad \neg R$	_____
12	$\neg P \lor \neg Q$	_____

1.8.7. $P \lor Q \vdash (P \to Q) \to Q$

1	$P \lor Q$	_____
2	$\quad \neg Q$	_____
3	$\quad\quad P \lor Q$	_____
4	$\quad\quad \neg\neg P \lor Q$	_____
5	$\quad\quad \neg P \to Q$	_____
6	$\quad\quad \neg\neg P$	_____
7	$\quad\quad P$	_____
8	$\quad\quad P \land \neg Q$	_____
9	$\quad\quad \neg(P \to Q)$	_____
10	$\neg Q \to \neg(P \to Q)$	_____
11	$(P \to Q) \to Q$	_____

1.8.8. $\vdash (P \to Q) \to [(P \to \neg Q) \to \neg P]$

1	$\quad P \to Q$	_____
2	$\quad\quad P \to \neg Q$	_____
3	$\quad\quad\quad P$	_____
4	$\quad\quad\quad P \to \neg Q$	_____
5	$\quad\quad\quad \neg Q$	_____
6	$\quad\quad\quad P \to Q$	_____
7	$\quad\quad\quad Q$	_____
8	$\quad\quad \neg P$	_____
9	$\quad (P \to \neg Q) \to \neg P$	_____
10	$(P \to Q) \to [(P \to \neg Q) \to \neg P]$	_____

Exercises 9–10: Logical Implication and Conclusive Deductions
Determine whether the following **implication** *claims are true. Then determine whether the* **deductions** *given are* **conclusive**. *Carefully point out where and how a rule of inference is being misused.*

1.8.9. $(P \to Q) \to R,\ \neg R \vDash \neg Q$

1	$(P \to Q) \to R$	Prem
2	$\neg R$	Prem
3	$(P \wedge Q) \to R$	Exp 1
4	$\neg (P \wedge Q)$	MT 3,2
5	$\neg P \wedge \neg Q$	DeM 4
6	$\neg Q$	Simp 5

1.8.10. $P \wedge Q \to R,\ R \to S \vDash P \wedge \neg S \to \neg Q$

1	$P \wedge Q \to R$	Prem
2	$R \to S$	Prem
3	Q	Spsn for NI
4	$P \wedge \neg S$	Spsn for CP
5	P	Simp 4
6	Q	Reit 3
7	$P \wedge Q$	Conj 5,6
8	$P \wedge Q \to R$	Reit 1
9	R	MP 8,7
10	$R \to S$	Reit 2
11	S	MP 10,9
12	$\neg S$	Simp 4
13	$\neg Q$	NI 3-12
14	$P \wedge \neg S \to \neg Q$	CP 4-13

Exercises 11–13: True or False
Are the following statements true or false? Explain your answer.

1.8.11. *Proof by Contradiction* is one of the oldest and most important forms of reasoning used in mathematics.

1.8.12. *Proof by Contradiction* can be used to prove propositions having any logical form.

1.8.13. *Contraposition* is one of the main forms of *indirect proof.*

Exercises 14–16: Negating Mathematical Propositions
Use Replacement Rules governing negated sentences to work the following.

1.8.14. *An element is a member of the union $A \cup B$ if and only if it is a member of A or a member of B.* What must be done to show that an element is *not* a member of $A \cup B$? What *Replacement Rule* is involved?

1.8.15. *Functions f and g are inverses if and only if for all x and y in the proper domains $g(f(x)) = x$ and $f(g(y)) = y$.* What must be done to show f and g are *not* inverses? What *Replacement Rule* is involved?

1.8.16. *A series is conditionally convergent if and only if it is convergent but not absolutely convergent.* If you know that a series is *not* conditionally convergent, what might it be? What *Replacement Rules* justify your conclusion?

Exercises 17–24: Deductions

Construct deductions, **where possible***, for the following problems, using* **only the Int-Elim Rules** *discussed up to this point. If a deduction still cannot be constructed, explain why not.*

1.8.17. $P \wedge \neg P \vdash Q$ **1.8.21.** $\vdash (\neg P \to P) \to P$

1.8.18. $Q, \neg(P \wedge Q) \vdash \neg P$ **1.8.22.** $P \wedge \neg Q \vdash \neg(P \to Q)$

1.8.19. $P \to (Q \wedge R), Q \to S, \neg S \vdash \neg P$ **1.8.23.** $\neg(P \to Q) \vdash P \wedge \neg Q$

1.8.20. $P \wedge Q \to R \vdash P \wedge \neg R \to \neg Q$ **1.8.24.** $(P \to Q) \to P \vdash P$

Exercises 25–31: More Deductions

Show that the following deduction claims hold, using any Inference Rules available so far or as specified.

1.8.25. $P \vee Q, \neg P \vdash Q$ [Prove this directly, using *Cndnl*.]

1.8.26. $\neg(P \wedge Q), Q \leftrightarrow R \vdash P \to \neg R$

1.8.27. $P \to R, Q \to \neg R \vdash P \to \neg Q$

1.8.28. $P \wedge Q \to R \dashv\vdash (P \to R) \vee (Q \to R)$

1.8.29. $P \vee Q, P \to R, Q \to R \vdash R$

1.8.30. $\neg Q \vee \neg R, P \to Q \wedge R \vdash \neg P$

1.8.31. $\vdash P \vee \neg P$

Exercises 32–37: Relations Among Inference Rules

Explore the connections between the following Inference Rules.

1.8.32. *Redundancy of Contraposition*

Show that *Conpsn* can be eliminated from PL's Natural Deduction System with no loss of deductive power, provided it has *MP* and *CP* along with *NI* and *NE*, i.e., show that $\mathbf{P} \to \mathbf{Q} \dashv\vdash \neg\mathbf{Q} \to \neg\mathbf{P}$ using these inference rules.

1.8.33. *Proof by Contradiction and Double Negation*

Show that *NE* can be eliminated from PL's Natural Deduction System with no loss of deductive power, provided it has *NI* and *DN Elimination* ($\neg\neg\mathbf{P} \vDash \mathbf{P}$). Hence, intuitionists dispute *DN Elimination* as well as *NE*.

1.8.34. *Law of Non-Contradiction and Law of Excluded Middle*

Show $\neg(\mathbf{P} \wedge \neg\mathbf{P}) \vdash \mathbf{P} \vee \neg\mathbf{P}$, given *DeM* and *DN Elimination* ($\neg\neg\mathbf{P} \vDash \mathbf{P}$). Without *DN*, what's the most you can conclude from $\neg(\mathbf{P} \wedge \neg\mathbf{P})$?

1.8.35. *De Morgan's Laws and Double Negation*

Using the first rule of *DeM* and *DN*, show that the second rule also holds: $\neg(\mathbf{P} \vee \mathbf{Q}) \dashv\vdash \neg\mathbf{P} \wedge \neg\mathbf{Q}$.

1.8.36. *Intuitionist Scruples Relaxed?*
According to intuitionists, $\neg\neg P \not\vdash P$. However, using the intuitionistically
valid *DN Introduction* ($\mathbf{P} \models \neg\neg\mathbf{P}$) or *NI*, show that $\neg P \vdash\!\vdash \neg\neg\neg P$. Thus,
negative sentences are very different from positive sentences for intuitionists.

1.8.37. *More Inference Rules*
Should the argument forms of Examples 3 and 4 be added to PL's Natural
Deduction System? Why or why not?

Exercises 38–39: Deducing Everyday Arguments
*Symbolically formulate and deduce the following arguments, using the capital
letters indicated for the positive component sentences.*

1.8.38. If the FBI operates a Sting operation, it will be Encouraging crime.
If the FBI does not operate a Sting operation, crime will Increase. Law and
order are upheld if and only if crime is not Encouraged and crime does not
Increase. Therefore Law and order cannot be upheld.   [S, E, I, L]

1.8.39. The weather is Rainy if it is Spring. If the weather is Rainy and the
roof has not been Fixed, water will Leak into the house. It is Spring, and
there is no Leak. Therefore, the roof has been Fixed.   [R, S, F, L]

Exercises 40–42: Irrationality Proofs
Work the following mathematical proofs using proof by contradiction.

1.8.40. Give another proof by contradiction for the irrationality of $\sqrt{2}$, as
follows. Begin as in the text, but then consider the number of factors of 2 in
the prime factorizations of the right and left sides of equation ($*$).

1.8.41. Prove that $\sqrt{3}$ is irrational. Use an argument similar to the proof of
the irrationality of $\sqrt{2}$. Would such an argument also show $\sqrt{4}$ is irrational?

1.8.42. Prove that $\sqrt[3]{2}$ is irrational.

Exercises 43–46: The Pigeonhole Principle
The following problems involve the Pigeonhole Principle.

1.8.43. Using *Proof by Contradiction* in an informal paragraph-style proof,
prove the *Pigeonhole Principle*: if m objects are distributed to n containers
for $m > n$, then at least one container will hold more than one object.

1.8.44. Explain why there must be more than 50 New York City residents
having exactly the same number of hairs on their heads (though you may
be hard pressed to locate two of them!). Use the fact that there are at most
150,000 hairs on any one person's head.

1.8.45. *Party Handshakes*
  a. Show that at a party attended by more than one person, at least two of
     them shook hands with the same number of people.
  b. Five couples attend a party and shake hands (once) with those they've
     not met before. Excluding the host, they all shook hands with a different
     number of people. How many hands did the host's friend shake? How
     many hands did the host shake?

1.9 Inference Rules for OR

To complete PL's Natural Deduction System, we still need inference rules for ∨. As before, we'll have both *Elimination* and *Introduction Rules*, as well as *Replacement Rules* for expanding or contracting formulas.

1.9.1 *Elimination Rule for* OR: *Disjunctive Syllogism*

We can draw a conclusion from a disjunction in several ways. The simplest and most widely used inference rule for this is *Disjunctive Syllogism (DS)*. It proceeds by *exclusion*, by *ruling out an alternative*.

There are four basic forms of *DS*, whose schema is as follows:

$$DS \qquad \begin{array}{c|c} \mathbf{P \vee Q} \\ \mathbf{\neg P} \\ \hline\hline \mathbf{Q} \end{array} \quad \begin{array}{c|c} \mathbf{P \vee Q} \\ \mathbf{\neg Q} \\ \hline\hline \mathbf{P} \end{array} \quad \begin{array}{c|c} \mathbf{\neg P \vee Q} \\ \mathbf{P} \\ \hline\hline \mathbf{Q} \end{array} \quad \begin{array}{c|c} \mathbf{P \vee \neg Q} \\ \mathbf{Q} \\ \hline\hline \mathbf{P} \end{array}$$

We can show that these rules are sound with a truth table (see Exercise 1) or by demonstrating that any conclusion *DS* generates can be deduced via other rules (see Exercise 2).

Disjunctive Syllogism is essentially *Modus Ponens* in the language of disjunction. This is easy to see by looking at the third form of *DS*.

✢ **Example 1.9.1**
Show that **P → Q, P ⊢ Q** without using *MP*.

Solution
The following proof diagram establishes the claim.

1	**P → Q**	Prem
2	**P**	Prem
3	**¬P ∨ Q**	Cndnl 1
4	**¬¬P**	DN 2
5	**Q**	DS 3, 4

Disjunctive Syllogism is applied in a wide variety of everyday contexts. Causal inductive inferences in natural science are like this rule, except scientists usually don't know all possible alternatives. Detectives use *DS* to decide who did what to whom and why (see Exercise 42). Sherlock Holmes' *modus operandi* was "Eliminate all other factors, and the one which remains must be the truth." And, to solve a Sudoku puzzle, one uses *DS* over and over.

In mathematics, *DS* is widely used. For instance, to solve an equation subject to side conditions, you first find all possible solutions and then rule out those that don't satisfy the problem's constraints (see Exercise 45).

The following example uses *DS* as its main proof strategy, but *NI* is first used to show that one disjunct fails to hold.

❖**Example 1.9.2**

Show that $P \to Q \wedge R$, $P \vee S$, $\neg Q \vdash S$.

Solution

The following proof diagram establishes the claim.

1	$P \to Q \wedge R$	Prem
2	$P \vee S$	Prem
3	$\neg Q$	Prem
4	P	Spsn for NI
5	$P \to Q \wedge R$	Reit 1
6	$Q \wedge R$	MP 5, 4
7	Q	Simp 6
8	$\neg Q$	Reit 3
9	$\neg P$	NI 4–8
10	S	DS 2, 9

1.9.2 Elimination Rule for OR: Proof by Cases

A second *Elimination Rule*, traditionally called *Constructive Dilemma*, is nearly as important as *DS*, but it's a bit more complex. Mathematicians know this proof strategy as *Proof by Cases*, so we'll use that term and abbreviate it as *Cases*. Arguments constructed according to this rule derive a sentence **R** on the basis of a disjunction **P** ∨ **Q** together with subproofs of **R** from **P** and **R** from **Q**. *Proof by Cases* gives us another suppositional rule.

In schematic form, *Cases* proceeds as follows:

Cases

The reason this rule is called *Constructive Dilemma* can be explained in behavioral terms. It presents a dilemma in which each alternative leads to the same, possibly unsavory, conclusion.

Here's a classic example from Greek philosophy. Euathlus (E) is trained as a lawyer by Protagoras (P), with the agreement that E will pay P for his training as soon as he wins his first case. When E decides not to become a lawyer, he is taken to court by P, who argues that if E wins he must pay according to their contract, while if E loses he must pay according to the court's ruling. Thus, E must pay up. E faces a real dilemma, which it seems he can't escape. However, if E learned his lessons well, he could launch a creative counterattack along the same lines (see Exercise 41).

The reason for the name *Proof by Cases* should be obvious—you prove **R** by supposing in turn the two alternatives **P** and **Q**, showing that *in each case* **R** follows. Since one of these two cases must hold, you may conclude **R**.

Cases is a sound rule of inference. If the only two cases both give the same result, then that result must follow, even though we may not know which case holds. Given *CP*, we can argue that *Cases* is sound by showing that **P** ∨ **Q**, **P** → **R**, **Q** → **R** ⊨ **R** (see Exercise 6). We can also demonstrate its soundness by showing that any argument *Cases* validates can be deduced using inference rules already accepted (see Exercise 37).

We'll illustrate this rule with two examples, starting with one from PL. Using *Cases*, we can prove one direction of the ∨-form of *Cndnl*. The other direction follows from rules that introduce disjunctions (see Exercise 23).

✤ Example 1.9.3

Show ¬**P** ∨ **Q** ⊢ **P** → **Q** without using *DS* or *Cndnl*, using *Cases* instead.

Solution

- In the following deduction, the first subproof proves the conclusion's contrapositive, since that fits better with the negative case under consideration.
- The second subproof deduces the conditional as given.

1	¬**P** ∨ **Q**	Prem
2	¬**P**	Spsn 1 for Cases
3	¬**Q**	Spsn for CP
4	¬**P**	Reit 2
5	¬**Q** → ¬**P**	CP 3–4
6	**P** → **Q**	Conpsn 5
7	**Q**	Spsn 2 for Cases
8	**P**	Spsn for CP
9	**Q**	Reit 7
10	**P** → **Q**	CP 8–9
11	**P** → **Q**	Cases 1, 2–6, 7–10

Formal deductions argue each case separately; the final conclusion depends upon both jointly. An informal *Cases* proof may proceed in a less obvious way, using phrases like *on the one hand* and *on the other hand* to indicate the different cases. Or you may find only one case proved in detail, particularly if the second case proceeds similarly. You may also see the phrase *without loss of generality* (or its abbreviation *wolog*) *suppose such-and-such is the case.* The reason behind these practices is that when cases are analogous, it isn't necessary to repeat the deduction. Treat this as an invitation to work the second subproof for yourself. You'll occasionally find a wrinkle in the argument that costs you some time and effort and makes you wonder why the writer thought the subproofs were so similar. Filling in the gaps keeps life interesting and mathematicians honest!

Proof by Cases is a favorite proof strategy in many fields of mathematics. Two different kinds of mathematical proofs use cases. The first is a genuine application of *Cases*. This occurs when a concept can be divided into different cases. The absolute value function, for instance, is defined as a split function, and the numerical relation *greater than or equal to* is composed of two alternatives. The next example illustrates this kind of argument.

❖ **Example 1.9.4**

Prove that $|x| \geq 0$ for all real numbers.

Solution

The absolute value function is defined by $|x| = \begin{cases} x : & \text{if } x \geq 0, \\ -x : & \text{if } x < 0. \end{cases}$

To prove the given result, we'll take cases based upon the definition.

· For $x \geq 0$, $|x| = x$, so in this case the result holds by substitution: $|x| \geq 0$.

· On the other hand, if $x < 0$, $|x| = -x$.

Since $x < 0$, $-x > 0$.

Thus, $|x| > 0$, so certainly $|x| \geq 0$. (This requires the *Addition Rule*, discussed below.)

· Hence, in all cases $|x| \geq 0$.   ■

Proof by Cases is a valuable technique to consider when the disjuncts present *exclusive alternatives*—evens and odds; rationals and irrationals; negatives, zero, and positives. If being a member of each subclass entails the desired result, then the conclusion holds without exception.

The disjunctions used in these situations can be thought of as instances of the *Law of Excluded Middle*—*n is even or n is not even, x is rational or it is not rational*, etc. A *Cases* argument based on these alternatives proves a sentence R on the basis of $P \vee \neg P$, whether or not it has been expressly stated. Since this sentence is a tautology, it is not really an additional premise, though it should be stated for the argument to be complete.

LEM appears here in an essential way, generating the cases to be investigated. In fact, there are many situations in which such a tautology is a rather natural way to make a derivation. This divide-and-conquer approach breaks a proof up into manageable pieces by focusing attention on distinct subclasses. Introducing an instance of *LEM* may demand creativity and insight (what's the best way to split things up?), but it's a standard mathematical maneuver.

So far, we've excluded tautologies from our Natural Deduction System because they're rarely used in mathematics. But having found a natural way in which they occur, we'll now make an exception. An instance of the *Law of Excluded Middle* may be asserted anywhere in a proof, citing *LEM* as the reason. As noted in Section 1.8 (see Exercise 1.8.31), such a sentence can be proved from no premises, using rules already available, so adopting this rule doesn't increase the power of our deduction system, only its efficiency.

The rule schema for *LEM* is the following:

LEM

$$\mathbf{P} \vee \neg\mathbf{P}$$

A second type of proof involving cases has a different logical basis and character because the cases are hierarchically related, becoming ever more inclusive (prime numbers and positive integers; integers, rationals, and real numbers). Proofs like this are interesting because the final case proves the entire proposition, not just one part of it. Why, then, are the earlier cases considered? Because the final case often needs the earlier results for its proof.

For example, to prove $a^m \cdot a^n = a^{m+n}$ for rational exponents, you first prove it for positive integers, then for integers, and finally for rational numbers (see Section 3.2). The latter proofs appeal to the law for positive integers or integers. Although the same formal result is concluded for these different cases, the final conclusion isn't validated by *Cases*, but simply uses a proposition proved for a more limited class of objects to establish the same result in a broader context. Such proofs are not in the *Proof by Cases* mold, but they are an important proof technique involving cases. Mathematicians call this proof strategy *bootstrapping*, because you gradually pull yourself up by your bootstraps, as it were, proving the general case from the special ones.

1.9.3 *Introduction Rule for* Or: *Addition*

The simplest *Introduction Rule* for $\vee$ seems useless at first glance. The *Addition Rule* (*Add*) permits you to conclude $\mathbf{P} \vee \mathbf{Q}$ from either $\mathbf{P}$ or $\mathbf{Q}$.

The schema for *Add* is the following:

Add

$$\mathbf{P} \qquad\qquad\qquad \mathbf{Q}$$
$$\overline{\mathbf{P} \vee \mathbf{Q}} \qquad\qquad\qquad \overline{\mathbf{P} \vee \mathbf{Q}}$$

Why would anyone want to conclude $\mathbf{P} \vee \mathbf{Q}$ if they already know one of the disjuncts? Example 4 is a case where this is needed. We had the inequality $|x| > 0$ but needed $|x| \geq 0$, which is the disjunction $|x| > 0 \vee |x| = 0$. *Addition* warrants the further obvious inference.

The following argument for the commutativity of $\vee$ illustrates the use of *Add* in a formal *Proof by Cases* deduction.

✤**Example 1.9.5**

Show that $\mathbf{P} \vee \mathbf{Q} \vdash \mathbf{Q} \vee \mathbf{P}$.

Solution

· Note first (as in Example 1.5.2) that only one direction needs to be proved. For if we establish $\mathbf{P} \vee \mathbf{Q} \vdash \mathbf{Q} \vee \mathbf{P}$, $\mathbf{Q} \vee \mathbf{P} \vdash \mathbf{P} \vee \mathbf{Q}$ follows: both show that it is valid to interchange $\mathbf{P}$ and $\mathbf{Q}$ in a disjunction.

· The proof of $\mathbf{Q} \vee \mathbf{P}$ from $\mathbf{P} \vee \mathbf{Q}$ goes as follows:

1	$\mathbf{P} \vee \mathbf{Q}$	Prem
2	$\quad \mathbf{P}$	Spsn 1 for Cases
3	$\quad \mathbf{Q} \vee \mathbf{P}$	Add 2
4	$\quad \mathbf{Q}$	Spsn 2 for Cases
5	$\quad \mathbf{Q} \vee \mathbf{P}$	Add 4
6	$\mathbf{Q} \vee \mathbf{P}$	Cases 1, 2–3, 4–5

1.9.4 *Introduction Rule for* OR*: Either-Or*

A second *Introduction Rule* is known as *Either-Or (EO)*. Whereas *Addition* is used when one of $\mathbf{P}$ or $\mathbf{Q}$ is the case and you want to conclude the disjunction $\mathbf{P} \vee \mathbf{Q}$, *Either-Or* is the rule to use when you want to conclude a disjunction but either disjunct might be true. This happens frequently, so *EO* is used more often in mathematical proofs than *Add*.

Either-Or proceeds in the following way. Since $\mathbf{P} \vee \mathbf{Q}$ is the case whenever either disjunct is true, $\mathbf{P} \vee \mathbf{Q}$ holds even if one of them is not the case. In fact, if you can show that one of the disjuncts *must be* the case whenever the other one *isn't*, then the disjunction follows.

Schematically we have the following:

EO

Let's look at why this procedure is sound. The subproof in the left-hand scheme proves $\neg\mathbf{P} \to \mathbf{Q}$ (via *CP*), and since $\neg\mathbf{P} \to \mathbf{Q} \vDash \mathbf{P} \vee \mathbf{Q}$ (via *Cndnl* and *DN*), it also proves $\mathbf{P} \vee \mathbf{Q}$. Analogously, $\neg\mathbf{Q} \to \mathbf{P} \vDash \mathbf{P} \vee \mathbf{Q}$.

Note that *EO* concludes $\mathbf{P} \vee \mathbf{Q}$ from a subargument based on the negation of just *one* of the disjuncts. You do not also need a subargument based on the negation of the other disjunct. Which disjunct to negate depends upon what else you have to work with—one negation may combine more easily with the premises than the other. In this case, the *Forward Method of Proof Analysis* helps you choose the best strategy. If you don't immediately see which negation to begin with, try them both and see which one works better.

To illustrate *Either-Or*, we'll look at a formal example from PL and an informal one from mathematics.

❖Example 1.9.6

Show that $\neg(\mathbf{P} \wedge \mathbf{Q}) \vdash \neg\mathbf{P} \vee \neg\mathbf{Q}$ without using *DeM*.

Solution

The following proof diagram establishes the claim. Note in line 5 that we cite a previous example that we chose not to make into an inference rule.

1	$\neg(\mathbf{P} \wedge \mathbf{Q})$	Prem
2	$\neg\neg\mathbf{P}$	Spsn for EO
3	$\neg(\mathbf{P} \wedge \mathbf{Q})$	Reit 1
4	$\mathbf{P}$	DN 2
5	$\neg\mathbf{Q}$	Example 1.8.3 3,4
6	$\neg\mathbf{P} \vee \neg\mathbf{Q}$	EO 2–5

❖Example 1.9.7

Prove that $ab = 0 \leftrightarrow a = 0 \vee b = 0$.

Solution

Since the main connective is $\leftrightarrow$, we'll use *BI*. The two subproofs use rules for introducing and eliminating disjunctions (see Exercise 43).
We assume as already known that $x \cdot 0 = 0 = 0 \cdot x$ for any real number x.
Proof:

    $(\rightarrow)$ First suppose $ab = 0$.

        If $a \neq 0$, multiplying $ab = 0$ by $1/a$ yields $b = (1/a) \cdot 0 = 0$. ✓

    $(\leftarrow)$ Now suppose $a = 0 \vee b = 0$.

        If $a = 0$, then $ab = 0 \cdot b = 0$.

        If $b = 0$, a similar result follows.

    Thus, $ab = 0$.

  Therefore $ab = 0 \leftrightarrow a = 0 \vee b = 0$. ■

1.9.5 Replacement Rules Involving Disjunctions

Section 1.5 gave three *Replacement Rules* for conjunction: *Commutation*, *Association*, and *Idempotence*. These same rules hold for disjunction.

They are schematized as follows:

Comm $(\vee)$	$\mathbf{P} \vee \mathbf{Q} :: \mathbf{Q} \vee \mathbf{P}$
Assoc $(\vee)$	$\mathbf{P} \vee (\mathbf{Q} \vee \mathbf{R}) :: (\mathbf{P} \vee \mathbf{Q}) \vee \mathbf{R}$
Idem $(\vee)$	$\mathbf{P} \vee \mathbf{P} :: \mathbf{P}$

In addition, two sorts of *Distribution Replacement Rules* govern how conjunctions and disjunctions can be expanded or contracted. Even though these are less obvious than the other rules, they have algebraic analogues that we're familiar with (see Exercises 1.3.6–7).

There are four forms here, which are schematized as follows:

$$\mathbf{P} \wedge (\mathbf{Q} \vee \mathbf{R}) :: (\mathbf{P} \wedge \mathbf{Q}) \vee (\mathbf{P} \wedge \mathbf{R}) \qquad \textit{Dist} \ (\wedge \ \textbf{over} \ \vee)$$
$$(\mathbf{P} \vee \mathbf{Q}) \wedge \mathbf{R} :: (\mathbf{P} \wedge \mathbf{R}) \vee (\mathbf{Q} \wedge \mathbf{R}) \qquad \textit{Dist} \ (\wedge \ \textbf{over} \ \vee)$$
$$\mathbf{P} \vee (\mathbf{Q} \wedge \mathbf{R}) :: (\mathbf{P} \vee \mathbf{Q}) \wedge (\mathbf{P} \vee \mathbf{R}) \qquad \textit{Dist} \ (\vee \ \textbf{over} \ \wedge)$$
$$(\mathbf{P} \wedge \mathbf{Q}) \vee \mathbf{R} :: (\mathbf{P} \vee \mathbf{R}) \wedge (\mathbf{Q} \vee \mathbf{R}) \qquad \textit{Dist} \ (\vee \ \textbf{over} \ \wedge)$$

EXERCISE SET 1.9

Exercises 1–7: Soundness of Inference Rules
Show that the following Int-Elim Rules are sound.

1.9.1. *Disjunctive Syllogism* (*DS*)
a. $\mathbf{P} \vee \mathbf{Q}, \ \neg\mathbf{P} \vDash \mathbf{Q}$ c. $\neg\mathbf{P} \vee \mathbf{Q}, \ \mathbf{P} \vDash \mathbf{Q}$
b. $\mathbf{P} \vee \mathbf{Q}, \ \neg\mathbf{Q} \vDash \mathbf{P}$ d. $\mathbf{P} \vee \neg\mathbf{Q}, \ \mathbf{Q} \vDash \mathbf{P}$

1.9.2. *Disjunctive Syllogism and Modus Ponens*
Show that $\mathbf{P} \vee \mathbf{Q}, \ \neg\mathbf{P} \vdash \mathbf{Q}$ without using *DS*. This result, along with Example 1, establishes the equivalence of *DS* and *MP*.

1.9.3. *Forms of Disjunctive Syllogism*
Show the following, using only the *first form* of *DS* plus other inference rules.
a. $\mathbf{P} \vee \mathbf{Q}, \ \neg\mathbf{Q} \vdash \mathbf{P}$ b. $\neg\mathbf{P} \vee \mathbf{Q}, \ \mathbf{P} \vdash \mathbf{Q}$ c. $\mathbf{P} \vee \neg\mathbf{Q}, \ \mathbf{Q} \vdash \mathbf{P}$

1.9.4. *Law of Excluded Middle* (*LEM*): $\vDash \mathbf{P} \vee \neg\mathbf{P}$

1.9.5. *Addition* (*Add*)
a. $\mathbf{P} \vDash \mathbf{P} \vee \mathbf{Q}$ b. $\mathbf{Q} \vDash \mathbf{P} \vee \mathbf{Q}$

1.9.6. *Proof by Cases* (*Cases*)
a. Show that $\mathbf{P} \vee \mathbf{Q}, \ \mathbf{P} \to \mathbf{R}, \ \mathbf{Q} \to \mathbf{R} \vDash \mathbf{R}$.
b. Use part *a*, along with *CP*, to explain why *Cases* is sound.

1.9.7. *Either-Or*
Show that $\neg\mathbf{P} \to \mathbf{Q} \vdash \mathbf{P} \vee \mathbf{Q}$ using *LEM*, *Add*, *Cases*, and *MP*. Thus, the left-hand form of *EO* is sound.

Exercises 8–9: Completing Deductions
Fill in the reasons for the following deductions.

1.9.8. $P \to (Q \to R), \ \neg R \vdash \neg P \vee \neg Q$

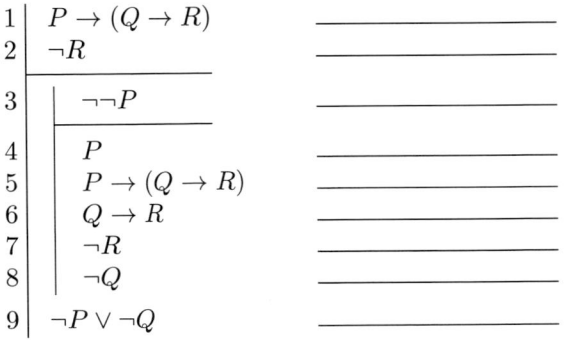

1	$P \to (Q \to R)$	_____
2	$\neg R$	_____
3	$\neg\neg P$	_____
4	P	_____
5	$P \to (Q \to R)$	_____
6	$Q \to R$	_____
7	$\neg R$	_____
8	$\neg Q$	_____
9	$\neg P \vee \neg Q$	_____

1.9.9. $P \vee Q \vdash (P \to Q) \to Q$

1	$P \vee Q$	_____
2	$P \to Q$	_____
3	$P \vee Q$	_____
4	P	_____
5	$P \to Q$	_____
6	Q	_____
7	Q	_____
8	Q	_____
9	Q	_____
10	$(P \to Q) \to Q$	_____

Exercises 10–12: Logical Implication and Conclusive Deductions

*Determine whether the following **implication** claims are true. Then determine whether the **deductions** given are **conclusive**. Carefully point out where and how a rule of inference is being misused.*

1.9.10. $\neg P \wedge Q \to R,\ R \to P \vDash P \vee Q$

1	$\neg P \wedge Q \to R$	Prem
2	$R \to P$	Prem
3	$\neg P$	Spsn for EO
4	$\neg P \wedge Q$	Add 3
5	$\neg P \wedge Q \to R$	Reit 1
6	R	MP 5, 4
7	$R \to P$	Reit 2
8	P	MP 7, 6
9	$P \vee Q$	Add 8
10	$P \vee Q$	EO 3–9

1.9.11. $P \vee Q \to \neg R,\ Q \vee R \vDash \neg R$

1	$P \vee Q \to \neg R$	Prem
2	$Q \vee R$	Prem
3	Q	Spsn for NI
4	$P \vee Q$	Add 3
5	$P \vee Q \to \neg R$	Reit 1
6	$\neg R$	MP 5, 4
7	$Q \vee R$	Reit 2
8	$\neg Q$	DS 7, 6
9	Q	Reit 3
10	$\neg Q$	NI 3–9
11	$\neg R$	DS 2, 10

1.9.12. $P \to Q$, $R \to \neg Q$, $P \vee \neg R \vDash Q$

1	$P \to Q$	Prem
2	$R \to \neg Q$	Prem
3	$P \vee \neg R$	Prem
4	P	Spsn 1 for Cases
5	$P \to Q$	Reit 1
6	Q	MP 5, 4
7	$\neg P$	Spsn 2 for Cases
8	$P \vee \neg R$	Reit 3
9	$\neg R$	DS 8, 7
10	$R \to \neg Q$	Reit 2
11	Q	MT 10, 9
12	Q	Cases 3; 4–6, 7–11

Exercises 13–15: True or False?
Are the following statements true or false? Explain your answer.

1.9.13. *Disjunctive Syllogism* is a widely used inference rule in everyday argumentation.

1.9.14. The only tautology included in PL's Natural Deduction System is the *Law of Non-Contradiction*.

1.9.15. The *Int-Elim Rules* in PL's Natural Deductions System form a complete set of inference rules for PL.

Exercises 16–19: Contraction/Absorption Rules
Show that the following results hold and explain in words what they mean regarding contraction/absorption (proceeding left to right).

1.9.16. $\mathbf{P} \wedge (\mathbf{P} \vee \mathbf{Q}) \vdash \mathbf{P}$ **1.9.18.** $\mathbf{P} \wedge (\mathbf{Q} \vee \neg \mathbf{Q}) \vdash \mathbf{P}$

1.9.17. $\mathbf{P} \vee (\mathbf{P} \wedge \mathbf{Q}) \vdash \mathbf{P}$ **1.9.19.** $\mathbf{P} \vee (\mathbf{Q} \wedge \neg \mathbf{Q}) \vdash \mathbf{P}$

Exercises 20–21: Simplifying Expressions
Using the relevant Replacement Rules as well as any results from Exercises 16–19, show that the following hold.

1.9.20. $P \vee (P \wedge \neg Q) \vdash P$ **1.9.21.** $P \wedge (\neg P \vee Q) \vdash P \wedge Q$

Exercises 22–36: Deductions
Show that the following deduction claims hold.

1.9.22. $P \vee Q$, $P \to Q \vdash Q$ [Prove this directly, *without* using *NE*.]

1.9.23. $P \to Q \vdash \neg P \vee Q$ [Prove this *without* using *Cndnl*.]

1.9.24. $P \vee Q$, $P \to R$, $R \to \neg P \vdash Q$

1.9.25. $P \vee Q$, $P \to R$, $Q \to S \vdash R \vee S$

1.9.26. $P \to R$, $Q \to \neg R, \vdash \neg P \vee \neg Q$

1.9.27. $P \vee Q$, $\neg(P \wedge R)$, $R \vdash \neg Q$ **1.9.31.** $P \vee Q$, $P \vee \neg Q \vdash P$
1.9.28. $\neg P \vee R$, $\neg Q \vee \neg R \vdash \neg P \vee \neg Q$ **1.9.32.** $P \leftrightarrow (Q \leftrightarrow P) \vdash Q$
1.9.29. $Q \rightarrow R \vdash P \vee Q \rightarrow P \vee R$ **1.9.33.** $(P \rightarrow Q) \vee R \vdash P \rightarrow (Q \vee R)$
1.9.30. $\vdash (P \rightarrow Q) \vee (Q \rightarrow P)$ **1.9.34.** $\vdash (P \leftrightarrow Q) \vee (P \vee Q)$
1.9.35. $P \rightarrow ((Q \rightarrow R) \vee (\neg Q \rightarrow S)) \vdash P \rightarrow R \vee S$
1.9.36. $P \wedge Q \rightarrow R$, $P \rightarrow R \vee Q \vdash P \rightarrow R$

Exercises 37–40: Inference Rule Connections
Show that the following relations hold between various inference rules.

1.9.37. *Cases and Other Inference Rules*
Show that if *Cases* is dropped from our Natural Deduction System there is
no loss of deductive power: show $\mathbf{P} \vee \mathbf{Q}$, $\mathbf{P} \rightarrow \mathbf{R}$, $\mathbf{Q} \rightarrow \mathbf{R} \vdash \mathbf{R}$.

1.9.38. *Either-Or and Law of Excluded Middle*
Show how the rule *Either-Or* can be used to deduce *LEM*. Hence, *EO* is
unacceptable to intuitionist logicians.

1.9.39. *NE from NI, LEM, and DS*
Show that the consequences generated by *NE* from its given setup can also
be obtained by *NI* in combination with *LEM* and *DS*.

1.9.40. *Proof by Contradiction, Addition, and Disjunctive Syllogism*
Using the rules *Add* and *DS*, show that $\mathbf{Q} \wedge \neg \mathbf{Q} \vdash \mathbf{R}$ for any $\mathbf{R}$.

1.9.41. *Constructive Dilemma*
Develop the classic example of *Euathlus vs. Protagoras* further, only this time
make the dilemma turn out favorable to Euathlus.

1.9.42. *Detective Reasoning: Use Or rules to help solve the following case.*
Security was breached at Bank One and a large sum of money taken. Police
have assembled the following clues. Track down all the culprits with a formal
deduction, using T, C, L, V, W, S, and D for the positive atomic sentences.
 1. The suspects are a Teller, a Computer programmer, a Loan officer, and
    a Vice-President.
 2. If it was the Computer programmer or the Teller, then one of the others
    was also involved.
 3. If it was the Teller or the Vice-President, then it did not occur on the
    Weekend.
 4. If Security was bypassed, then it was either the Computer programmer
    or the Vice-President.
 5. If the break-in was Discovered later, then Security was bypassed.
 6. The break-in was on the Weekend, but it was not Discovered until later.

Exercises 43–47: Mathematical Proofs
Work the following mathematical proofs, as indicated.

1.9.43. Analyze the logical structure of the proof given in Example 7, pointing out which *Int-Elim Rules* for $\vee$ are used and where.

1.9.44. Prove $n^2 + n + 1$ *is odd for all natural numbers n.*
[Hint: use some natural classification of natural numbers and prove the result for each type of number.]

1.9.45. Using the result of Example 7, solve $x^2 - x - 6 = 0$, subject to the constraint that $x \geq 0$. Construct a step-by-step argument, noting where PL inference rules are being used.

1.9.46. Using the result of Example 7, find the solution set of all ordered pairs (x, y) such that $x(1 - y^2) = 0$ and $(x + 2)y = 0$. Construct a step-by-step argument, noting where PL inference rules are being used.

1.9.47. *Absolute Value Proofs*
Using *Cases*, where appropriate, prove for any real numbers a and b that the following results hold.

a. $|-a| = |a|$ c. $a \leq |a|$
b. $-|a| \leq a$ d. $|a + b| \leq |a| + |b|$

Exercises 48–50: Exploring *Exclusive-Or* Rules
Work the following problems related to $\veebar$.

1.9.48. Show that the following inference rule is sound: $\mathbf{P} \veebar \mathbf{Q}$, $\mathbf{P} \vDash \neg \mathbf{Q}$. Is this rule sound if $\vee$ is substituted for $\veebar$?

1.9.49. Show that the following conversion rule holds: $\mathbf{P} \veebar \mathbf{Q}$, $\mathbf{P} \to \mathbf{R}$, $\mathbf{Q} \to \mathbf{S}$, $\mathbf{R} \veebar \mathbf{S} \vDash (\mathbf{R} \to \mathbf{P}) \veebar (\mathbf{S} \to \mathbf{Q})$. Is a similar rule sound for $\vee$?

1.9.50. Which of the *Int-Elim Rules* for $\vee$ are sound if $\vee$ is replaced by $\veebar$? Give reasons to support your answer.

Chapter 2
First-Order Logic

2.1 Symbolizing Sentences

First-Order Logic (FOL) extends *Propositional Logic* by taking into account the inner logical structure of sentences as well as their combinatorial connections. In this chapter we'll learn how to read and write first-order sentences, we'll look at FOL's semantics (truth and consequences), and we'll select natural deduction inference rules for constructing proofs.

We could easily get tangled up in technical details here. As our goal is not to master the fine points of *First-Order Logic* but to understand how logic undergirds mathematical proof strategies, our treatment of FOL will be less formal than that of a typical logic text.

2.1.1 *Propositional Logic: Complete But Deficient*

We noted earlier that our Natural Deduction System for PL is both sound and complete. You may wonder, then, why PL's extension to FOL is necessary. It's because *Propositional Logic* is still deficient in certain respects. PL's completeness is tied to its truth-functional combinatorial character.

PL is *deductively complete*, in the sense that any consequence of a set of premises that logically follows due to the truth-functional forms of the sentences involved can be proved from them using PL's inference rules.

PL is also *expressively complete*. The set of logical connectives we selected— $\{\neg, \wedge, \vee, \rightarrow, \leftrightarrow\}$—is *complete*, in the sense that a sentence formulated using any truth-functional connective whatsoever is logically equivalent to one involving only these connectives. In fact, even fewer connectives suffice. We'll argue this pleasantly surprising result later in the text (see Section 7.5).

Taken together, we can say that PL's syntactic capabilities are sufficient to articulate anything that can be said *using truth-functional connectives*, and that PL's deductive capabilities can prove anything that follows from a set of premises *by virtue of the logical structure supplied by these connectives*.

Thus, these completeness results, while significant, are system dependent. In an absolute sense, *Propositional Logic* is incomplete. It cannot serve as the full underlying logic for mathematical argumentation—or everyday reasoning, for that matter. PL's expressive and deductive capabilities are limited.

To put it another way, *Propositional Logic* does the job it was designed for, but it doesn't do everything that needs to be done. Forks or chopsticks work fine for solid foods, but soup requires a spoon. Many intuitively valid argument forms will appear invalid when viewed from the standpoint of PL, because its focus is too coarse to recognize all the logical components. This problem is rectified by extending *Propositional Logic* to *First-Order Logic*.

© Springer Nature Switzerland AG 2019

C. Jongsma, *Introduction to Discrete Mathematics via Logic and Proof*,
Undergraduate Texts in Mathematics,
https://doi.org/10.1007/978-3-030-25358-5_2

2.1.2 *The Need For FOL: PL's Inadequacy*

Propositional Logic doesn't probe the logical structure of atomic sentences, so it cannot determine whether that contributes anything to the overall logical composition of a compound sentence. Consequently, inferences depending upon the logical form of the premises' most basic constituents will be missed by PL. To illustrate this, we can choose almost any mathematical argument.

✤ Example 2.1.1

Show that the following is invalid when rendered as an argument in PL:
For all real numbers, $x < y$ if and only if $x+r = y$ for some positive number r; $e < 3$;   therefore, $e + r = 3$ for some positive number r.

Solution

· Using PL sentence forms, we would write this as $P \leftrightarrow Q$, R; therefore S. Statements Q and S differ, even though they are closely related.
· This is not a valid PL argument though the given argument *is* valid. The conclusion logically follows from the premises because of their internal logical structure, as we will be able to demonstrate later. Logical connectives enter into the argument but PL is insufficient to establish its validity.

The logical structure residing within atomic mathematical sentences comes from how *equals* and the quantifiers *some* and *all* are used. In order for our logical system to deal with such structure, we'll need to develop a theory of identity and a theory of logical quantifiers. We'll also have to take into account other internal structures—the more strictly linguistic components— in order to fully symbolize the logical structure of a sentence.

2.1.3 *The Symbolic Vocabulary of First-Order Logic*

Compared with sentences we use every day, mathematical sentences are highly symbolic. These symbols make mathematics less accessible to some, so it's good to avoid mathematical notation when words serve as well. Nevertheless, mathematical symbolism simplifies expressions, is essential for concise exposition, and even aids reasoning. Imagine how difficult algebra would be using only words—as was done prior to the seventeenth century!

FOL takes symbolic representation to the extreme. Transcribing a sentence into first-order notation transforms it into a string of symbols, erasing all traces of natural language. Even mathematics textbooks don't go this far, but taking this approach, for now, will help us better understand the internal logical structure of sentences, a prerequisite for making deductions with them.

Just as in mathematics, *First-Order Logic* uses symbols to denote *objects*, *operations*, *functions*, and *relations*, but it also uses them for *properties*, such as numbers being rational, which mathematics usually states in words. FOL's alphabet also includes logical symbolism, such as connectives from PL and quantifier symbols. Punctuation is done with parentheses, just like in PL.

We will use lowercase letters, either constants $(a, b; m, n)$ or variables (x, y, z), to represent individual objects. Properties will be symbolized by uppercase letters $(A, B; P, Q)$, as will relations (R, S). Pa and Qx indicate that object a has property P and that x has property Q. Rab and $Sxyz$ denote that a and b are in relation R and that x, y, z are in relation S.

The standard symbol for the *universal quantifier* is $\forall$, which is read *for all*. Think of it as an upside-down A, the first letter of All. $\forall x Qx$ asserts that property Q holds for all objects x of a certain sort.

The *existential quantifier* symbol $\exists$ is a backward E (as in $\exists$xists). $\exists x Qx$ asserts the existence of an object x with property Q. It can be read as *there is an x such that Qx* or *for some x Qx*.

A third quantifier symbol, $\exists!$, is read *there exists a unique*. The *unique existence quantifier* can be defined in terms of $\exists$ and logical connectives or in terms of both $\exists$ and $\forall$ along with logical connectives (see Section 2.4). We could therefore eliminate it without any loss of expressive power, but we'll retain it to keep symbolic formulations of uniqueness statements simpler.

2.1.4 *Preparing to Read and Write FOL Sentences*

Let's now see how to read and write FOL sentences. Think of FOL as an artificial foreign language used for logically analyzing sentences. If you feel at home in algebra or catch on quickly to using a computer language, FOL syntax won't be difficult. You may write some confused-looking sentences at first, but with practice, you'll soon become fluent with first-order notation.

To formulate propositions of a mathematical theory in a specialized first-order language, we'll need an *interpretation key* to make the *overall context* and *specific meaning* of the symbols clear.

Such a key first of all identifies the intended *universe of discourse U*, the *non-empty set of objects* being considered, along with their *properties, operations, functions*, and *relations*. Variables and quantifiers are then interpreted relative to U. Universal statements become assertions about all the objects in U; existential statements claim the existence of an object in U.

Second, an *interpretation key* assigns meaning to symbols for the various mathematical entities mentioned. It's not crucial to include symbols having a familiar meaning, but we must stipulate meanings for any nonstandard symbols. *Logical connectives will always have a fixed interpretation.*

Given a universe of discourse and a key, we can combine the symbols to construct FOL sentences. Writing a sentence in a first-order format requires both an intuitive knowledge of how to compose well-formed sentences and a knack for analyzing sentences' internal logical structure. Constructing first-order sentences that mean what we want them to say takes practice.

It's easier to go the other way, translating FOL sentences into informal mathematical statements. Given a sentence and an interpretation key, we can first translate the sentence into a stilted literal formulation and then reformulate it in a more idiomatic natural-language equivalent. This process is similar to translating a sentence from a foreign language into one's native tongue.

2.1.5 Reading Singly-Quantified Sentences

We'll begin our study of *First-Order Logic* formulations by considering a few examples that involve a single quantifier.

✤ Example 2.1.2

Translate the sentence $\forall n(Pn \wedge n \neq 2 \to On)$ into mathematical English, using the following interpretation key:

KEY: $U = \{1, 2, 3, \ldots\}$

Pn: n is prime On: n is odd

Solution

· As an abstract first-order statement, we can read this as *For all n, if Pn and n is not equal to 2, then On*.

· With the interpretation key, we can give this a more meaningful reading. The inner sentence is *If n is prime and n is not equal to 2, then n is odd.* This is being asserted in a universal way: *For all positive integers n, if n is prime and is not equal to 2, then n is odd.*

· A more idiomatic translation of the sentence would be *All prime numbers different from 2 are odd.*

✤ Example 2.1.3

Translate $\neg \exists x(x \neq 0 \wedge x + x = x)$ into mathematical English, using the following interpretation key:

KEY: $U = \mathbb{R}$, the set of all real numbers

Solution

· All mathematical symbols here are to be taken in their usual sense.

· This sentence is a negated existential sentence: *It is not the case that there exists a real number x such that $x \neq 0$ and $x + x = x$.*

· To indicate that the inner sentence is governed by the existential quantifier, the words *such that* are inserted after the quantifier phrase. This wording is typical of existential sentences; *such that* is not used after a universal quantifier (see the last example).

· We can push the negation further into the sentence to give it a more natural reading: *There are no real numbers x such that $x \neq 0$ and $x + x = x$.*

· An even more compact idiomatic translation is *No non-zero real numbers x satisfy $x + x = x$.* Or, putting it completely in words, *No non-zero real numbers remain unchanged when added to themselves.*

2.1.6 Writing Singly-Quantified Sentences

In transcribing a sentence from mathematical English into the formal language of *First-Order Logic*, we'll use the following approach:

1) Stipulate a universe of discourse for the objects under consideration.
2) Give an interpretation key for the objects, properties, operations, functions, and relations mentioned.

3) Determine the logical structure of the sentence and write it in a way that clearly indicates its logical structure.
4) Translate the sentence into the symbolism of FOL.

The next two examples illustrate this procedure.

✦ Example 2.1.4

Analyze the logical structure of the sentence *If a function is differentiable, then it is continuous*. Then symbolize it with FOL notation.

Solution

· This sentence is less complex than the last example, though it may seem more difficult since we're now translating into logical symbolism.
· We need to tease out the underlying logical structure of this sentence.
· We'll first specify a domain of discourse. Since this sentence usually occurs in an introductory calculus course, we'll assume that context. The objects under consideration, then, are real-valued functions of a real variable.
· Two properties of functions are mentioned: being differentiable and being continuous. These could be spelled out by definitions involving limits, but that's not needed here. We'll simply choose one-place predicate letters to stand for these properties.

KEY: U = the set of all real-valued functions of a real variable
　　Df: f is differentiable 　　　　Cf: f is continuous

· Although the proposition seems to refer to a single function, its obvious intent is more general, namely, *Every differentiable function is continuous*. This makes the universal quantifier explicit, but it still hides the conditional nature of the inner sentence. A formulation that makes both aspects apparent is the wordier *For every real-valued function f of a real variable, if f is differentiable, then f is continuous*.
· This now makes the logical structure clear.

$FORMULATION$: $\forall f(Df \to Cf)$

The key to transcribing sentences into FOL symbolism, as this example illustrates, is to recognize the underlying logical form of the sentence. Mathematical propositions formulated in ordinary English often leave some logical structure implicit. If you analyze and retranslate the sentence, verbally or mentally, into a form that better reveals its logical deep structure, you're more than half-way done. This is an art that improves with practice.

✦ Example 2.1.5

Transcribe the second premise and the conclusion of Example 1 into FOL notation: $e < 3$, and *there is a positive real number r such that $e + r = 3$*.

Solution

· Writing the second premise is trivial: it's just as given: $e < 3$.
· A universe of discourse compatible with Example 1's conclusion is the set of positive real numbers, $\mathbb{R}^+$. This U, however, makes it impossible to formulate the argument's first premise, which refers to all real numbers.

· So we'll choose $\mathbb{R}$ as our universe of discourse and restrict our attention
to positive real numbers manually, as it were, with standard symbolism
inside the sentence. Our interpretation key, therefore, consists of simply
stipulating a universe of discourse.

   KEY: $U = \mathbb{R}$, the set of real numbers

· The conclusion claims the existence of a real number that is positive and
whose sum with e is 3. Both of these must hold, so we conjoin them.

· The two sentences are thus written as follows, using standard notation:

   FORMULATION: *Premise*: $e < 3$;   *Conclusion*: $\exists r(r > 0 \wedge e + r = 3)$

2.1.7 Using Restricted Quantifiers

Mathematicians rarely stipulate a universe of discourse. Sometimes U is
understood from the context, but often the type of objects being considered
is symbolized within the sentence itself, much like we did in the last example to specify a positive addend. As this leads to longer, more cumbersome
sentences, we can use *restricted quantifiers* to simplify our formulations. For
example, instead of $\exists r(r > 0 \wedge e + r = 3)$, we can write $(\exists r > 0)(e + r = 3)$,
where $\exists r > 0$ can be read as *there is a positive real number* r. Restricted-
quantifier notation is especially useful when mathematical sentences involve
multiple quantifiers (see Example 10 and Exercise 18).

A restricted quantifier could have been used in Example 4, too. Its formu-
lation would be $(\forall f \in \mathcal{D})\,Cf$, where Cf still means f is continuous, $\mathcal{D}$ stands
for the set of differentiable real-valued functions, and $\forall f \in \mathcal{D}$ means *for all
functions in the set* $\mathcal{D}$.

Let's generalize the last examples. A universal sentence that asserts $P(x)$[1]
for all objects x in a set S is of the form $(\forall x \in S)P(x)$, while an existential
sentence that says some x in S satisfies $P(x)$ is of the form $(\exists x \in S)P(x)$. Put
into expanded form, the two inner sentences are quite different. The universal
sentence becomes the quantified *conditional* $\forall x(x \in S \rightarrow P(x))$, while the
existential sentence becomes the quantified *conjunction* $\exists x(x \in S \wedge P(x))$.

This is typical. *Universal sentences usually contain the connective* $\rightarrow$ *while
existential sentences involve* $\wedge$. It is important to keep these two straight and
not interchange the two main interior connectives in the expanded format.

Using a restricted quantifier avoids the problem of choosing a connective,
because none is present—it gets submerged in the restricted quantifier. There
may be times, though, when you'll need to expand a restricted quantifier in
order to work further with the given sentence. Knowing how to treat these
two different cases will help you take advantage of the convenient notation
of restricted quantification while being aware of the logical structure of the
sentences they abbreviate.

---

[1] We will use $P(x)$, with parentheses, to symbolize a *statement*, possibly complex. $P(x)$
might indicate that x has a *property* P, but it need not—that can be symbolized as Px.

2.1.8 *Multiply-Quantified Sentences*

So far our examples have contained only a single quantifier, but mathematical sentences typically have more than one. When all the quantifiers are the same, there is little problem translating them, but when both universal and existential quantifiers occur, the potential for confusion increases.

✤**Example 2.1.6**

Translate $\forall x \forall y (x \cdot y = 0 \leftrightarrow x = 0 \lor y = 0)$ into mathematical English. Let the universe of discourse be $\mathbb{R}$, the set of real numbers.

Solution

Since there are two universal quantifiers, x and y are completely general. In words, this sentence says *the product of two real numbers is zero if and only if at least one of them is zero.*

✤**Example 2.1.7**

Translate $\forall n (On \leftrightarrow \exists k (n = 2k + 1))$ into mathematical English.
  KEY: $U = \{0, 1, 2, \ldots\} = \mathbb{N}$, the set of natural numbers
      On: n is odd

Solution

· This universally quantified biconditional could be taken as the definition of being odd. A literal translation is *for all natural numbers n, n is odd if and only if there is some natural number k such that n is equal to $2k + 1$.*
· Putting this into more idiomatic English: *a natural number is odd if and only if it is one more than an even number.*

The two different quantifiers in the last example were fairly easy to parse correctly, but this is not always the case.

✤**Example 2.1.8**

Translate $\forall x \exists \bar{x} (x + \bar{x} = 0 = \bar{x} + x)$ into mathematical English.
  KEY: $U = \mathbb{R}$, the set of real numbers

Solution

· The key only identifies a universe of discourse. All symbols should thus be taken in their usual sense relative to real-number arithmetic. The double equation abbreviates the conjunction $(x + \bar{x} = 0) \land (0 = \bar{x} + x)$.
· Read literally, this sentence says *For every real number x there is a real number $\bar{x}$ such that $x + \bar{x} = 0$ and $0 = \bar{x} + x$.*
· The key to putting this into good mathematical English is properly placing the quantifiers. To abbreviate the inner sentence, using standard terminology, let's call $\bar{x}$ an *additive inverse* of x, since their sum in either order yields the *additive identity* 0.
· Do we now translate the sentence as *Every real number has an additive inverse*, or as *There is an additive inverse for every real number?*

- These two options may seem to say the same thing, but there's an important difference. The first sentence says that any real number x has some associated $\bar{x}$ as its additive inverse. Strictly interpreted, the second sentence says some real number $\bar{x}$ is the additive inverse for *every* real number x. In the first case, $\bar{x}$ may change, depending on x; but in the second case $\bar{x}$ is the same for all x. The second sentence thus translates $\exists \bar{x} \forall x(x + \bar{x} = 0 = \bar{x} + x)$, where the quantifiers have been interchanged.
- The correct formulation for the given sentence, therefore, is *Every real number x has an additive inverse $\bar{x}$.*

Our final examples illustrate the reverse process of translating from informal mathematical English into the formalism of *First-Order Logic*. Again, this is more complicated when multiple quantifiers are involved. To check your work, carefully translate your answer back into an informal statement.

❖ **Example 2.1.9**

Give a formal definition of the number-theoretic relation *divides*: *d divides a, denoted by $d \mid a$, if and only if a is a multiple of d.*

Solution

- This definition is intended to hold for all integers, not only for two numbers, so our FOL sentence will begin $\forall d \forall a$, followed by the defining clause.
- The definition says $d \mid a$ *if and only if a is a multiple of d.* Rather than use a special symbol to denote the relation *is a multiple of*, this time we'll spell it out: *a is a multiple of d if and only if $a = m \cdot d$ for some m.*
- Proceeding further, the phrase *for some m* indicates an existential quantifier. We know from arithmetic that this m is unique (when $d \neq 0$), but since uniqueness isn't being asserted, we'll use an ordinary existential quantifier. The position and scope of $\exists m$ are not difficult to determine: $\exists m$ governs the equation $a = m \cdot d$.
- Putting all these things together, we obtain the following:

KEY: $U = \mathbb{Z}$, the set of integers

FORMULATION: $\forall d \forall a(d \mid a \leftrightarrow \exists m(a = m \cdot d))$

Our final example is the famous parallel postulate of Euclidean geometry, in the version made popular by John Playfair around 1800. This example also introduces the idea of variables of different *sorts*.

❖ **Example 2.1.10**

Put into FOL notation: *Given a line and a point not on the line, there is a unique line passing through the given point parallel to the given line.*

Solution

- There are two sorts of objects here: points and lines. To simplify matters, lowercase letters like l and m will stand for lines, and uppercase letters like P will denote points. Quantified variables are then restricted *by their case* instead of by using a restricted quantifier or a predicate qualifier.

· The universe of discourse contains two sub-universes or sorts of objects—points and lines. Since *Playfair's Postulate* is part of plane geometry, we'll take U to be all points and lines in some unspecified plane.

· Some lines are in the binary relation *is parallel to*, which we'll symbolize using the suggestive notation $\parallel$. Some points and lines are in the binary relation *lies on*. Thinking of lines as sets of points, we'll use the set-theoretic notation $P \in l$ to indicate that *point P lies on line l. Point P does not lie on line l* is then symbolized by $P \notin l$.

· The sentence needs quantifiers. The original sentence mentions *a line* and *a point*, but the obvious intent is to state something about *any* line and *any* point, so it will contain the universal quantifier $\forall$ twice. It also claims the existence of a unique line, so we'll use the quantifier $\exists!$.

· And the postulate needs some logical connectives. The phrase *not on* obviously involves negation. An *and* is present in the original sentence, but this merely indicates that two universal quantifiers are being successively asserted; it does not signify a genuine propositional connective.[2] We simply write one quantifier after the other, as $\forall l \forall P$.

· The sentence does involve a conjunction, though it is not indicated by *and*. The unique line must satisfy two properties simultaneously—it must pass through the given point, *and* it must be parallel to the given line.

· One more connective could appear in our formal version, because we still need to restrict point P to those not on line l. This is done for universal quantifiers, as noted earlier, by means of a conditional sentence with the restriction occurring in its antecedent. Alternatively, we can use a restricted quantifier to shorten our formulation.

· We're now ready to make a formal translation. Note the type and order of the quantifiers, and convince yourself that this is what's needed.

KEY: $U = \{x : x$ is a point or x is a line in some common plane$\}$
  $P \in l$: point P lies on line l $m \parallel l$: line m is parallel to line l

RESTRICTED FORMULATION: $\forall l (\forall P \notin l) \exists! m(P \in m \wedge m \parallel l)$

EXPANDED FORMULATION: $\forall l \forall P(P \notin l \rightarrow \exists! m(P \in m \wedge m \parallel l))$

The examples in this section begin to indicate the logical complexity of mathematical statements. The FOL versions we've seen look more complex than their corresponding English formulations, and they certainly appear more artificial. This is no doubt because you are more familiar with English, but it is also because of the logical complexity that informal sentences typically conceal. A sentence's logical structure does not become explicit until the sentence is fully analyzed and symbolized as we have done, using the tools of *First-Order Logic*. You will have to learn to live, at least temporarily, with the linguistic complexity that arises from using first-order formulations.

---

[2] *And* will indicate a logical connective, however, if the properties *being a point* and *being a line* are symbolized using a qualifying predicate letter. See Exercise 45.

EXERCISE SET 2.1

Exercises 1–4: Translating Singly-Quantified Sentences

*Translate each of the following sentences into informal mathematical English.
Then tell whether it is true or false under the given interpretation.*
  SYMBOLISM AND KEYS:
    $\mathbb{N}$ *is the set of natural numbers;* $\mathbb{Z}$ *is the set of integers;* $\mathbb{Q}$ *is the set of
    rational numbers; and* $\mathbb{R}$ *is the set of real numbers.*

2.1.1. $\forall x(x + 0 = x)$
  Key: $U = \mathbb{R}$

2.1.2. $\forall x(0 \div x = 0)$
  Key: $U = \mathbb{Q}$

2.1.3. $(\exists x \neq 3)(|x - 2| = 1)$
  Key: $U = \mathbb{Z}$

2.1.4. $\exists x(x > x^2)$
  Key: $U = \mathbb{N}$

Exercises 5–18: Translating Multiply-Quantified Sentences

*Translate each of the following sentences into informal mathematical English.
Then tell whether it is true or false under the given interpretation.*
  SYMBOLISM AND KEYS: $\mathbb{N}$, $\mathbb{Z}$, $\mathbb{Q}$, *and* $\mathbb{R}$ *are as for Exercises 1–4.*
  The symbol $\in$ *denotes set membership:* $x \in S$ *means* x *belongs to set* S.

2.1.5. $\forall x \forall y \exists z(x + z = y)$
  Key: $U = \mathbb{Z}$

2.1.6. $\forall x(x \neq 0 \rightarrow \exists y(x \cdot y = 1))$
  Key: $U = \mathbb{Q}$

2.1.7. $\forall x \forall y \forall z(x \cdot (y+z) = x \cdot y + x \cdot z)$
  Key: $U = \mathbb{R}$

2.1.8. $\forall x \forall y \forall z(x < y \rightarrow x \cdot z < y \cdot z)$
  Key: $U = \mathbb{R}$

2.1.9. $\forall x \exists y(y \leq x)$
  Key: $U = \mathbb{N}$

2.1.10. $\exists y \forall x(y \leq x)$
  Key: $U = \mathbb{N}$

2.1.11. $\forall x \exists y(y > x)$
  Key: $U = \mathbb{N}$

2.1.12. $\exists y \forall x(y > x)$
  Key: $U = \mathbb{N}$

2.1.13. $\forall x \forall y(x \notin \mathbb{Q} \land y \notin \mathbb{Q} \rightarrow x \cdot y \notin \mathbb{Q})$
  Key: $U = \mathbb{R}$

2.1.14. $\forall p(Pp \leftrightarrow \forall m \forall n(p = m \cdot n \rightarrow m = 1 \lor n = 1))$
  Key: $U = \mathbb{N}$; Pn: n is prime

2.1.15. $\forall a \forall b \forall p(Pp \land p \mid ab \rightarrow p \mid a \lor p \mid b)$
  Key: $U = \mathbb{N}$; Pn: n is prime; $x \mid y$: x divides y

2.1.16. $\forall P \forall Q(P \neq Q \rightarrow \exists! l(P \in l \land Q \in l))$
  Key: see Example 10

2.1.17. $\forall l \forall m(l \parallel m \leftrightarrow \neg \exists P(P \in l \land P \in m))$
  Key: see Example 10

2.1.18. $L = \lim\limits_{x \to a} f(x) \leftrightarrow (\forall \epsilon > 0)(\exists \delta > 0)(\forall x)(0 < |x-a| < \delta \rightarrow |f(x)-L| < \epsilon)$
  Key: $U = \mathbb{R}$

Exercises 19–31: Writing Singly-Quantified Sentences

*Determine whether each of the following sentences is true or false. Then
translate it from mathematical English into FOL notation. For each sentence,
choose a universe of discourse and interpret any nonstandard symbols.*

2.1.19. All equilateral triangles are equiangular.

2.1.20. No scalene triangle is isosceles.

2.1.21. Some rectangles are squares.

2.1.22. Some rectangles are not squares.

2.1.23. Not all triangles are congruent.

2.1.24. No odd number is composite.

2.1.25. Some prime numbers are even.

2.1.26. All isosceles right triangles are equilateral.

2.1.27. No real number satisfies the equation $x^2 + 1 = 0$.

2.1.28. Some real numbers do not satisfy the inequality $a^2 < a$.

2.1.29. Every natural number is either even or odd.

2.1.30. All natural numbers are both integers and rational numbers.

2.1.31. There is exactly one even prime number.

Exercises 32–33: True or False
Are the following statements true or false? Explain your answer.

2.1.32. Any consequence that logically follows from a set of premises can be proved from them using *Propositional Logic*'s inference rules.

2.1.33. Universal sentences typically contain an inner conjunction when formulated with the machinery of *First-Order Logic*.

Exercises 34–44: Multiply-Quantified Mathematical Sentences
Determine whether each of the following sentences is true or false. Then translate it from mathematical English into FOL notation. For each sentence, choose a universe of discourse and interpret any nonstandard symbols.

2.1.34. Real numbers multiplied together in either order give the same result.

2.1.35. For any two positive real numbers a and b, there is a natural number n such that $na > b$.

2.1.36. The sum of two rational numbers is rational.

2.1.37. All positive integers can be factored into a product of two primes.

2.1.38. Every natural number is smaller than some other natural number.

2.1.39. There is an integer that is less than all other integers.

2.1.40. Between any two distinct real numbers there is a rational number.

2.1.41. For integers a, b, and c, if $a \neq 0$, then $ab = ac$ if and only if $b = c$.

2.1.42. A function f is monotone increasing on the set of real numbers if and only if whenever $x_1 < x_2$, $f(x_1) < f(x_2)$.

2.1.43. If $0 \leq a_n \leq b_n$ for all n, then $\sum a_n$ converges if $\sum b_n$ does.

2.1.44. There are exactly two real solutions to $x^2 - 1 = 0$.

2.1.45. Formulate *Playfair's Parallel Postulate* (see Example 10) using symbols to stand for the properties of being a point and being a line, qualifying the variables inside the sentence. Compare your result with the formulations given earlier, which used variables of different sorts.

2.2 First-Order Logic: Syntax and Semantics

First-Order Logic intensifies the mathematical mode of expression you've probably used for years, making sentences fully symbolic. Summarizing the general structure of the sentences we've encountered so far, we can say:

> *Mathematical sentences affirm or deny that properties or relations hold of or among some or all objects of certain types.*

We could explore FOL's syntactic formation rules in detail,[3] but we'll only lay the basic groundwork needed to explain *First-Order Logic's* significance for mathematical proof construction.

2.2.1 *FOL Syntax*: *Open and Closed Sentences*

FOL sentences come in two varieties—open or closed, depending on how variables appear in them. A variable is *free* if and only if it does not lie within the scope of a quantifier; otherwise it is *bound*. *Open* sentences are ones that have a free variable. All occurrences of x in $\exists x(x+x = x)$ are bound. In $\forall z(z + \bar{z} = 0)$, z is bound, but $\bar{z}$ is free. As written, $\forall n Pn \wedge n \neq 2 \rightarrow On$ has one bound occurrence of n and two free occurrences. To quantify more occurrences, we must use parentheses. For example, $\forall n(Pn \wedge n \neq 2 \rightarrow On)$ is the universally quantified sentence of Example 2.1.2.

A *closed sentence* is one that is not open: it has no free variables. It makes a claim that is either true or false when its nonlogical symbols are assigned meanings by an interpretation key. Open sentences generally have no definite truth value, even after being interpreted.

Calling both open and closed formulas *sentences* is a conscious departure from earlier usage. According to Section 1.2's definition, only closed sentences count as sentences, since only they are true or false. However, in order to use PL's inference rules for open as well as closed sentences (in keeping with mathematical practice), we must stretch the meaning of a sentence to this new case. Open sentences are complete, if indefinite, statements.

✤**Example 2.2.1**

Give examples of closed and open sentences and tell whether they're true.

Solution
- $\forall x \exists \bar{x}(x + \bar{x} = 0 = \bar{x} + x)$ of Example 2.1.8 is a closed formula. It is true for ordinary addition when $\mathbb{R}$ is the universe of discourse, but it is false if $\mathbb{N}$ is the universe of discourse, for 1 has no natural number additive inverse.
- The subformulas $\exists \bar{x}(x + \bar{x} = 0 = \bar{x} + x)$ and $x + \bar{x} = 0 = \bar{x} + x$ are both open sentences, having free variable occurrences. Neither one is true or false if either $\mathbb{R}$ or $\mathbb{N}$ is the universe of discourse and $+$ is ordinary addition. However, if $U = \{0\}$ and $+$ is ordinary addition, both are true.
- This example shows that truth values for sentences are *interpretation-dependent*, something that's beneficial for logic, as we'll see next.

---

[3] We do this for the simpler system of PL in Section 3.3, where we look at *String Theory*.

2.2.2 *FOL Semantics: Interpretations and Truth Values*

Closed first-order sentences have an abstract generic meaning that only becomes concrete when we create an interpretation key. This may seem like a liability when symbolizing mathematical sentences, but it's actually an asset, both for logic and mathematics. Having multiple interpretations is important for FOL's semantics; it's also a feature of contemporary branches of mathematics, such as modern geometry, abstract algebra, and advanced analysis.

To probe FOL sentences' logical connections, we can't use truth tables, as we did for PL, assigning truth values to atomic sentences without regard for what they say. We now need to consider what the atomic components of sentences assert relative to one another. There may be logical relations among sentences that depend upon their internal logical structure, so some truth-value assignments may not be possible. For example, if $\forall x (0 \leq x)$ is true, then $0 \leq a$ must also be true for any constant a, regardless of the intended meaning of $\leq$ or the value of a. Similarly, if $\exists x (x \neq 0)$ is true, $\forall x (x = 0)$ must be false, because these atomic sentences contradict one another.

In assigning truth values to first-order sentences, then, we must know *what* the atomic sentences *abstractly assert* about a potential universe of discourse. We'll have to delve more deeply into sentence semantics than before, using interpretations, as we did at the outset (see Section 1.1).

2.2.3 *Interpreting Closed Sentences of FOL*

To define properties such as logical truth or relations such as logical implication, we'll consider all well-formed formulas as *uninterpreted—sentences abstracted from all particular nonlogical meaning.*

There are several phases to interpreting first-order sentences. First, we choose a non-empty *universe of discourse U*. We then *assign meanings* to symbols in a way that respects their category types—constants as distinguished objects in U, property symbols as properties of U's members, predicate symbols as relations, n-ary function symbols as functions of n variables, and so on. Logical symbols have a fixed meaning: $\wedge$, $\vee$, $\neg$, $\rightarrow$, and $\leftrightarrow$ signify standard truth-functional connectives; $=$ denotes the identity relation; and quantifiers $\forall$, $\exists$, and $\exists!$ are assigned their usual meaning relative to U.

Given this process, any closed sentence will have a unique truth value—it's true if and only if what it asserts is the case for that interpretation.

✦ **Example 2.2.2**

Give interpretations, with their truth values, for the following sentences:
$$0 < 1; \quad 1 < (1 + 1); \quad 1 \cdot 1 = 1; \quad \exists x (x \cdot x < x); \quad \forall x (x \neq 0 \rightarrow 0 < x).$$

Solution

· These are statements of ordinary arithmetic. To interpret them, we'll specify a universe of discourse and assign meanings to 0, 1, $+$, $\cdot$, and $<$.

· The usual meaning of these sentences is the following:

KEY: $U = \mathbb{N} = \{0, 1, 2, \ldots\}$

0: *zero*	1: *one*	
+: *addition*	· : *multiplication*	< : *less-than*

- We can abbreviate this by saying *Let the symbols have their usual meaning.* Writing everything out, though, highlights the fact that symbols do not have inherent meaning—they must be interpreted to communicate their intended meaning. That some symbols have a fixed meaning due to long and familiar usage should not obscure this point.
- Under this interpretation, *sentence four is false, while the others are true.*
- For a second interpretation, change the universe of discourse to $\mathbb{R}$, the set of all real numbers, and interpret the symbols as before, relative to $\mathbb{R}$.
- Now *the first four sentences are true and the last one is false.* The first three are obviously true, and the fourth is true because $\frac{1}{2} \cdot \frac{1}{2} < \frac{1}{2}$. A *counterexample* showing that the fifth sentence is false is given by letting $x = -1$: -1 is not zero, and -1 is also not greater than 0.

The next example illustrates the potent technique of *finite-universe interpretation.* Here we take a finite universe and specify which abstract properties and relations we want to consider true. The final result may look contrived (it is), but artificial constructions like this reveal important logical relationships.

✦ Example 2.2.3

Give a finite-universe interpretation for Example 2's sentences (repeated below) that makes the second one false and the others true.

$$0 < 1; \quad 1 < (1+1); \quad 1 \cdot 1 = 1; \quad \exists x(x \cdot x < x); \quad \forall x(x \neq 0 \rightarrow 0 < x).$$

Solution

- Since the only constants are 0 and 1, we'll construct an interpretation whose universe of discourse contains only two elements. We need at least two *distinct* elements denoted by 0 and 1, otherwise the second sentence will necessarily be true if the first one is.
- We'll now stipulate what we want to be true about 0 and 1 regarding $<$, $+$, and $\cdot$ so that all sentences will have the desired truth values.
- We want the first and third sentences to be true about 0 and 1. If $0 < 1$, this satisfies the last sentence as well, since U only contains 0 and 1.
- To make the fourth sentence true, we must either take $0 \cdot 0 < 0$ or $1 < 1$. Let's do the latter. Then to falsify the second sentence, we must choose $1 + 1 = 0$ and take $1 \not< 0$.
- Putting all of this together gives us the following:

    KEY: $U = \{0, 1\}$

    $<$: Let $0 < 1$, $1 < 1$ define the wedge-relation $<$. Thus, $1 \not< 0$.

    $+$: Let $0 + 0 = 0$, $0 + 1 = 0$, $1 + 0 = 0$, and $1 + 1 = 0$ define $+$.

      [Think of $+$ as always producing the smallest element in U, i.e., 0].

    $\cdot$: Let $0 \cdot 0 = 0$, $0 \cdot 1 = 1$, $1 \cdot 0 = 1$, and $1 \cdot 1 = 1$ define operation $\cdot$.

      [Think of $\cdot$ as always giving the larger of the two elements].

- This makes the second sentence false, since $1 \not< 0$, but the others are true.

2.2.4 *Interpreting Open Sentences of FOL*

One issue we must address is how to work with open sentences. As noted above, open sentences are typically neither true nor false—free variables have no assigned meaning. This is not a limitation of the interpretation process—*free variables are supposed to be indeterminate.*

But this puts us in a predicament: we need to assign truth values to open sentences in order for them to participate in FOL's semantics (implication, equivalence) and derivations. To resolve this issue, we'll introduce the idea of an *extended interpretation*, which makes open sentences determinate.

Suppose $P(x)$ is an open sentence with free variable x. Beginning as before, we choose a universe of discourse and assign meanings to the various symbols. This interpretation leaves $P(x)$ indeterminate because it has an uninterpreted variable x. So we'll next *extend the interpretation* by assigning x some object in the universe of discourse. This extended interpretation gives the sentence a definite meaning, so its truth can now be determined.

More generally, an *extended interpretation of an FOL sentence is an interpretation of the sentence together with a meaning assignment to all of its free variables.* Ordinarily, some of these extensions will yield true statements about the structure, while others will be false. An extended interpretation that makes an open sentence true is said to *satisfy* the sentence.

✤**Example 2.2.4**

Discuss the truth values of the following sentences in the ordinary theory of natural number arithmetic under different extended interpretations:
$$0 < 1; \quad w < (w + w); \quad 1 \cdot x = 1; \quad y \cdot y < y; \quad z \neq 0 \to 0 < z.$$

Solution

Here we're assuming the intended interpretation of Example 2 as our base. Let's see what effect different extensions of this interpretation have.

1) $0 < 1$ is *true* under any extension of this interpretation: there are no free variables to assign a meaning.
2) The truth value of $w < (w + w)$ is *indefinite*: it is *false* if w is interpreted as 0, but is *true* otherwise.
3) $1 \cdot x = 1$ also has an *indefinite* truth value: it is *true* if and only if x is assigned the value 1.
4) $y \cdot y < y$, though open, is *always false*: no natural number value can be assigned to y that will satisfy this sentence.
5) $z \neq 0 \to 0 < z$, on the other hand, is *always true*: 0 is the smallest natural number, so every extended interpretation here satisfies this sentence.

2.2.5 *Truth and Falsehood in FOL*

Using extended interpretations, we can now discuss FOL's semantics. The central notions of *true* and *false* for FOL sentences are defined as follows:

Definition 2.2.1: *Truth Values for FOL Sentences*

a) *A sentence is **true under an interpretation** if and only if it is true under every possible extension of the interpretation.*

b) *A sentence is **false under an interpretation** if and only if it is false under every possible extension of the given interpretation.*

c) *A sentence is **indeterminate under an interpretation** if and only if it is true under one extension and false under another.*

This definition merely sharpens our earlier notion of being true under an interpretation. For closed sentences, nothing new is added—they remain either true or false under an extended interpretation. But we now have a way to talk about open sentences being true or false relative to an interpretation.

When a sentence is true under an interpretation, the mathematical structure involved is called a *model*[4] for that sentence.

Definition 2.2.2: *Logically True/False, Logically Indeterminate*

*A sentence is **logically** ...*

a) ***true** if and only if it's true under all interpretations.*

b) ***false** if and only if it's false under all interpretations.*

c) ***indeterminate** otherwise.*

Intuitively, a sentence is logically true/false if and only if it is true/false irrespective of the meaning of its nonlogical terms. Thus, any extended interpretation always gives a sentence that is true/false. To decide whether a sentence is logically true/false, though, we do not have to examine all possible structures and interpretations. We only need to argue in some way that it *must be true/false of all structures under all extended interpretations.*

✤ **Example 2.2.5**

Determine the logical status of the following sentences.

a) $Px \vee \neg Px$

b) $\forall x Px \wedge \neg Pa$

c) $\forall x Px \rightarrow \exists x Px$

d) $\exists x \forall y Pxy$.

Solution

a) Any extended interpretation of $Px \vee \neg Px$ yields an instance of LEM, which is a tautology, so this sentence is logically true. Its universal closure $\forall x(Px \vee \neg Px)$ is logically true for the same reason.

b) $\forall x Px \wedge \neg Pa$ is logically false. If everything in universe U has property P, whatever that is, then any element a in U must also have property P. But since a doesn't have property P, not all elements do.

c) $\forall x Px \rightarrow \exists x Px$ is logically true. To show this, we must argue that $\exists x Px$ is true whenever $\forall x Px$ is. If $\forall x Px$ is true, then all members of universe U have property P, however P is interpreted. Certainly, then, some member of U has property P, i.e., $\exists x Px$ is true. (This depends upon the

---

[4] Unfortunately, logical usage of the term *model* is at odds with how applied mathematicians use this term. In logic a model is the mathematical *structure* satisfying the theory; in applied mathematics, the *theory* is the model for the situation.

fact, unstressed until now, that *only non-empty sets* are accepted—by convention—as *bona fide universes of discourse.*)

d) $\exists x \forall y Pxy$ is logically indeterminate: we can give it two interpretations, one making it true, and another making it false.

Let $U = \{2, 3, 4, \dots\}$ for both cases.

If Pxy means *x is a prime factor of y*, the sentence is false. Integers greater than 1 do have prime factors, but not the same one, i.e., while $\forall y \exists x Pxy$ is true, $\exists x \forall y Pxy$ is not.

The sentence $\exists x \forall y Pxy$ becomes true, however, if we take Pxy to mean $x \leq y$, for 2 is less than or equal to all members of U.

2.2.6 *Logical Implication in FOL*

In Section 1.1 we said that *an argument is valid if and only if every interpretation making its premises true also makes its conclusion true.* This continues to hold in *First-Order Logic* for closed sentences, but to admit open sentences into our arguments, we must assume our broader sense of interpretation.

Definition 2.2.3: *Logical Implication for FOL Sentences*
*A set of sentences $\mathcal{P}$ **logically implies** sentence **Q**, written $\mathcal{P} \models$ **Q**, if and only if every interpretation making $\mathcal{P}$ true also makes **Q** true.*

As before, we do not have to consider all possible extended interpretations to demonstrate logical implication. We can use a generic argument.

✦ **Example 2.2.6**
Justify the following implication claims, involving only closed sentences.
a) $\forall x(Px \rightarrow Qx), \exists x Px \models \exists x Qx$ b) $\forall y \exists x(x < y) \not\models \forall x(x \neq 0 \rightarrow 0 < x)$

Solution
These are closed sentences, so we only need to consider non-extended interpretations.

a) Suppose $\forall x(Px \rightarrow Qx)$ and $\exists x Px$ are both true for some universe of discourse U and some properties P and Q.
Then some element a in U has property P (premise 2): Pa is true.
Furthermore, since any element of U having property P also has property Q (premise 1), this holds for a: $Pa \rightarrow Qa$ is true.
Hence (by *Modus Ponens*) a must have property Q: Qa is true.
So, the conclusion $\exists x Qx$ is true whenever the premises are true. ✓

b) To show that this is invalid, we will construct a *counterargument*.
Let $U = \mathbb{Z}$, and take the usual meaning for the various symbols.
Then the first sentence is true, because for any integer y there is an integer x less than it (e.g., $x = y - 1$). But the second sentence is false, because -1 is different from 0 but not greater than 0. ✓
The following finite-universe interpretation gives an *alternative counterargument*: take $U = \{0, 1\}$, and let $0 < 0$ and $1 < 1$ be the only order relations. Then the premise is true while the conclusion is false. ✓

The next example presents an argument containing open sentences, something that occurs frequently in mathematical reasoning.

✤Example 2.2.7
Show that $x = 1 \lor x = -1, x \neq 1 \vDash x = -1$.

Solution
Suppose the two premises are true when x is assigned to some a in U.
Then, $a = 1 \lor a = -1$ (interpreting the first premise).
But $a \neq 1$ (similarly interpreting the second premise).
By process of elimination (by *Disjunctive Syllogism*), $a = -1$.
So $x = -1$ is true under any extended interpretation that makes both premises true, which means the argument is valid. ✓

The last two examples show that demonstrating validity has two parts: interpreting FOL symbolism and variables, and using PL reasoning on the interpreted statements. The reasoning used here with respect to quantifiers will be formalized in Sections 2.3 and 2.4.

2.2.7 *Logical Equivalence for FOL*

Logical equivalence can also be defined using extended interpretations, making $\mathbf{P} \boxminus \mathbf{Q}$ if and only if $\mathbf{P} \vDash \mathbf{Q}$ and $\mathbf{Q} \vDash \mathbf{P}$, as it was for PL.

Definition 2.2.4: *Logical Equivalence for FOL Sentences*
Sentences $\mathbf{P}$ *and* $\mathbf{Q}$ *are* **logically equivalent**, *written* $\mathbf{P} \boxminus \mathbf{Q}$, *if and only if every interpretation satisfying* $\mathbf{P}$ *satisfies* $\mathbf{Q}$, *and conversely.*

The next example illustrates this idea. The first equivalence is trivial but points out that the variables used to write quantified sentences are irrelevant. The second one introduces something we'll explore further in Section 2.4.

✤Example 2.2.8
Determine the truth of the following equivalence claims.
a) $\forall x Px \boxminus \forall y Py$ b) $\neg\forall x Qx \boxminus \forall x \neg Qx$

Solution
a) Since both $\forall x Px$ and $\forall y Py$ assert that every object in a universe of discourse has property P, they are alike true or false.
   Thus, these sentences are *are logically equivalent* ✓.
b) $\neg\forall x Qx$ and $\forall x \neg Qx$ are *not logically equivalent*, though people often talk as if they were. Moving the negation sign changes a sentence's meaning. *Not all numbers are odd* does not mean *All numbers are not odd*. If $U = \mathbb{N}$ and Qx means x *is odd*, the first sentence is true while the second is false. So $\neg\forall x Qx \nvDash \forall x \neg Qx$, making $\neg\forall x Qx \boxminus\!\!\!/ \ \forall x \neg Qx$, too. ✓
   The second sentence does imply the first, however. If everything in universe of discourse U lacks property Q, then no members of U have that property. So $\forall x \neg Qx \vDash \neg\forall x Qx$. ✓

2.2.8 *Multiple Interpretations and Foundations*

As noted, the key semantic ideas of *First-Order Logic* require an abstract point of view. Formulas must be able to have different meanings in order to deal with their truth and consequences. This formal approach is also important for contemporary foundations of mathematics, where multiple interpretations have clarified the logical structure of mathematical theories. To investigate logical relationships among the axioms of a theory, we must be able to vary the interpretation.

The concepts of *logical consistency*, *logical independence*, and *theory completeness* were defined for *Propositional Logic* in Section 1.3, but these apply to *First-Order Logic* as well: here they're based on the truth of interpretations instead of truth tables. A simple example will illustrate their usefulness.

❖**Example 2.2.9**

Investigate the notions of logical implication, consistency, and independence for the sentences in Example 2, again repeated for easy reference.
$$0 < 1; \quad 1 < (1+1); \quad 1 \cdot 1 = 1; \quad \exists x(x \cdot x < x); \quad \forall x(x \neq 0 \rightarrow 0 < x).$$

Solution

· The usual number-theoretic interpretation satisfies all but the fourth sentence. Thus, the *fourth* sentence is not a logical consequence of the rest. In fact, it is logically independent of the other four since there is also an interpretation that makes all of them true (see Exercise 7).

· Example 2's second interpretation made all of the sentences true except the last, so the first four do not imply the fifth. And, since all five sentences are logically consistent, the *fifth* is logically independent of the rest.

· Example 3 demonstrated that the *second* sentence doesn't follow from the rest; it, too, is logically independent of them.

· Other interpretations show that the *first* and *third* sentences are logically independent of the rest (see Exercise 49), making these sentences a completely independent set. While it may seem that the first sentence logically follows from the fifth one, an interpretation in which 0 and 1 denote the same object blocks such an inference.

2.2.9 *Abstract Formalism in Mathematics*

While logic and foundations benefit from an abstract viewpoint, does mathematics itself gain anything from such a perspective?

Let's first consider this philosophically. One might take mathematical sentences as intrinsically meaningless formulas. According to this outlook, called *formalism*, mathematicians merely play games with strings of symbols according to prescribed rules, much like one does in moving chess pieces.

The ideas of David Hilbert early in the twentieth century, introduced for developing proof theory, tended to encourage such an outlook. But most practicing mathematicians reject this view. Mathematics is more than manipulating symbols according to the rules of logic. Even in very abstract areas, research is directed both by general (though often abstract) mathematical intuitions and by knowledge of specific concrete models.

Nevertheless, there are mathematical reasons for allowing formulas to have multiple interpretations and for treating them at times as uninterpreted sentences. By noting similar results in different settings, mathematicians can abstract from the particularities of each structure and develop a general theory for the common features of all such structures. This approach makes economy of thought possible, because proving an abstract result demonstrates that it holds for all the interpretations or models of the theory.

This approach also makes mathematics more rigorous. By ignoring the specific meaning of a particular interpretation, mathematicians can guard against smuggling unwarranted assumptions into a proof. In fact, it was concern for deductive rigor that led mathematicians to adopt an abstract viewpoint in the late nineteenth century. After more than 2000 years, overlooked properties of the *betweenness* relation were recognized by Pasch, Hilbert, and others and were explicitly incorporated into modern axiomatic geometry.

If you proceed further in mathematics, you'll meet this abstract approach again and again. The theory of vector spaces (linear algebra), for example, applies to a wide variety of structures. The objects may be matrices or functions or n-tuples of numbers, and their operations may be quite different. Vector spaces vary greatly, yet they share a common vector space structure. We'll see a similar thing in Chapter 7 when we look at Boolean algebras.

EXERCISE SET 2.2

Exercises 1–6: FOL Syntax
In the problems below, interpret all standard symbols in the usual way, with letters from the front of the alphabet as constants. Then if the symbol strings are formulas, identify them as open or closed sentences and explain why.

2.2.1. $\exists x(x^2 < 0)$

2.2.2. $y \neq 0 \to \exists x(x < y)$

2.2.3. $|x| = x \vee -x$

2.2.4. $a^3 + 1 = 0 \to a = -1$

2.2.5. $\forall x \forall y \forall z(x \mid y \to x \mid yz)$

2.2.6. $\sin(\pi/2) = \sqrt{2}$

2.2.7. Give an interpretation for the sentences of Example 2, repeated here, that will make them all true. Thus, these sentences are logically consistent.

$$0 < 1; \quad 1 < (1+1); \quad 1 \cdot 1 = 1; \quad \exists x(x \cdot x < x); \quad \forall x(x \neq 0 \to 0 < x).$$

Exercises 8–10: True or False
Are the following statements true or false? Explain your answer.

2.2.8. Truth tables aren't useful for deciding the truth and consequences of FOL sentences.

2.2.9. A closed FOL sentence can be either true or false, depending on its interpretation.

2.2.10. Multiple interpretations are useful in mathematics as well as logic.

Exercises 11–17: Interpretations and Truth Values for Arithmetic

Determine the truth values of the following sentences for each interpretation. Let 0 have its usual meaning, and take s (the first letter of s̲uccessor) as a one-place function symbol. $\mathbb{N}$ and $\mathbb{Z}$ stand for the set of natural numbers and the set of integers respectively.

a. $\forall x(s(x) \neq 0)$ b. $\forall x \forall y(s(x) = s(y) \to x = y)$ c. $\forall y(y \neq 0 \to \exists x(y = s(x)))$

2.2.11. $U = \mathbb{N}$; $s(x) = x + 1$

2.2.12. $U = \mathbb{Z}$; $s(x) = x + 1$

2.2.13. $U = \mathbb{N}$; $s(x) = 2x$

2.2.14. $U = \mathbb{Z}$; $s(x) = 2x + 1$

2.2.15. $U = \mathbb{N}$; $s(x) = x^2$

2.2.16. $U = \mathbb{Z}$; $s(x) = x^2 + 1$

2.2.17. $U = \mathbb{N}$; $s(x) = \begin{cases} x + 1: & \text{if } x \text{ is even} \\ \dfrac{x+1}{2}: & \text{if } x \text{ is odd} \end{cases}$

Exercises 18–24: Geometric Interpretations and Truth Values

Determine the truth values of the following sentences for each interpretation. The terms 'point' and 'line' refer to objects of different sorts. Assume universe U contains all possible (interpreted) points and lines of some (Euclidean) plane, and that 'passes through' has the ordinary meaning of "contains."

a. At least one line passes through each pair of distinct points.

b. At most one line passes through each pair of distinct points.

c. Each line passes through at least two distinct points.

d. Each line misses (does not pass through) at least one point.

2.2.18. Point: point
Line: line

2.2.19. Point: point on the x-axis
Line: x-axis

2.2.20. Point: point with integer coordinates (integral lattice points)
Line: line

2.2.21. Point: point with integer coordinates (integral lattice points)
Line: line passing through the origin

2.2.22. Point: point with integer coordinates (integral lattice points)
Line: line passing through the origin with slope a rational number

2.2.23. Point: point with integer coordinates (integral lattice points)
Line: rectangle with sides parallel to the axes, integral lattice point vertices

2.2.24. Point: point with non-zero integer coordinates
Line: circle centered at the origin, passing through some point, with radius length a positive whole number

Exercises 25–31: Algebraic Interpretations and Truth Values

*Determine the truth value of the following four sentences for each interpretation. Let e denote a constant and * a binary operation. $\mathbb{N}$ and $\mathbb{Q}$ stand for the set of natural numbers and the set of rational numbers respectively.*

a. $\forall x \forall y \forall z((x * y) * z = x * (y * z))$ b. $\forall x(e * x = x = x * e)$

c. $\forall x \exists y(x * y = e = y * x)$ d. $\forall x \forall y(x * y = y * x)$

2.2.25. $U = \mathbb{N}$
  $e = 0;\quad a * b = a + b$
2.2.26. $U = \mathbb{N}$
  $e = 0;\quad a * b = \min\{a, b\}$
2.2.27. $U = \mathbb{Q}$
  $e = 0;\quad a * b = \dfrac{a + b}{2}$

2.2.28. $U = \mathbb{Q}$
  $e = 1;\quad a * b = a \cdot b$
2.2.29. $U = \mathbb{N}$
  $e = 0;\quad a * b = |a - b|$
2.2.30. $U = \mathbb{N}$
  $e = 0;\quad a * b = \begin{cases} a : & a \neq 0 \\ b : & a = 0 \end{cases}$

2.2.31. $U = $ all six rotations and flips of an equilateral triangle onto itself
  $e = 360°$ rotation; $a * b = $ motion a followed by motion b

Exercises 32–35: Truth Values for Abstract FOL Sentences
Determine whether the following sentences are logically true, logically false, or logically indeterminate. Explain your answers, using the notions of interpretations and models (finite or infinite).

2.2.32. $\forall x \forall y (Pxy \lor Pyx)$
2.2.33. $\forall x (Px \rightarrow \neg Px)$

2.2.34. $\forall x [Px \land \neg Qx \leftrightarrow \neg(Px \rightarrow Qx)]$
2.2.35. $\forall x (Px \land Qx) \rightarrow \exists x Px \land \exists x Qx$

Exercises 36–41: Truth Values for Mathematical FOL Sentences
Determine whether the following mathematical sentences are logically true, logically false, or logically indeterminate. Explain your answers, using the notions of interpretations and models (finite or infinite).
*Note: do **not** assume that familiar symbols/words have their usual meanings!*

2.2.36. $1 < 2$
2.2.37. $x = 1 \rightarrow x^2 = 1$
2.2.38. $x = 0 \land x = 1 \rightarrow 0 = 1$

2.2.39. $\forall x (x = 0 \lor x \neq 0)$
2.2.40. $\exists x (x^2 - 2 = 0)$
2.2.41. No right angles are angles.

Exercises 42–44: Aristotelian Logic and Logical Implication
Show that the following syllogistic argument forms are valid using arbitrary interpretations and PL forms of reasoning.

2.2.42. $\exists x(Px \land Qx),\ \forall x(Qx \rightarrow Rx) \vDash \exists x(Px \land Rx)$
2.2.43. $\forall x(Px \rightarrow Qx),\ \forall x(Rx \rightarrow \neg Qx) \vDash \forall x(Px \rightarrow \neg Rx)$
2.2.44. $\exists x(Px \land Qx),\ \forall x(Rx \rightarrow \neg Qx) \vDash \exists x(Px \land \neg Rx)$

Exercises 45–47: Implication and Equivalence for FOL Sentences
Show that the following implications and equivalences hold using arbitrary interpretations and PL forms of reasoning.

2.2.45. $\forall x(Px \lor Qx),\ \exists x(\neg Qx) \vDash \exists x(\neg Px)$
2.2.46. $\forall x(Px \leftrightarrow Qx) \vDash \exists x Px \rightarrow \exists x Qx$
2.2.47. $\forall x(Px \rightarrow Qx \land Rx) \vDash\!\dashv \forall x[(Px \rightarrow Qx) \land (Px \rightarrow Rx)]$

2.2.48. *Order of Quantifiers*
 a. Determine the truth value of $\forall x \exists y(y \leq x)$ and $\exists y \forall x(y \leq x)$ for the following interpretations. Interpret $\leq$ as usual.
  i. $U = \mathbb{N}$ ii. $U = \mathbb{Z}$ iii. $U = \mathbb{Q}$ iv. $U = \mathbb{R}$
 b. Given your answers to part *a*, which formula makes a stronger claim, $\exists y \forall x Pxy$ or $\forall x \exists y Pxy$? Explain.

Exercises 49–50: Logical Independence
The following problems deal with logical independence and consistency.

2.2.49. *Independence of Example 2's Sentences*
 a. Show that the first sentence of Example 2 is independent of the others.
 b. Show that the third sentence of Example 2 is independent of the others.

2.2.50. *Consistency and Independence*
 a. Are the three arithmetic sentences given prior to Exercises 11–17 consistent? Are they independent of one another? Cite any relevant results from your work on those problems.
 b. Are the four geometric sentences given prior to Exercises 18–24 consistent? Are they independent of one another? Cite any relevant results from your work on those problems.
 c. Are the four algebraic sentences given prior to Exercises 25–31 consistent? Are they independent of one another? Cite any relevant results from your work on those problems.

2.3 Rules for Identity and Universal Quantifiers

First-Order Logic's Natural Deduction System will modify and extend that of *Propositional Logic*. All earlier rules remain in force, except that now they'll apply to both *open and closed sentences*. This allows us to make normal mathematical arguments, such as the following: $x = 1 \lor x = -1$, $x \not< 0$, $-1 < 0$; *therefore* $x = 1$ (see Exercise 1).

In addition, we'll have rules governing the identity relation $=$ and the quantifiers $\forall$, $\exists$, and $\exists!$. As before, our main rules will be *Int-Elim Rules*, but we'll also have some *Replacement Rules* for negating quantified sentences.

In this section we'll look at the inference rules for identity and the universal quantifier. The rules for existential sentences and the *Replacement Rules* will be covered in the next section.

2.3.1 *Rules of Inference For Identity*: *Substitution*

Logic for mathematics should include rules for identity. In elementary settings, an equation[5] says two expressions represent the *same quantity*, but *equals* is important in advanced fields as well. Identity also occurs in nonmathematical contexts, so we'll treat the theory of identity as part of *First-Order Logic* and will always interpret $=$ to mean *is identical to*.

Intuitively, objects are identical if and only if they can't be distinguished, i.e., if and only if they have exactly the same features. This characterization is known as *Leibniz's Law of Indiscernibility*. More precisely, we can say that if $\mathbf{t_1}$ and $\mathbf{t_2}$ are *terms* denoting objects in a universe of discourse, then $\mathbf{t_1} = \mathbf{t_2}$

---

[5] Mathematicians distinguish between an *equation* and a *mathematical identity*. The former is what logic calls an identity. The latter is an equation that holds for *all* values in a certain domain, i.e., an equation's *universal closure*.

if and only if given any formula $\mathbf{P}(\cdot\,\mathbf{x}\,\cdot)$, $\mathbf{P}(\cdot\,\mathbf{t_2}\cdot)$ holds whenever $\mathbf{P}(\cdot\,\mathbf{t_1}\cdot)$ does, and conversely. Here, as before, $\mathbf{P}(\cdot\,\mathbf{x}\,\cdot)$ represents an FOL *statement*, and $\cdot\,\mathbf{x}\,\cdot$ indicates an occurrence of $\mathbf{x}$ in the statement; this occurrence is replaced by $\mathbf{t_1}$ and $\mathbf{t_2}$ in the formulas $\mathbf{P}(\cdot\,\mathbf{t_1}\cdot)$ and $\mathbf{P}(\cdot\,\mathbf{t_2}\cdot)$.

Definition 2.3.1: *Leibniz's Law of Indiscernibility for Identity*
$\mathbf{t_1} = \mathbf{t_2}$ *if and only if* $\mathbf{P}(\cdot\,\mathbf{t_1}\cdot) \mathrel{\vDash\!\dashv} \mathbf{P}(\cdot\,\mathbf{t_2}\cdot)$ *for all statements* $\mathbf{P}$.

In theory, this definition enables us to determine when two terms denote the same object. In practice, though, we'll usually use an *Introduction Rule* based upon this law. And in some cases (for example, in *Set Theory*), equality will be asserted by means of an axiom or a theorem that gives an independent mathematical criterion for when two objects are identical.

Leibniz's Law of Indiscernibility can also be used for drawing conclusions from an identity. If you know $\mathbf{P}(\cdot\,\mathbf{t_1}\cdot)$ and $\mathbf{t_1} = \mathbf{t_2}$, you may conclude $\mathbf{P}(\cdot\,\mathbf{t_2}\cdot)$. In fact, you may substitute $\mathbf{t_2}$ for $\mathbf{t_1}$ as often as you wish. This rule, which we'll call *Substitution of Equals (Sub)*, is schematized as follows:

$$
\textit{Sub} \quad
\begin{array}{|l}
\mathbf{P}(\cdot\,\mathbf{t_1}\cdot) \\
\mathbf{t_1} = \mathbf{t_2} \\
\hline\hline
\mathbf{P}(\cdot\,\mathbf{t_2}\cdot)
\end{array}
$$

2.3.2 Equivalence Properties of Identity

Substitution of Equals is the main *Elimination Rule* for identity. In addition, there is an *Introduction Rule* and two other *Int-Elim Rules*. These three rules are convenient for making deductions, though they do not characterize identity to the same extent that *Leibniz's Law of Indiscernibility* does. Many other relations, known as *equivalence relations*, also satisfy these three properties.

The *Law of Identity (Iden)* is based on the *reflexive* property of $=$. Since each object is identical with itself, you may write $\mathbf{t} = \mathbf{t}$ on any line of a deduction. No premises are needed for this conclusion. This law's soundness should be clear, but it may also be demonstrated by appealing to *Leibniz's Law of Indiscernibility* (see Exercise 10a).

Schematically, this rule of inference is as follows:

$$
\textit{Iden} \quad
\begin{array}{|l}
\\
\hline\hline
\mathbf{t} = \mathbf{t} \qquad [\mathbf{t} \text{ any term}]
\end{array}
$$

The *Law of Identity* is seldom used in mathematics, though there are a few occasions where it seems indispensable. Its most important use in elementary mathematics is to create rigorous arguments proving mathematical identities.[6] To show $\tan^2 x + 1 = \sec^2 x$, for instance, we ordinarily begin

---

[6] Here the more specialized mathematical sense of *identity* is meant. See note 5.

with one side of the identity and generate a sequence of algebraically equivalent expressions until we arrive at the other side (see below). Arguments are composed of sentences, however, not a list of identical terms. This standard approach can be made rigorous by beginning with an identity via *Iden* and repeatedly using *Sub* to change one side of the equation into the next expression.

Besides being reflexive, all equivalence relations share two additional properties: *symmetry*, and *transitivity* (see Exercise 1.3.10). *Transitivity of Identity* (*Trans*) concludes $t_1 = t_3$ from $t_1 = t_2$ and $t_2 = t_3$. This can be considered a special case of *Sub* (see Exercise 8b).

Schematically, we have:

$$\begin{array}{|l} t_1 = t_2 \\ t_2 = t_3 \\ \hline\hline t_1 = t_3 \end{array} \qquad \textbf{\textit{Trans}}$$

The soundness of this rule is obvious: if one object is denoted by both t_1 and t_2 and another one by t_2 and t_3, then these objects are identical since t_2 names them both. Hence t_1 and t_3 name the same object: $t_1 = t_3$. This rule can be derived via *Leibniz's Law of Indiscernibility* (see Exercise 10c).

The inference rule *Symmetry of Identity* (*Sym*) concludes $t_2 = t_1$ from $t_1 = t_2$. Like *Trans*, it is both an *Introduction* and *Elimination Rule*.

Schematically, we have:

$$\begin{array}{|l} t_1 = t_2 \\ \hline\hline t_2 = t_1 \end{array} \qquad \textbf{\textit{Sym}}$$

This rule's soundness is also obvious, but it can be demonstrated using *Leibniz's Law of Indiscernibility* (Exercise 10b) or *Sub* (Exercise 8a).

While $=$ is clearly symmetrical, at times it gets treated as if it means *yields* or *produces*, which converts it into a one-directional relation. The distributive law $a(b + c) = ab + ac$ is often thought to signify something different from its reversed form, $ab + ac = a(b + c)$. Moving from left to right, the first is an expansion rule, while the latter is a rule for factoring. The legitimacy of these reverse procedures is captured by the symmetry of equality.

Sym can be used to generalize *Sub*, so you can substitute a term for its equal no matter how equality is asserted. *Trans* can also be generalized: if t_1 and t_2 appear in either order on line i while t_2 and t_3 appear in either order on line j, you may conclude $t_1 = t_3$ on line k (see Exercise 9a). This can be used to justify Euclid's first axiom, for instance, which says that *things equal to the same thing are equal to each other* (see also Exercise 6.3.39). When arguing your deductions in these generalized ways, you may simply cite *Sub* or *Trans*, though strictly speaking, you should transform the identity using *Sym* before applying *Sub* or *Trans* to obtain the conclusion.

2.3.3 *Using Equations in Mathematical Proofs*

The rules for identity underlie the way we string equations together in an argument. We'll look at how to do this and what to avoid.

✦**Example 2.3.1**

Expand the polynomial $(x+2)(x+3)$ in a step-by-step fashion.

Solution

We can multiply these binomials as follows:
$$(x+2)(x+3) = x(x+3) + 2(x+3)$$
$$= x^2 + 3x + 2x + 6$$
$$= x^2 + 5x + 6.$$

The equations in this example could have been written all on one line, as $(x+2)(x+3) = x(x+3) + 2(x+3) = x^2 + 3x + 2x + 6 = x^2 + 5x + 6$, but doing so fails to highlight that $(x+2)(x+3) = x^2 + 5x + 6$. Stringing equations together vertically indicates the same sequence of equalities but better shows that the top-left expression equals the bottom-right one. A formal argument would use *Trans* repeatedly (see Exercise 9b).

This example illustrates the proper way to prove an equation. Begin with the expression on one side, transform it into an equal expression, and continue until you arrive at the final expression on the other side of the equation.

Unfortunately, a *bad proof strategy* that violates this advice is sometimes picked up in elementary algebra and trigonometry. That approach begins with the identity to be proved and works with *both sides at once* until arriving at an equation known to be true. Such a (*bad!*) argument might look like:
$$\tan^2 x + 1 = \sec^2 x$$
$$\sin^2 x / \cos^2 x + 1 = 1/\cos^2 x$$
$$\sin^2 x + \cos^2 x = 1.$$

This approach is *misleading* even when all the individual equations are true, because it is no longer clear what's given, what's been proved, and what still needs to be argued. In fact, it starts by asserting the very thing that needs proof! Such a procedure easily leads to invalid arguments. Here's a cute "proof" that $1 = 2$, argued in the same way:
$$1 = 2 \qquad \text{[identity to be proved]}$$
$$0 = 0 \qquad \text{[multiply by 0—a legitimate operation]}$$

Getting a true final conclusion doesn't qualify as a valid demonstration of the original statement. The sequence of equations generated must be reversible for it to be conclusive, in which case the reversed sequence is the proof.

A modified version of this approach is acceptable, however. If you keep the forward and backward directions separate, you can work with both sides of an equation, though not by creating a sequence of transformed equations. Start with one side of the equation in the top-left corner and with the other side in the bottom-right corner. Then, finding values equal to the first side, list them downwards. List values equal to the other side upwards. Once you reach a common value, you have a proof. Taking the above example:

$$\tan^2 x + 1 = (\sin^2 x / \cos^2 x) + 1$$
$$\downarrow = (\sin^2 x + \cos^2 x) / \cos^2 x \qquad \text{[add fractions]}$$
$$= 1/ \cos^2 x \qquad\qquad \text{[substitute } \sin^2 x + \cos^2 x = 1]$$
$$\vdots \qquad \|$$
$$\uparrow = 1/ \cos^2 x \qquad \text{[substitute the definition for } \sec x]$$
$$= \sec^2 x$$

This dual approach is legitimate. In fact, since you may not know at the outset which side of an equation will be easier to work with, it makes sense to use both a forward and backward approach to generate your proof. But still, *don't organize your argument as a sequence of transformed equations starting with what you want to prove!*

2.3.4 *Universal Instantiation:* ∀ *Elimination*

$\forall \mathbf{xP(x)}$ asserts that $\mathbf{P(x)}$ holds for every object $\mathbf{x}$ in the universe of discourse U. If $\mathbf{t}$ denotes an object in U, $\mathbf{P(t)}$, obtained by substituting $\mathbf{t}$ for $\mathbf{x}$ everywhere in $\mathbf{P(x)}$, must be true if $\forall \mathbf{xP(x)}$ is. The *Elimination Rule* that justifies this inference is *Universal Instantiation*. Schematically, we have:

$$\left. \begin{array}{c} \forall \mathbf{xP(x)} \\ \hline\hline \mathbf{P(t)} \end{array} \right. \qquad \text{[} \mathbf{t} \text{ any term]} \qquad\qquad \textbf{\textit{UI}}$$

Such a rule is sound: it's valid to *instantiate* $\forall \mathbf{xP(x)}$ to $\mathbf{t}$, obtaining $\mathbf{P(t)}$.

Think of $\forall \mathbf{xP(x)}$ as asserting a *grand conjunction*, with one conjunct $\mathbf{P(t)}$ for every $\mathbf{t}$ in U. According to the inference rule *Simp*, generalized to handle any number of conjuncts, we can conclude any $\mathbf{P(t)}$, no matter what $\mathbf{t}$ denotes. *UI* is thus an FOL version of *Simp*.

UI only applies to whole sentences, not sub-sentences. For instance, it is invalid to conclude $\neg Pa$ from $\neg\forall xPx$. Even if not every object has property P, we have no reason to conclude that a particular object a fails to have it.

2.3.5 *The Role of UI in Mathematical Proofs*

UI is used repeatedly in mathematics, though it is usually not made explicit. To prove that a result holds for some object a, we typically instantiate previous results, known to hold universally for a certain class of objects, to that object, because a belongs to the class being considered.

For example, to prove that $\triangle ABC \cong \triangle A'B'C'$, we take a congruency criterion such as SAS and apply it to the instance being considered. Similarly, if a is an odd number, we can conclude that a^2 is also odd because all odd numbers have odd squares.

Two simple formal mathematical arguments illustrate how *UI* gets used, along with some earlier rules for identity.

❖**Example 2.3.2**

Show that $\forall x(s(x) > 0)$, $1 = s(0) \vdash 1 > 0$. Here s might denote the successor function, which takes each natural number to the next one; 0 the number zero; and 1 the number $s(0)$.

Solution

The following proof diagram establishes the claim. Note that step 4 could be omitted by using a generalized version of *Sub*.

1	$\forall x(s(x) > 0)$	Prem
2	$1 = s(0)$	Prem
3	$s(0) > 0$	UI 1
4	$s(0) = 1$	Sym 2
5	$1 > 0$	Sub 3, 4

❖**Example 2.3.3**

Show that $\forall x(x * e = x)$, $\forall x(x^{-1} * x = e) \vdash e^{-1} = e$, where e is a constant, $*$ is a binary operation, and $(\ )^{-1}$ denotes a function that assigns (an inverse) a^{-1} to each element a in the universe of discourse.

Solution

The following proof diagram establishes the claim. This time we'll give an abridged demonstration, using generalized versions of our inference rules for $=$. Constructing a longer proof is left as an exercise (see Exercise 13).

1	$\forall x(x * e = x)$	Prem
2	$\forall x(x^{-1} * x = e)$	Prem
3	$e^{-1} * e = e$	UI 2
4	$e^{-1} * e = e^{-1}$	UI 1
5	$e^{-1} = e$	Trans 3, 4

2.3.6 Universal Generalization: ∀ Introduction

Many mathematical theorems are law-like statements that hold for entire classes of objects. To prove them we must be able to deduce sentences of the form $\forall \mathbf{x} \mathbf{P}(\mathbf{x})$. To demonstrate that $\mathbf{P}(\mathbf{x})$ holds for all $\mathbf{x}$ in some universe $\mathbf{U}$, we must show that $\mathbf{x}$ belonging to $\mathbf{U}$ entails $\mathbf{P}(\mathbf{x})$. If $\mathbf{U}$ is small enough, it might be possible to do this for each member individually. But in most cases, this is hopeless or impossible—mathematical theories usually have infinite models. Hence the time-worn adage, *you can't prove a result with examples*.

The problem with using specific cases to argue the truth of a universal proposition is that each such argument may depend upon the *particularities* of the object chosen and so not be valid for other members of the universe. On the other hand, if it *were* possible to find an object **a** that was completely representative—a *typical generic element of U*, as it were—then an argument given for this object should work in general and be considered a proof.

In reality, there are no *generic brand-x* objects. Every object has features that distinguish it from others. Nevertheless, if an argument uses only results that are true of everything in **U**, it would be a *proof by generic example*. This proof strategy is captured by the rule *Universal Generalization* (*UG*).

This inference rule says that if a sentence $\mathbf{P(a)}$ can be shown to hold for an arbitrary representative individual **a**, where nothing is asserted of **a** except what follows from its being a member of **U**, then we may conclude $\forall\mathbf{xP(x)}$.

Schematically, we have:

$$\begin{array}{ll} \mathbf{P(a)} & \text{[\textbf{a} an arbitrary constant]} \\ \hline\hline \forall\mathbf{xP(x)} & \end{array} \qquad \textit{UG}$$

A proof by *UG* requires **a** to be an arbitrary element of **U**, but we can't say this inside FOL. Using **a** as a constant automatically means, according to FOL's semantics, that **a** is a *member* of U. We can guarantee that **a** denotes an *arbitrary* member by surveying the written argument to make sure nothing has been asserted about **a** that can't be said about every member of **U**.

Informal arguments using *UG*, though, do start by saying something like *let* **a** *be an element of* U. This signifies that **a** is intended to be perfectly general. There is nothing special about the letter **a**, of course. In practice you'll probably use whatever letter helps you remember the sort of object you're considering. Mathematical proofs often use **x** in a dual role, both as the arbitrary individual to be argued about and as the replacing variable. In formal proofs, though, we'll distinguish these two roles with separate letters.

❖ Example 2.3.4

Show that $\forall\mathbf{x(Px \to Qx)}, \forall\mathbf{xPx} \vdash \forall\mathbf{xQx}$.

Solution

This is established by the following proof diagram.
Note that in line 3 we instantiate line 1 to an arbitrary element **a** of the universe of discourse, so that line 6 can generalize on it.

$$\begin{array}{ll} 1 \mid \forall\mathbf{x(Px \to Qx)} & \text{Prem} \\ 2 \mid \forall\mathbf{xPx} & \text{Prem} \\ \hline 3 \mid \mathbf{Pa \to Qa} & \text{UI 1} \\ 4 \mid \mathbf{Pa} & \text{UI 2} \\ 5 \mid \mathbf{Qa} & \text{MP 3, 4} \\ 6 \mid \forall\mathbf{xQx} & \text{UG 5} \end{array}$$

The next example instantiates more than one universally quantified variable at a time, something that would officially require several applications of *UI* to legitimize. We can also universally generalize more than one variable at a time, which will shorten other deductions. Informal mathematical arguments use these abbreviation maneuvers frequently.

❖**Example 2.3.5**

Show that $\forall x \forall y (x + y = y + x)$, $\forall x (x + 0 = x) \vdash \forall x (0 + x = x)$.

Solution

This claim is established by the following proof diagram.

$$
\begin{array}{ll}
1 & \forall x \forall y (x + y = y + x) \qquad \text{Prem} \\
2 & \forall x (x + 0 = x) \qquad\qquad\; \text{Prem} \\
\hline
3 & a + 0 = a \qquad\qquad\qquad \text{UI 2} \\
4 & a + 0 = 0 + a \qquad\qquad\; \text{UI 1 (twice)} \\
5 & 0 + a = a \qquad\qquad\qquad \text{Sub 3, 4} \\
6 & \forall x (0 + x = x) \qquad\qquad \text{UG 5}
\end{array}
$$

The soundness of *UG* should be intuitively clear. If we know that a typical element **a** of the universe of discourse U satisfies statement **P(x)**, then this is true of all elements in U.

Here, again, it may help to think of $\forall \mathbf{x} \mathbf{P}(\mathbf{x})$ as a grand conjunction. Having shown by a generic argument that each conjunct holds, we can conjoin them to obtain the universal statement. *UG* is thus the FOL counterpart of *Conj*.

2.3.7 *The Role of UG in Mathematical Proofs*

Like other inference rules, *UG* is often tacitly assumed. It can be helpful to keep it in mind, however, when constructing proofs of mathematical propositions in the form $\forall \mathbf{x} \mathbf{P}(\mathbf{x})$. *Begin by supposing that* **a** *is an object of the sort being considered.* Draw a diagram to visualize the situation, and assess what you know about such objects. As you do these things, you're overcoming the psychological barrier of getting started. To continue your proof, use a *Backward-Forward* proof analysis. Analyze how to prove **P(a)**, given the logical form of the sentence. Take stock of what you know and what you still want to show. If it's possible to split the domain into several types of objects, try proof by *Cases* to generate the result for each type of object before generalizing.

UG is also useful when the statement to be proved is a restricted universal sentence, i.e., when it makes a claim about some *subset* of the universe of discourse. The following example illustrates how to proceed in such a case.

❖**Example 2.3.6**

Analyze the beginning and overall proof strategy to use in proving the proposition *All positive real numbers have real square roots.*

Solution

· Assuming that this is a result in real-number arithmetic, the appropriate universe of discourse is $U = \mathbb{R}$, the set of real numbers. Asserting that numbers have square roots is equivalent to saying they are squares of numbers.

· Two slightly different formal proof strategies are available to us here, depending on whether we use a restricted quantifier to formulate the sentence.

a) In expanded form, this sentence is $\forall x(x > 0 \rightarrow \exists y(y^2 = x))$.
To start a proof of this, we would assume that r is an arbitrary real
number (for UG) and then prove (via CP) that $r > 0 \rightarrow \exists y(y^2 = r)$.

b) Using a restricted quantifier, this sentence is $(\forall x > 0)(\exists y)(y^2 = x)$.
To prove this, using a version of UG, we would suppose that r is
an arbitrary positive real number, and we would then try to deduce
$\exists y(y^2 = r)$.

· In either case, having proved the inner sub-sentence, we would universally
generalize via UG to complete the proof.

· An informal proof follows the second strategy. We would compress the two
suppositions and say, without using a new letter for the instance, *Suppose
x is a positive real number*. After showing that x has a square root, the
proof would be considered complete, implicitly applying UG.

EXERCISE SET 2.3

2.3.1. Give a formal deduction for the following argument, citing any PL
rules of inference used: $(x = 1) \vee (x = -1)$, $x \not< 0$, $-1 < 0$ $\vdash$ $x = 1$.

Exercises 2–3: True or False
Are the following statements true or false? Explain your answer.

2.3.2. To prove an identity, you manipulate both sides of the equation until
a true identity results.

2.3.3. To prove $\forall x P(x)$, you show $P(a)$ holds for all instances a.

Exercises 4–5: Completing Deductions
Fill in the reasons for the following deductions.

2.3.4. $a = c$, $b = d$ $\vdash$ $Rab \leftrightarrow Rcd$

1	$a = c$	_____
2	$b = d$	_____
3	Rab	_____
4	$a = c$	_____
5	Rcb	_____
6	$b = d$	_____
7	Rcd	_____
8	Rcd	_____
9	$a = c$	_____
10	Rad	_____
11	$b = d$	_____
12	Rab	_____
13	$Rab \leftrightarrow Rcd$	_____

2.3.5. $\forall x(Px \lor Qx)$, $\forall x(\neg Qx \lor Rx) \vdash \forall x(Px \lor Rx)$

1	$\forall x(Px \lor Qx)$	_____
2	$\forall x(\neg Qx \lor Rx)$	_____
3	$\quad \neg Pa$	_____
4	$\quad\ \forall x(Px \lor Qx)$	_____
5	$\quad\ Pa \lor Qa$	_____
6	$\quad\ Qa$	_____
7	$\quad\ \forall x(\neg Qx \lor Rx)$	_____
8	$\quad\ \neg Qa \lor Ra$	_____
9	$\quad\ Ra$	_____
10	$Pa \lor Ra$	_____
11	$\forall x(Px \lor Rx)$	_____

Exercises 6–7: Logical Implication and Conclusive Deductions

Determine whether the following **implication** *claims are true. Then determine whether the* **deductions** *given are* **conclusive**. *Carefully point out where and how a rule of inference is being misused.*

2.3.6. $\forall xPx \to \forall yQy$, $\neg Qb \vDash \neg \forall xPx$

1	$\forall xPx \to \forall yQy$	Prem
2	$\neg Qb$	Prem
3	$Pa \to Qb$	UI 1
4	$\neg Pa$	MT 3, 2
5	$\neg \forall xPx$	UG 4

2.3.7. $\forall x \forall y(x \neq y \to Pxy)$, $\forall x \forall y(x = y \to Qxy) \vDash \forall x \forall y(\neg Pxy \to Qxy)$

1	$\forall x \forall y(x \neq y \to Pxy)$	Prem
2	$\forall x \forall y(x = y \to Qxy)$	Prem
3	$\quad \neg Pxy$	Spsn for CP
4	$\quad\ \forall x \forall y(x \neq y \to Pxy)$	Reit 1
5	$\quad\ \neg(x \neq y)$	MT 4, 3
6	$\quad\ x = y$	DN 5
7	$\quad\ \forall x \forall y(x = y \to Qxy)$	Reit 2
8	$\quad\ Qxy$	MP 7, 6
9	$\neg Pxy \to Qxy$	CP 3-8
10	$\forall x \forall y(\neg Pxy \to Qxy)$	UG 9

Exercises 8–13: Rules for Identity

The following problems explore properties of inference rules for identity.

2.3.8. Prove the following properties of identity using *Substitution of Equals* and the *Law of Identity*.

a. *Symmetry:* $\mathbf{t_1} = \mathbf{t_2} \vdash \mathbf{t_2} = \mathbf{t_1}$

b. *Transitivity:* $\mathbf{t_1} = \mathbf{t_2}$, $\mathbf{t_2} = \mathbf{t_3} \vdash \mathbf{t_1} = \mathbf{t_3}$

2.3.9. *Generalized Transitivity of Identity*
a. How many different generalized forms of *Trans* are there? Pick one other than the official version and derive it, using *Trans* along with *Sym*.
b. Use *Trans* to show the following: $x_1 = x_2$, $x_2 = x_3$, $x_3 = x_4 \vdash x_1 = x_4$.

2.3.10. *Identity is an Equivalence Relation*
Using *Leibniz's Law of Indiscernibility*, show that identity is an equivalence relation. That is, show:
a. *Reflexive Law*: $\mathbf{t} = \mathbf{t}$
b. *Symmetric Law*: $\mathbf{t_1} = \mathbf{t_2} \vdash \mathbf{t_2} = \mathbf{t_1}$
c. *Transitive Law*: $\mathbf{t_1} = \mathbf{t_2}$, $\mathbf{t_2} = \mathbf{t_3} \vdash \mathbf{t_1} = \mathbf{t_3}$

2.3.11. *Distinguishing Objects*
a. Prove, using rules for identity, that two objects are distinct if some property P distinguishes them, i.e., show Pa, $\neg Pb \vdash a \neq b$.
b. Conversely, using *Leibniz's Law of Indiscernibility*, determine what follows when $a \neq b$.

2.3.12. Prove that Px, $x = a \lor x = b \vdash Pa \lor Pb$.

2.3.13. Construct an expanded version of the proof given in Example 3, using the rules of inference precisely as given.

Exercises 14–18: Proving Mathematical Identities
Prove the following identities in algebra and trigonometry, handling equations in the proper way. You may assume the rules of arithmetic, definitions of trigonometric functions, and the basic identity $\sin^2 x + \cos^2 x = 1$.

2.3.14. $(a - b)(a + b) = a^2 - b^2$ **2.3.16.** $(x+y)^3 = x^3 + 3x^2 y + 3xy^2 + y^3$
2.3.15. $\tan x / \sin x = \sec x$ **2.3.17.** $\cot^2 x = \csc^2 x - 1$
2.3.18. Symbolically formulate and then prove: *The difference between two consecutive squares of integers is the sum of these integers.*

Exercises 19–20: Deducing Logical Truths
Using the rules of inference for identity and the universal quantifier, deduce the following logical truths.
2.3.21. $\forall x \forall y \forall z (x = y \land x = z \to y = z)$ **2.3.22.** $\forall x \forall y \forall z (x \neq z \to x \neq y \lor y \neq z)$

Exercises 21–22: Aristotelian Logic and Derivations
Deduce the following arguments, which represent certain syllogistic forms from Aristotelian Logic, using FOL's inference rules for universal sentences.
2.3.21. $\forall x (\mathbf{P}x \to \mathbf{Q}x)$, $\forall x (\mathbf{Q}x \to \mathbf{R}x) \vdash \forall x (\mathbf{P}x \to \mathbf{R}x)$
2.3.22. $\forall x (\mathbf{P}x \to \mathbf{Q}x)$, $\forall x (\mathbf{R}x \to \neg \mathbf{Q}x) \vdash \forall x (\mathbf{P}x \to \neg \mathbf{R}x)$

Exercises 23–28: Deductions
Deduce the following, using FOL's inference rules for universal sentences.
2.3.23. $\forall x (Px \to Qx)$, $\neg Qa \vdash \neg Pa$
2.3.24. $\forall x (Px \to Qx) \vdash \forall x (Px \land Rx \to Qx)$
2.3.25. $\forall x (\neg (Px \land Qx))$, $\forall x Px \vdash \forall x (\neg Qx)$

2.3.26. $\forall x(Px \vee Qx),\ \forall x(Px \rightarrow \neg Rx)\ \vdash\ \forall x(Rx \rightarrow Qx)$

2.3.27. $\forall x Px \vee \forall x Qx \vdash \forall x(Px \vee Qx)$

2.3.28. $\forall x(Px \rightarrow Qx) \vdash \forall x Px \rightarrow \forall x Qx$

Exercises 29–30: Interderivability
Show the following interderivability results, using FOL's inference rules.

2.3.29. $\forall x(Px \wedge Qx)\ \dashv\vdash\ \forall x Px \wedge \forall x Qx$

2.3.30. $\forall x(Px \rightarrow Qx \wedge Rx)\ \dashv\vdash\ \forall x((Px \rightarrow Qx) \wedge (Px \rightarrow Rx))$

Exercises 31–35: Mathematical Deductions
Prove the following, using FOL's inference rules.

2.3.31. Show that if $*$ denotes a binary operation, then the following general results hold. Thus, if $*$ stands for addition or subtraction, we obtain Euclid's second and third axioms *Equals added to/subtracted from equals are equal.*
   a. $a = b \vdash a * c = b * c$
   b. $a = b \wedge c = d \vdash a * c = b * d$

2.3.32. Show that $a < b \rightarrow a + c < b + c$ for any real numbers a, b, and c. You may use the results of Exercise 31 as well as the *definition of less than* $[x < y \leftrightarrow x + z = y$ for some $z > 0]$, *associativity of addition* $[(x + y) + z = x + (y + z)]$, and *commutativity of addition* $[x + y = y + x]$.

2.3.33. Show that $\forall x \forall y(x \leq y \vee y \leq x),\ \forall x \forall y \forall z(x \leq y \wedge y \leq z \rightarrow x \leq z),\ a \leq b,\ a \not\leq c \vdash c \leq b$.

2.3.34. Show that $\forall x \forall y \forall z(x \otimes (y \oplus z) = (x \otimes y) \oplus (x \otimes z)),\ \forall x(x \oplus 0 = x),\ \forall x(x \oplus x' = 1),\ \forall x(x \otimes 1 = x),\ \forall x(x \otimes x' = 0) \vdash \forall x(x \otimes x = x)$, where 0 and 1 are constants, $\oplus$ and $\otimes$ are binary operations, and ( )$'$ is a unary operation or function.

2.3.35. Show that $\forall x(\min x \leftrightarrow \forall y(x \leq y)),\ \forall x \forall y(x \leq y \wedge y \leq x \rightarrow x = y) \vdash \min a \wedge \min b \rightarrow a = b$. In words, if a set has a minimum for a relation $\leq$ that is *antisymmetric* (i.e., satisfies the second premise), it must be unique.

2.4 Rules for Existential Quantifiers

This section gives *Int-Elim Rules* for $\exists$ as well as some *Replacement Rules*. We'll also briefly digress to consider a related technique, the *Method of Analysis*. With this, FOL's Natural Deduction System will be complete.

2.4.1 *Existential Generalization*

Section 2.3 cautioned you not to prove results using examples. Nevertheless, *UG* is, in a sense, *proof by generic example*, so we need to qualify that advice. And, for the case at hand, we need to reject it altogether. To prove an existential sentence $\exists \mathbf{x P(x)}$, our main proof strategy is *precisely to use an example*. If $\mathbf{P(t)}$ holds for some *instance* $\mathbf{t}$, we may then conclude $\exists \mathbf{x P(x)}$. This mode of inference is called *Existential Generalization (EG)*.

Schematically, we have:

$$\left| \begin{array}{l} \mathbf{P(t)} \qquad [\text{t any term}] \\ \rule{4cm}{0.4pt} \\ \mathbf{\exists x P(x)} \end{array} \right. \qquad\qquad \textit{EG}$$

To apply *EG*, we must do two things—locate a candidate **t**, and show that **t** satisfies **P(x)**. We may then conclude **∃xP(x)**. This is a sound rule of inference. If **P(t)** is true for some **t** in the universe of discourse, then **∃xP(x)** is certainly true, given the meaning of ∃.

Just as it helps to think of a universal sentence as a grand conjunction, so it helps to think of an existential sentence **∃xP(x)** as a *generalized disjunction*. Saying that there is an object **x** satisfying **P(x)** is like asserting that object **a** satisfies **P(x)**, or object **b** does, or ... object **t** does, etc. *EG* is the FOL version of *Add*, i.e., if one of the disjuncts is the case, so is the full disjunction.

❖**Example 2.4.1**
   Show how *EG* is used in *Geometry* and *Number Theory*.

   Solution
   · To prove in *Geometry* that there is a line m parallel to a given line l and passing through a point P off l, we construct such a line by means of congruent, alternate interior angles. This proves the existential claim.
   · To deduce that an odd number m times an odd number n is odd, we must show that there is an integer k such that $mn = 2k + 1$.
     If $m = 2i + 1$ and $n = 2j + 1$, then $mn = (2i + 1)(2j + 1) = 2(2ij + i + j) + 1$, so $k = 2ij + i + j$ is the number needed.
   · In each case, *EG* validates the existential result, though it usually remains below the surface.

   The following example uses *EG* in a formal setting. It also illustrates an important point about *EG* that is easily misunderstood. In passing from **P(t)** to **∃xP(x)**, *you do not need to replace every occurrence of* **t** *in* **P(t)** *by* **x**. In fact, doing so may not yield the sentence **∃xP(x)** you want to prove, for **P(x)** may already contain an occurrence of **t**. *EG* allows you to existentially generalize from a formula containing **t** to one where **t** has been replaced by **x** *any number of times*. If you use the *Backward Method of Proof Analysis* and remember what needs proving, you'll be able to identify the appropriate sentence **P(x)** and see how to apply *EG*.

❖**Example 2.4.2**
   Show that $\forall x(x < s(x)) \vdash \exists x(x < s(0))$.   Here 0 is a constant, < is a binary relation, and s is a (successor) function.

   Solution
   · The following short proof diagram gives our argument.
     To prove $\exists x(x < s(0))$ via *EG*, we take $P(x)$ to be $x < s(0)$, not $0 < s(x)$ or $x < s(x)$. In step 3, then, we only replace the first occurrence of 0 by x.

· *Note*: we could also conclude $\exists x(0 < s(x))$ or $\exists x(x < s(x))$ in step 3, but we don't, because that's not what we want. This highlights the value of the *Backward Method of Proof Analysis*! Generalize to what you need.

$$
\begin{array}{lll}
1 & \forall x(x < s(x)) & \text{Prem} \\
\hline
2 & 0 < s(0) & \text{UI 1} \\
3 & \exists x(x < s(0)) & \text{EG 2}
\end{array}
$$

2.4.2 *EG and the Method of Analysis*

Existential Generalization proofs often require some ingenuity. Sometimes an instance **t** is easy to find; at other times it may be extremely difficult. The existence of everywhere-continuous, nowhere-differentiable functions is a case in point. Before Weierstrass constructed such functions in 1872, mathematicians thought continuous functions were nearly everywhere differentiable.

Often, most of the work in developing a proof by *EG* consists of figuring out which object is a viable candidate, even though *that* isn't strictly part of the deduction. An argument that fails to explain how one arrived at the particular instance being used is pedagogically unsatisfying.

✤**Example 2.4.3**

Analyze the method of showing that $2x^2 + x - 6 = 0$ has a positive solution.

Solution

· The usual process of showing that $2x^2 + x - 6 = 0$ has a positive solution is interesting from a logical viewpoint. Essentially, *we assume there is a solution* (denoted by x) and then argue in a *logically forward direction* from the equation to determine its value.

· Finding (say, by factoring) that $x = \frac{3}{2}$, we then turn around and show that $\frac{3}{2}$ actually is a solution—we substitute $\frac{3}{2}$ into the equation and calculate $2(\frac{3}{2})^2 + \frac{3}{2} - 6 = 0$. The lowly *check* is what proves the existential claim.

This procedure—assume what needs to be proved and argue forward to what follows from it—is known as the *Method of Analysis*. It seems completely wrong. Why would we assume the proposition we want to prove? But remember, the resulting conclusion of such an argument is only a *necessary* condition for the proposition. Analysis is a fruitful tool of mathematical discovery, but it does not produce a proof. The *Method of Analysis* must be followed by the *method of synthesis*, i.e., by a deductive proof of the proposition. This might occur merely by reversing the argument, showing that the necessary condition is also *sufficient*; or it might use the conclusion in some other way, as in the above example, to show that the proposition is satisfied.

The *Method of Analysis* has a long and distinguished history. It was first used in ancient Greek geometry, but it was later reinterpreted and made the basis of elementary algebra by Viète (1591), Descartes (1637), and others in their theory of equations. We'll see shortly that it is important for unique existence proofs as well as existence proofs.

2.4.3 *Existential Instantiation*

Existential Instantiation (EI), the *Elimination Rule* for ∃, is the most complex quantifier *Int-Elim Rule*. It proceeds as follows. If $\exists \mathbf{x} \mathbf{P}(\mathbf{x})$ is true, then some object satisfies $\mathbf{P}(\mathbf{x})$, though we don't know which one. Temporarily using an unassigned constant $\mathbf{a}$ to name this object, we prove a sentence $\mathbf{Q}$ that does not depend on which name was used. $\mathbf{Q}$ then follows from $\exists \mathbf{x} \mathbf{P}(\mathbf{x})$.

Schematically, we have:

$$
\begin{array}{|l}
\exists \mathbf{x} \mathbf{P}(\mathbf{x}) \\
\quad \begin{array}{|l}
\mathbf{P}(\mathbf{a}) \qquad [\mathbf{a} \text{ an arbitrary constant}] \\
\mathbf{Q} \qquad\; [\mathbf{a} \text{ not in } \mathbf{Q}] \qquad\qquad \textbf{\textit{EI}} \\
\end{array} \\
\hline
\mathbf{Q}
\end{array}
$$

To use *EI*, make sure that no unwarranted results about $\mathbf{a}$ are assumed in the argument and that the conclusion $\mathbf{Q}$ does not mention the name $\mathbf{a}$. Then $\mathbf{Q}$ may be exported one proof-level out as a conclusion of $\exists \mathbf{x} \mathbf{P}(\mathbf{x})$.

We can gain a more intuitive understanding of *EI* by again considering existential sentences as disjunctions. Doing so turns *EI* into the FOL version of *Cases*. Starting with the grand disjunction $\exists \mathbf{x} \mathbf{P}(\mathbf{x})$, if we can deduce $\mathbf{Q}$ from a generic disjunct $\mathbf{P}(\mathbf{a})$, we can then conclude $\mathbf{Q}$ in the main argument.

Like *Cases*, *EI* is a sound rule, though demonstrating this is complicated, since it's a suppositional FOL rule.

✚ **Example 2.4.4**

Illustrate *EI*'s use in the *Number Theory* argument of Example 1.

Solution

· Example 1 looked at part of a proof that the product of odd numbers is odd. According to the definition, n *is odd* $\leftrightarrow \exists i(n = 2i + 1)$. So if n is odd, $\exists i(n = 2i + 1)$ by *BE*. We can next let i_0 name the integer asserted to exist and argue in terms of that: suppose $n = 2i_0 + 1$. Any conclusion $\mathbf{Q}$ following from this then also follows by *EI* from $\exists i(n = 2i + 1)$.

· Informal mathematical arguments don't bother to introduce i_0—they use the symbol i both as a variable and as the name of the instance, essentially dropping the existential quantifier as the argument continues. A more formal approach uses i for the variable and i_0 for the arbitrary constant denoting the instance, but the proof procedure is roughly the same.

The next example argues an immediate inference from *Aristotelian Logic* (see Example 2.2.6a) to illustrate the formal use of *EI*. Note that when a premise set includes both an existential and a universal sentence, you should *first instantiate the existential sentence so that you know what value to instantiate the universal sentence to*. This is good advice for creating informal proofs as well.

✤**Example 2.4.5**
   Show that $\forall x(Px \rightarrow Qx)$, $\exists x Px \vdash \exists x Qx$.

Solution
   The following proof diagram establishes this claim.

1	$\forall x(Px \rightarrow Qx)$	Prem
2	$\exists x Px$	Prem
3	$\quad Pa$	Spsn for EI
4	$\quad\quad \forall x(Px \rightarrow Qx)$	Reit 1
5	$\quad\quad Pa \rightarrow Qa$	UI 4
6	$\quad\quad Qa$	MP 5, 3
7	$\quad\quad \exists x Qx$	EG 6
8	$\exists x Qx$	EI 2, 3-7

2.4.4 Simplifying Negated Quantified Sentences

UG and *EG* are direct ways to prove universal and existential sentences. An indirect proof strategy—*Proof by Contradiction*—is also available. We assume the quantified sentence's negation and deduce contradictory sentences.

To use negated quantified sentences as suppositions, we need to be able to draw conclusions from them. Two *Replacement Rules* cover this.

If $\neg\forall\mathbf{x}\mathbf{P}(\mathbf{x})$ is true, then not all elements of **U** satisfy $\mathbf{P}(\mathbf{x})$, i.e., at least one does not satisfy $\mathbf{P}(\mathbf{x})$. Thus, $\exists\mathbf{x}(\neg\mathbf{P}(\mathbf{x}))$ is true. The converse also holds. If $\exists\mathbf{x}(\neg\mathbf{P}(\mathbf{x}))$ is true, then some object of **U** fails to satisfy $\mathbf{P}(\mathbf{x})$, and so $\neg\forall\mathbf{x}\mathbf{P}(\mathbf{x})$ is true. This shows that $\neg\forall\mathbf{x}\mathbf{P}(\mathbf{x}) \vDash\dashv \exists\mathbf{x}(\neg\mathbf{P}(\mathbf{x}))$, which gives us *Universal Negation* (*UN*), the rule for negating universal statements.

Existential Negation (*EN*) tells how to negate existential sentences. An argument like that just given shows $\neg\exists\mathbf{x}\mathbf{P}(\mathbf{x}) \vDash\dashv \forall\mathbf{x}(\neg(\mathbf{P}(\mathbf{x}))$ (see Exercise 9).

We thus have the following negation *Replacement Rules*:

UN	$\neg\forall\mathbf{x}\mathbf{P}(\mathbf{x}) :: \exists\mathbf{x}(\neg\mathbf{P}(\mathbf{x}))$
EN	$\neg\exists\mathbf{x}\mathbf{P}(\mathbf{x}) :: \forall\mathbf{x}(\neg\mathbf{P}(\mathbf{x}))$

The soundness of these *Replacement Rules* can be demonstrated by deducing each equivalent from its mate using only *Int-Elim Rules* for quantified sentences and *Proof by Contradiction* (see Exercise 10). We could leave them out of FOL's Natural Deduction System with no loss of deductive power, but we'll include them since they make it easier to work with negated sentences.

Another way to think about these rules is to consider quantified sentences as generalized conjunctions and disjunctions. Then *UN* and *EN* are the FOL counterparts of the *DeM Replacement Rules*. Both rules treat negated quantifiers in the same way: *move* ¬ *past the quantifier and change the quantifier*. To simplify the result further, you apply PL *Negation Replacement Rules* to the negated inner sentence. The next two examples illustrate this process.

❖ **Example 2.4.6**
   a) Negate $\forall x(Px \to Qx)$ and simplify the result.
   b) Negate $(\forall x \in \mathcal{P})Qx$, the restricted-quantifier formulation of part a.

Solution
   a) By *UN*, $\neg\forall x(Px \to Qx)$ simplifies to $\exists x(\neg(Px \to Qx))$.
      Using *Neg Cndnl* to simplify this further gives $\exists x(Px \land \neg Qx)$.
   b) $\neg(\forall x \in \mathcal{P})Qx$ is equivalent to $(\exists x \in \mathcal{P})(\neg Qx)$, for if not all x in $\mathcal{P}$ have
      property Q, then some x in $\mathcal{P}$ must fail to have property Q.
      Note that *we do not negate the quantifier restriction* $x \in \mathcal{P}$. We instead
      negate the additional properties x *has*. Restricted quantifiers are thus
      negated in the same way as ordinary quantifiers (see also Exercise 33).
      Our final result is a restricted-quantifier form of part a's conclusion:
      $(\exists x \in \mathcal{P})(\neg Qx)$ abbreviates $\exists x(Px \land \neg Qx)$.

❖ **Example 2.4.7**
   Negate and simplify Playfair's *Euclidean Parallel Postulate*: *for any line l*
   and any point P not on l there is a line m through P that is parallel to l.

Solution
   · The negated sentence is symbolized as follows (see Example 2.1.10):
      $\neg\forall l\forall P(P \notin l \to \exists m(P \in m \land m \parallel l))$
   · We can now pass to the following logical equivalents, moving $\neg$ inward:
      $\exists l\exists P(\neg(P \notin l \to \exists m(P \in m \land m \parallel l)))$ via *UN*, twice
      $\exists l\exists P(P \notin l \land \neg\exists m(P \in m \land m \parallel l))$ via *Neg Cndnl*
   · Translating this sentence back into mathematical English, we have:
      there is a line l and a point P not on l such that no line m passing through
      P is parallel to l.
   · We have transformed the negation somewhat, but we can go further:
      $\exists l\exists P(P \notin l \land \forall m(\neg(P \in m \land m \parallel l)))$ via *EN*
      $\exists l\exists P(P \notin l \land \forall m(P \notin m \lor m \nparallel l))$ via *DeM*
      $\exists l\exists P(P \notin l \land \forall m(P \in m \to m \nparallel l))$ via *Cndnl*
   · Translating this final sentence back into mathematical English gives us the
      following final form for the negation of *Playfair's Postulate*:
      there is a line l and a point P not on l such that all lines m passing through
      P are not parallel to l.

   The last example illustrates the issue of deciding which sentence is the
simplest negation of the original. Generally, the most natural form of an
existential sentence is a quantified conjunction, and the most natural form of
a universal sentence is a quantified conditional (see Section 2.1).

   In mathematics, you'll usually negate a sentence without going through
a series of equivalent sentences—the problem of which negation to choose
among several may not arise. Instead, you may face a different problem. How
can you be sure that your negation is correct? If there's any uncertainty, you
should carefully formalize the sentence and negate it as done above.

2.4.5 *Deductions Using Replacement Rules*

We'll now look at two examples that use *UN* and *EN* in formal deductions. These will also involve the *Int-Elim Rules* for existential sentences. Note in the first one that *EI* is once again used prior to *UI*.

❖**Example 2.4.8**

Show that $\forall x(Px \wedge Qx)$, $\neg\forall x(Px \wedge Rx) \vdash \neg\forall x(Px \rightarrow Rx)$.

Solution

The following proof diagram establishes this claim.

1	$\forall x(Px \wedge Qx)$	Prem
2	$\neg\forall x(Px \wedge Rx)$	Prem
3	$\exists x(\neg(Px \wedge Rx)$	UN 2
4	$\neg(Pa \wedge Ra)$	Spsn for EI
5	$\neg Pa \vee \neg Ra$	DeM 4
6	$\forall x(Px \wedge Qx)$	Reit 1
7	$Pa \wedge Qa$	UI 6
8	Pa	Simp 7
9	$\neg Ra$	DS 5, 8
10	$Pa \wedge \neg Ra$	Conj 8, 9
11	$\exists x(Px \wedge \neg Rx)$	EG 10
12	$\exists x(\neg(Px \rightarrow Rx))$	Neg Cndnl 11
13	$\neg\forall x(Px \rightarrow Rx)$	UN 12
14	$\neg\forall x(Px \rightarrow Rx)$	EI 3, 4-13

The last example used a counterexample to prove its conclusion, as does the next one. However, each of these could have been proved instead using *Proof by Contradiction* (see Exercise 14).

❖**Example 2.4.9**

Show that $\forall x(0 \neq s(x)) \vdash \neg\forall y\exists x(y = s(x))$, where s is a function.

Solution

· To prove the conclusion, we'll deduce its equivalent, $\exists y\forall x(y \neq s(x))$. This is an existential sentence, so we need an a to make $\forall x(a \neq s(x))$ true. But this is immediate, given the premise: take $a = 0$.

· This yields the following formal proof diagram.

1	$\forall x(0 \neq s(x))$	Prem
2	$\exists y\forall x(y \neq s(x))$	EG 1
3	$\exists y(\neg\exists x(y = s(x)))$	UN 2
4	$\neg\forall y\exists x(y = s(x))$	EN 3

2.4.6 *Uniqueness Assertions in Deductions*

To prove a sentence of the form $\exists!\,\mathbf{x}\mathbf{P}(\mathbf{x})$, we need to find an instance $\mathbf{t}$ that satisfies $\mathbf{P}(x)$ and show that $\mathbf{t}$ is the only object that does. This gives us our final *Replacement Rule*, which we'll call *Unique Existence* (*Uniq Exis*).

$$\exists!\,\mathbf{x}\mathbf{P}(\mathbf{x}) :: \exists\mathbf{x}\mathbf{P}(\mathbf{x}) \wedge \forall\mathbf{x}\forall\mathbf{y}(\mathbf{P}(\mathbf{x}) \wedge \mathbf{P}(\mathbf{y}) \to \mathbf{x} = \mathbf{y}) \qquad \textit{Uniq Exis}$$

The elimination procedure for $\exists!$ is much like that for $\exists$. Given $\exists!\,\mathbf{x}\mathbf{P}(\mathbf{x})$, you first conclude $\exists\mathbf{x}\mathbf{P}(\mathbf{x})$ by means of *Uniq Exis* and *Simp*. Then, using *EI*, you suppose that $\mathbf{a}$ is the (unique) instance and argue in terms of it. In informal proofs, though, you will go directly from the unique existence claim to supposing that $\mathbf{a}$ is the unique instance satisfying $\mathbf{P}(\mathbf{x})$. Whatever is properly proved from $\mathbf{P}(\mathbf{a})$ then follows from $\exists!\,\mathbf{x}\mathbf{P}(\mathbf{x})$.

To prove $\exists!\,\mathbf{x}\mathbf{P}(\mathbf{x})$, you deduce $\exists\mathbf{x}\mathbf{P}(\mathbf{x})$ and $\forall\mathbf{x}\forall\mathbf{y}(\mathbf{P}(\mathbf{x}) \wedge \mathbf{P}(\mathbf{y}) \to \mathbf{x} = \mathbf{y})$ separately and then conjoin them, i.e., you show that some object $\mathbf{x}$ satisfies $\mathbf{P}(-)$ and that if $\mathbf{x}$ and $\mathbf{y}$ denote any objects satisfying $\mathbf{P}(-)$, then $\mathbf{x} = \mathbf{y}$.

Although we can prove these conjuncts in either order, the most intuitive approach is to show that there is such an object before proving it is unique. Nevertheless, this common-sense approach is often the less fruitful course of action. By instead supposing that $\mathbf{x}$ and $\mathbf{y}$ satisfy $\mathbf{P}(-)$ (usually without supposing that they're distinct), we may not only find that there is at most one object, but also what that object must be.

At times, by supposing only that $\mathbf{x}$ satisfies $\mathbf{P}(-)$ (the *Method of Analysis*), you may be able to determine what the object *must be*—say, that $\mathbf{x} = \mathbf{t}$. This proves uniqueness (the second conjunct), but it also gives you an instance $\mathbf{t}$ to check for satisfying the existence clause.

❖**Example 2.4.10**

Prove that the additive identity for real-number arithmetic is unique, i.e., show that $\exists!z\forall x(x + z = x \wedge z + x = x)$ (see also Exercise 34a).

Solution

Let's begin with existence. 0 is an identity because $x + 0 = x = 0 + x$.
To prove uniqueness, suppose z_1 and z_2 each satisfy the identity equations.
Then $z_1 = z_1 + z_2 = z_2$, and so $z_1 = z_2$.   ∎

This sort of argument occurs repeatedly in mathematics for identities. Proofs for the uniqueness of inverses proceed similarly (see Exercise 34b).

2.4.7 *Formal vs. Informal Proofs*

We'll close this lesson with an example that illustrates the striking difference between an informal proof and a formal one.

❖**Example 2.4.11**

Prove the transitive law *if $a < b$ and $b < c$, then $a < c$*, given the laws for arithmetic and the definition *$x < y$ if and only if there is a positive real number z such that $x + z = y$*.

Solution

Suppose $a < b$ and $b < c$. Then there are positive real numbers p_1 and p_2 such that $a + p_1 = b$ and $b + p_2 = c$. Substituting, we get $(a + p_1) + p_2 = c$. This gives us $a + (p_1 + p_2) = c$. Since $p_1 + p_2$ is positive, $a < c$.   ∎

❖ **Example 2.4.12**

Rework the last example using a formal proof diagram.

Solution

· The formal proof goes as follows:

1	$\forall x \forall y (x < y \leftrightarrow \exists z(z > 0 \wedge x + z = y))$	Prem
2	$\forall x \forall y (x > 0 \wedge y > 0 \rightarrow x + y > 0)$	Prem
3	$\forall x \forall y \forall z ((x + y) + z = x + (y + z))$	Prem
4	$a < b \wedge b < c$	Spsn for CP
5	$a < b$	Simp 4
6	$\forall x \forall y (x < y \leftrightarrow \exists z(z > 0 \wedge x + z = y))$	Reit 1
7	$a < b \leftrightarrow \exists z(z > 0 \wedge a + z = b)$	UI 6 (2×)
8	$\exists z(z > 0 \wedge a + z = b)$	BE 7, 5
9	$b < c$	Simp 4
10	$b < c \leftrightarrow \exists z(z > 0 \wedge b + z = c)$	UI 6 (2×)
11	$\exists z(z > 0 \wedge b + z = c)$	BE 10, 9
12	$z_1 > 0 \wedge a + z_1 = b$	Spsn for EI
13	$z_2 > 0 \wedge b + z_2 = c$	Spsn for EI
14	$z_1 > 0$	Simp 12
15	$z_2 > 0$	Simp 13
16	$z_1 > 0 \wedge z_2 > 0$	Conj 14, 15
17	$\forall x \forall y (x > 0 \wedge y > 0 \rightarrow x + y > 0)$	Reit 2
18	$z_1 > 0 \wedge z_2 > 0 \rightarrow z_1 + z_2 > 0$	UI 17 (2×)
19	$z_1 + z_2 > 0$	MP 18, 16
20	$a + z_1 = b$	Simp 12
21	$b + z_2 = c$	Simp 13
22	$(a + z_1) + z_2 = c$	Sub 20, 21
23	$\forall x \forall y \forall z ((x + y) + z = x + (y + z))$	Reit 3
24	$(a + z_1) + z_2 = a + (z_1 + z_2)$	UI 23 (2×)
25	$a + (z_1 + z_2) = c$	Sub 22, 24
26	$z_1 + z_2 > 0 \wedge a + (z_1 + z_2) = c$	Conj 19, 25
27	$\exists z(z > 0 \wedge a + z = c)$	EG 26
28	$\exists z(z > 0 \wedge a + z = c)$	EI 8, 11, 12-27
29	$a < c \leftrightarrow \exists z(z > 0 \wedge a + z = c)$	UI 6 (2×)
30	$a < c$	BE 29, 28
31	$a < b \wedge b < c \rightarrow a < c$	CP 4-30
32	$\forall x \forall y \forall z (x < y \wedge y < z \rightarrow x < z)$	UG 31 (3×)

· Our premises include the *definition of* < (line 1); the fact that *the sum of positive numbers is positive* (line 2); and the *associative law for addition* (line 3). These turn out to be sufficient.
· Although this deduction is long, it has been shortened by combining multiple universal instantiations (steps 7, 10, 18, 24, 29), the two existential instantiation subproofs (steps 12-28), and the multiple universal generalizations (step 32). Despite its length, your knowledge of logic should enable you to follow this proof without difficulty.

2.4.8 *Logic's Contribution to Mathematics*

We've explored *Propositional Logic* and *First-Order Logic* in a partly formal fashion, using symbols for sentences, connectives, predicates, quantifiers, and so on, both to formulate inference rules and to work examples and exercises.

There are good reasons for this degree of formality. The first is that logic needs to focus on logical form in order to analyze patterns of valid inference and construct conclusive deductions.

A second reason is that most mathematical arguments are too complex to logically analyze and formalize until you're familiar with all the connectives, quantifiers, and inference rules. But with FOL's full deduction system at our disposal, we're finally able to tackle genuine mathematical arguments. As you saw from the last example, however, even a simple mathematical argument becomes unbearably long and complicated when its full logical detail is disclosed. Informal proofs take for granted many details that go beyond what's needed to communicate their key ideas.

On the other hand, logic helps us see that *it is possible in principle to make deductions logically rigorous*. Over the past century and a half, logic has progressed to where mathematical logicians now believe that FOL can formalize any mathematical proof (though some advocate extending FOL to a higher order logic, where one can quantify over subsets of the universe of discourse as well as its elements). While this significant achievement is still underappreciated by many, current theorem-proving software programs have taken advantage of these developments. The use and value of computers for constructing proofs will undoubtedly only increase as time goes on.

But to return to an earlier point, the fact that rigor *can be* attained does not mean that it *should be*. A high degree of rigor is important for some foundational concerns and for developing automated deduction systems, but at some point in everyday mathematical arguments there is a trade-off between logical rigor and clarity. Constructing *natural deduction proofs* with component subproofs helps to exhibit the main parts of a deduction, but the accumulation of logical details eventually obscures the argument for humans.

The main benefit of having studied *Propositional Logic* and *First-Order Logic* is that you now know the inference rules underlying proof techniques used all the time in mathematics and other fields. Being familiar with these inference rules, with the overall strategy of the *Backward-Forward Method of*

Proof Analysis, and with using subproofs to obtain key results as intermediate steps should help you decide what proof strategies might be fruitful.

Using these tools effectively will still take lots of practice. The logic you've learned should function unobtrusively below the surface as you study other areas of discrete mathematics. You'll eventually forget what terms like *Modus Tollens* and *Existential Instantiation* refer to, and you may not be able to create formal proofs on the spot. But having absorbed the key ideas of logic, you should be well equipped to follow the inferential maneuvers in proofs constructed by others and to map out a strategy for making your own proofs.

EXERCISE SET 2.4

Exercises 1–2: Completing Deductions
Fill in the reasons for the following deductions.

2.4.1. $\exists x Px,\ \forall x(Qx \to \neg Px) \vdash \neg\forall x Qx$

1	$\exists x Px$	_____
2	$\forall x(Qx \to \neg Px)$	_____
3	$\quad Pa$	_____
4	$\quad\quad \forall x(Qx \to \neg Px)$	_____
5	$\quad\quad Qa \to \neg Pa$	_____
6	$\quad\quad \neg Qa$	_____
7	$\quad\quad \exists x(\neg Qx)$	_____
8	$\quad\quad \neg\forall x Qx$	_____
9	$\neg\forall x Qx$	_____

2.4.2. $\neg\exists x(Px \wedge Qx),\ \exists x(Rx \wedge Qx) \vdash \neg\forall x(Rx \to Px)$

1	$\neg\exists x(Px \wedge Qx)$	_____
2	$\exists x(Rx \wedge Qx)$	_____
3	$\quad Ra \wedge Qa$	_____
4	$\quad\quad \neg\exists x(Px \wedge Qx)$	_____
5	$\quad\quad \forall x(\neg(Px \wedge Qx))$	_____
6	$\quad\quad \neg(Pa \wedge Qa)$	_____
7	$\quad\quad \neg Pa \vee \neg Qa$	_____
8	$\quad\quad Qa$	_____
9	$\quad\quad \neg Pa$	_____
10	$\quad\quad Ra$	_____
11	$\quad\quad Ra \wedge \neg Pa$	_____
12	$\quad\quad \exists x(Rx \wedge \neg Px)$	_____
13	$\exists x(Rx \wedge \neg Px)$	_____
14	$\exists x(\neg(Rx \to Px))$	_____
15	$\neg\forall x(Rx \to Px)$	_____

Exercises 3–4: Logical Implication and Conclusive Deductions
*Determine whether the following **implication** claims are true. Then determine whether the **deductions** given are **conclusive**. Carefully point out where and how a rule of inference is being misused.*

2.4.3. $\forall x Px \rightarrow \exists y Qy, \ \neg\forall y Qy \vDash \forall x(\neg Px)$

1	$\forall x Px \rightarrow \exists y Qy$	Prem
2	$\neg\forall y Qy$	Prem
3	$Pa \rightarrow \exists y Qy$	UI 1
4	$\quad Pa \rightarrow Qb$	Spsn for EI
5	$\quad\quad \neg\forall y Qy$	Reit 2
6	$\quad\quad \neg Qb$	UI 5
7	$\quad\quad \neg Pa$	MT 4, 6
8	$\quad\quad \neg\exists x Px$	EG 7
9	$\neg\exists x Px$	EI 3, 4–8
10	$\forall x(\neg Px)$	EN 9

2.4.4. $\exists x Px, \ \forall y(Py \vee Qy) \vDash \neg\forall y Qy$

1	$\exists x Px$	Prem
2	$\forall y(Py \vee Qy)$	Prem
3	Pa	EI 1
4	$Pa \vee Qa$	UI 2
5	$\neg Qa$	DS 4, 3
6	$\exists y(\neg Qy)$	EG 5
7	$\neg\forall y Qy$	UN 6

Exercises 5–8: True or False
Are the following statements true or false? Explain your answer.

2.4.5. You can never prove anything with examples.

2.4.6. *Int-Elim Rules* for $\forall$ and $\exists$ are counterparts to PL rules for $\wedge$ and $\vee$.

2.4.7. The *Method of Analysis* is used to prove that the numbers found in solving an equation satisfy the equation.

2.4.8. FOL provides the tools for making mathematical arguments rigorous.

Exercises 9–10: Soundness of UN and EN
Work the following problems related to the Replacement Rules UN and EN.

2.4.9. Show the soundness of *EN* by explaining why the two sentence forms involved are logically equivalent.

2.4.10. Show the following without using *UN* or *EN*:
  a. $\neg\forall x Px \vdash \exists x(\neg Px)$ b. $\neg\exists x Px \vdash \forall x(\neg Px)$

Exercises 11–15: Deductions
Deduce the following, using FOL's inference rules.

2.4.11. $\forall x(Px \lor Qx),\ \exists x(\neg Px) \vdash \exists x Qx$

2.4.12. $\exists x(\neg(Px \land Qx)),\ \forall x Px \vdash \exists x(\neg Qx)$

2.4.13. $\exists x(Px \land Qx) \vdash \exists x Px \land \exists x Qx$

2.4.14. $\forall x(Px \land Qx),\ \neg\forall x(Px \land Rx) \vdash \neg\forall x(Px \to Rx)$ Prove this using *NI*.

2.4.15. $\exists x Px,\ \forall x(Px \to x = b \lor x = c) \vdash Pb \lor Pc$

Exercises 16–17: Aristotelian Logic and Derivations
Deduce the following, which represent syllogistic forms of argument.

2.4.16. $\exists \mathbf{x}(\mathbf{Px} \land \mathbf{Qx}),\ \forall \mathbf{x}(\mathbf{Qx} \to \mathbf{Rx}) \vdash \exists \mathbf{x}(\mathbf{Px} \land \mathbf{Rx})$

2.4.17. $\exists \mathbf{x}(\mathbf{Px} \land \mathbf{Qx}),\ \forall \mathbf{x}(\mathbf{Rx} \to \neg\mathbf{Qx}) \vdash \exists \mathbf{x}(\mathbf{Px} \land \neg\mathbf{Rx})$

Exercises 18–19: Interderivability
Deduce the following interderivability results.

2.4.18. $\exists x(Px \lor Qx) \dashv\vdash \exists x Px \lor \exists x Qx$

2.4.19. $\exists x(Px \to Qx) \dashv\vdash \forall x Px \to \exists x Qx$

2.4.20. Argue for the validity of the following equivalences. Thus, each quantifier can be defined in terms of the other one and negation, if so desired.
  a. $\exists \mathbf{x} \mathbf{Px} \dashv\vdash \neg\forall \mathbf{x}(\neg\mathbf{Px})$ b. $\forall \mathbf{x} \mathbf{Px} \dashv\vdash \neg\exists \mathbf{x}(\neg\mathbf{Px})$

Exercises 21–26: Logical Equivalences and Negations
Determine logical equivalents for the following negations. Simplify the results using Replacement Rules from PL to put them in their most natural form.

2.4.21. $\neg\exists x(Px \land Qx)$ **2.4.24.** $\neg\forall x(Px \to Qx)$

2.4.22. $\neg\exists x(Px \leftrightarrow Qx)$ **2.4.25.** $\neg\forall x \forall y \exists z(x + z = y)$

2.4.23. $\neg\forall x(Px \land Qx)$ **2.4.26.** $\neg\forall x \forall y \exists m(N(m) \land mx > y)$

Exercises 27–30: Negating Formal Definitions
*On the supposition that the antecedent of each biconditional does **not** hold, explain what can be concluded (via NBE); that is, **negate** the defining condition (the clause following the double arrow). Fully simplify your answer.*

2.4.27. $E(a) \leftrightarrow \exists k(a = 2k)$ **2.4.29.** $x < y \leftrightarrow \exists z(z > 0 \land x+z = y)$

2.4.28. $a \mid b \leftrightarrow \exists m(a = mb)$ **2.4.30.** $S \subseteq T \leftrightarrow \forall x(x \in S \to x \in T)$

Exercises 31–32: Negating Informal Definitions
Formulate the following definitions, identifying a universe of discourse and nonstandard symbols. Then negate the defining condition and simplify.

2.4.31. A positive integer greater than 1 is *prime* if and only if it has no positive integer factors except itself and 1.

2.4.32. $\{x_n\}$ is a *Cauchy Sequence* if and only if for every positive real number ϵ there is a natural number N such that the absolute difference $|x_m - x_n|$ is less than ϵ for all m and n greater than N.

2.4.33. *Negating Restricted and Unrestricted Existentials*
  a. Formulate $(\exists \mathbf{x} \in \mathcal{P})\mathbf{Qx}$ without using a restricted quantifier.
  b. Negate and simplify both sentences in part *a*. Then tell why these represent the same sentence.

Exercises 34–36: Formal and Informal Deductions
Construct deductions for the following mathematical results, as instructed.
Take Examples 11 *and* 12 *as models for how much logical detail to supply.*

2.4.34. *Identities and Inverses*
  a. Give a formal proof to show that the additive identity for arithmetic is unique, i.e., deduce $\exists! z \forall x (x + z = x \wedge z + x = x)$ (see Example 11).
  b. Give an informal argument to show that the additive inverse of any real number is unique.

2.4.35. Prove the following argument in two ways, first by giving an informal argument, and second by giving a formal argument: *All rational numbers are algebraic. All transcendental numbers are not algebraic. Some real numbers are transcendental. Therefore, some real numbers are irrational.*

2.4.36. Using the definition for being odd, $\forall x (Ox \leftrightarrow \exists y (x = 2y + 1))$, plus any laws of algebra, construct a formal proof of the fact that *the product of two odd numbers is odd.* The argument is sketched in Examples 1 and 4.

Chapter 3
Mathematical Induction and Arithmetic

3.1 Mathematical Induction and Recursion

First-Order Logic supplies two methods for proving universal sentences—the direct method of *Universal Generalization* and the indirect method of *Proof by Contradiction*. *Proof by Mathematical Induction* (*PMI*) is a strategy that can be used when the universal quantifier ranges over the set of natural numbers. This is a proof technique found throughout mathematics, though it originates in natural number arithmetic.

Our focus in this section is on the most basic form of mathematical induction. We'll explain its overall strategy, give an intuitive argument for its soundness, illustrate it with examples from various fields of mathematics, and show how it's related to *Recursive Definition*. Later sections look at some variations and extensions of *PMI*, investigate its theoretical basis in *Peano Arithmetic*, and explore the notion of divisibility for the natural numbers.

3.1.1 *Introduction to Mathematical Induction*

To introduce *Proof by Mathematical Induction*, we'll analyze the *Tower of Hanoi* (Figure 3.1), a favorite game of mathematicians, computer scientists, and neuro-psychologists because of the thought process involved. This one-person game[1] is played by moving discs, one at a time, from a starting peg to a terminal peg, using a third peg for temporary holding, never placing a larger disc on top of a smaller one.

Fig. 3.1 Tower of Hanoi

Playing the game isn't difficult when there are only a few discs, but it becomes harder with a large number of discs. And determining the least number of moves needed to complete the game in general may seem impossible.

We'll show that the game can be played, and we'll determine the optimal number of moves, regardless of how many discs are used. Before you read the analysis below, try the game yourself with three or four discs. If you don't have the game at hand, play it online or with different sized coins, using dots on a piece of paper to represent the pegs.

Game Analysis
   1) For one disc, the game takes one move.

---

[1] Invented by mathematician and Fibonacci aficionado E. Lucas in 1883, versions of this game and a discussion of its history can be found on numerous web sites.

© Springer Nature Switzerland AG 2019

C. Jongsma, *Introduction to Discrete Mathematics via Logic and Proof*,
Undergraduate Texts in Mathematics,
https://doi.org/10.1007/978-3-030-25358-5_3

2) Two discs can be transferred in a minimum of three moves:
  i) move the small disc to the temporary peg;
  ii) place the large disc on the terminal peg; and
  iii) put the small disc back on top of the large one.
3) Three discs require a minimum of seven moves (try it).
n) How many moves if you start with four discs? with five? in general?

To determine the minimum number of moves needed, let's tabulate our results and look for an emerging pattern.

Number of Discs	Minimal Number of Moves
1	1
2	3
3	7
4	?
n	m_n

This approach is only viable with a good-sized sample of tabulated values. However, playing more games to produce enough data from which to conjecture the general relationship between the number of discs n and the minimal number of moves m_n is time-consuming and error-ridden, since it's easy to lose track of what you're doing and repeat earlier moves. What we need in addition to concrete data is insight into how the minimal number of moves can be calculated from what we've done. How can we argue (a thought experiment now) that the above data is correct and then extrapolate to get further correct values? Let's start with $n = 2$. To move two discs, move the top one to the temporary peg; then move the other disc to the terminal peg; finally, move the small disc back on top of the other one—a total of three moves minimum.

For three discs, first move the top two discs to the temporary holding peg (we know this can be done as prescribed, so move both at once and register your count as 3); then move the bottom disc to the terminal peg (move 4); and finally, move the two smaller discs onto that disc (3 more moves). The total number of moves needed for the whole process is 7.

Generalizing, at each stage we can move the top n discs to the temporary peg in some m_n moves, move the bottom disc to the terminal peg in one move, and then move the n discs back on top of the large disc in another m_n moves. The total moves for $n + 1$ discs is thus $2m_n + 1$ moves.

This recursive procedure is captured by the following equations:

$$m_1 = 1$$
$$m_{n+1} = 2m_n + 1$$

This does not explicitly define m_n in terms of n, but at least it convinces us that the game can be played with any number of discs, and it generates enough accurate values to help us determine the pattern. Beginning with $n = 1$, we get the following table of values:

Discs	Minimal Moves
1	**1**
2	$2 \cdot 1 + 1 \;=\; \mathbf{3}$
3	$2 \cdot 3 + 1 \;=\; \mathbf{7}$
4	$2 \cdot 7 + 1 \;=\; \mathbf{15}$
5	$2 \cdot 15 + 1 \;=\; \mathbf{31}$
$\vdots$	$\vdots$
n	$\mathbf{m_n}$
$n+1$	$\mathbf{2 \cdot m_n + 1}$

Tower of Hanoi Moves

A slight modification of our formulas makes the doubling more prominent.

$$m_1 + 1 = 2$$
$$m_{n+1} + 1 = 2m_n + 2 = 2(m_n + 1).$$

This sequence proceeds by doubling from 2, giving powers of 2: 2, 4, 8, ... In general, therefore, $m_n + 1 = 2^n$, or $m_n = 2^n - 1$.

A three-step argument proves that this general formula is correct:
1) $2^n - 1$ is the minimal number of moves for $n = 1$ discs.
2) If $m_k = 2^k - 1$ is the minimum number of moves for k discs, $k + 1$ discs require $m_{k+1} = 2m_k + 1 = 2(2^k - 1) + 1 = 2^{k+1} - 1$ moves.
3) From steps 1 and 2, we conclude that $m_n = 2^n - 1$ for every n. The value of m_n at each stage depends on the previous value, starting from the first. This formula thus holds for all n.

3.1.2 *Proof by Mathematical Induction*

Proof by Mathematical Induction is the specialized method of proving a proposition $P(n)$ for all counting numbers n by the three-step process just described. We'll schematize it and name each step for future reference.

Proof by Mathematical Induction
To prove $\forall \mathbf{n} \mathbf{P}(\mathbf{n})$ when $U = \{1, 2, \ldots\}$, the set of counting numbers:
1) *Base case*: *prove* $\mathbf{P}(\mathbf{1})$.
2) *Induction step*: *assume* $\mathbf{P}(\mathbf{k})$ for an arbitrary number k; *prove* $\mathbf{P}(\mathbf{k+1})$.
3) *Conclusion*: *conclude* $\forall \mathbf{n} \mathbf{P}(\mathbf{n})$.

To use a fanciful analogy, visualize an infinite line of dominoes D_n waiting to be knocked down. To get all dominoes to fall, you push the lead domino D_1 (prove the base case), and you make sure the dominoes are lined up so each domino D_k knocks over the next one D_{k+1} (prove the induction step). If this is done, all dominoes will fall (draw the universal conclusion).

Or, think of proof by induction as climbing an infinitely tall ladder. You can reach every rung R_n (the conclusion) if you start on the bottom rung R_1 (the base case) and have some way to move from each rung R_k to the one right above it R_{k+1} (the induction hypothesis).

We'll give a more detailed analysis of *PMI* in Section 3.3, but for now, note that steps one and two are needed to guarantee the validity of the conclusion. If either is missing, you can prove false statements (see Exercises 12 and 13).

For now we'll concentrate on how *Proof by Mathematical Induction* gets used. Example 1's result was known already to Archimedes around 250 B.C., but it was independently rediscovered by a precocious young Gauss about two millennia later. Gauss, the story goes, used the idea to add up the first 100 numbers, quickly solving a problem his teacher had hoped would keep the class busy for a while.

✤ Example 3.1.1

Prove Gauss's classic formula for summing up the following series:
$$1 + 2 + \cdots + n = \frac{n(n+1)}{2} \text{ for all counting numbers } n.$$

Solution

· We'll give two arguments for this result, starting with an informal proof that matches how Gauss calculated his answer of 5050.
  Proof #1:
  · Note that the first and last terms of the series add up to $n + 1$, as do the second term and the second-to-last term, etc.
  · Let's list the series twice, once in increasing order $(1 + 2 + \cdots + n)$ and once in decreasing order $(n + (n - 1) + \cdots + 1)$.
    Adding first terms, second terms, etc. gives n sums, all equal to $n + 1$.
  · Half of this total is the value of the original series: $\dfrac{n(n+1)}{2}$. ∎

· This clever proof is informative, but it lacks rigor, due to the missing terms indicated by the ellipsis. On the other hand, it does show how the sum's value arises. Our second argument, using *PMI*, assumes this value as given.
  Proof #2:

  1) *Base case*
  $$1 = \frac{1 \cdot 2}{2}. \checkmark$$

  2) *Induction step*
  Suppose $1 + 2 + \cdots + k = \dfrac{k(k+1)}{2}$.   Indn Hyp

  Then $1 + \cdots + k + (k+1) = \dfrac{k(k+1)}{2} + (k+1)$ Sub

  $$= \frac{k(k+1) + 2(k+1)}{2}$$ Algebra

  $$= \frac{(k+1)(k+2)}{2}. \checkmark$$ Factoring

3) *Conclusion*
   Therefore, $1 + 2 + \cdots + n = \dfrac{n(n+1)}{2}$. ∎ *PMI*

Following the advice of Section 2.3 about how to organize a proof with equations, our induction-step argument began with one side of the equation and moved toward the other side, substituting the induction hypothesis (when $n = k$) to obtain a new expression equal to the original one.

❖ **Example 3.1.2**
Find and prove a formula for the odd-number series $1 + 3 + \cdots + (2n - 1)$, used by Galileo around 1600 in analyzing the motion of falling bodies.

Solution
The sequence of partial sums for this series, corresponding to successive values of n, is $1, 4, 9, 16, \ldots$, evidently a sequence of squares.
So we'll conjecture that $1 + 3 + \cdots + (2n - 1) = n^2$.
We'll give three proofs of this formula, exhibiting different degrees of rigor.
Proof #1:

The first proof illustrates concretely how the Pythagoreans originally discovered and deduced it about 500 BC. Adding successive odd numbers can be shown by arranging each new odd term as a sort of carpenter's square of dots/pebbles around the earlier configuration. This produces a square of n^2 objects. ∎

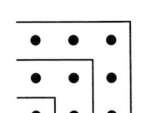

Proof #2:
Now we'll argue this based on the result of Example 1.
Since $1 + 2 + \cdots + n = \dfrac{n(n+1)}{2}$, $\ 2 + 4 + \cdots + 2n = n^2 + n$.
Subtracting n from this last series, 1 from each term, we get our result:
$1 + 3 + \cdots + (2n - 1) = n^2$. ∎
Proof #3:
Our final argument uses *Proof by Mathematical Induction*.
1) *Base case*
   $1 = 1^2$. ✓
2) *Induction step*
   Suppose $1 + 3 + \cdots + (2k - 1) = k^2$.   Indn Hyp
   Then $1 + \cdots + (2k - 1) + (2k + 1) = k^2 + 2k + 1$ Sub
                                $= (k + 1)^2$. ✓   Factoring
3) *Conclusion*
   Therefore, $1 + 3 + \cdots + (2n - 1) = n^2$ for all n. ∎ PMI

The next example is a fairly simple result from algebra, but it illustrates how mathematical induction can be used to prove an inequality.

❖ **Example 3.1.3**
Show for any real numbers a and b that if $a \leq b$, then $na \leq nb$ for all counting numbers n. Assume the additivity property $x \leq y \rightarrow x + z \leq y + z$.

Solution

We'll use a combination of *Universal Generalization* and *Proof by Mathematical Induction* here. Suppose a and b are any real numbers (for *UG*) and that $a \leq b$. To show $na \leq nb$, we'll use induction on n.

Proof:

1) *Base case*
$$1 \cdot a = a \leq b = 1 \cdot b. \checkmark$$

2) *Induction step*

Suppose $ka \leq kb$.	Indn Hyp
Then $(k+1)a = ka + a$	Algebra
$\leq ka + b$	Add Prop $(a \leq b)$
$\leq kb + b$	Add Prop $(ka \leq kb$ by hyp$)$
$\leq (k+1)b. \checkmark$	Factoring

3) *Conclusion*

Therefore, $na \leq nb$ for all n.   ∎   PMI

Although *Mathematical Induction* is now an essential feature of mathematics, it wasn't explicitly recognized until rather late. The ancient Greeks never used it. Some induction-like reasoning can be found in the work of a few medieval Arabic and Jewish mathematicians, but the first formulation of *Mathematical Induction* occurs in the work of the seventeenth-century French mathematician Blaise Pascal in connection with his investigation of what we now call *Pascal's Triangle* (see Section 4.4).

3.1.3 Recursive Definitions

Proof by Mathematical Induction often goes hand-in-hand with *Recursive Definition*. To define some operation, relation, or property, you may be able to define it for all numbers by some logical equivalence, as we did earlier with the notion of being odd (Example 2.1.8) and being divisible (Example 2.1.10). Sometimes, though, you need to define a notion piece-meal, first for 1, then for 2, and so on. Such a process would never terminate, so we use an inductive process to define it successively-all-at-once. This is known as *Recursive Definition*. The meaning/value of the defined notion for each counting number depends upon its meaning/value for earlier numbers.

Simple recursive definitions have two parts, corresponding to the first two parts of *PMI*: an *initialization step/base case* defines the concept for 1, and a *recursion step* tells how it is defined for any number $k + 1$ in terms of its predecessor k. Such a process uniquely defines the concept. We state an informal version of this claim in the following *Recursion Theorem*. Its truth is intuitively clear, but proving it rigorously requires an argument we can't do at this point.[2]

---

[2] For a discussion of this theorem, see Leon Henkin's April 1960 article *On Mathematical Induction* in **The American Mathematical Monthly**.

Theorem: *Recursive Definitions Are Well Defined*

> *If a concept is defined recursively, first for $n = 1$ and then for $n = k+1$ in terms of its meaning for $n = k$, then the concept is uniquely defined for all counting numbers.*

To show how recursive definitions are made and used, we'll present a recursive definition for exponentiation and then prove a basic law for exponents (see Exercises 24–26 and 3.2.30–34 for other laws).

✤**Example 3.1.4**

Define exponentiation a^n for real numbers a and counting numbers n.

Solution
- Let a be an arbitrary real number. Then
  1) $a^1 = a$, and
  2) $a^{k+1} = a^k \cdot a$ for all k.
- This defines a^n for any counting number n. The first power of a real number a is the number itself, and any later power is the product of the preceding power with the number a.
- Starting with step 1 and using step 2 repeatedly, we can generate any positive integer power of a. The first two of these are:

$$a^2 = a^1 \cdot a = a \cdot a \qquad \text{the product of two } a\text{'s}$$
$$a^3 = a^2 \cdot a = a \cdot a \cdot a \quad \text{the product of three } a\text{'s}$$

✤**Example 3.1.5**

Using the recursive definition for exponentiation and the *Associative Laws* for addition and multiplication, prove the law of exponents $a^m \cdot a^n = a^{m+n}$, where a is a real number and both m and n are counting numbers.

Solution

Proof:
Suppose a is any real number and m is any counting number. For such an a and m, we'll prove the law for any counting number n by mathematical induction.

1) *Base case*

$a^m \cdot a^1 = a^m \cdot a$	Defn part 1, Sub
$= a^{m+1}.$ ✓	Defn part 2, Sub

2) *Induction step*

Suppose $a^m \cdot a^k = a^{m+k}$.   Indn Hyp

Then $a^m \cdot a^{k+1} = a^m \cdot (a^k \cdot a)$	Defn part 2, Sub
$= (a^m \cdot a^k) \cdot a$	Assoc for $\cdot$
$= a^{m+k} \cdot a$	Ind Hyp, Sub
$= a^{(m+k)+1}$	Defn part 2, Sub
$= a^{m+(k+1)}.$ ✓	Assoc for $+$

3) *Conclusion*

Therefore, $a^m \cdot a^n = a^{m+n}$ for all counting numbers n by *PMI*. ✓

Since a and m are arbitrary, the conclusion holds for any real number a and any counting numbers m and n by *Universal Generalization*. ∎

For our last example, we'll give a rigorous definition for finite series like those used in Examples 1 and 2.

✤**Example 3.1.6**

Give a recursive definition for the finite series $\sum_{i=1}^{n} a_i = a_1 + a_2 + \cdots + a_n$.

Solution

Let a_i be any real numbers, $i = 1, 2, \ldots, n$. Then

1) $\sum_{i=1}^{1} a_i = a_1$, and

2) $\sum_{i=1}^{k+1} a_i = \sum_{i=1}^{k} a_i + a_{k+1}$.

This definition allows us to replace $1 + 2 + \cdots + n$ in Example 1 with $\sum_{i=1}^{n} i$ and the series $1 + 3 + \cdots + (2n - 1)$ in Example 2 with $\sum_{i=1}^{n} (2i - 1)$.

3.1.4 *Aside on Proof Style*

This may be a good time to comment on evolving proof styles. While our proof analyses in the rest of the text will occasionally identify key logical rules of inference employed in a proof, we'll no longer put our arguments into the standard proof-diagram format that we developed for logic. We're now beginning to move toward a more conventional mathematical style, occasionally using what we know about logic to help us *choose a proof strategy* for constructing an argument. This allows us to keep our proofs simpler and not spell out all the logical details (recall Example 2.4.13). We may at times still use a two-column format of sorts like above for clarity (this will be especially true in Section 3.4, when we develop *Peano Arithmetic* in a careful axiomatic fashion), but we'll mostly cite as our reasons *some mathematical result* or *algebra* or the name of a technique like *factoring*. Eventually, when the inferences are obvious, we will omit giving reasons for our steps. You may do likewise in working problems from the exercise sets as the text proceeds and as you gain confidence in constructing proofs.

If you write your proofs in an informal paragraph style, merging your *Backward-Forward Proof Analysis* with your deduction, be sure to use words like *we know* to indicate results known to hold and words like *we want to show* or *we need to prove* to indicate results you'd like to use but still don't have. Without such indicators, your proofs will become chaotic and confuse your reader. That person may be you—if you need to go back to an earlier part of your argument, you may not recognize which results you've established and which ones still need to be argued. For this reason, mathematics textbooks and instructors usually insist that, in the end, no matter how they were discovered and first written up, proofs should be presented in a completely forward direction, a practice we modeled in the first two chapters on logic.

EXERCISE SET 3.1

Exercises 1–5: Triangular Numbers

Square numbers, of the form n^2, are so named because they can be represented by an $n \times n$ square array of dots. **Triangular numbers** are the numbers $1, 3, 6, 10, \ldots$, which can be pictured with a triangular array of dots of size $T_n = 1 + 2 + 3 + \cdots + n$. These numbers were introduced in Example 1.

3.1.1. *Sums of Triangular Numbers*
Prove that the sum of two successive triangular numbers, $T_n + T_{n+1}$, is a perfect square in the following ways:
  a. First determine what square this triangular sum is. Illustrate your result by joining dot diagrams in the shape of right triangles for $n = 1, 2, 3$.
  b. Prove your general result algebraically using UG, adding the triangular number formulas provided by Example 1.
  c. Prove your general result using mathematical induction. Use the recursive relationship that generates triangular numbers: $T_{k+1} = T_k + (k + 1)$.

3.1.2. Prove that $\sum_{i=1}^{n} T_i = \dfrac{n + 2}{3} T_n$ using mathematical induction.

3.1.3. Prove that $8T_n + 1 = (2n + 1)^2$ in the following ways:
  a. Geometrically for the cases $n = 2, 3$, putting four paired triangular numbers around a central unit square.
  b. Algebraically, using the formulas provided by Example 1.
  c. Using PMI, based on the recursive relationship $T_{k+1} = T_k + (k + 1)$.

3.1.4. *Squares of Triangular Numbers*
Prove that $T_n^2 = \sum_{i=1}^{n} i^3$ using PMI and the recursive relationship $T_{k+1} = T_k + (k + 1)$.

3.1.5. Prove that $T_{n+1}^2 - T_n^2 = (n + 1)^3$ in the following ways.
  a. Algebraically, using the formulas provided by Example 1.
  b. Algebraically, using the result of Exercise 4.
  c. Using PMI, based on the recursive relationship $T_{k+1} = T_k + (k + 1)$.

Exercises 6–9: Divisibility and Mathematical Induction

Prove the following divisibility results for all counting numbers n using PMI. Recall that $a \mid b$ (a divides b) means $b = ma$ for some integer m.

3.1.6. *Factors of $x^n - 1$*
  a. $2 \mid 3^n - 1$
  b. $3 \mid 4^n - 1$
  c. Conjecture and prove a divisibility result for 4: $4 \mid$ _____ .
  d. Conjecture and prove a divisibility result for m: $m \mid$ _____ .
  e. Find a zero of the polynomial $x^n - 1$. According to the *Factor Theorem*, what's a factor of $x^n - 1$? How does this match your result in part d?

3.1.7. $3 \mid n^3 + 2n$

3.1.8. $5 \mid n^5 - n$

3.1.9. Show that $n^5/5 + n^3/3 + 7n/15$ is an integer for all counting numbers.

Exercises 10–13: Exploring Induction and Recursion
Work the following, which explore aspects of mathematical induction.

3.1.10. Explain in your own words why induction is a valid form of mathematical argumentation. Is this a form of proof you have used before?

3.1.11. Explain in your own words how *Recursive Definition* defines a concept for all counting numbers.

3.1.12. Is the proposition $n^2 - n + 17$ *is prime for all counting numbers* true or false? If it is true, prove it using mathematical induction. If it is false, give a counterexample and explain why such a proof breaks down.

3.1.13. What's wrong with the following proof that *birds of a feather flock together*, i.e., all the birds in a flock of any size are of the same species.
Proof:
Let n denote the size of a flock of birds. We'll prove the result by induction for any counting number n.
1) If a flock has only one member, obviously every member is of the same species, so the result holds when $n = 1$.
2) Suppose, now, that the result holds for any flock of size k. Then it must also hold for any flock of size $k + 1$, too.
   Consider a flock of size $k + 1$. One bird flies off, leaving a flock of size k. All the birds in this subflock are then of the same species.
   The wayward bird returns, and another one flies off. This also leaves a subflock of size k, so this group of birds are of the same species. But the birds that never flew off are of the same species as the ones that did.
   Hence, all birds in the full flock of size $k + 1$ are of the same species.
3) Therefore, by induction, all flocks contain birds of a common species.   ∎

Exercises 14–16: True or False
Are the following statements true or false? Explain your answer.

3.1.14. *Proof by Mathematical Induction* is a logical inference rule for concluding universal sentences in mathematics.

3.1.15. *Recursive Definitions* are for mathematical definitions what *Proofs by Mathematical Induction* are for mathematical proofs.

3.1.16. *Proof by Mathematical Induction* is often paired with *Recursive Definition* in developing mathematics.

Exercises 17–19: Finite Power Series
Prove the following formulas for series of powers using PMI.

3.1.17. $\displaystyle\sum_{i=1}^{n} i^2 = \frac{n(n+1)(2n+1)}{6}$

3.1.18. $\displaystyle\sum_{i=1}^{n} i^3 = \left[\frac{n(n+1)}{2}\right]^2 = \left(\sum_{i=0}^{n} i\right)^2$ [this connects to Exercise 4]

3.1.19. $\displaystyle\sum_{i=1}^{n} i^4 = \frac{n(n+1)(2n+1)(3n^2 + 3n - 1)}{30}$

Exercises 20–23: Finite Series of Products and Their Inverses

Prove the following, using mathematical induction.

3.1.20. $\sum\limits_{i=1}^{n} i(i+1) = \dfrac{n(n+1)(n+2)}{3}$

3.1.21. $\sum\limits_{i=1}^{n} (2i-1)(2i+1) = \dfrac{n(4n^2+6n-1)}{3}$

3.1.22. $\sum\limits_{i=1}^{n} \dfrac{1}{(2i-1)(2i+1)} = \dfrac{n}{2n+1}$

3.1.23. Determine and prove a formula for summing $\sum\limits_{i=1}^{n} \dfrac{1}{i(i+1)}$.
Hint: try $n = 1, 2, 3$ to conjecture the formula.

Exercises 24–26: Laws of Exponents for Counting Numbers

Prove the following for counting-number exponents. You may use Example 4, Example 5, and any general laws for addition or multiplication.

3.1.24. $(a^m)^n = a^{m \cdot n}$ **3.1.25.** $a^n \cdot b^n = (a \cdot b)^n$

3.1.26. $a^n/a^m = a^{n-m}$, $m < n$.   *Hint*: do induction on n; generalize on m.

Exercises 27–28: Factorials

Work the following problems on factorials. Recall that n! denotes the product of all positive integers from 1 to n inclusive.

3.1.27. *Recursive Formula for Factorials*
  a. Give a recursive definition of $n!$.
  b. Using part a, show that $2! = 2 = 1 \cdot 2$ and $3! = 6 = 1 \cdot 2 \cdot 3$. What familiar formula for $n!$ results from your recursive definition?

3.1.28. Prove $\sum\limits_{i=1}^{n} i \cdot i! = (n+1)! - 1$ using the recursive definition of factorial from Exercise 27 and mathematical induction.

Exercises 29–30: Inequalities

Prove the following results, using mathematical induction where appropriate.
3.1.29. If $0 < a < 1$, then $a^n < 1$ for any counting number n. You may assume the basic order properties of $<$.

3.1.30. *Bernoulli's Inequality*
Prove $(1+b)^n \geq 1 + bn$ for counting numbers n, where $b > -1$, $b \in \mathbb{R}$. You may use any basic results about exponentiation and the order relation $\leq$. Point out where $b > -1$ enters into your proof.

3.2 Variations on Induction and Recursion

At times *PMI* may seem like the natural strategy to try for making a proof, but if the base case doesn't match what's needed or the induction step fails to go through, you may need another form of induction. We'll explore variants of *PMI* in this section, as well as another version of recursive definition.

3.2.1 Modified Proof by Mathematical Induction

As we saw in Section 3.1, *Proof by Mathematical Induction* begins with the case $n = 1$ and, after the induction step, concludes a result $P(n)$ for all counting numbers n. Sometimes, though, the base case we want isn't $n = 1$. Perhaps $P(n)$ holds for all natural numbers, including 0.[3] If so, either we can prove $P(0)$ separately, or we can make $n = 0$ our base case. *Modified Proof by Mathematical Induction* (*Mod PMI*) is the latter strategy.

Mathematical induction can be argued for a variety of integer base cases. If we want to prove a result, for example, for all n-sided polygons, we would begin with $n = 3$, because taking $n < 3$ makes no sense. Whatever the starting point n_1, if the induction step shows how to pass from $n = k$ to $n = k + 1$ for all integers $k \geq n_1$, we can conclude $P(n)$ for all $n \geq n_1$.

Modified Proof by Mathematical Induction (*Mod PMI*)
    Suppose $U = \mathbb{Z}$, the set of integers.
    1) *Modified base case*: <u>prove $\mathbf{P(n_1)}$</u> for some integer $\mathbf{n_1}$.
    2) *Modified induction step*: <u>assume $\mathbf{P(k)}$</u> for $\mathbf{k \geq n_1}$; <u>prove $\mathbf{P(k + 1)}$</u>.
    3) *Modified conclusion*: <u>conclude $(\forall \mathbf{n} \geq \mathbf{n_1})\mathbf{P(n)}$</u>.

Our first example illustrates modified induction, using a mix of backward and forward argumentation, typical of informal mathematical arguments.

✤**Example 3.2.1**
    Show that $n^3 < n!$ for all $n \geq 6$.

Solution
Our proof uses a modified base case. To establish the induction step, it helps to prove another inequality first, which can be formulated as a separate proposition. It, too, can be proved via *Mathematical Induction* (see Exercise 2), but we'll use some basic results that permit us to avoid it.
Proof:
    1) *Base case*
        · $6^3 < 6!$; i.e., $216 < 720$. ✓
        · Note that 6 is the best we can do for n_1: $125 > 120$.
          While the result holds for $n = 0$, it fails for $n = 1, \ldots, 5$.
    2) *Induction step*
        · Suppose that $k^3 < k!$ for $k \geq 6$.
        · We want to show that $(k + 1)^3 < (k + 1)!$.
        · We'll use a backward argument here; a forward argument would appear unmotivated. The first part is reversible, so we'll use *iff* here (standard mathematical shorthand for *if and only if*).

$$(k + 1)^3 < (k + 1)! \qquad \text{Desired Conclusion}$$
$$\text{iff} \quad (k + 1)^3 < k!\,(k + 1) \qquad \text{Factoring}$$
$$\text{iff} \quad (k + 1)^2 < k! \qquad \text{Canceling } (k \neq -1)$$

---

[3] We'll discuss our reasons for choosing to include 0 in $\mathbb{N}$ in Sections 3.4 and 5.3.

- We know $k^3 < k!$ by our induction hypothesis.
  If we can also show that $(k+1)^2 < k^3$, combining these two inequalities will prove $(k+1)^2 < k!$, our desired inequality.
- Thus, we only need $(k+1)^2 < k^3$ for $k \geq 6$ to conclude the proof.
- *This result* actually holds for $k \geq 3$, as we'll prove. Two subproofs illustrate the different paths we can take. We'll look at both of them.

Subproof #1:

$$k^3 > (k+1)^2 \qquad \text{Desired Conclusion}$$
$$\text{iff} \qquad k^3 - k^2 - 2k > 1 \qquad \text{Algebra}$$
$$\text{iff} \quad k(k+1)(k-2) > 1 \qquad \text{Factoring}$$

- But since $k \geq 3$, each of these left-hand-side factors is larger than 1, and so the product is also larger than 1 (see also Exercise 1). ✓

Subproof #2:

- In this subproof, we'll argue the above result by first replacing one k in k^3 by 3, which is less than k, and then showing that $3k^2 > (k+1)^2$.
- If this holds, the original inequality will, too.

$$k^3 > (k+1)^2 \qquad \text{Desired Conclusion}$$
$$\text{if} \qquad 3k^2 > (k+1)^2 \qquad \text{Since } k \geq 3,\ k^3 \geq 3k^2$$
$$\text{iff} \qquad 2k^2 - 2k > 1 \qquad \text{Algebra}$$
$$\text{iff} \qquad 2k(k-1) > 1 \qquad \text{Factoring}$$

- But for $k \geq 3$, $2k > 1$ and $k - 1 > 1$ (you can prove *these* by mathematical induction, too, if you wish), so $2k(k-1) > 1$. ✓

3) *Conclusion*
   By *Mod PMI*, $n^3 < n!$ for all $n \geq 6$. ∎

As this example shows, if you need a non-trivial result in the middle of an induction argument, you may first need to prove *that* proposition, which may need its own induction argument. Whether or not a proof requires multiple inductions partly depends upon how much you know before you start.

3.2.2 *Mathematical Induction and Integer Arithmetic*

To prove a proposition $P(n)$ not merely for all integers greater than some initial integer but for *all integers*, we typically proceed as follows. We use induction to prove $P(n)$ for all natural numbers (or we use induction to show $P(n)$ for all positive integers and prove $P(0)$ separately), and then we use the induction *result* just proved to show $P(n)$ for negative integers $n \in \mathbb{Z}^-$. Although we could develop a reversed form of induction for negative integers, this is not normally done, because the result already proved should enable us to prove the general result and avoid a further induction proof.

For example, we can extend the additive law for exponents (Example 3.1.5) to integer exponents. To do this, we first need a definition for negative exponents. Here, too, we can avoid induction (recursion) by building upon the recursive definition for counting-number exponents.

❖ **Example 3.2.2**

Define zero and negative exponents.

Solution

The standard definitions, chosen so the laws of exponents will continue to hold for the new values (see Example 4 and Exercises 30–34), are as follows:

Definition: *If a is a non-zero real number, then*

    1) $a^0 = 1$, *and*

    2) $a^{-n} = 1/a^n$ *for positive integers n.*

Since a^n is already defined for positive integers (see Example 3.1.4) and never yields 0 (since $a \neq 0$), our symbolism is well defined.

In the next two examples we'll first show that the definition just given for a^{-n} holds for all integers n, and then we'll get a start on demonstrating the addition law for all integral exponents.

❖ **Example 3.2.3**

Show that $a^{-n} = 1/a^n$ for all integers n.

Solution

Proof:

Assuming the above definition, we'll argue by cases to avoid induction.

  Case 1: $n \in \mathbb{Z}^+$

    $a^{-n} = 1/a^n$. ✓   Defn of Negative Exponents

  Case 2: $n = 0$

    $a^{-0} = a^0$ Arithmetic

       $= 1$ Defn of Zero Exponent

       $= 1/1$ Arithmetic

       $= 1/a^0$. ✓   Defn of Zero Exponent, Sub

  Case 3: $n \in \mathbb{Z}^-$

  Let p be the positive natural number associated with n: i.e., $n = -p$.

    $a^{-n} = a^p$ Sub, Arithmetic

       $= 1/(1/a^p)$ Reciprocals, Algebra

       $= 1/a^{-p}$ Defn of Negative Exponents, Sub

       $= 1/a^n$. ✓   Sub

Thus, $a^{-n} = 1/a^n$ for all integers n.   ■

❖ **Example 3.2.4**

Prove $a^m \cdot a^n = a^{m+n}$ for any natural number m and any integer n.

Solution

Proof:

We'll consider two cases. The second appeals to a result in the exercises that requires only earlier results.

  Case 1: both m and n are natural numbers

    We proved this for counting-number exponents in Example 3.1.5.

    When $m = 0$ or $n = 0$, the equation to check is trivial since $a^0 = 1$. ✓

Case 2: m is a natural number and n is a negative integer
   Let p be the positive natural number associated with n: i.e., $n = -p$.

$$\begin{aligned} a^m \cdot a^n &= a^m \cdot a^{-p} && \text{Sub} \\ &= a^m \cdot (1/a^p) && \text{Defn of Negative Exponents} \\ &= a^m/a^p && \text{Meaning of Division} \\ &= a^{m-p} && \text{See Exercise 33} \\ &= a^{m+(-p)} && \text{Meaning of Subtraction} \\ &= a^{m+n}. \checkmark && \text{Sub} \end{aligned}$$

Since we've shown the result for all possible cases, $a^m \cdot a^n = a^{m+n}$ for all natural numbers m and all integers n. ∎

3.2.3 *Strong Proof by Mathematical Induction*

We've seen how to modify *PMI* by using a different *base case*. Now we'll look at a variation that modifies the *induction step*. This version is needed when $P(k + 1)$ doesn't relate well to its predecessor $P(k)$ but connects to earlier cases. In such situations, the standard induction hypothesis is too restrictive. Since earlier $P(n)$ can be considered established prior to $n = k + 1$, any of them should be available for demonstrating $P(k+1)$, not just $P(k)$. There is then no real reason to focus on successors here. We therefore formulate the induction clause in terms of numbers less than some arbitrary k.

   Schematically, we have the following:

Strong Proof by Mathematical Induction (*Strong PMI*)
   1) *Base case*: <u>prove **P(1)**</u>.
   2) *Strong Induction step*: <u>assume **P(n)**</u> for **1 ≤ n < k**; <u>prove **P(k)**</u>.
   3) *Conclusion*: <u>conclude ∀n**P(n)**</u>.

   This procedure can be modified in an obvious way to deal with all natural numbers or all integers from some point on, just as before. We'll illustrate this with a number theory result drawn from Euclid's ***Elements*** (VII.31).

❖**Example 3.2.5**
   Prove that *every number greater than or equal to* 2 *has a prime divisor*.

Solution
   Recall that $n \geq 2$ *is prime if and only if there is no pair of strictly smaller numbers a and b such that $a \cdot b = n$*.
Proof #1: *proof by mathematical induction*
   We begin our induction at $n = 2$ and use strong induction since prime factors don't relate well to immediate predecessors.
   1) *Base case*
      The number 2 has itself as a prime divisor. ✓
   2) *Induction step*
      Suppose all numbers $n < k$ have prime divisors.
      Then k is either prime or not prime.

    Case 1: k is prime
      Then k has itself as a prime divisor. ✓
    Case 2: k is not prime (k is composite)
      If k is not prime, then $k = a \cdot b$, for $2 \leq a, b < k$.
      The strong induction hypothesis applies to a, so a has a prime divisor.
      But any divisor of a is a divisor of k; so k has a prime divisor. ✓
    Thus, in all cases, k has a prime divisor. ✓

3) *Conclusion*
    All integers greater than or equal to 2 have prime divisors. ∎

Proof #2: Euclid's infinite descent argument
  Note: for Euclid, numbers (multitudes) begin at 2, so 1 is not a number.
  If a number A is composite, then some smaller number B divides it.
  If B is prime, the proof is done.
  If not, then B has some smaller number C that divides it.
  Again, if C is prime the proof is done, since C divides A.
  But if not, and if repeating this process never yields a prime divisor, there
  will be an unending sequence of successively smaller composite divisors of
  A, which is impossible for numbers.
  So, eventually there must be a prime factor of A. ∎

3.2.4 Well-Ordering Principle and Infinite Descent

Euclid's argument, just given, assumes a result descriptively called the *Least-Number Principle* (*every non-empty subset of the natural numbers has a least number*), but it is officially called the *Well-Ordering Principle*. This result can be proved by *Contradiction* using *Strong Mathematical Induction*.

Theorem 3.2.1: *Well-Ordering Principle*
  Every non-empty set of natural numbers has a least number.

Proof:
· Suppose to the contrary that S is a non-empty set of natural numbers with no least number.
  Then S can't contain 0, the smallest of all natural numbers.
· Let P denote the complement of S, all numbers *not* in S. Then P contains 0.
· Let $P(n)$ be the proposition n *is in* P. Then $P(0)$ is true.
  Now suppose $P(n)$ is true for all $n < k$, i.e., all $n < k$ belong to P.
  Then these n are not in S.
  But then k isn't in S, either, or it would be the least number in S.
  So k is in P; that is, $P(k)$ holds.
· By strong induction, all natural numbers are in P, i.e., S is empty.
  This contradicts our initial supposition.
· Therefore every non-empty set of natural numbers has a least number. ∎

    The *Well-Ordering Principle* validates *Proof by Infinite Descent*, an indirect proof technique used as an alternative to *Proof by Mathematical Induction* by Euclid and others and popularized especially by Fermat in his work on Diophantine number theory. An *Infinite Descent* argument supposes that a result holds for an arbitrary natural number n and proves that it must then hold for a smaller natural number m. Repeating this argument would yield an infinite sequence of strictly decreasing natural numbers, an impossibility. So there must not be any natural number satisfying the proposition.

For example, to paraphrase Euclid's proof in Example 5, if we suppose an arbitrary composite number n has no prime factor, then there must be a smaller composite number m (a proper divisor of n), also with no prime factor. As this would lead to an infinite descent of composite numbers, all composite numbers must have a prime factor. A version of the proof that $\sqrt{2}$ *is irrational* also uses this technique (see Exercise 49).

3.2.5 *Modified Recursive Definitions*

Mod PMI and *Strong PMI* correspond to certain forms of *Recursive Definition*. Recursive definitions may begin with an integer other than 1, and they may define a concept for a number in terms of its meaning for one or more preceding numbers, whether or not they immediately precede the number under consideration. A good example of such a definition is the one for the *Fibonacci sequence*. Since each term of the sequence is defined in terms of the two preceding values, the base case must give the first two values of the sequence. This is variously given as $F_0 = 0$, $F_1 = 1$ or $F_1 = 1$, $F_2 = 1$.

❖ **Example 3.2.6**
The Fibonacci sequence $1, 1, 2, 3, 5, 8, 13, 21, \cdots$ begins with the pair of numbers $F_1 = 1$, $F_2 = 1$, and each term F_n thereafter is generated as the sum of the two preceding ones. Give a recursive definition of this sequence.

Solution
Let F_n stand for the nth term in the Fibonacci sequence. Then
1) $F_1 = 1$, $F_2 = 1$; and
2) $F_{n+2} = F_n + F_{n+1}$ for $n \geq 1$.

Exploring the properties and the prolific applications of the Fibonacci sequence and related sequences could be a full-time occupation. We know many things about such sequences. We'll state one here (in a form that can be generalized) and include some others in the exercises (see Exercises 35–47). In Section 3 we'll derive a rather surprising formula for F_n.

❖ **Example 3.2.7**
Show that every third Fibonacci number is even, i.e., show $F_3 \mid F_{3n}$.

Solution
Proof:
We'll prove this using ordinary induction on n.
1) *Base case*
   $F_3 \mid F_3$, i.e., $2 \mid 2$. ✓

2) *Induction step*

Suppose F_{3k} is even: $2 \mid F_{3k}$. 　　　　　　Indn Hyp

Then $F_{3(k+1)} = F_{3k+3}$ 　　　　　　　　　　Algebra

$\qquad\qquad = F_{3k+1} + F_{3k+2}$ 　　　　　　　Defn Fib Seq

$\qquad\qquad = F_{3k+1} + (F_{3k} + F_{3k+1})$ 　　Defn Fib Seq

$\qquad\qquad = 2F_{3k+1} + F_{3k}$. 　　　　　　　Algebra

Thus, being the sum of two even numbers, $F_{3(k+1)}$ is even. ✓

3) *Conclusion*

Therefore, F_{3n} is even for all positive integers. ∎

EXERCISE SET 3.2

Exercises 1–4: Inequalities via Induction

Prove the following, using Mod PMI and any basic results from arithmetic.

3.2.1. $m \cdot n > 1$ for all natural numbers $m, n > 1$.

3.2.2. $n^3 > (n+1)^2$ for all natural numbers $n \geq 3$.

3.2.3. $n! > 2^n$ for natural numbers $n \geq 4$.

3.2.4. $2^n > n^2$ for all natural numbers $n \geq 5$.

Exercises 5–6: Geometry and Induction

Prove the following geometric results using induction.

3.2.5. Determine and prove a formula giving the maximum number of intersection points for a collection of two or more lines.

3.2.6. Formulate and prove a formula for the sum in degrees of the interior angles of a convex (non-indented) polygon. Is this true for all polygons?

Exercises 7–9: True or False

Are the following statements true or false? Explain your answer.

3.2.7. *Strong PMI* proves results for all integers, not only natural numbers.

3.2.8. *Mod PMI* differs from *PMI* in its base case.

3.2.9. *Proof by Infinite Descent* is a reversed-induction version of *PMI*.

Exercises 10–13: Finite Arithmetic Sequences

*An **arithmetic sequence** $a_0, a_1, \ldots, a_n$ is a sequence such that each term is a constant difference more than the preceding one.*

3.2.10. Give a recursive definition for an arithmetic sequence whose initial term is $a_0 = a$ and whose constant difference is d.

3.2.11. Conjecture and prove a formula for the general term a_n of an arithmetic sequence in terms of the first term $a_0 = a$ and the constant difference d.

3.2.12. Explain why successive odd numbers form an arithmetic sequence. What is a_0? What is d? What is a_{25}? Which a_n is 99?

3.2.13. A job initially pays \$10.50 per hour. If an employee receives a 50 cent raise every eight months, how much will she earn per hour after 15 years?

Exercises 14–16: Finite Arithmetic Series

A **finite arithmetic series** is a sum $S_n = \sum\limits_{i=0}^{n} a_i$ in which the terms a_i form an arithmetic sequence.

3.2.14. Determine the sum of all even numbers less than or equal to 100.

3.2.15. Determine the sum of all odd numbers less than 100.

3.2.16. Determine and prove a general formula for the sum S_n of a finite arithmetic series in terms of the first term $a_0 = a$ and the common difference d.

Exercises 17–20: Adding Up Powers

Prove the following summation formulas using mathematical induction.

3.2.17. $\sum\limits_{i=0}^{n} 2^i = 2^{n+1} - 1$ **3.2.18.** $\sum\limits_{i=0}^{n} 3^i = \dfrac{3^{n+1} - 1}{2}$

3.2.19. Based on the pattern in Exercises 17 and 18, conjecture a formula for $\sum\limits_{i=0}^{n} 4^i$. Test it when $n = 2, 3, 4$, and then prove your result.

3.2.20. Generalize the results of Exercises 17–19: conjecture a formula for $\sum\limits_{i=0}^{n} r^i$ and then prove your result. Need r be a natural number in the formula?

Exercises 21–24: Finite Geometric Sequences

A **geometric sequence** is a sequence of numbers $a_0, a_1, \ldots, a_n$ such that each new term is a constant multiple r of the last one.

3.2.21. Give a formal recursive definition for a geometric sequence whose first term is $a_0 = a$ and whose constant multiple is r.

3.2.22. Conjecture and prove a formula for the general term a_n of a geometric sequence in terms of the first term $a_0 = a$ and the constant multiple r.

3.2.23. List the first few terms of the geometric sequence that starts with $a_0 = 32$ and proceeds by repeated halving. What is a_{100} for this sequence?

3.2.24. A principal of A_0 dollars is deposited in an account that yields $100r$ percent interest per period, compounded once per period. Prove that the amount A accumulated after t periods is $A(t) = A_0(1 + r)^t$. Tell why annual accumulated amounts form a geometric sequence.

Exercises 25–26: Finite Geometric Series

A **finite geometric series** is a sum $S_n = \sum\limits_{i=0}^{n} a_i$ whose terms a_i form a geometric sequence.

3.2.25. Determine a formula for S_n in terms of $a_0 = a$ and r. *Hint*: Express S_n using only a and r; then factor out a, multiply the rest by $\dfrac{1 - r}{1 - r}$, and simplify. Prove your formula using induction. Alternatively, apply Exercise 20.

3.2.26. Use the formula developed in Exercise 25 to determine the sum of the first 12 terms of a geometric series whose first term is 32 and whose common multiple is $1/2$. (*Caution:* a_0 is the *first* term.)

Exercises 27–29: Finite Products

*The **finite product** $a_1 \cdot a_2 \cdots a_n$ is denoted in compact form by $\prod_{i=1}^{n} a_i$.*

3.2.27. Give a recursive definition of $\prod_{i=1}^{n} a_i$.

3.2.28. Using *Mod PMI*, prove that $\prod_{i=2}^{n} \left(1 - \dfrac{1}{i}\right) = \dfrac{1}{n}$ for all $n \geq 2$.

3.2.29. Using *Mod PMI*, prove that $\prod_{i=2}^{n} \left(1 - \dfrac{1}{i^2}\right) = \dfrac{n+1}{2n}$ for all $n \geq 2$.

Exercises 30–34: Laws of Exponents for Integers

Prove the following results using Example 3.1.5, Exercises 3.1.24–26, Examples 2–4 above, or any basic results of arithmetic or algebra not related to exponents. You may also use any result that appears above the one you are working. Use induction where it seems appropriate.

3.2.30. $a^m \cdot a^n = a^{m+n}$, where m and n are any integers.

3.2.31. $(a^m)^n = a^{m \cdot n}$, where m and n are any integers.

3.2.32. $(a \cdot b)^n = a^n \cdot b^n$, where n is any integer.

3.2.33. $a^n / a^m = a^{n-m}$, where m and n are any natural numbers.

3.2.34. $a^n / a^m = a^{n-m}$, where m and n are any integers.

Exercises 35–42: Fibonacci Sequence and Divisibility Results

Prove the following results about the Fibonacci sequence (see Example 6).

3.2.35. F_n and F_{n+1} are relatively prime (no common divisors except 1).

3.2.36. F_{3n+1} and F_{3n+2} are both odd numbers.

3.2.37. F_{4n} is divisible by 3.

3.2.38. F_{5n} is divisible by 5.

3.2.39. F_{6n} is divisible by 8.

3.2.40. $F_{m+n} = F_{n-1}F_m + F_n F_{m+1}$ for $n \geq 2$, $m \in \mathbb{N}$.

3.2.41. Note the results of Example 7 and Exercises 37–39. On the basis of the emerging pattern, conjecture a result for F_{mn}. Test your conjecture on some $m = 7$ terms; if correct, prove your conjecture. Exercise 40 may help.

3.2.42. Which Fibonacci terms are prime? What subscripts must a Fibonacci prime have? Is the converse also true? Why or why not? (Note: it is not yet known whether there are infinitely many Fibonacci primes.)

Exercises 43–47: Fibonacci Sequence, Squares, and Products

Prove (and find, where indicated) the following formulas for the Fibonacci sequence (see Example 6). Not all need mathematical induction.

3.2.43. $1^2 + 1^2 = 2$, $1^2 + 2^2 = 5$, $\ldots$; $F_n^2 + F_{n+1}^2 = $?? .

3.2.44. $1^2+1^2 = 2$, $1^2+1^2+2^2 = 6$, $1^2+1^2+2^2+3^2 = 15, \ldots$; $\sum\limits_{i=1}^{n} F_i^2 = \;$?? .

3.2.45. $F_{n+1}^2 = F_n F_{n+2} + (-1)^n$.

3.2.46. $F_n F_{n+3} = F_{n+2}^2 - F_{n+1}^2$.

3.2.47. Every positive integer is the sum of distinct Fibonacci numbers.

3.2.48. *Binary Representation of natural numbers*

  a. Prove that every positive integer m can be uniquely expressed as the sum of distinct binary powers (powers in the form 2^n).

  b. Explain why all numbers m satisfying $0 \leq m < 2^n$ can be uniquely represented in base two notation by an n-place binary digit (bit) numeral string $b_{n-1}b_{n-2}\cdots b_1 b_0$, where each b_i equals 0 or 1 and $m = b_{n-1}\cdot 2^{n-1} + b_{n-2}\cdot 2^{n-2} + \cdots + b_1 \cdot 2^1 + b_0 \cdot 2^0$.

  c. Can the result in parts *a* and *b* be generalized to other bases? How?

3.2.49. Prove that $\sqrt{2}$ is irrational using *Proof by Infinite Descent*. *Hint*: begin a *Proof by Contradiction* without assuming that the fraction representative m/n for $\sqrt{2}$ is in reduced form (see Section 1.8).

3.3 Recurrence Relations; Structural Induction

Inductive reasoning occurs throughout mathematics because natural numbers occur everywhere. A similar form of reasoning is employed in inductive structures that don't involve numbers. This is of special interest to logicians and computer scientists, who often encounter such phenomena. Before looking at this broader context of induction, however, we'll briefly explore how to determine closed formulas for recursively defined sequences of numbers, whose discovery at times seems mysterious. We'll only touch on some elementary matters here—more can be found in books devoted to this topic.

3.3.1 *Solving Recurrence Relations*

Proof by Mathematical Induction can prove the correctness of closed formulas for sequences that are initially defined using a base value and a recurrence relation. But where do these formulas come from? Must the pattern be intuited from a short list of initial values as we did for the Tower of Hanoi? Or are there more systematic methods for generating such formulas?

Solving recurrence relations is like solving differential equations in calculus or difference equations in linear algebra—we first develop a general solution and then find a particular solution satisfying a given initial condition. Here we solve a recurrence equation to express a_n in terms of n and a_0, obtaining a closed-formula solution when the particular base value for a_0 is specified.

There are some standard procedures for solving recurrence relations, but as with differential equations, it's easy to run into a problem that's difficult to solve. We'll present a few examples that illustrate how closed formulas can be found in some elementary ways.

❖**Example 3.3.1**

Determine a closed formula for the sequence $\{a_n\}$ recursively defined by
1) *Base case*: $a_0 = 4$.
2) *Recurrence equation*: $a_k = a_{k-1} + 2k$ for $k \geq 1$.

Solution

· The first few numbers in this sequence are 4, 6, 10, 16, 24. The sequence begins with 4 and continues by successively adding 2, 4, 6, 8, ...
· To express a_n in terms of n and the initial term 4, we'll list the sequence in expanded form so the arithmetic doesn't hide the operations involved.

$$a_1 = 4 + 2$$
$$a_2 = (4 + 2) + 4 = 4 + (2 + 4) = 4 + 2(1 + 2)$$
$$a_3 = [4 + (2 + 4)] + 6 = 4 + (2 + 4 + 6) = 4 + 2(1 + 2 + 3)$$
$$a_4 = [4 + (2 + 4 + 6)] + 8 = 4 + (2 + 4 + 6 + 8) = 4 + 2(1 + 2 + 3 + 4)$$

· The pattern is now obvious: $a_n = 4 + 2(1 + 2 + \cdots + n)$
· We already know how to add the first n counting numbers (Example 3.1.1), so our closed formula is $a_n = 4 + 2[n(n + 1)/2] = n^2 + n + 4$. This formula agrees with the terms listed above; mathematical induction proves that it holds in general (see Exercise 4b).

· Since our recurrence equation defines a_k in terms of a single multiple of a_{k-1}, we can use another procedure to develop our formula. This time we'll work backwards from a_n instead of forward from a_0.

$$a_n - a_{n-1} = 2n$$
$$a_{n-1} - a_{n-2} = 2(n - 1)$$
$$\vdots$$
$$a_1 - a_0 = 2$$

· We could expand these expressions by successive substitutions to put everything in terms of a_n, n, and a_0 (see Exercise 4a), but we'll instead add all of these equations together.
· The left-hand sum forms a telescoping series (a_{n-1} cancels $-a_{n-1}$, etc.), leaving $a_n - a_0 = 2 + \cdots + 2(n - 1) + 2n = 2(1 + \cdots + (n - 1) + n)$.
· Again, knowing the sum of successive counting numbers, this simplifies to the closed formula mentioned above: $a_n = n^2 + n + 4$.

❖**Example 3.3.2**

Determine a closed formula for the sequence $\{a_n\}$ recursively defined by
1) *Base case*: $a_0 = 1$
2) *Recurrence equation*: $a_k = 2a_{k-1} + 3$ for $k \geq 1$.

Solution

· For this sequence, let's again list the first few terms: 1, 5, 13, 29, 61. This isn't a well-known sequence, but since repeated doubling occurs due to the recurrence equation, perhaps the formula contains a power of 2.
· Let's find the pattern by using *backwards substitution*, relating specific terms to earlier ones. (Medieval mathematicians did something similar to this in some of their work with particular sequences.)

- We'll calculate a_4's value with this method.

$$a_4 = 2a_3 + 3$$
$$= 2(2a_2 + 3) + 3 = 4a_2 + 3 + 6$$
$$= 4(2a_1 + 3) + 3 + 6 = 8a_1 + 3 + 6 + 12$$
$$= 8(2a_0 + 3) + 3 + 6 + 12$$
$$= 16a_0 + 3 + 6 + 12 + 24$$

- Putting this result in factored form, we have

$$a_4 = 2^4 a_0 + 3(1 + 2 + 4 + 8) = 16a_0 + 3 \cdot 15 = 16 + 45 = 61. \checkmark$$

A closed formula for the general term therefore seems to be

$$a_n = 2^n a_0 + 3(1 + 2 + \cdots + 2^{n-1}) = 2^n + 3(2^n - 1) = 2^{n+2} - 3.$$

- We needed to know how to evaluate a finite geometric series to arrive at our final formulation, but this gives us a formula to test, using *Proof by Mathematical Induction*: $a_n = 2^{n+2} - 3$ (see Exercise 5b). As suspected, this involves a power of 2.

Our final example looks at a famous sequence defined by a *second-order linear homogeneous recurrence relation*, where each term is a linear combination of the two preceding terms. We won't develop the general theory of solving such relations, but our solution method works for others, too.

❖ Example 3.3.3

Find a closed formula for terms F_n of the Fibonacci sequence (see Example 3.2.6), defined here by $F_0 = 0$, $F_1 = 1$, $F_{n+2} = F_n + F_{n+1}$ for $n \geq 0$.

Solution

- We might try to use the telescoping sum approach of Example 1, but this leads to a formula for summing Fibonacci numbers (see Exercise 6a), not to a closed formula for F_n.
- Working backwards from F_n as in Example 2 expresses F_n as linear combinations of pairs of earlier terms, but the constants that arise are themselves Fibonacci numbers, whose formula we still need, and the process ends in the original recurrence equation $F_n = F_{n-2}F_1 + F_{n-1}F_2$ or the even more obvious $F_n = F_{n-1}F_0 + F_n F_1$ (see Exercise 6c).
- So we need a new approach. First note that several sequences might satisfy the recurrence equation, depending on what the base-case values are. Note also that if $\{a_n\}$ and $\{b_n\}$ are general solutions, so is any combination $\{c_1 a_n + c_2 b_n\}$: $c_1 a_{n+2} + c_2 b_{n+2} = (c_1 a_n + c_2 b_n) + (c_1 a_{n+1} + c_2 b_{n+1})$.
- To find an $\{a_n\}$ that solves the recurrence relation, two paths are typically taken: develop a *generating function* for the Fibonacci sequence (i.e., a formal power series whose coefficients are the terms of the sequence); or make an intelligent guess at the general form of F_n. The latter is a bit less motivated, but as it requires less background machinery, we'll sketch that solution here. A number of details are left for the reader (see Exercise 7a).
- It's well known that the ratio of Fibonacci terms F_{n+1}/F_n tends toward the limit $\phi \approx 1.618 \cdots$ known as the *golden ratio*. The information we'll ex-

tract from this is that the Fibonacci sequence is approximately a geometric sequence. This gives us a general formula to try.

· So, applying the *Method of Analysis*, suppose that some sequence of powers $F_n = r^n$ satisfies the recurrence relation: $r^{n+2} = r^n + r^{n+1}$.
Then $r^n(r^2 - r - 1) = 0$.
The viable solutions here are $r_1 = \dfrac{1 + \sqrt{5}}{2}$ and $r_2 = \dfrac{1 - \sqrt{5}}{2}$.

· Conversely, these roots satisfy the Fibonacci recurrence equation, and any linear-combination sequence $\{c_1 r_1^n + c_2 r_2^n\}$ does as well.

· To determine which one of these yield the Fibonacci sequence, we'll equate the values for the base cases: $0 = c_1 r_1^0 + c_2 r_2^0$, and $1 = c_1 r_1 + c_2 r_2$.
Solving this simple system of equations yields $c_2 = -c_1$ and $c_1 = 1/\sqrt{5}$.

Thus, $F_n = \dfrac{(1 + \sqrt{5})^n - (1 - \sqrt{5})^n}{2^n \sqrt{5}} . \checkmark$

· It's somewhat surprising and interesting that while the Fibonacci sequence contains only integers, the closed formula for its terms involves irrational numbers, the golden ratio $\phi = \dfrac{1 + \sqrt{5}}{2}$ and $1 - \phi = \dfrac{1 - \sqrt{5}}{2}$.

3.3.2 *Structural Induction*: *Theory of Strings*

The natural number system $\mathbb{N}$ can be defined as an inductive structure—starting with 0, we can continue through the entire set by going from one number to the next. We'll develop this idea in detail in Section 3.4. Numerical substructures can also be defined recursively; in fact, every recursive sequence of numbers defines such a structure (see Exercises 20–22). Our focus in the rest of this section, though, will be on recursively defined *non-numerical structures*, such as *strings* and *well-formed formulas*.

The *theory of strings* is an important foundational topic in logic, computer science, and linguistics. It deals with syntactically correct formulas, something that computer programmers know all too well from experience is critical to successfully compiling their work.

Informally, *strings* are finite ordered lists (sequences) of characters taken from some *alphabet*. The symbol ε will denote the *empty string*, a string with no characters. This string simplifies and completes the theory of strings, playing much the same role here that 0 does in arithmetic.

Definition 3.3.1: *Finite Strings Over an Alphabet*
*If alphabet $\mathcal{A}_p = \{a_1, a_2, \ldots, a_p\}$ is a set of distinct characters, $\mathcal{A}_p^*$, the set of all **finite strings/words** over $\mathcal{A}_p$, is defined as follows:*
1) *Base case: ε is a string;   [think of ε as an honorary member of $\mathcal{A}_p^*$]*
2) *Recursion step: if s is a string, then so is the extension sa_i for any a_i in $\mathcal{A}_p$ appended to s; and*
3) *Closure clause: all strings result from applying the recursion step finitely many times, starting with ε, i.e., $s = \varepsilon s_1 s_2 \cdots s_n$ for s_i in $\mathcal{A}_p$.*

Given the intended interpretation of ε as the empty string, we can assert:
1) ε is not a character $(\varepsilon \neq a_i)$;
2) $\varepsilon a_i = a_i$ for all characters a_i in $\mathcal{A}_p$.

Thus, from the base case and the recursion step, we can conclude that all characters $a_i = \varepsilon a_i$ are strings. Compound strings are constructed by appending characters to the right end of a string of characters, which the closure clause says is how all strings can be formed. In fact, string representation is unique, which we'll prove to illustrate *structural induction*, a proof technique based on a *recursive definition*, like the one just given.

Theorem 3.3.1: *Unique Representation of Strings*
A string s over $\mathcal{A}_p = \{a_1, a_2, \ldots, a_p\}$ is either the empty string ε or has a unique representation $s_1 s_2 \cdots s_n$, where each $s_i = a_j$ for some j.

Proof:
Base case: If $s = \varepsilon$, the claim is trivially true. ✓
Induction step: Suppose (*induction hypothesis*) that s is either ε or has a unique representation as a string of characters from $\mathcal{A}_p$, and consider any extension sa_i for some i.
Case 1: If $s = \varepsilon$, then $sa_i = \varepsilon a_i = a_i$, which is unique. ✓
Case 2: Suppose s is uniquely represented by a string of characters. Then sa_i is uniquely represented by a string of characters, too. ✓
Conclusion: Thus, every string s is either the empty string ε or has a unique representation $s = s_1 s_2 \cdots s_n$, where all $s_i = a_j$ for some j. ∎

Proof by structural induction in general follows the way in which a structure is recursively defined. If $\mathcal{S}$ is an inductive structure whose elements are generated by some procedure from a set of base cases, an inductive proof that a result $\mathbf{P(x)}$ holds for all $\mathbf{x}$ in $\mathcal{S}$ proceeds as follows:

Proof by Structural Induction
1) *Base case*: prove $\mathbf{P(b_i)}$ for all base cases $\mathbf{b_i}$.
2) *Induction step*: assume $\mathbf{P(x)}$ for arbitrary $\mathbf{x}$ in $\mathcal{S}$, and prove $\mathbf{P(y)}$ for any $\mathbf{y}$ generated from $\mathbf{x}$ by the recursive procedure.
3) *Conclusion*: conclude $(\forall \mathbf{x} \in \mathcal{S})\mathbf{P(x)}$.

Our recursive definition of strings allows us to append a character to a string. We'll now extend this idea to define a concatenation operation on strings.

Definition 3.3.2: *String Concatenation*
*Let $\mathcal{A}_p^*$ be the set of finite strings over alphabet $A_p = \{a_1, a_2, \ldots, a_p\}$. Then the **string concatenation** operation $\cdot$ is defined for strings s and t by:*
1) *Base case: $s \cdot \varepsilon = s = \varepsilon s$; and*
2) *Recursion step: $s \cdot (ta_i) = (s \cdot t)a_i$.*

❖**Example 3.3.4**

Using the recursive definitions for strings and string concatenation, show
that concatenation is associative for characters: $a_1 \cdot (a_2 \cdot a_3) = (a_1 \cdot a_2) \cdot a_3$.

Solution

First note that $a_i \cdot a_j = a_i \cdot (\varepsilon a_j) = (a_i \cdot \varepsilon) a_j = a_i a_j$.
Thus, $a_1 \cdot (a_2 \cdot a_3) = a_1 \cdot (a_2 a_3) = (a_1 \cdot a_2) a_3 = a_1 a_2 a_3$, and
$(a_1 \cdot a_2) \cdot a_3 = (a_1 \cdot a_2) \cdot \varepsilon a_3 = ((a_1 \cdot a_2) \cdot \varepsilon) a_3 = (a_1 \cdot a_2) a_3 = a_1 a_2 a_3$.
Therefore, $a_1 \cdot (a_2 \cdot a_3) = a_1 a_2 a_3 = (a_1 \cdot a_2) \cdot a_3$. ✓

Associativity of concatenation for characters justifies writing any product of characters as a string of those characters in order, dropping product
symbols and parentheses, as just illustrated. Generalizing, the concatenation
product of two strings is a string: if $s = s_1 s_2 \cdots s_m$ and $t = t_1 t_2 \cdots t_n$, then
$s \cdot t = s_1 s_2 \cdots s_m t_1 t_2 \cdots t_n$ (see Exercise 24). Concatenated strings, no matter
how complex the product, can be rewritten as a simple list of characters.

Concatenation is thus an associative operation on strings (see Exercise 25), but it is not commutative if the alphabet has at least two char-
acters. Also, the empty string ε satisfies identity results such as $\varepsilon \cdot x = x$
and $x \cdot y = \varepsilon \leftrightarrow x = \varepsilon \wedge y = \varepsilon$. The set $\mathcal{A}_p^*$ of all finite strings over $\mathcal{A}_p$ thus
forms an algebraic structure of sorts under concatenation.[4]

Once an inductive structure has been defined, functions can be defined on
them, though care must be taken to make sure they are well defined. The
length of a string is recursively defined as follows:

Definition 3.3.3: *Length of a String*
  *The **length of a string** in $\mathcal{A}_p^*$ is defined by:*
  1) *Base case:* $\ell(\varepsilon) = 0$; *and*
  2) *Recursion step:* $\ell(sa_i) = \ell(s) + 1$ *for any string s.*

This definition means that $\ell(a_i) = 1$, $\ell(a_i a_j) = 2$, etc. If $s = s_1 s_2 \cdots s_n$
for s_i in $\mathcal{A}_p$, then $\ell(s) = n$. This is well defined because s has a unique
representation as a string of characters by *Theorem 1*.

Proposition 3.3.1: *Length of Concatenated Strings*
  If s and t are strings, $\ell(s \cdot t) = \ell(s) + \ell(t)$.

Proof:
  We'll outline the steps and leave the details for Exercise 26.
  The proof is by induction on t, following $\mathcal{A}_p^*$'s recursive definition.
  Base case: take $t = \varepsilon$.
  Induction step: assume it holds for arbitrary t; prove it holds for all ta_i.
  To prove this, use the recursion steps from the definition of concatenation
  and the definition of length. ∎

---

[4] On axiomatizing the theory of strings, see the article *String Theory* by John Corcoran
et al. in the December, 1974 issue of ***The Journal of Symbolic Logic***.

3.3.3 *Structural Induction*: *Well-Formed Formulas*

We built up the syntax of *Propositional Logic* piece-by-piece in Chapter 1. Given sentences P and Q, we formed compound sentences using truth-functional connectives $\neg, \wedge, \vee, \rightarrow, \leftrightarrow$. For complex sentences, we used parentheses to guarantee that they were uniquely constructed/readable, but we also adopted priority conventions so we could write sentences more compactly.

Let's now make this process more mathematically precise. Suppose we have a base collection of sentence letters $P_1, P_2, \ldots$ for representing (atomic) propositions. We represent compound propositions by joining sentences using truth-functional connectives, initially basic sentences, but then also any sentences generated from them via the connectives. Not all strings of PL symbols represent propositions—they must be *well-formed formulas*. To guarantee unique readability of these formulas, we'll use parentheses around any compound sentence since it may occur in even more complex sentences.

Definition 3.3.4: *Well-Formed Formulas of Propositional Logic*
Let $\mathcal{A} = \{(,), \neg, \wedge, \vee, \rightarrow, \leftrightarrow, P_1, P_2, \ldots\}$ [i.e., left and right parentheses, connective symbols, and basic proposition symbols], and consider finite strings over this alphabet of PL symbols.
*Strings that are **well-formed formulas** (**wffs**) are defined as follows:*
1) *Base case: all P_i are well-formed formulas;*
2) *Recursion step: if **P** and **Q** are well-formed formulas, then so are $(\neg \mathbf{P})$, $(\mathbf{P} \wedge \mathbf{Q})$, $(\mathbf{P} \vee \mathbf{Q})$, $(\mathbf{P} \rightarrow \mathbf{Q})$, and $(\mathbf{P} \leftrightarrow \mathbf{Q})$; and*
3) *Closure clause: all wffs result from applying the recursion step a finite number of times, starting with basic proposition symbols P_i.*

❖**Example 3.3.5**
Which of the following are well-formed formulas for *Propositional Logic?*
$$P_3 \vee (P_2; \quad) \leftrightarrow (\wedge; \quad (\rightarrow P_1 P_2; \quad (P_1); \quad P_1 \vee \neg P_2; \quad (P_1 \rightarrow ((\neg P_2) \wedge P_3))$$

Solution
The first three words are clearly gibberish; also, all compound well-formed formulas must begin and end with a parenthesis (see Exercise 28b), while these don't. The fourth isn't well formed because it contains no connective. The fifth word looks good, but only because of our earlier priority conventions. It is not well formed, again because it lacks enclosing parentheses. Adding these would still not produce a well-formed formula, though, because no parentheses enclose $\neg P_2$. Only the last word is a well-formed formula. The production graph in Figure 3.2 shows how it is constructed according to our recursion procedure.

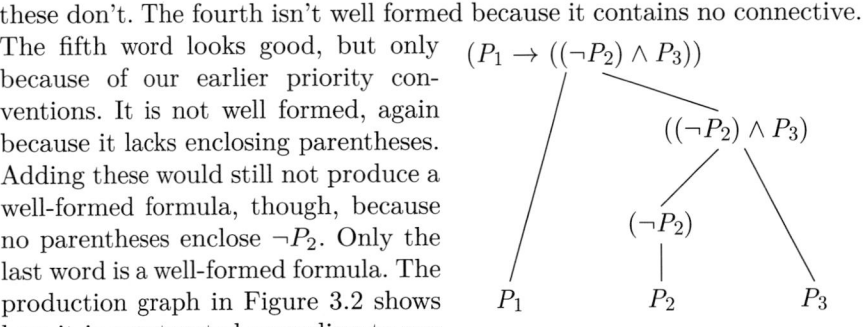

Fig. 3.2 Production graph for Example 3.3.5

We can now use the recursive definition of wffs to prove results about them or to define functions on them like a truth-value function (see Exercise 31). We'll illustrate this with the following results, beginning with two simple lemmas about the occurrence of parentheses. These set us up to prove the *Unique Composition Theorem*.

Lemma 3.3.1: *Balanced Parentheses in Wffs*

Every wff has an even number of parentheses, each left parenthesis balanced by a unique right parenthesis.

Proof:
  This result can be proved inductively (see Exercise 28), as follows:
  1) *Base case*: show that all base wffs P_i satisfy the lemma; and
  2) *Induction step*: supposing **P** and **Q** are wffs satisfying the lemma, show that wffs generated from them using PL connectives also satisfy it. ∎

An *initial segment of a string* $s_1 s_2 \cdots s_n$ is a substring $s_1 s_2 \cdots s_k$ where $k \leq n$. A *proper initial segment of a string* is one where $k < n$. The next lemma gives an important property of proper initial segments of wffs.

Lemma 3.3.2: *Parentheses in Proper Initial Segments of Wffs*

Every proper initial segment of a wff has more left than right parentheses. Consequently, no proper initial segment of a wff is itself a wff.

Proof:
  We'll prove this inductively.
  1) *Base case*: Since no P_i has a proper initial segment, the lemma is vacuously satisfied.
  2) *Induction step*: suppose **P** and **Q** are wffs satisfying the lemma.
    · The proper initial segments of $(\neg \textbf{P})$ are the substrings $\underline{(}$, $\underline{(\neg}$, and $\underline{(\neg \textbf{P}_0}$ for $\textbf{P}_0$ a proper initial segment of **P**.
    · Since $\textbf{P}_0$ already has more left parentheses by supposition, each of these has an excess of left parentheses.
    · Now suppose $*$ represents a binary propositional connective. The proper initial segments of $(\textbf{P} * \textbf{Q})$ are the substrings $\underline{(}$, $\underline{(\textbf{P}_0}$, $\underline{(\textbf{P}}$, $\underline{(\textbf{P}*}$, $\underline{(\textbf{P} * \textbf{Q}_0}$, and $\underline{(\textbf{P} * \textbf{Q}}$, for $\textbf{P}_0$ a proper initial segment of **P** and $\textbf{Q}_0$ a proper initial segment of **Q**.
    · Since by supposition both $\textbf{P}_0$ and $\textbf{Q}_0$ contain more left parentheses than right, while both **P** and **Q** contain equal amounts of them, these substrings each contain an excess of left parentheses.
  3) *Conclusion*: the lemma is satisfied by all wffs. ∎
  As a consequence, no wff has a proper initial segment that is a wff, since wffs have a balanced number of left and right parentheses. ∎

With these lemmas, we're ready to state and prove that wffs are uniquely composed, often called the *Unique Readability Theorem*. As strings, wffs already have unique representations by *Theorem* 1. This theorem goes further to assert that wffs can be constructed in only one way.

Theorem 3.3.2: *Unique Composition of Well-Formed Formulas*
A well-formed formula **F** *is exactly one of the following*:
a) *A unique propositional symbol* P_i;
b) *A unique negation* $(\neg \mathbf{P})$ *of a wff* **P**; *or*
c) *A unique compound* $(\mathbf{P} * \mathbf{Q})$ *for wffs* **P** *and* **Q** *and* $*$ *one of the binary connectives* $\wedge, \vee, \rightarrow, $ *or* $\leftrightarrow$.

Proof:
· By definition, **F** must be one of these possibilities, and these alternatives are distinct. What still needs proving is **F**'s unique composition.
· The P_i are distinct, so if $\mathbf{F} = P_i$, its composition is unique.
· Suppose **P** has a unique composition. If $\mathbf{F} = (\neg\mathbf{P})$, it must be a unique negation. For suppose $(\neg\mathbf{P}) = (\neg\mathbf{Q})$. Then, dropping off the parentheses and the negation signs, we have $\mathbf{P} = \mathbf{Q}$, which has a unique composition. Thus, **F** does as well.
· To show that there is a unique way to compose $\mathbf{F} = (\mathbf{P} * \mathbf{Q})$ from uniquely composed wffs **P** and **Q**, suppose $(\mathbf{P} * \mathbf{Q}) = (\mathbf{R} \circ \mathbf{S})$, where **R** and **S** are uniquely composed wffs and $*$ and $\circ$ are any binary connectives.
· Dropping the leading left parentheses, we have $\mathbf{P} * \mathbf{Q}) = \mathbf{R} \circ \mathbf{S})$.
· If $*$ and $\circ$ occur in the same spots of these strings, then $\mathbf{P} = \mathbf{R}$ and $\mathbf{Q} = \mathbf{S}$, which means **F** is uniquely composed.
· Else, without loss of generality, suppose $*$ occurs before $\circ$.
· Then wff **P** is a proper initial segment of **R**, which is impossible.
· Hence, $\mathbf{P} = \mathbf{R}$ and $\mathbf{Q} = \mathbf{S}$: **F** has a unique composition as $(\mathbf{P} * \mathbf{Q})$. ∎

❖**Example 3.3.6**
Prove that wffs have the same number of left parentheses as connective symbols.

Solution
This result is easy to prove inductively.
1) *Base case*: wffs P_i have 0 left parentheses and 0 connective symbols.
2) *Induction step*: suppose wffs **P** and **Q** have the same number of left parentheses as connective symbols.
   · Since $(\neg\mathbf{P})$ has 1 more left parenthesis and 1 more connective symbol than **P**, it has the same number of left parentheses as connective symbols.
   · $(\mathbf{P} * \mathbf{Q})$ also has 1 more left parenthesis and 1 more connective symbol than the equal sums of those respectively in **P** and **Q**, so it has the same number of left parentheses as connective symbols.
3) *Conclusion*: all wffs have the same number of left parentheses as connective symbols. ∎

While our recursive definition for wffs gives us an unambiguous notation to represent them, the proliferation of parentheses can make long formulas cumbersome and difficult to decipher. That's why we usually drop the outermost parentheses and adopt priority conventions for dropping others.

In Section 1.2, we saw another approach to eliminating parentheses, one proposed by the Polish logician Łukasiewicz in 1924. Polish notation pre-fixes the operators in front of the formulas being connected, using letters to indicate the connectives instead of the traditional symbols. Thus, N, K, A, C, E (mostly the initial letters for relevant Polish words) are used in place of $\neg, \wedge, \vee, \rightarrow, \leftrightarrow$ respectively. This doesn't lead to more readable formulas, but it streamlines notation and is used by some computer programming languages. Hewlett Packard used reverse Polish notation in its calculators starting in the 1970s. This requires a little practice to become familiar with it but offers a pleasantly efficient parenthesis-free way to make calculations. Exercise 32 explores this system of notation further.

✤**Example 3.3.7**

Write the following well-formed formulas (given in *Polish Notation*) in conventional logical symbolism: $NCP_3NKP_2P_1$; $AP_3CNP_2NP_1$.

Solution

Start from the right side and proceed left. Upon encountering an operation symbol, perform it on the subformula(s) just gotten. This yields the subformulas $P_2 \wedge P_1$, $\neg(P_2 \wedge P_1)$, $P_3 \rightarrow \neg(P_2 \wedge P_1)$, $\underline{\neg(P_3 \rightarrow \neg(P_2 \wedge P_1))}$; and $\neg P_1$, $\neg P_2$, $\neg P_2 \rightarrow \neg P_1$, $\underline{P_3 \vee (\neg P_2 \rightarrow \neg P_1)}$.

More can be done with strings and formulas, but this would take us deeper into the technicalities of logic than we intend to go.

EXERCISE SET 3.3

Exercises 1–3: Closed Formulas for Series of Powers
Find and prove closed formulas for the following series, as indicated. These results are used in introductory calculus to integrate simple power functions.

3.3.1. *Sum of Successive Counting Numbers*

a. Verify the identity $n^2 = \sum_{i=1}^{n} i^2 - \sum_{i=1}^{n} (i-1)^2$. Then use it to derive

$$\sum_{i=1}^{n} i = \frac{n(n+1)}{2}. \quad \textit{Hint: } \sum_{i=1}^{n}(i-1)^2 = \sum_{i=1}^{n} i^2 - 2\sum_{i=1}^{n} i + \sum_{i=1}^{n} 1.$$

b. Compare your work in part *a* with Example 3.1.1's solution.

3.3.2. *Sum of Successive Squares*

a. Verify the identity $n^3 = \sum_{i=1}^{n} i^3 - \sum_{i=1}^{n} (i-1)^3$. Then use it to derive

$\sum_{i=1}^{n} i^2 = \dfrac{n(n+1)(2n+1)}{6}$. Use the expansion method suggested in Exercise 1, along with the identity $(i-1)^3 = i^3 - 3i^2 + 3i + 1$.

b. Use *Mathematical Induction* to prove $\sum_{i=1}^{n} i^2 = \dfrac{n(n+1)(2n+1)}{6}$.

3.3.3. *Sums of Successive Powers*

a. Describe the method of deriving a closed formula for $\sum_{i=1}^{n} i^k$ as exhibited for $k = 1, 2$ in Exercises 1–2. Could this algebraic method be used to develop formulas for higher powers k? What would be involved?

b. Prove $\sum_{i=1}^{n} i^3 = \dfrac{n^4 + 2n^3 + n^2}{4} = \left(\dfrac{n(n+1)}{2}\right)^2$ using induction.

Exercises 4–7: Closed Formulas for Sequences
The following exercises concern the sequences in Examples 1–3.

3.3.4. *Closed Formula for Example 1*
a. Solve the recurrence relation for the sequence of Example 1 defined by $a_0 = 4$, $a_k = a_{k-1} + 2k$ by working backwards from a_n to a_0 as in Example 2, using successive substitutions to put a_n in terms of n and a_0.
b. Prove by *Mathematical Induction* that $a_n = n^2 + n + 4$.

3.3.5. *Closed Formula for Example 2*
a. Solve the recurrence relation for the sequence of Example 2 defined by $a_0 = 1$, $a_k = 2a_{k-1} + 3$ by creating a telescoping series as in Example 1 to put a_n in terms of n and a_0.
b. Prove by *Mathematical Induction* that $a_n = 2^{n+2} - 3$.

3.3.6. *Solving the Fibonacci-Sequence Recurrence Relation (Example 3)*
a. Show that if a telescoping-series approach is used (as in Example 1) to solve the Fibonacci recurrence relation, the result is $F_{n+2} - 1 = \sum_{i=0}^{n} F_i$.
b. Using *Mathematical Induction*, prove that $\sum_{i=0}^{n} F_i = F_{n+2} - 1$.
c. Show that if a backwards-substitution approach is used (as in Example 2) to solve the Fibonacci recurrence relation, the coefficients that arise are themselves Fibonacci numbers, leading finally to the defining equations $F_n = F_{n-2}F_1 + F_{n-1}F_2$ and $F_n = F_{n-1}F_0 + F_n F_1$.

3.3.7. *A Closed Formula for the Fibonacci Sequence*
a. Fill in the solution details in Example 3 for the approach finally taken: show why if $\{a_n\}$ and $\{b_n\}$ satisfy the Fibonacci recurrence equation, so does any $\{c_1 a_n + c_2 b_n\}$; why if $r^{n+2} = r^n + r^{n+1}$, then the viable solutions are $r_1 = (1+\sqrt{5})/2$ and $r_2 = (1-\sqrt{5})/2$; and why the linear-combination sequence $\{c_1 r_1^n + c_2 r_2^n\}$ that satisfies the Fibonacci base-case values has $c_2 = -c_1$ and $c_1 = 1/\sqrt{5}$, yielding $F_n = [(1+\sqrt{5})^n - (1-\sqrt{5})^n]/(2^n \sqrt{5})$. Explain why these results imply that F_n is the Fibonacci sequence.
b. Assuming $\lim\limits_{n\to\infty} \dfrac{F_{n+1}}{F_n}$ exists, show that it is the golden ratio $\phi = \dfrac{1+\sqrt{5}}{2}$.

Exercises 8–17: Formulas for Other Recursive Sequences
Find closed formulas for the following sequences in the ways suggested. Then prove your formulas for a_n using Mathematical Induction. Closed formulas for finite arithmetic and geometric series are needed for some problems.

3.3.8. $a_0 = 5;\ a_k = a_{k-1} + 2 - k$

Use the telescoping-series approach of Example 1 to determine a_n.

3.3.9. $a_0 = 2;\ a_{k+1} = 3a_k$

List the first four terms in expanded format to intuit the formula for a_n.

3.3.10. $a_0 = 3/4;\ a_{k+1} = 2^{k+1}a_k$

List the first four terms in expanded format to intuit the formula for a_n.

3.3.11. $a_0 = 1/2;\ a_{k+1} = (k+1)a_k$

Use backwards substitution to determine a_n. Check for $n = 0, 1, 2, 3$.

3.3.12. $a_0 = 4;\ a_k = 3a_{k-1} - 5$

Use backwards substitution to determine a_n. Check for $n = 0, 1, 2, 3$.

3.3.13. $a_0 = 0,\ a_k = \dfrac{2 + a_{k-1}}{3}$

Use backwards substitution to determine a_n. Check for $n = 0, 1, 2, 3$.

3.3.14. $a_0 = 0,\ a_k = a_{k-1} + k^2 - k$

Use the telescoping-series approach of Example 1.

3.3.15. $a_0 = 0,\ a_1 = 1,\ a_{k+2} = 3a_{k+1} - 2a_k$

List several terms to intuit the formula for a_n.

3.3.16. $a_1 = 1,\ a_2 = 3,\ a_{k+2} = a_k + a_{k+1}$ [the Lucas sequence]

Use the method of Example 3 and check your formula for $n = 0, 1, 2, 3$.

3.3.17. $a_0 = 2,\ a_1 = 4,\ a_{k+2} = a_{k+1} + 2a_k$

List several terms to intuit the formula for a_n. Then solve the second-order recurrence equation using the method of Example 3 and check your formula.

Exercises 18–19: True or False

Are the following statements true or false? Explain your answer.

3.3.18. Sequences defined by linear recurrence equations can always be solved by backwards substitution.

3.3.19. Strings of PL symbols are well-formed formulas if they have the same number of left parentheses, right parentheses, and connective symbols.

Exercises 20–22: Inductive Substructures of Number Systems

3.3.20. Determine what substructure of $\mathbb{N}$ has elements defined by the following recursive definition: $a_0 = 1,\ a_k = k^2 a_{k-1}$.

3.3.21. Determine what substructure of $\mathbb{Z}$ is defined by the following definition: $a_0 = 2,\ a_{2k+1} = -a_{2k} + 1,\ a_{2k+2} = a_{2k} + 3$ for $k \geq 0$.

3.3.22. Action $n * m$ is defined recursively for natural numbers n and integers m by $0 * m = m$ and $(k+1) * m = (k * m)/2$, yielding rational numbers.

 a. Prove that $n * m = m/2^n$ for all n. Use the recursive definition given and any results from ordinary arithmetic.

 b. Prove that if a and b are two rational numbers of the form $n * m$ for some natural number n and integer m, then so is $a + b$ and $a - b$.

 c. Prove that $(n_1 * m_1) \cdot (n_2 * m_2) = (n_1 + n_2) * (m_1 \cdot m_2)$. Thus, rational numbers in the form $n * m$ are closed under multiplication.

Exercises 23–27: String Concatenation

Work the following using the recursive definition of string concatenation.

3.3.23. *String Concatenation Identity*
Prove that $\varepsilon \cdot x = x$ for all strings over alphabet $A_p = \{a_1, a_2, \ldots a_p\}$. Thus, ε is the (two-sided) identity element for concatenation of strings.

3.3.24. *String Concatenation*
a. Prove for any string s and any characters a_i, that $s \cdot a_i = s a_i$ and $(a_1 \cdot a_2) \cdot (a_3 \cdot a_4) = a_1 a_2 a_3 a_4 = a_1 \cdot (a_2 a_3 \cdot a_4)$, extending Example 4.
b. Prove that any strings $s = a_1 a_2 \cdots a_m$ and $t = b_1 b_2 \cdots b_n$ can be concatenated to yield $s \cdot t = st = a_1 a_2 \cdots a_m b_1 b_2 \cdots b_n$.
c. Prove that all concatenated string products can be written as strings.

3.3.25. *Concatenation is Associative but not Commutative*
a. Prove that $(x \cdot y) \cdot z = x \cdot (y \cdot z)$ for any strings x, y, and z.
b. Tell why concatenation is not commutative in general.

3.3.26. *Proposition 1: Length of Concatenated Strings*
Prove that if s and t are strings, $\ell(s \cdot t) = \ell(s) + \ell(t)$.

3.3.27. *Character Occurrences in Strings*
a. Give a recursive definition for how often a character occurs in a string.
b. Prove a sum theorem for the number of occurrences of a character in concatenated strings.

Exercises 28–34: Well-Formed Formulas and Words

The following deal with PL wffs and words constructed in formal languages.

3.3.28. *Lemma 1: Balanced Parentheses in Wffs*
a. Prove by induction that *every wff has an even number of parentheses, each left parenthesis balanced by a unique right parenthesis.*
b. Prove that *all compound PL wffs start with a left parenthesis and end with a matched right parenthesis.*

3.3.29. Determine whether each of the following strings is a wff. If it is, give a production graph; if it is not, explain why not.
a. $(P_1 \wedge \neg P_2)$
b. $(P_1 \wedge ((\neg P_2) \to P_1) \vee P_3)$
c. $((P_1 \to (\neg P_2)) \wedge (P_3 \vee (\neg P_1)))$
d. $((P_1 \to (P_2 \wedge P_3)) \leftrightarrow ((P_1 \wedge P_2) \to (P_1 \wedge P_3)))$

3.3.30. Suppose P, Q, R, and S denote PL wffs as defined in this section, and let X and Y denote strings formed from PL's alphabet.
a. If string $P \wedge Q$ is identical to string $R \wedge S$, must $P = R$ and $Q = S$? Explain.
b. Can string $(P \wedge Q)$ be identical to string $(X \vee Y)$? Explain. Does your answer contradict the *Unique Composition Theorem*?

3.3.31. Given a truth-value assignment $a(P_i) = T$ or $a(P_i) = F$ for all atomic formulas P_i, recursively define the truth-value function v that extends a in the normal way, i.e., define $v(\mathbf{P})$ for all wffs $\mathbf{P}$ as usual, extending $v(P_i) = a(P_i)$.

3.3.32. *Polish Notation for PL Formulas*
a. Let N, K, A, C, E indicate the negation, conjunction, alternation (disjunction), conditional, and biconditional connectives respectively, and prefix these to the propositions they operate on. If $P_1, P_2, \ldots$ are propositional characters, recursively define well-formed formulas in *Polish Notation*.
b. Where does the main connective appear in a wff using this notation?
c. Determine whether the following are wffs in *Polish Notation*. Then write them in standard notation. $AP_1CP_1P_2$; $EP_1P_2NP_1$; $EKP_1P_2NP_2$.
d. To determine whether a string of characters in *Polish Notation* is a wff, use the following procedure. Begin with a count of 1 and proceed from left to right within a formula. Add 1 for each binary connective symbol, add 0 for the unary connective symbol, and subtract 1 for each propositional character. What final count does a wff have? Test your claim out on the strings in part c; then argue your result using an inductive argument.
e. Write $5 \times ((2 + 3) - 1) + 7 \times 4 - 6$ using reverse Polish notation.

3.3.33. *Words in the Balanced-Parentheses Language*
Suppose $A = \{(,)\}$, a set with a left and right parenthesis, and let $\mathcal{L}$ consist of all words formed as strings of balanced parentheses, in the following way:
Base case: ε, the empty string, is a word.
Recursion step: if v and w are words, then so are (v) and vw.
Closure clause: all words are produced using the *recursion step* finitely many times, starting from the base case.
a. Show that $(())$, $(()())()$, and $((()())(()))$ all belong to $\mathcal{L}$.
b. Formulate and inductively prove a proposition that captures the idea of words in $\mathcal{L}$ having balanced parentheses.
c. Show that the $(()(())))(()$ and $()(()((()))))()$ are not words in $\mathcal{L}$.
d. State and argue for the correctness of a procedure that decides whether a string of left and right parentheses is a word in $\mathcal{L}$. Begin at the left-most parenthesis and work your way forward in the string with your procedure.

3.3.34. *Hofstadters's MIU System*
Hofstadter's MIU System[5] has alphabet $\{M, I, U\}$ and four recursive rules governing word/string production (see Exercises 1.5.19–24). If the sole base-case word for this system is MI, prove the following:
a. Every word in the MIU system begins with an M and has no other M.
b. MI^k is a word in this system for every $k = 2^n$, where I^k abbreviates a list of k Is.
c. MI^{10} and MI^{11} are words in this system.
d. Is MI^3 a word in this system? MI^6? Explain.
e. *Further Exploration*: What sorts of words can be produced from MI via the given rules? Can you find any necessary and sufficient conditions for a word to be produced in the MIU system? Use your work in parts b–d; see also Exercise 1.5.24.

---

[5] This is treated in *A Simple Decision Procedure for Hofstadter's MIU-System* by Swanson and McEliece in **The Mathematical Intelligencer** 10 (2), Spring 1988, pp. 48-9.

3.4 Peano Arithmetic

Proof by Mathematical Induction is an inductive proof strategy designed for the natural number structure $\mathbb{N}$. Elementary mathematics merely assumes the existence of the natural number system, focusing on how calculations are made. More advanced branches of mathematics take the procedures and facts of arithmetic for granted, as we have, too, up to this point. But in order to delve into the connection between mathematical induction and the natural number system, we must explore arithmetic more theoretically.

We have two options here. We can either accept arithmetic as foundational and look for axioms to characterize it as fully as possible, or we can take arithmetic as a theory that depends upon something even more basic, such as logic or set theory. We'll look at a version of the second alternative in Section 5.3; here we'll develop the first option.

3.4.1 $\mathbb{N}$ as an *Inductive Structure*

Let's begin by briefly analyzing what it means for $\mathbb{N}$ to be an inductive structure. This will give us some background for axiomatizing arithmetic.

✤ **Example 3.4.1**
  Exhibit $\mathbb{N}$ as an inductive structure.

Solution
  · Natural numbers answer the question "how many?" Since 0 can be an answer to this (though it took millennia for this to be recognized), we'll include it as a natural number. For us, $\mathbb{N} = \{0, 1, 2, \ldots\}$.
  · Natural numbers also have an intimate connection to one another, indicated by their position in the order used for counting: each number after 0 is one more than the preceding number in the counting sequence. Let's use $S(n)$ to denote the number $n + 1$ that succeeds n in the list.
  · This gives us a recursive way to present $\mathbb{N}$, where S is a successor operator:
    1) *Base case*: 0 is a natural number;   [i.e., 0 is a member of $\mathbb{N}$]
    2) *Recursion step*: if n is a natural number, then so is $S(n)$; and
    3) *Closure clause*: if n is a natural number, then it can be obtained by applying S a finite number of times (n, actually), starting with 0.

This example gives us an informal description of $\mathbb{N}$, but it lacks rigor because the concept of a natural number seems already embedded in the idea of repeatedly applying the successor operator S to 0. The axiomatization given below will help us avoid this circularity in the closure clause.

3.4.2 *Historical Context of Axiomatic Arithmetic*

Axiomatic arithmetic probably sounds like an oxymoron. After all, arithmetic deals with applying computational algorithms like long division to numbers expressed in our base-ten place-value system. Where are the axioms and definitions and proofs in all this?

If this is your response, it's understandable. Arithmetic is taught with little concern for proof because young children don't need (and can't follow) deductive arguments when they first learn computational techniques. Most civilizations have treated arithmetic as a collection of specialized calculation procedures. Algorithmic mathematics became part of theoretical mathematics only about 150 years ago when mathematicians began investigating the nature of number systems in more depth and organizing them deductively.

In the nineteenth century, for a variety of technical and educational reasons, mathematicians became concerned with the axiomatic basis and deductive structure of their discipline. In attempting to establish a rigorous foundation for calculus, they also reconsidered the foundations of algebra and arithmetic. Leading mathematicians mounted an *arithmetization* program in analysis to banish informal ideas coming from geometry and physics. In the end, they defined real numbers in terms of set theory and rational numbers, rational numbers in terms of set theory and integers, and integers in terms of set theory and natural numbers. Some mathematicians went even further and grounded the natural numbers in set theory. Others accepted the natural numbers as an appropriate starting point and axiomatized them instead.

Two mathematicians who investigated the natural number system were Richard Dedekind, who in 1888 used ideas from set theory as a foundation, and, independently, Giuseppe Peano, who in 1889 took the natural numbers as primitive and formulated a list of axioms for them. Now known as the *Peano Postulates*, they might better be called the *Dedekind-Peano Postulates*, since Dedekind was the first person to isolate the key properties of the natural numbers. Peano, pictured in Figure 3.3, developed arithmetic deductively as well, something Dedekind did not do,

Fig. 3.3 G. Peano

so natural number arithmetic is appropriately called *Peano Arithmetic*.

3.4.3 *Postulates for Peano Arithmetic*

Very young children learn to count by reciting number words in a certain order, touching or pointing to the objects they're counting. They may not know how the numbers relate to one another nor what they mean about the quantity of objects being counted, but they learn to start with *one* and continue with a definite progression of words, almost like a nursery rhyme. The purpose of counting becomes clearer to them over time.

Children also learn that the sequence of counting numbers never stops, that by counting long enough and using new unit names like *thousand*, they can keep counting forever. They learn, too, how to use counting (at first on their fingers) to do simple arithmetic: adding is done by counting forward, subtracting by counting backward. Because other operations build on addition and subtraction, counting forms the ultimate basis for all of arithmetic.

Peano's approach to arithmetic parallels this way of learning about numbers and computation. He takes the notion of the natural numbers occurring in ordered succession as the basis for developing arithmetic into a deductive theory. Surprisingly, he found that four simple postulates, stating some obvious properties of the natural numbers, together with a fifth closure postulate, formed a sufficient axiomatic foundation for arithmetic.

We'll state these postulates in ordinary mathematical English and then write them formally using logical symbolism. When we formalize the *Peano Postulates* as axioms, our list of postulates shrinks to three axioms, because two of them are covered by the way FOL expresses sentences.[6]

1) 0 is a natural number.
2) Every natural number has a unique successor.
3) 0 is not the successor of any natural number.
4) Distinct natural numbers have distinct successors.
5) *Axiom of Mathematical Induction* (*Closure Axiom*)

Peano Postulates

The natural numbers are thus 0 and all its successors, each number after 0 having both a unique successor and a unique predecessor.

Note that the language for *Peano Arithmetic* has two non-logical symbols: the constant 0 denotes the least natural number,[7] and S represents the successor function. $S(x)$, which we'll abbreviate as Sx, signifies the unique successor of x in the usual ordering of $\mathbb{N}$.[8] These are the only primitive notions we need; other ideas will be introduced via definitions.

Using this first-order language, we can formalize the first four *Peano Postulates*. The first postulate—0 is a natural number—disappears, because this is presupposed by FOL notational conventions. The constant symbol 0 necessarily stands for some distinguished member of $\mathbb{N}$, our universe of discourse.

The second postulate, that successors are unique, is also unnecessary. If S represents a (successor) function, then Sx is defined for all x and this output is unique. Thus, this postulate is also omitted in a formal listing of the *Peano Postulates*.

The third postulate, that 0 is not the successor of any natural number, does need to be postulated. It can be formulated by either of the following:

$$\neg \exists n(Sn = 0) \qquad \text{or} \qquad \forall n(Sn \neq 0).$$

We'll choose the latter because it is a universal sentence, but the former immediately follows from it.

Axiom 3.4.1: *Successors Are Non-Zero*
$\forall n(Sn \neq 0)$

---

[6] When arithmetic is developed within set theory, all five postulates come into play.

[7] There is no unanimity among mathematicians about whether $\mathbb{N}$ should include 0 or start with 1. Even Peano waffled on this. We're including 0 for reasons already mentioned. We'll use $\mathbb{N}^+$ to represent the set of positive natural numbers.

[8] We're consciously avoiding the notation $x+1$ (or even x^+) for the successor of x since we've yet to define addition, which will be done in terms of the successor function S.

The fourth postulate, which says that distinct numbers have distinct successors, can be formulated by either of the following logical equivalents:

$$\forall m \forall n(m \neq n \to Sm \neq Sn) \qquad \text{or} \qquad \forall m \forall n(Sm = Sn \to m = n).$$

We'll choose the second positive formulation as our axiom, which says that predecessors of successors are unique.

Axiom 3.4.2: *Predecessors of Successors Are Unique*
$\forall m \forall n(Sm = Sn \to m = n)$

3.4.4 *The Axiom of Mathematical Induction*

The fifth *Peano Postulate* is a closure axiom for $\mathbb{N}$, considered as an inductive structure. As noted above, the *Axiom of Mathematical Induction* says that all natural numbers are generated by starting with 0 and repeatedly applying the successor function, i.e., by counting on from 0. Put in terms of sets, if $\mathcal{P}$ is the subset of natural numbers that contains 0 and contains the successor Sk of every number k it contains, then $\mathcal{P} = \mathbb{N}$.

We can express the *Axiom of Mathematical Induction* in first-order notation (which doesn't contain symbols for the universe $\mathbb{N}$ or any subset $\mathcal{P}$) by letting $P(n)$ stand for the proposition n *belongs to* $\mathcal{P}$. The postulate then becomes *if $P(0)$ holds, and if $P(Sk)$ holds whenever $P(k)$ does, then $P(n)$ holds for all natural numbers n.*

More formally, here is the official version of the *Induction Axiom*:

Axiom 3.4.3: *Axiom of Mathematical Induction*
$P(0) \wedge \forall k(P(k) \to P(Sk)) \to \forall n P(n)$

Since this is intended to hold for any formula $P(n)$ in the language of arithmetic, we actually have a collection of axioms, called an *axiom schema*, instead of a single axiom. Moreover, since the successor of natural number k is $k + 1$ (as it will be officially, once we define addition), we'll eventually replace Sk by $k + 1$ in our formulation, yielding the following more familiar version of the axiom:

Induction Axiom, Additive Form: $P(0) \wedge \forall k(P(k) \to P(k + 1)) \to \forall n P(n)$.

We'll use this more familiar form of the axiom to explain its relation to *Proof by Mathematical Induction* below, but we'll revert to using the official form when we begin rigorously developing *Peano Arithmetic*.

The truth of the *Axiom of Induction* isn't hard to see. Suppose the antecedent of the axiom. Then $P(0)$ holds. But so does $P(1) = P(0 + 1)$, because $P(k + 1)$ holds whenever $P(k)$ does. Similarly, since $P(1)$ holds, so does $P(2)$; and so on. $P(n)$ must therefore hold for every natural number n.

Given this argument, shouldn't we consider the *Axiom of Mathematical Induction* a theorem? This way of thinking, while understandable, is flawed. Here's why. Axiomatizing *Peano Arithmetic* delineates its models. Prior to *Axiom 3*, we've only established that a model $\mathbb{N}$ for *Peano Arithmetic* contains a non-successor element 0 and that predecessors of successors are unique. We still don't know, for example, whether all non-zero numbers are successors, nor, if they are, whether they are eventual successors of 0 or something else.

As a matter of fact, the results just mentioned *don't follow* from the first two axioms. It's relatively easy to concoct interpretations in which *Axiom* 1 and *Axiom* 2 are true but these fail, and so *Axiom* 3 is false of these models (see Exercise 32). Thus, while we can show that any number of successors of 0 are natural numbers, we can't claim that this process generates all of $\mathbb{N}$.

To make our above argument rigorous, we might try to use *Proof by Mathematical Induction*. But keep in mind that we're now developing *Peano Arithmetic* deductively from an axiomatic basis, so we can only use FOL's inference rules and the axioms of the theory. As noted in Section 3.1, *Proof by Mathematical Induction* is not a rule of inference from logic. It still needs justification, which we'll do shortly by appealing to the *Axiom of Mathematical Induction*. So we can't use it to prove the axiom that justifies it. Since *Axiom* 3 is nevertheless true of $\mathbb{N}$, we're adopting it as an axiom.

3.4.5 *Proof by Mathematical Induction Revisited*

To understand the connection between *Proof by Mathematical Induction* and the *Axiom of Mathematical Induction*, let's diagram how induction proceeds.

$$
\begin{array}{rl|l}
n_1 & P(0) & \cdots \\
n_2 & \quad\begin{array}{|l} P(k) \end{array} & \text{Spsn for CP} \\
 & \quad\begin{array}{|l} \\ \vdots \\ \\ \end{array} & \\
n_3 & \quad P(k+1) & \cdots \\
n_4 & P(k) \to P(k+1) & \text{CP } n_2\text{-}n_3 \\
n_5 & \forall k(P(k) \to P(k+1)) & \text{UG } n_4 \\
 & \quad \vdots & \\
? & \forall n P(n) & \text{???}
\end{array}
$$

The base case (we're starting at 0 to line up with $\mathbb{N}$) is line n_1, and the induction step occurs in lines n_2–n_3. This allows us to conclude first the conditional $P(k) \to P(k+1)$ in line n_4 and then its universal generalization $\forall k(P(k) \to P(k+1))$ in line n_5. But that doesn't yield the conclusion $\forall n P(n)$.

As argued above, we could continue by instantiating to $k = 0$, which via *MP* will prove $P(1)$, then $P(2)$, etc. Repeating this process k_0 times, we can prove $P(k_0)$ for any natural number k_0. However, we'll never be able to prove $P(n)$ for every natural number n in this way—it would require an infinitely long proof. Nor will we be able to generalize to get $\forall n P(n)$, since all $P(k_0)$ are particular cases. So, we're at an impasse with this approach.

A second way to continue our proof is to conjoin sentences n_1 and n_5 to get $P(0) \land \forall k(P(k) \to P(k+1))$ (line n_6). This still won't yield $\forall n P(n)$, though, unless we can somehow combine this conjunction with another proposition to produce it. But this is just what the *Axiom of Mathematical Induction* gives us: $[P(0) \land \forall k(P(k) \to P(k+1))] \to \forall n P(n)$ (line n_7). By asserting this axiom and applying *MP*, we can now conclude what's wanted (line n_8).

$$
\begin{array}{lll}
n_1 & P(0) & \cdots \\
& \vdots & \\
n_5 & \forall k(P(k) \to P(k+1)) & \text{UG } n_4 \\
n_6 & P(0) \land \forall k(P(k) \to P(k+1)) & \text{Conj } n_1, n_5 \\
n_7 & [P(0) \land \forall k(P(k) \to P(k+1))] \to \forall n P(n) & \text{Ax Indn} \\
n_8 & \forall n P(n) & \text{MP } n_7, n_6 \\
\end{array}
$$

The three-step mathematical-induction proof technique thus finds its justification in FOL's *Inference Rules* and in the *Axiom of Mathematical Induction*. Having shown this, we may legitimately use mathematical induction to deduce propositions in *Peano Arithmetic*. This makes induction proofs more manageable than if we were to use *Axiom* 3 directly.

All the variations on mathematical induction—*PMI*, *Mod PMI*, and *Strong PMI*—can be justified by the *Axiom of Mathematical Induction* (see Exercise 35), so we may use any form of *Proof by Mathematical Induction* in *Peano Arithmetic*'s deductions.

3.4.6 *Peano Arithmetic: Addition*

We'll now demonstrate how *Peano Arithmetic* can be rigorously developed from its axiomatic foundation. You may find this both frustrating and exhilarating, because you have little to work with at the outset—only the *Peano Postulates* and *First-Order Logic*. You'll have to continually check that what you're using in your proofs is justified by what you've already proved, rather than by what you know from earlier courses. But achieving each new goal can give you a sense of deductive accomplishment—it's like developing a computer method for doing complicated computations starting with machine level manipulations of 0s and 1s.

Let's start by once again listing the three *Peano Postulates*, including a successor version of mathematical induction.

Axiom 3.4.1: *Zero is not a Successor*
  $\forall n(Sn \neq 0)$

Axiom 3.4.2: *Predecessors of Successors Are Unique*
  $\forall m \forall n(Sm = Sn \to m = n)$

Axiom 3.4.3: *Axiom of Mathematical Induction*
  $P(0) \land \forall k(P(k) \to P(Sk)) \to \forall n P(n)$

Proof Technique: Mathematical Induction (Successor Version)
  From $P(0)$ and a deduction of $P(Sk)$ from $P(k)$ for an arbitrary natural number k, conclude $\forall n P(n)$.

Amazingly, the *Peano Postulates* are all that's needed to *prove the laws* supporting computational arithmetic—the *Commutative Laws, Associative Laws, Distributive Laws,* and *Cancellation Laws* that govern the ordinary binary operations of addition and multiplication.

The successor function is thus the ultimate basis of *Peano Arithmetic.* We'll define addition recursively in terms of it and prove that addition has the properties we know. We'll then use addition and the successor function to define multiplication. Other operations and relations can also be defined, building on what's available up to that point (see Exercises 20–31).

We can prove a few things about simple successors (see Exercises 3 and 5), but we'll introduce addition so we can prove and use $Sm = m + 1$. Note in the results below that all relevant leading universal quantifiers have been left off to simplify the formulations for readability.

Definition 3.4.1: *Addition of Natural Numbers (Successor Version)*
a) $m + 0 = m$
b) $m + Sk = S(m + k)$ *for any natural number* k

In this definition, equation a defines adding 0 to a number, while equation b says that adding a *successor* of a number is the successor of an earlier sum. These two equations give the official definition of addition. To put the second equation in a more recognizable form, we'll define 1 and prove an elementary proposition that does not require mathematical induction.

Definition 3.4.2: *Definition of* 1
$1 = S0$

Proposition 3.4.1: *Successors as Sums*
$Sm = m + 1$

Proof:
$$\begin{aligned} Sm &= S(m + 0) && \text{Defn of Addn, Eqn } a; \text{ Sub} \\ &= m + S0 && \text{Defn of Addn, Eqn } b \\ &= m + 1 \quad \blacksquare && \text{Defn of 1; Sub} \end{aligned}$$

This simple result allows us to rewrite the recursive definition for addition in the following more standard formulation. We'll label this a definition, though it's really a proposition following from *Definition* 1 and *Proposition* 1.

Definition 3.4.3: *Addition of Natural Numbers (Standard Version)*
a) $m + 0 = m$
b) $m + (k + 1) = (m + k) + 1$

Note in the second equation that we do *not* write $m + k + 1$ without parentheses. Addition is a *binary operation*, so only two numbers can be combined at a time. Only after a sum is obtained can it be added to another number. As we proceed in *Peano Arithmetic*, we must be *very picky about the use of parentheses* so we don't assume a result that hasn't yet been proved. Be on your guard against smuggling results into your arguments that aren't warranted by the premises. Ironically, developing deductive arguments for arithmetic can be difficult for advanced mathematics students, because they know too much!

Having the result of *Proposition* 1, we can put the *Axiom of Induction* and its counterpart *Proof by Mathematical Induction* into its usual additive format (see above). *This is the form we'll mostly use from this point on.*

We can now prove some key propositions about addition. We'll work some results here and leave others for the exercises (see Exercises 11–13).

Proposition 3.4.2: 0 *is an Additive Identity*
$$m + 0 = m = 0 + m$$

Proof:
We know from the definition of addition that $m + 0 = m$, so all we need to prove is $0 + m = m$.

1) *Base case*
   $0 + 0 = 0.$ ✓   Defn of Addn, Eqn *a*
2) *Induction step*
   Suppose $0 + k = k$.   Indn Hyp
   Then $0 + (k + 1) = (0 + k) + 1$ Defn of Addn, Eqn *b*
   $\qquad\qquad = k + 1.$ ✓   Sub
3) *Conclusion*
   Thus, $0 + m = m$ for all m. ∎   Indn

We actually know more than that 0 is *an* identity for addition of natural numbers. In Example 2.4.11 we showed that there was *at most one* additive identity for ordinary arithmetic, using an argument requiring nothing more than the rules for identity. Our argument here proves that there is *at least one* additive identity. Together they demonstrate that 0 is the unique additive identity for *Peano Arithmetic*.

Looking at the last proposition in a slightly different light, we can say that 0 *commutes* with all natural numbers. Showing that any two natural numbers commute will be left as an exercise (see Exercise 11). As a second step toward getting that result, though, we'll show that 1 also commutes with everything. This result is needed to prove the general case.

Proposition 3.4.3: *Commutativity of 1*
$$1 + m = m + 1$$

Proof:
We'll again argue by mathematical induction.

1) *Base case*
   $1 + 0 = 1 = 0 + 1.$ ✓   Propn 2, UI ($m = 1$)
2) *Induction step*
   Suppose $1 + k = k + 1$.   Indn Hyp

$$\text{Then } 1 + (k+1) = (1+k)+1 \qquad \text{Defn of Addn, Eqn } b$$
$$= (k+1)+1. \checkmark \qquad \text{Sub}$$

3) *Conclusion*

$$\text{Thus,} \qquad 1 + m = m + 1 \text{ for all } m. \quad \blacksquare \qquad \text{Indn}$$

These last two propositions begin to show how recursive definitions and mathematical induction work together to prove the fundamental computational laws of arithmetic. This is less trivial than it looks. In developing an axiomatic theory from scratch, the subject matter doesn't tell you *when* you have to prove *what*, nor which results you should take as axiomatic. It may be that in trying to prove a certain result, you get stuck and need to prove another result first so you can use it in your proof. This is what would have happened if we had tried to deduce *Proposition* 3 before *Proposition* 2. This sort of thing can happen when you know a number of true results in a given theory and you want to organize them deductively into a series of propositions. For instance, if we were to try to prove the *Commutative Law*, having already done two special cases in the above propositions, we would discover that our proof could use other results that haven't been proved (see Exercise 11b).

As you construct proofs on your own, you'll quickly sense that the laws of arithmetic are logically intertwined and that the order in which you prove them may make a difference. Once someone finds a way to organize them, proofs become more straightforward. Whenever this is the case, you should suspect that serious work went on behind the scenes to arrange the results so that they could be readily demonstrated one after the other.

We'll finish exploring addition by proving the *Associative Law*—which, as our proof makes clear, could have been done before *Proposition* 2. Equation b in the recursive definition of addition is a special instance of this law.

Proposition 3.4.4: *Associative Law for Addition*
$$(l + m) + n = l + (m + n)$$

Proof:

We'll use induction on n and generalize on the first two variables.

1) *Base case*

$$(l + m) + 0 = l + m \qquad \text{Defn Addn Eqn } a$$
$$= l + (m + 0). \checkmark \qquad \text{Defn Addn Eqn } a$$

2) *Induction step*

$$\text{Suppose} \quad (l+m)+k = l+(m+k). \qquad \text{Indn Hyp}$$
$$\text{Then } (l+m)+(k+1) = ((l+m)+k)+1 \qquad \text{Defn Addn Eqn } b$$
$$= (l+(m+k))+1 \qquad \text{Sub}$$
$$= l+((m+k)+1) \qquad \text{Defn Addn Eqn } b$$
$$= l+(m+(k+1)). \checkmark \qquad \text{Defn Addn Eqn } b$$

3) *Conclusion*

$$\text{Hence,} \qquad (l+m)+n = l+(m+n) \text{ for all } n. \qquad \text{Indn}$$

Generalizing, we can conclude $(l+m)+n = l+(m+n)$. $\blacksquare$

3.4.7 *Peano Arithmetic*: *Multiplication*

Having seen how to treat addition, we can be brief in introducing multiplication. Like addition, multiplication is defined recursively. And once multiplication is available, we can go further and introduce exponentiation. We already did this in a rigorous fashion in Sections 3.1 and 3.2 to illustrate *Mathematical Induction* proofs, so we won't repeat it here.

Definition 3.4.4: *Multiplication* (*Standard Version*)
  a) $m \cdot 0 = 0$
  b) $m \cdot (k + 1) = m \cdot k + m$

Based on this definition, the recursive definition of addition, the definition of 1, and the above propositions for addition, results analogous to the ones argued for addition can be proved for multiplication; for example,

$$m \cdot 0 = 0 = 0 \cdot m, \qquad \text{and} \qquad m \cdot 1 = m = 1 \cdot m.$$

Having just proved propositions similar to these, the proofs for these and other results involving multiplication will be left for Exercises 14–19.

EXERCISE SET 3.4

Note: *In proving propositions of* **Peano Arithmetic***, be sure not to use a result unless it is a Peano Postulate, a definition, or something previously proved. Put your proof in a two-column format as in the text, using parentheses around sums and products. Quantifiers are specified for clarity.*

Exercises 1–2: Baby-Steps Arithmetic
Prove the following basic arithmetic results.

3.4.1. Using the definitions $2 = SS0$ and $4 = SSSS0$ along with the recursive definition for addition, show that $2 + 2 = 4$.

3.4.2. Using the above definitions and Exercise 1 along with properties for addition and the recursive definition for multiplication, show that $2 \cdot 2 = 4$.

Exercises 3–7: Natural Numbers, Successors, and Sums
Use the Peano Postulates, the Successor Version of Proof by Mathematical Induction, and the Definition of Addition to prove the following.

3.4.3. $\forall n (Sn \neq n)$

3.4.4. $\forall m \forall n (m + n = n \rightarrow m = 0)$
Hint: use induction on n and generalize on m.

3.4.5. $\forall n (n = 0 \lor \exists! m (n = Sm))$

3.4.6. $\forall l \forall m \forall n (l + n = m + n \rightarrow l = m)$
Hint: use induction on n and generalize on m and l.

3.4.7. $\forall m \forall n (m + n = 0 \rightarrow m = 0 \land n = 0)$
Hint: start by using Exercise 5 to show $n = 0$.

Exercises 8–10: True or False

Are the following statements true or false? Explain your answer.

3.4.8. Arithmetic was axiomatized already in ancient times by the Greeks.

3.4.9. The *Axiom of Induction* is needed to justify *Proof by Mathematical Induction*.

3.4.10. The *Peano Postulates* were demonstrated by Dedekind before Peano formulated them as postulates for arithmetic.

Exercises 11–13: Properties of Addition

Prove the following propositions for addition.

3.4.11. *Commutative Law for Addition*
  a. Using any results from the section, prove that addition is commutative: $\forall m \forall n (m + n = n + m)$. *Hint*: use induction on n and generalize on m.
  b. Examine your proof in part *a*. Did you use the *Associative Law*? Can the *Commutative Law* be proved prior to *Proposition* 4? *Note*: a degree of associativity is already present in the recursive definition of addition.

3.4.12. *Cancellation Laws for Addition*
Prove the following, using any results mentioned up to this point (do *not* subtract equals). Did you use commutativity of addition for your proof?
  a. $\forall l \forall m \forall n (l + n = m + n \rightarrow l = m)$ b. $\forall l \forall m \forall n (l + m = l + n \rightarrow m = n)$

3.4.13. *Dichotomy Law*
$\forall m \forall n \exists d (m + d = n \lor n + d = m)$

Exercises 14–19: Properties of Multiplication

Prove the following propositions for multiplication.

3.4.14. *Multiplicative Properties of* 0
  a. $\forall m (m \cdot 0 = 0 = 0 \cdot m)$ b. $\forall m \forall n (m \cdot n = 0 \rightarrow m = 0 \lor n = 0)$

3.4.15. *Identity Property for Multiplication*
$\forall m (1 \cdot m = m = m \cdot 1)$

3.4.16. *Distributive Laws*
  a. $\forall l \forall m \forall n (l \cdot (m + n) = l \cdot m + l \cdot n)$ b. $\forall l \forall m \forall n ((l + m) \cdot n = l \cdot n + m \cdot n)$

3.4.17. *Associative Law for Multiplication*
$\forall l \forall m \forall n ((l \cdot m) \cdot n = l \cdot (m \cdot n))$

3.4.18. *Commutative Law for Multiplication*
$\forall m \forall n (m \cdot n = n \cdot m)$

3.4.19. *Cancellation Laws for Multiplication*
  a. $\forall l \forall m \forall n (l \neq 0 \land l \cdot m = l \cdot n \rightarrow m = n)$ b. $\forall l \forall m \forall n (l \neq 0 \land m \cdot l = n \cdot l \rightarrow m = n)$

Exercises 20–31: Order Properties of the Natural Numbers

The order-relation $\leq$ is defined for $\mathbb{N}$ by $\forall m \forall n (m \leq n \leftrightarrow \exists d (m + d = n))$. Prove the following results for $\leq$.

3.4.20. $\forall m (m \leq m)$

3.4.21. $\forall l \forall m \forall n (l \leq m \land m \leq n \rightarrow l \leq n)$

3.4.22. $\forall m (m \leq m + 1)$

3.4.23. $\forall n (0 \leq n)$

3.4.24. $\forall n (n \leq 0 \rightarrow n = 0)$

3.4.25. $\forall m \forall n (m \leq n \land n \leq m \rightarrow m = n)$

3.4.26. $\forall m \forall n (m \leq n \lor n \leq m)$

3.4.27. $\forall n (n \leq 1 \rightarrow n = 0 \lor n = 1)$

3.4.28. $\forall m \forall n (m \leq n \leq m + 1 \rightarrow n = m \lor n = m + 1)$

3.4.29. $\forall m \forall n (m \leq n + 1 \leftrightarrow m \leq n \lor m = n + 1)$

3.4.30. $\forall l \forall m \forall n (l \leq m \rightarrow l + n \leq m + n)$

3.4.31. $\forall l \forall m \forall n (l \leq m \rightarrow l \cdot n \leq m \cdot n)$

Exercises 32–33: Metalogical Exploration of Peano Arithmetic

The following problems explore how the Peano Postulates restricts its models.

3.4.32. *Models of Peano Arithmetic's Axioms 1 and 2*

  a. Find a model $\mathcal{N}$ of the first two axioms of *Peano Arithmetic* in which there are numbers besides 0 that are not successors of another number.

  b. Find a model $\mathcal{N}$ of the first two axioms of *Peano Arithmetic* in which all numbers except 0 are successors of some number but in which not every number is an eventual successor of 0.

  c. Find a model $\mathcal{N}$ of the first two axioms of *Peano Arithmetic* that has a subset $\mathcal{P}$ containing 0 as well as the successor of every element it contains, yet which is not $\mathcal{N}$. What does this say about *Axiom 3*'s relation to *Axioms 1* and 2?

  d. Can a finite set be a model of the first two axioms of *Peano Arithmetic*? Why or why not?

3.4.33. *Logical Independence of Peano Postulates*

  a. Using the method of models, show that *Axiom 1* is not logically implied by *Axioms 2* and 3. Thus, *Axiom 1* is independent of *Axioms 2* and 3.

  b. Using the method of models, show that *Axiom 2* is not logically implied by *Axioms 1* and 3. Thus, *Axiom 2* is independent of *Axioms 1* and 3.

3.4.34. *Mathematical Induction and Well-Ordering*

The *Well-Ordering Principle* (*every non-empty subset S of the natural numbers* $\mathbb{N}$ *has a least element*) was proved in Section 3.3 using *Strong Mathematical Induction*. Assuming the *Well-Ordering Principle* prove the *Axiom of Mathematical Induction* (without using *induction*, of course). Thus, the *Well-Ordering Principle* is logically equivalent to the *Axiom of Induction*.

3.4.35. *Variants of Mathematical Induction*

Formulate and prove the following, using the *Axiom of Induction*:

  a. A *Modified Axiom of Induction*

  b. A *Strong Axiom of Induction*

3.5 Divisibility

The *Peano Postulates* and recursive definitions for addition, multiplication, and exponentiation give us an algebraic structure in which we can do ordinary arithmetic. In the last Exercise Set, we also defined the order-relation $\leq$ and stated key results holding for it (see Exercises 3.4.20–31).

We'll now turn to *divisibility* and the related notions of *greatest common divisor* and *least common multiple*. We'll explore some basic divisibility results, both to round out our discussion of natural number arithmetic and to provide background material for our later discussion of modular arithmetic in Section 6.4. We'll prove these in a more informal paragraph style.

3.5.1 *Divisibility*: *Definition and Properties*

The natural number system $\mathbb{N}$ remains our overall context, though for some results the integers $\mathbb{Z}$ will be a more convenient universe of discourse (as indicated below) because subtraction is a full operation there. To be logically rigorous, we need to develop the basic theory of integer arithmetic, but we'll postpone this until we can make use of some set-theoretic ideas introduced in Section 6.3. Any results about the integers used here are ones learned early in school. Without listing list all of them, we'll carefully state and prove these and others after we've defined $\mathbb{Z}$ in Section 6.4.

One important result we need is the *Division Algorithm*, which allows us to divide one integer by another when there are remainders. This captures the ordinary process of long division. We'll prove this theorem in Section 6.4.

Theorem 3.5.1: *Division Algorithm*
   If an integer n is divided by a positive integer d, there results a unique quotient q and a unique remainder r such that $n = qd + r$ with $0 \leq r < d$.

When the remainder r is 0 in a division $n \div d$, the divisor d divides n exactly. We'll define this relationship for all integers. Note that $|$ denotes a *relation*, not an operation like $\div$.

Definition 3.5.1: *Divides*; *Divisor*, *Multiple*
   Suppose d, m, n are integers. Then
   a) *d **divides** n, denoted by $d \mid n$, if and only if $n = md$ for some m;*
   b) *d is a **divisor** of n (and n is a **multiple** of d) if and only if $d \mid n$.*

❖**Example 3.5.1**
   Show that $3 \mid 12$, $-4 \mid 20$, and $5 \nmid 9$; does $0 \mid 0$?

Solution
   · $3 \mid 12$ because $12 = 4 \cdot 3$.
   · $-4 \mid 20$ because $20 = -5 \cdot -4$.
   · $5 \nmid 9$ because no integral multiple of 5 yields 9; $9 \div 5$ leaves remainder 4.
   · $0 \mid 0$ by our definition (though you can't divide *by* 0) because $0 = 1 \cdot 0$;
   in fact, $d \mid 0$ for all integers d because $0 = 0 \cdot d$.

Divisibility satisfies a number of basic properties. These are listed in the next proposition, where again all the variables range over $\mathbb{Z}$.

Proposition 3.5.1: *Properties of Divisibility*
a) $1 \mid a$; $a \mid 0$.
b) *If $a > 0$ and $c \mid a$, then $c \le a$.*
c) *Reflexivity*: $a \mid a$.
d) *Anti-symmetry*: $a \mid b \wedge b \mid a \to a = \pm b$.
e) *Transitivity*: $a \mid b \wedge b \mid c \to a \mid c$.
f) *Divisibility of sums and differences*: $a \mid b \wedge a \mid c \to a \mid (b \pm c)$.
g) *Divisibility of (scalar) products*: $a \mid b \to a \mid mb$ *for any m.*
h) *Divisibility of linear combinations*: $a \mid b \wedge a \mid c \to a \mid (m_1 b + m_2 c)$.

Proof:
We'll prove three of these and leave the others as exercises.
b) Suppose $a > 0$ and $c \mid a$.
    If $c \le 0$, then $c \le a$ because $0 < a$.
    If $c > 0$, then $a = qc$ for $q \ge 1$.
    Thus, $c \le a$.   ∎
d) Suppose $a \mid b$ and $b \mid a$.
    Then $a = mb$ and $b = na$.
    Thus, $a = mna$, and so $mn = 1$.
    Either $m = 1 = n$, in which case $a = b$;
    or $m = -1 = n$, in which case $a = -b$.
    Hence, $a = \pm b$.   ∎
f) Suppose $a \mid b \wedge a \mid c$.
    Then $b = m_1 a$ and $c = m_2 a$.
    So $b \pm c = m_1 a \pm m_2 a = (m_1 \pm m_2)a$.
    Thus, $a \mid (b \pm c)$.   ∎

3.5.2 *Common Divisors and Common Multiples*

Common divisors (common multiples) and greatest common divisors (least common multiples) are defined as follows. Here $\mathbb{N}$ is our universe of discourse.

Definition 3.5.2: *Common Divisor; Greatest Common Divisor*
Suppose a, b, c, d are natural numbers.
a) *c is a **common divisor** of a and b if and only if $c \mid a \wedge c \mid b$.*
b) *d is the **greatest common divisor** of a and b (not both 0), denoted by $d = \gcd(a, b)$, if and only if d is a common divisor of a and b and whenever c is a common divisor of a and b, then $c \le d$.*

Definition 3.5.3: *Common Multiple; Least Common Multiple*
Suppose a, b, l, m are natural numbers.
a) *m is a **common multiple** of a and b if and only if $a \mid m \wedge b \mid m$.*
b) *l is the **least common multiple** of a and b, denoted by $l = \text{lcm}(a, b)$, if and only if l is a common multiple of a and b and whenever m is a common multiple of a and b, then $l \le m$.*

✤ **Example 3.5.2**

a) Find the common divisors of 12 and 20; of 30 and 42; of 15 and 22.
b) Find the common multiples of 12 and 20; of 30 and 42; of 15 and 22.
c) How are the common divisors related to $\gcd(a, b)$ for the pairs given? How are the common multiples related to the $\operatorname{lcm}(a, b)$?
d) How are $\gcd(a, b)$ and $\operatorname{lcm}(a, b)$ related for the pairs given?

Solution

a) The common divisors of 12 and 20 are $1, 2, 4$; so $\gcd(12, 20) = 4$.
   The common divisors of 30 and 42 are $1, 2, 3, 6$; so $\gcd(30, 42) = 6$.
   The only common divisor of 15 and 22 is 1; so $\gcd(15, 22) = 1$.
b) The common multiples of 12 and 20 are $60, 120, \ldots$; $\operatorname{lcm}(12, 20) = 60$.
   The common multiples of 30 and 42 are $210, 420, \ldots$; $\operatorname{lcm}(30, 42) = 210$.
   The common multiples of 15 and 22 are $330, 660, \ldots$; $\operatorname{lcm}(15, 22) = 330$.
c) From the definition we know that $\gcd(a, b)$ is the largest common divisor. But more than this seems to be true, as we can see in these examples— each common divisor divides the greatest common divisor. Similarly, all common multiples are multiples of the least common multiple.
d) Clearly, it seems $\gcd(a, b) \mid \operatorname{lcm}(a, b)$, but this isn't the best we can do in relating these two. See if you can discover what more is true.

We can easily find common divisors and the greatest common divisor for two numbers a and b by listing their divisors separately, comparing lists, and picking out common divisors and the $\gcd(a, b)$. Finding divisors is time-consuming, however, so we'll introduce a more efficient method.

3.5.3 *The Euclidean Algorithm*

Euclid put forward a procedure for calculating the greatest common divisor, now called the *Euclidean Algorithm*, in *Propositions* VII.1–2 of his ***Elements***. The ancient Chinese used an equivalent method to help reduce common fractions to lowest form. We'll look at both of these procedures.

✤ **Example 3.5.3**

The Chinese *Reciprocal Subtraction Method* for calculating the greatest common divisor of two numbers subtracts the smaller number from the larger until it no longer yields a positive remainder. The last remainder is then subtracted from the other number until the same thing happens. This process is repeated until the remainder equals the number being subtracted, which produces the greatest common divisor.

Let's use this to calculate $\gcd(36, 128)$ and $\gcd(15, 49)$.

Solution

Here is the reciprocal subtraction procedure in schematic form.

①$128 \to 92 \to 56 \to 20$; ③$20 \to 4$ ①$49 \to 34 \to 19 \to 4$; ③$4 \to 1$
②$36 \to 16$; ④$16 \to 12 \to 8 \to 4$ ②$15 \to 11 \to 7 \to 3$; ④$3 \to 2 \to 1$

Thus, $\gcd(36, 128) = 4$. Thus, $\gcd(15, 49) = 1$.

Essentially, long division is repeatedly subtracting one number from another, yielding a remainder in the end. In the last example, $128 \div 36$ leaves remainder 20; $36 \div 20$ leaves remainder 16; etc. The Chinese *Reciprocal Subtraction Method* gradually divides the larger number by the smaller, then divides the smaller number by the remainder, and so on, until the final nonzero remainder equals the divisor. The *Euclidean Algorithm* does this more compactly, as illustrated below, except that we divide until the remainder is 0, the last divisor being the greatest common divisor.

✤ Example 3.5.4

Use the *Euclidean Algorithm* to calculate $\gcd(36, 128)$ and $\gcd(15, 49)$.

Solution

$128 \div \underline{36} = 3, \text{rem.} \underline{20}; \quad \underline{36} \div \underline{20} = 1, \text{rem.} \underline{16}; \quad \underline{20} \div \underline{16} = 1, \text{rem.} \underline{4};$
$\quad \underline{16} \div \underline{4} = 4, \text{rem.} 0; \quad \text{so } \gcd(36, 128) = 4.$

$49 \div \underline{15} = 3, \text{rem.} \underline{4}; \quad \underline{15} \div \underline{4} = 3, \text{rem.} \underline{3}; \quad \underline{4} \div \underline{3} = 1, \text{rem.} \underline{1};$
$\quad \underline{3} \div \underline{1} = 3, \text{rem.} 0; \quad \text{so } \gcd(15, 49) = 1.$

The Chinese *Reciprocal Subtraction Method* and the *Euclidean Algorithm* seem mysterious at first. The reason they work—the crux of the proof that these methods yield $\gcd(a, b)$—lies in the following *Lemma*.

Lemma 3.5.1: *Common GCD in Stages of the Euclidean Algorithm*
If $n = qd + r$, then $\gcd(n, d) = \gcd(d, r)$.

Proof:
· Suppose $n = qd + r$. [We'll prove that every common divisor of n and d is a common divisor of d and r, and conversely.]
· Let c be a common divisor of n and d. Then $c \mid (n - qd)$, i.e., $c \mid r$.
So c is a common divisor of d and r. ✓
· Conversely, suppose c is a common divisor of d and r.
Then $c \mid (qd + r)$, i.e., $c \mid n$.
So c is a common divisor of d and n. ✓
· Since their sets of common divisors are the same, $\gcd(n, d) = \gcd(d, r)$. ∎

Theorem 3.5.2: *Euclidean Algorithm Outcome*
The *Euclidean Algorithm* applied to integers a and b, not both 0, produces $\gcd(a, b)$.

Proof:
· Since the *Euclidean Algorithm* yields strictly decreasing remainders, after a finite number of steps the remainder will be 0, and the preceding remainder will be $\gcd(a, b)$.
· For suppose the procedure applied to $a \le b$ yields, after n steps:
$b \div a = q_1, \text{rem.} r_1; \quad a \div r_1 = q_2, \text{rem.} r_2; \quad r_1 \div r_2 = q_3, \text{rem.} r_3; \quad \cdots$
$r_{n-1} \div r_n = q_{n+1}, \text{rem.} 0, \quad a > r_1 > \cdots > r_n > 0.$
· Then by *Lemma 1*, $\gcd(a, b) = \gcd(a, r_1) = \gcd(r_1, r_2) = \cdots = \gcd(r_{n-1}, r_n) = \gcd(r_n, 0) = r_n.$ ∎

Making substitutions in the equations associated with the *Euclidean Algorithm* will show that $\gcd(a, b)$ is an integer linear combination of a and b. To illustrate, we'll follow up on the first part of Example 4 and then prove that this relation always holds.

✤ **Example 3.5.5**

Find a linear combination of 36 and 128 yielding $\gcd(36, 128) = 4$.

Solution

From Example 4, we can write the following sequence of equations:

$$128 = 3 \cdot 36 + 20; \qquad 20 = 128 - 3 \cdot 36$$
$$36 = 1 \cdot 20 + 16; \qquad 16 = 36 - 1 \cdot 20 = 36 - (128 - 3 \cdot 36) = 4 \cdot 36 - 128$$
$$20 = 1 \cdot 16 + 4; \qquad 4 = 20 - 1 \cdot 16 = (128 - 3 \cdot 36) - (4 \cdot 36 - 128)$$
$$16 = 4 \cdot 4 + 0. \qquad\qquad = 2 \cdot 128 - 7 \cdot 36 = 256 - 252.$$

Thus, $\gcd(36, 128) = 4 = 2 \cdot 128 - 7 \cdot 36.$ ✓

3.5.4 *Properties of the Greatest Common Divisor*

Based on how the *Euclidean Algorithm* works, we can conjecture the following theorem. This result probably isn't the first one that comes to mind when thinking about what a greatest common divisor is, but it's a highly versatile tool for proving results that involve the greatest common divisor.

Theorem 3.5.3: *The GCD is the Least Positive Linear Combination*
The $\gcd(a, b)$ is the least positive linear combination $ma + nb$ for integers m and n.

Proof:
· Consider all possible linear combinations of a and b. The collection of all such positive sums (there are many) has a least element by the *Well-Ordering Principle*, say, $d = ma + nb$. We'll show that $d = \gcd(a, b)$.
· We first show that d is a common divisor.
By the *Division Algorithm*, $a = qd + r$ for $0 \leq r < d$.
Then $r = a - qd = a - q(ma + nb)$, a linear combination less than d.
Since d is the smallest positive linear combination, $r = 0$.
Thus, $a = qd$, so $d \mid a$. ✓
· Similarly, $d \mid b$. So d is a common divisor of a and b. ✓
· Now let c denote any positive common divisor of a and b.
Then $c \mid (ma + nb)$, i.e., $c \mid d$, making $c \leq d$.
Thus, $d = \gcd(a, b)$ by the definition. ∎

Corollary 3.5.3.1: *GCD Generates all Linear Combinations*
The set of multiples of $d = \gcd(a, b)$ is the same as the set of all linear combinations $ma + nb$ of a and b.

Proof:
Linear combinations of a and b are multiples of d because $d \mid a$ and $d \mid b$. ✓
By *Theorem 3*, all multiples of d are also linear combinations of a and b. ∎

A fragment of *Theorem* 3's proof (stated next) is often taken as the definition of $\gcd(a, b)$, because it treats everything in terms of divisibility instead of involving the order relation $\leq$. We used the more intuitive notion of *greatest* in our definition, leaving this characterization as a consequence.

Corollary 3.5.3.2: *Divisibility Characterization of GCD*
$d = \gcd(a, b)$ *if and only if* $d > 0$, $d \mid a$, $d \mid b$, *and* $c \mid d$ *for all common divisors* c *of* a *and* b.

Proof:
The forward direction follows from the definition of $\gcd(a, b)$ and the theorem just proved, since c divides any linear combination of a and b. ✓
The backward direction follows from the definition of $\gcd(a, b)$ and the divisibility property of *Proposition* 1b. ∎

A similar divisibility characterization holds for $\operatorname{lcm}(a, b)$.

Proposition 3.5.2: *Divisibility Characterization of LCM*
$l = \operatorname{lcm}(a, b)$ *if and only if* $l > 0$, $a \mid l$, $b \mid l$, *and* $l \mid q$ *for all common multiples* q *of* a *and* b.

Proof:
The *Division Algorithm* can be used to show this. See Exercise 22b.

The last general property of greatest common divisors we'll consider is the relationship hinted at in Example 2, part d. We'll again use the linear combination characterization of $\gcd(a, b)$ to prove this result.

Theorem 3.5.4: *GCD × LCM = Product*
If $a, b \geq 1$, *then* $\gcd(a, b) \cdot \operatorname{lcm}(a, b) = a \cdot b$.

Proof:
· Let $d = \gcd(a, b)$, and let $l = \dfrac{ab}{d}$. We'll show $l = \operatorname{lcm}(a, b)$.
Since $d \mid a$ and $d \mid b$, $a = md$ and $b = nd$ for positive integers m and n.
Thus, $\dfrac{ab}{d} = na$ and $\dfrac{ab}{d} = mb$, so $l = \dfrac{ab}{d}$ is a multiple of both a and b. ✓
· To show l is the *least* common multiple, let q be any positive multiple of a and b.
We'll show $q \geq l$ by proving $\dfrac{q}{l}$ is a positive integer.
$$\frac{q}{l} = \frac{qd}{ab} = \frac{q(sa + tb)}{ab} \quad \text{for } d = sa + tb.$$
$$= \frac{q}{b}s + \frac{q}{a}t, \text{ which is an integer since } q \text{ is a multiple of } a \text{ and } b.$$
But since q and l are positive, $\dfrac{q}{l}$ is a positive integer.
Thus, $\dfrac{q}{l} \geq 1$, making $q \geq l$, which implies $l = \operatorname{lcm}(a, b)$. ∎

✤**Example 3.5.6**
Verify that $\gcd(56, 472) \cdot \operatorname{lcm}(56, 472) = 56 \cdot 472$.

Solution
After three divisions, the *Euclidean Algorithm* yields gcd(56, 472) = 8 (see Exercise 17b).
Listing multiples of 472 and checking whether they are integer multiples of 56 shows that lcm(56, 472) = 3304 (see Exercise 15b).
Checking: $8 \cdot 3304 = 56 \cdot 472$ (see Exercise 23c).

3.5.5 *Relatively Prime Natural Numbers*

Throughout the text we've used *prime numbers* in examples and exercises, even defining them in the solution of Example 3.2.5. Prime number factorizations can help determine gcd(a, b) and lcm(a, b), but showing how will be left for the exercises (see Exercise 23a).

In the rest of this section we'll define the relation of being *relatively prime* (or *coprime*) and prove a few results about this that follow from *Theorem 3*'s linear-combination characterization of gcd(a, b).

Definition 3.5.4: *Relatively Prime Numbers*
*Positive integers a and b are **relatively prime** if and only if* gcd(a, b) = 1.

The relation of being relatively prime is satisfied by pairs of distinct primes, but it also holds for other numbers.

✦**Example 3.5.7**
Show that the following pairs of numbers are relatively prime: {5, 17}; {18, 521}. Are 156 and 221 relatively prime?

Solution
The first two pairs of numbers are relatively prime: gcd(5, 17) = 1 and gcd(18, 521) = 1. However, gcd(156, 221) = 13, so the last two numbers are not relatively prime (see Exercise 24).

Proposition 3.5.3: *Relatively Prime Numbers' Linear Combination*
Numbers a and b are relatively prime if and only if $ma + nb = 1$ *for some integers m and n.*

Proof:
· If a and b are relatively prime, gcd(a, b) = 1.
  By *Theorem 3*, $ma + nb = 1$ for some integers m and n. ✓
· Conversely, if $ma + nb = 1$ for some integers m and n, this is obviously the smallest positive linear combination of a and b.
  By *Theorem 3*, gcd(a, b) = 1, so a and b are relatively prime. ■

Being relatively prime interacts in some important ways with divisibility. For instance, if relatively prime numbers a and b both divide a number c, then so does their product (see Exercise 25a). And if $d \mid ab$ while d and a are relatively prime, then $d \mid b$. This is a slight generalization of a result known as *Euclid's Lemma*, which is *Proposition* VII.30 in Euclid's **Elements**. The proof of this below, however, is more modern, drawing again upon the linear-combination characterization of $\gcd(a, b)$.

Theorem 3.5.5: *Euclid's Lemma* **(*Generalized*)**
   If $d \mid ab$ but d and a are relatively prime, then $d \mid b$.

Proof:
   · Suppose that $d \mid ab$ with d and a being relatively prime.
   · Then $ma + nd = 1$ by the last proposition.
   · Multiplying this by b yields $b = mab + nbd$, which is divisible by d.   ∎

Corollary 3.5.5.1: *Euclid's Lemma*
   If p is prime and $p \mid ab$ but $p \nmid a$, then $p \mid b$.

Proof:
   See Exercise 26.

EXERCISE SET 3.5

Exercises 1–3: True or False
Are the following statements true or false? Explain your answer.

3.5.1. For $n > 0$, $\gcd(n, 0) = n$ and $\operatorname{lcm}(n, 0) = 0$.

3.5.2. For natural numbers a and b, not both 0, $\gcd(a, b) \mid \operatorname{lcm}(a, b)$.

3.5.3. For natural numbers a and b, if $\operatorname{lcm}(a, b) = ab$, then a and b are relatively prime.

Exercises 4–8: Divisibility Properties
The following problems explore the notion of divisibility.

3.5.4. *Division and* 0
   a. Can $d = 0$ in the definition of divides?
   b. Why is the condition that a and b are not both 0 inserted in the definition of the greatest common divisor?

3.5.5. Show the following divisibility results:
   a. $15 \mid 240$ b. $-7 \mid 343$ c. $22 \nmid 6712$

3.5.6. *Verifying Proposition* 1
   a. Show that $7 \mid 112$, $112 \mid 1904$, and $7 \mid 1904$. What property of divisibility does this illustrate?
   b. Show that $13 \mid 156$ and $13 \mid 117$. Then show that $13 \mid (156 \pm 117)$.
   c. Show that $24 \mid 192$ and $24 \mid (15 \cdot 192)$.
   d. Show that $18 \mid 396$ and $18 \mid 216$. Then show that $18 \mid (3 \cdot 396 \pm 4 \cdot 216)$.

3.5.7. *Proving Proposition* 1
Let a, b, c and m be any integers. Prove the following properties:

a. $1 \mid a; \ a \mid 0$

b. $a \mid a$

c. $a \mid b \land b \mid c \to a \mid c$

d. $a \mid b \to a \mid mb$

e. $a \mid b \land a \mid c \to a \mid (m_1 b + m_2 c)$

3.5.8. *Divisibility and Multiples*
Prove that if $c \neq 0$, $ac \mid bc \leftrightarrow a \mid b$.

Exercises 9–13: Divisibility Properties and Decimal Numerals

Show that the following divisibility properties hold, where $\underline{abc}$ is the decimal numeral $100a + 10b + c$ and $\underline{bc}$ is $10b + c$.

3.5.9. *Divisibility by 2*
$2 \mid \underline{abc}$ if and only if $2 \mid c$.

3.5.10. *Divisibility by 3*
$3 \mid \underline{abc}$ if and only if $3 \mid (a + b + c)$. *Hint:* $100 = 99 + 1$, etc.

3.5.11. *Divisibility by 4*
$4 \mid \underline{abc}$ if and only if $4 \mid \underline{bc}$.

3.5.12. *Divisibility by 5*
$5 \mid \underline{abc}$ if and only if $5 \mid c$.

3.5.13. *Divisibility by 6*
$6 \mid \underline{abc}$ if and only if $2 \mid \underline{abc}$ and $3 \mid \underline{abc}$.

3.5.14. Show that $6 \mid n^3 - n$ for any integer n. *Hint:* factor $n^3 - n$.

Exercises 15–19: Calculating GCD and LCM

Calculate the greatest common divisor and least common multiple as indicated in the following problems.

3.5.15. *Calculating GCD and LCM via Lists*
List the following numbers' divisors and multiples and use those lists to determine $\gcd(a, b)$ and $\operatorname{lcm}(a, b)$.

  a. $a = 25$, $b = 70$ b. $a = 56$, $b = 472$ c. $a = 810$, $b = 2772$

3.5.16. *Calculating GCD via the Chinese Reciprocal Subtraction Method*
Determine $\gcd(a, b)$ using the Chinese *Reciprocal Subtraction Method*.

  a. $a = 25$, $b = 70$ b. $a = 56$, $b = 472$ c. $a = 810$, $b = 2772$

3.5.17. *Calculating GCD via the Euclidean Algorithm*
Determine $\gcd(a, b)$ using the standard *Euclidean Algorithm*.

  a. $a = 25$, $b = 70$ b. $a = 56$, $b = 472$ c. $a = 810$, $b = 2772$

3.5.18. *Calculating LCM via Theorem 4*
Determine $\operatorname{lcm}(a, b)$ using the product relationship of *Theorem 4*.

  a. $a = 25$, $b = 70$ b. $a = 56$, $b = 472$ c. $a = 810$, $b = 2772$

3.5.19. *GCD is a Linear Combination*
Using your work from Exercise 17, determine a linear combination so that $\gcd(a, b) = ma + nb$ for the following:

  a. $a = 25$, $b = 70$ b. $a = 56$, $b = 472$ c. $a = 810$, $b = 2772$

Exercises 20–22: Proofs of GCD Results

Prove the following results:

3.5.20. *Theorem 3*

a. Fill out the proof of *Theorem 3* by showing that $d \mid b$, which was said to follow similarly to how we proved $d \mid a$.

b. Show that if $\gcd(a, b) = ma + nb$, then $\gcd(m, n) = 1$.

3.5.21. *Factoring and the GCD*

a. Prove that $\gcd(ma, mb) = m \cdot \gcd(a, b)$.

b. Explain how the result of part *a* can be used to modify the Chinese *Reciprocal Subtraction Method* and the *Euclidean Algorithm*, so that they calculate $\gcd(a, b)$ for smaller numbers than the original ones. Then illustrate this by calculating $\gcd(252, 462)$ via the *Euclidean Algorithm*.

c. Prove that if $\gcd(a, b) = d$, then $\gcd\left(\dfrac{a}{d}, \dfrac{b}{d}\right) = 1$.

d. Put part *c* in your own words and explain why you think it should be true.

3.5.22. *Corollary 3.2 and Proposition 2*

a. Fill in the details for *Corollary 3.2*'s proof to show that *if $d = \gcd(a, b)$, then $d > 0$ and $c \mid d$ for all common divisors c of a and b.*

b. Prove *Proposition 2*.

Exercises 23–27: Primes and Relatively Prime Numbers

The following problems deal with primes and relatively prime numbers.

3.5.23. *Prime Factorization, GCD, and LCM*

a. Use that fact that each positive integer has a unique prime factorization to outline a method for calculating both $\gcd(a, b)$ and $\text{lcm}(a, b)$.

b. Using the method from part *a*, calculate $\gcd(24, 132)$ and $\text{lcm}(24, 132)$. Check that $\gcd(24, 132) \cdot \text{lcm}(24, 132) = 24 \cdot 132$.

c. Using the method from part *a*, calculate $\gcd(56, 472)$ and $\text{lcm}(56, 472)$. Check that $\gcd(56, 472) \cdot \text{lcm}(56, 472) = 56 \cdot 472$.

3.5.24. *Example 7*

a. Show that $\gcd(5, 17) = 1$.

b. Show that if p and q are distinct primes, then p and q are relatively prime.

c. Use some method to show that $\gcd(18, 521) = 1$; that $\gcd(156, 221) = 13$.

3.5.25. *Relatively Prime Numbers*

Prove the following, using any characterization of $\gcd(a, b)$.

a. $(a \mid c \wedge b \mid c) \wedge \gcd(a, b) = 1 \rightarrow ab \mid c$

b. $\gcd(a, b) = 1 \wedge \gcd(a, c) = 1 \rightarrow \gcd(a, bc) = 1$

c. $\gcd(a, b) = 1 \rightarrow \gcd(ma, b) = \gcd(m, b)$

d. $\gcd(a, b) = 1 \leftrightarrow \gcd(a^2, b^2) = 1$

3.5.26. Prove *Corollary 5.1, Euclid's Lemma*

If p is prime and $p \mid ab$ but $p \nmid a$, then $p \mid b$.

3.5.27. *Euclid IX.30, 31*

a. Prove that *if an odd number divides an even number, then it divides half of that number.*

b. Prove that *if an odd number is relatively prime with a number, then it is relatively prime with its double.*

Chapter 4
Basic Set Theory and Combinatorics

Chapter 3 looked at counting in a theoretical way, seeing how it formed the conceptual basis for *Peano Arithmetic*. This chapter continues the theme of counting, but in a more practical and less formal fashion—looking at ways to count collections of things, both ordered arrangements and combinations where order is irrelevant. Since it's sets of things that get counted, we'll begin with some elementary *Set Theory*, a topic that will be continued on a more advanced level in the next chapter. As before, we'll occasionally comment on proof strategy as we proceed, where appropriate.

4.1 Relations and Operations on Sets

Set Theory's ideas and terminology are indispensable for understanding many branches of mathematics and parts of computer science. In fact, some tout it as the ultimate foundation for all of mathematics. Here it forms the theoretical background needed for our study of *Combinatorics*, the subfield of mathematics that deals with counting combinations of things.

Sets arise both from aggregation (collecting things into a unified whole) and from classification (selecting things with a common defining property). We find sets in everyday life (a set of dishes, a collection of toys) as well as in science and mathematics (a family of cats, the set of prime numbers). Nomadic tribes kept track of their herds by making one-to-one correspondences between flocks of animals and objects stored in a pouch.

Despite sets having been useful since ancient times, a *theory* of sets is only about 150 years old. Treating categories or collections as sets to be manipulated, as conceptual objects in their own right, was not part of mathematics until the late nineteenth century. Only then was a genuine mathematical role for sets discovered—as a tool for understanding infinity. We'll discuss this important connection as well as its historical context in Chapter 5. Here we'll explore *Set Theory* as a basic foundation for topics in *Discrete Mathematics*.

4.1.1 *The Idea, Notation, and Representation of Sets*

Any definite collection of anything whatsoever forms a *set*. This isn't a definition; it merely uses the synonym *collection* to indicate what a set is. In fact, no definition will be given. We'll take *set* as a primitive term characterized by its use—formally, by the axioms that govern set relations and operations (see Section 6.3). Everyday experience helps us comprehend its core meaning, but *Set Theory* sharpens and extends our intuition. We can use other words as well—*class*, *group*, *family*, and so on—but the basic idea is the same: a set is a multiplicity of distinct objects collected into a single conceptual unit.

© Springer Nature Switzerland AG 2019

C. Jongsma, *Introduction to Discrete Mathematics via Logic and Proof*,
Undergraduate Texts in Mathematics,
https://doi.org/10.1007/978-3-030-25358-5_4

How a set's *elements* or *members* are related to one another by some ordering
or how they can be calculated with is irrelevant from a purely set-theoretic
viewpoint. Basic *Set Theory* only considers which things *belong to* what sets.

To formulate statements about sets, we'll ordinarily use capital letters
to indicate sets and lowercase letters to stand for their individual members,
though there will be occasional exceptions. The symbol $\in$ indicates set membership. Thus, if P denotes the set of prime numbers, $3 \in P$ asserts 3 *is a
prime number*; $x \in \mathbb{Q}$ says x *belongs to the set of rational numbers* $\mathbb{Q}$.

A set can be specified in two ways: by listing its elements or by specifying
a property shared by all and only those elements in the set. If the sets are
small enough, their members can be listed between braces. For example, the
set of primes less than ten is denoted by $\{2, 3, 5, 7\}$. Sometimes an ellipsis is
used to help list the elements, provided the pattern is clear from those that
are present. Thus, the first one hundred counting numbers can be written as
$\{1, 2, 3, \cdots, 99, 100\}$ and the entire set of natural numbers as $\{0, 1, 2, 3, \cdots\}$.

We can think of this *set-roster notation* as asserting a disjunction that
identifies a set's members. Thus, $S = \{a_1, a_2, \ldots, a_n\}$ indicates $x \in S$ if and
only if $x = a_1 \vee x = a_2 \vee \cdots \vee x = a_n$.

Set-descriptor notation specifies sets by means of a proposition that the
set's elements satisfy. For example, $\{x : x = 2n \text{ for some } n \in \mathbb{N}\}$ identifies the
set of all even natural numbers. The notation $S = \{x : P(x)\}$ denotes the set
whose elements satisfy $P(x)$. It can be taken as equivalent to $x \in S \leftrightarrow P(x)$.

We'll see below that *restricted set-descriptor notation* is needed for rigorously presenting a set. The notation $S = \{x \in U : P(x)\}$ indicates that S consists of all those elements inside U that satisfy $P(x)$.[1] Thus, the set of prime
natural numbers can be denoted by $\{x \in \mathbb{N} : x \text{ is prime}\}$. $S = \{x \in U : P(x)\}$
is equivalent to the membership claim $x \in S \leftrightarrow x \in U \wedge P(x)$.

To illustrate theorems in *Set Theory*, we'll often make use of diagrams.
These play the same role that geometric figures do in *Geometry*: they help
us see the truth of a proposition and follow an argument, but they're not a
substitute for a proof. *Set Theory* diagrams are called *Venn diagrams* after the
late nineteenth-century English logician John Venn, who used them in logic.
Somewhat similar devices had been used earlier by other mathematicians.[2]

Venn diagrams typically contain two or three circles located inside a rectangle (Figure 4.1). The outer rectangle indicates a universe of discourse U,
and the circles represent particular sets within it. When arbitrary sets are
intended, the circles are drawn as overlapping to allow all possible relations
among the sets. Shared regions do not automatically indicate shared members, they only permit that possibility. Shading will indicate a specific region,

---

[1] In Section 5.3 we'll see why we should restrict sets to those that can be formed inside
already existing sets.

[2] Venn attributes circle diagrams to Euler. Euler may have gotten the idea from his
teacher, Jean Bernoulli, who in turn may have been indebted to his collaborator Leibniz
(the linkage isn't clear). Leibniz used them to exhibit logical relations among various
classes. Versions prior to Venn, though, were less general and less versatile than his.

and the existence of an element within a set can be indicated by placing its
symbol within the appropriate region.

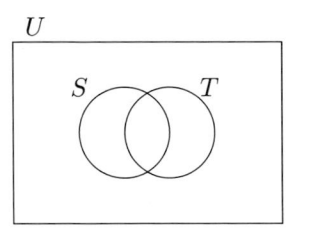 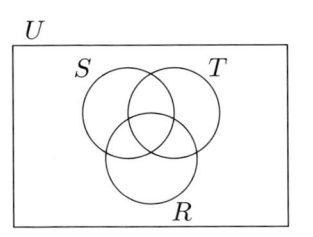

Fig. 4.1 Venn Diagrams

4.1.2 *Equal Sets and Subsets*

Because sets are completely determined by which members they have, two
sets are equal if and only if they contain exactly the same elements. This
gives us the following axiom or definition,[3] which we'll state formally.

Definition 4.1.1: *Equality for Sets*
$$S = T \leftrightarrow \forall x(x \in S \leftrightarrow x \in T)$$

$S = T$ can be demonstrated, according to this definition, by taking an
arbitrary element x and proving the biconditional $x \in S \leftrightarrow x \in T$. This
in turn can be done by two subproofs: supposing $x \in S$, prove $x \in T$; then
supposing $x \in T$, prove $x \in S$. At times, however, we may be able to chain
biconditionals together to deduce the necessary connection.

Arguing $x \in S \to x \in T$ and $x \in T \to x \in S$ in set-equality subproofs
amounts to showing that the first set is contained in the second one as a
subset. This leads to the next definition and to our first proposition.

Definition 4.1.2: *Subset and Superset Inclusions*
 a) ***Subset***: $S \subseteq T \leftrightarrow \forall x(x \in S \to x \in T)$
 b) ***Superset***: $T \supseteq S \leftrightarrow S \subseteq T$

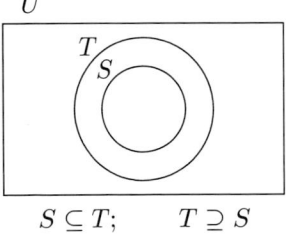

$$S \subseteq T; \qquad T \supseteq S$$

---

[3] There is more here than meets the eye, because we already have a fixed meaning for
equals (see Section 2.3). The forward part of this definition follows from FOL's inference
rules for identity and so can be proved. The backward part, however, must be asserted
as an axiom (see Section 5.3). Taken together, though, they specify how equality is used
in *Set Theory*, so we'll treat it here as a definition of set equality.

To show that S is a subset of T, you must show that each element in S is also in T. We'll use this proof procedure as well as the one for set equality to prove the first proposition—doing this in more detail here than usual to indicate the steps involved. Supply the logical reasons for the argument in the blanks given. Once you see what you need to do in such proofs, you can abbreviate them and write them more informally.

Proposition 4.1.1: *Equality and Subset Inclusion*[4]
$S = T \leftrightarrow S \subseteq T \wedge T \subseteq S$

Proof:
   Our proposition is a biconditional sentence, so we'll use _____ .
      · First suppose $S = T$.   _____
         Then $\forall x(x \in S \leftrightarrow x \in T)$.   _____
            Suppose that x is any element of S.   _____
            Then x must be an element of T.   _____
         Hence, $S \subseteq T$.   _____
         Similarly $T \subseteq S$.
         And so $S \subseteq T \wedge T \subseteq S$. ✓   _____
      · Conversely, suppose $S \subseteq T \wedge T \subseteq S$.   _____
         Then $\forall x(x \in S \to x \in T)$ and $\forall x(x \in T \to x \in S)$.   _____
         So $(x \in S \to x \in T) \wedge (x \in T \to x \in S)$, for any x.   _____
         This means $\forall x(x \in S \leftrightarrow x \in T)$.   _____
         Thus, $S = T$. ✓   _____
   Hence, $S = T \leftrightarrow S \subseteq T \wedge T \subseteq S$.   ■   _____

Sets are usually proved equal by the definition, using arbitrary elements of the sets to establish equality. However, at times it will be possible to remain up on the set level, working with subsets rather than elements. Then *Proposition* 1 may come in handy. Occasionally, it may also be possible to show that two sets are equal by working with set identities.

The subset relationship is sometimes confused with set membership. This is due to fuzzy thinking—subsets are not elements. Be careful to keep these two concepts distinct. The number 2 is an element of the set P of prime numbers; it is not a subset of P. On the other hand, the set P of prime numbers is a subset of the natural number system $\mathbb{N}$; it is not an element of $\mathbb{N}$. The potential for confusion on this score is increased when we consider sets whose elements are themselves sets (see Section 4.2).

According to *Proposition* 1, whenever $S \subseteq T$ and $T \subseteq S$, then $S = T$. In technical terms (see Section 7.1), this means $\subseteq$ is an *antisymmetric* relation. Additionally, like $\leq$ for numbers (another such relation), the $\subseteq$ relation is *not symmetric*, that is, it is not true that whenever $S \subseteq T$, then $T \subseteq S$. The subset relation does have two other basic properties, however. The first is completely trivial; the second is less so but should be fairly obvious.

---

[4] Note that this and later propositions are intended as universal statements, though we've omitted the quantifiers $\forall S \, \forall T$ in the interest of readability.

Proposition 4.1.2: *Reflexive Law for Inclusion*
   $S \subseteq S$

Proof:
   If $x \in S$, then $x \in S$.   ■

Proposition 4.1.3: *Transitive Law for Inclusion*
   $R \subseteq S \wedge S \subseteq T \rightarrow R \subseteq T$

Proof:

This is a conditional sentence, so we'll use *CP*.
Suppose $R \subseteq S \wedge S \subseteq T$.
Using the *Method of Backward Proof Analysis*
[this is *crucial*—else you may get lost in the
"givens" and not get started right], note that
we want to show that $R \subseteq T$.
This is done by proving that if $x \in R$, then
$x \in T$, too.

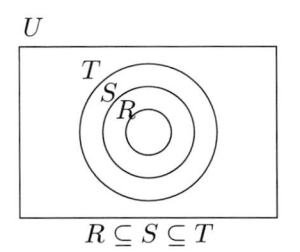

$R \subseteq S \subseteq T$

Again using *CP* as our proof strategy, we now start by supposing $x \in R$.
Since $R \subseteq S$, $x \in S$ by the definition of being a subset.
But $S \subseteq T$; so $x \in T$, too.
Thus, whenever $x \in R$, $x \in T$, so $R \subseteq T$.   ■

Our definition of subset inclusion allows for the possibility that the two sets
are equal, but there is also a more restricted notion of inclusion. *Proper inclusion* occurs when the subset is strictly smaller than the superset. This relation
is also transitive, but it is not reflexive or symmetric (see Exercises 10–12).

Definition 4.1.3: *Proper Subset Inclusion*
   $S \subset T \leftrightarrow S \subseteq T \wedge S \neq T$

Among all possible sets, one set is unique. This is the *empty set*, denoted
by $\emptyset$, which plays the role in *Set Theory* that the number 0 plays in *Arithmetic*
(see Exercises 13–16). And, like the number 0, it may give some trouble when
it's first encountered—how can something be a set if it doesn't have any
elements? It might help to think of sets more concretely as containers; an
empty set would be a container having no objects inside it.

Definition 4.1.4: *Empty Set*
   $\emptyset = \{x : x \neq x\}$

This definition asserts that $x \in \emptyset \leftrightarrow x \neq x$. Since the defining condition
for $\emptyset$ is a contradiction, it follows that $\forall x(x \notin \emptyset)$, i.e., $\emptyset$ has no elements.

Proofs for the first few propositions were rather detailed, but this won't
continue—it would make our deductions too long and would obscure the main
points of the proof. Get in the habit, though, of reading a proof with a pencil
and paper to fill in any missing details you need. Knowing which logical
strategies are available to prove a proposition from the relevant definitions
and earlier results should give you insight into what's going on in a proof.
Proposition 4's proof will give you some practice at this (see Exercise 21).

Proposition 4.1.4: *Empty Set Inclusion*
$\emptyset \subseteq S$

Proof:
Suppose that $x \notin S$.
But $x \notin \emptyset$, too.
Thus, $\emptyset \subseteq S$. ∎

4.1.3 Intersection and Union

The two most basic operations on sets are *intersection* and *union*. The intersection of two sets contains the elements common to both; the union contains all the elements in the two sets together. The intersection $S \cap T$ and union $S \cup T$ of sets S and T can be indicated by shading regions in a Venn diagram.

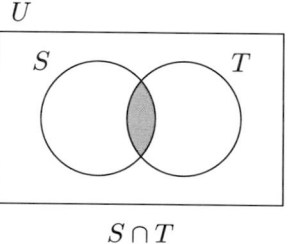
$S \cap T$

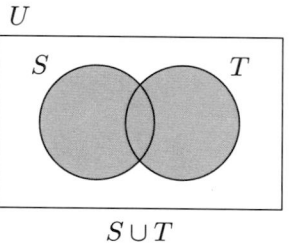
$S \cup T$

Definition 4.1.5: *Intersection*
$S \cap T = \{x : x \in S \wedge x \in T\}$

Intersection is a genuine binary operation, so $S \cap T$ is always defined. This is true even when they have no overlap, when they're *disjoint*. This can be drawn by leaving the center region unshaded or by separating the circles.

Definition 4.1.6: *Disjoint Sets*
S and T are **disjoint** if and only if $S \cap T = \emptyset$.

Definition 4.1.7: *Union*
$S \cup T = \{x : x \in S \vee x \in T\}$

It's clear from the definitions just given that intersection and union parallel the logical operations of conjunction and disjunction. This correspondence reveals itself more fully in the laws governing these operations, which are counterparts of the *Replacement Rules* for $\wedge$ and $\vee$. These laws are most easily proved by employing this correspondence; if the associated *Replacement Rules* from *Propositional Logic* are not used, their proofs can be tedious and lengthy. Drawing Venn diagrams for the sets being equated in the way suggested above may help you follow the arguments.

Proposition 4.1.5: *Idempotence Laws for Intersection and Union*
a) $S \cap S = S$
b) $S \cup S = S$

Proof:
  a) Let x be an arbitrary element.
    By the definition for intersection, $x \in S \cap S \leftrightarrow x \in S \wedge x \in S$.
    But by *Idem*, we can replace $x \in S \wedge x \in S$ with $x \in S$.
    Substituting, $x \in S \cap S \leftrightarrow x \in S$.
    But this means that $S \cap S = S$ by the definition for set equality.
  b) Note that part *b* differs from part *a* only in the operation involved.
    Replacing $\cap$ by $\cup$ and $\wedge$ by $\vee$ in *a*'s argument yields a proof for *b*. ∎

The way in which the second half of the last proposition was proved suggests that a *Duality Principle* may be at work in *Set Theory*: replace $\cap$ with $\cup$, and conversely, and you have a new proposition with a new proof. The following propositions seem to offer further confirmation of this, but does such a principle really hold? We'll return to this question later.

Proposition 4.1.6: *Commutative Laws for Intersection and Union*
  a) $S \cap T = T \cap S$
  b) $S \cup T = T \cup S$

Proof:
  a) $x \in S \cap T \leftrightarrow x \in S \wedge x \in T$ [Defn intersection]
        $\leftrightarrow x \in T \wedge x \in S$ [Comm of $\wedge$]
        $\leftrightarrow x \in T \cap S$. ∎   [Defn intersection]
  b) See Exercise 18a.

Proposition 4.1.7: *Associative Laws for Intersection and Union*
  a) $R \cap (S \cap T) = (R \cap S) \cap T$
  b) $R \cup (S \cup T) = (R \cup S) \cup T$

Proof:
See Exercises 17 and 18b.

Proposition 4.1.8: *Distributive Laws for Intersection and Union*
  a) $R \cap (S \cup T) = (R \cap S) \cup (R \cap T)$
  b) $R \cup (S \cap T) = (R \cup S) \cap (R \cup T)$

Proof:
See Exercise 19.

Proposition 4.1.9: *Absorption Laws and Subset Ordering*
  a) $S \cap T \subseteq S$; $S \cap T \subseteq T$
  b) $R \subseteq S$ and $R \subseteq T$ if and only if $R \subseteq S \cap T$.
  c) $S \subseteq S \cup T$; $T \subseteq S \cup T$
  d) $R \subseteq T$ and $S \subseteq T$ if and only if $R \cup S \subseteq T$.

Proof:

We'll sketch proofs for the first two parts and leave the others as exercises (see Exercise 18cd).

a) This is essentially the set-theoretic counterpart of the *PL* rule *Simp*:
$$x \in S \wedge x \in T \vDash x \in S, \quad x \in S \wedge x \in T \vDash x \in T.$$

b) R is contained in both S and T if and only if all its elements are in both S and T, that is, if and only if $R \subseteq S \cap T$. ∎

The first two parts of the last proposition say that $S \cap T$ is the largest set contained in both S and T. Similarly, parts *c* and *d* say that $S \cup T$ is the smallest set containing both S and T. These results are important for the theory of ordering collections of sets using the subset relation. We'll explore these and related matters in a more algebraic setting in Chapter 7.

4.1.4 *Set Difference and Set Complement*

Given sets S and T, we can form their union and intersection, but we can also take their set difference. Then, given a superset U, we can define set complement relative to U in terms of set difference.

Definition 4.1.8: *Set Difference, Set Complement*

a) $S - T = \{x : x \in S \wedge x \notin T\}$.

b) *Let U be any set. The **complement** of S inside U is $\overline{S} = U - S$.*

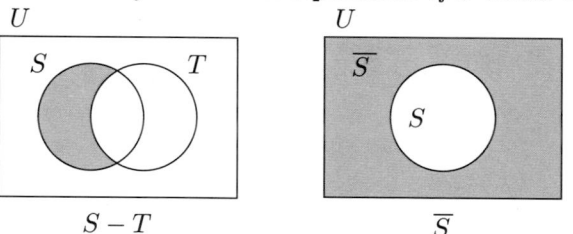

$$S - T \qquad\qquad \overline{S}$$

Set difference does not require $S \supseteq T$, but set complements are taken only when $U \supseteq S$. The following proposition relates complements to unions and intersections. Note once again how *Replacement Rules* play a crucial rule in the proof.

Proposition 4.1.10: *De Morgan's Laws for Set Complement*

If S and T are sets whose complements are taken with respect to a common set U, then

a) $\overline{S \cap T} = \overline{S} \cup \overline{T}$;

b) $\overline{S \cup T} = \overline{S} \cap \overline{T}$.

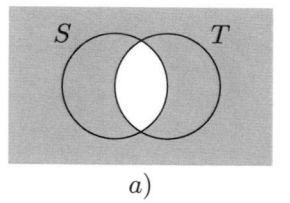

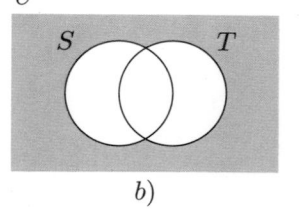

$$a) \qquad\qquad\qquad\qquad b)$$

Proof:

a) Shading in both $\overline{S \cap T}$ and $\overline{S} \cup \overline{T}$ gives diagram a) above, which makes it seem plausible that the complement of the intersection is the union of the complements.

The argument for this claim goes as follows:

$$
\begin{aligned}
x \in \overline{S \cap T} &\leftrightarrow x \in U \wedge x \notin (S \cap T) & \text{[Defn complement]}\\
&\leftrightarrow x \in U \wedge \neg(x \in S \wedge x \in T) & \text{[Defn intersection]}\\
&\leftrightarrow x \in U \wedge (x \notin S \vee x \notin T) & \text{[DeM for neg $\wedge$]}\\
&\leftrightarrow (x \in U \wedge x \notin S) \vee (x \in U \wedge x \notin T) & \text{[Distributive Law]}\\
&\leftrightarrow x \in (U - S) \vee x \in (U - T) & \text{[Defn set differ]}\\
&\leftrightarrow x \in (U - S) \cup (U - T) & \text{[Defn union]}\\
&\leftrightarrow x \in \overline{S} \cup \overline{T} & \text{[Defn complement]}
\end{aligned}
$$

Therefore, $\overline{S \cap T} = \overline{S} \cup \overline{T}$. ■ [Defn set equality]

b) See Exercise 20.

EXERCISE SET 4.1

You do not need to use a two-column format nor only logical rules of inference as reasons to work the following problems, but use logic to guide your proof strategy. Illustrate your results using Venn diagrams when appropriate.

4.1.1. *Venn diagrams*

a. A certain company presented a set of data having four categories (A, B, C, and D) using a Venn diagram, as shown. Explain why this diagram is deficient.

b. Modify the diagram in some way to show all possible regions.

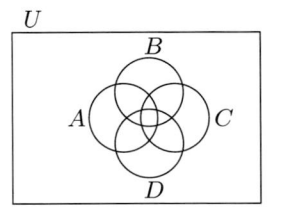

Exercises 2–5: Illustrating Basic Operations

Let $U = \{x \in \mathbb{N} : x \leq 30\}$, $E = \{x \in U : x \text{ is even}\}$, $O = \{x \in U : x \text{ is odd}\}$, and $P = \{x \in U : x \text{ is prime}\}$. Determine the following sets:

4.1.2. *Intersections*

a. $O \cap E$ b. $O \cap P$ c. $E \cap P$

4.1.3. *Unions*

a. $O \cup E$ b. $O \cup P$ c. $E \cup P$

4.1.4. *Complements Relative to U*

a. $\overline{E}$ b. $\overline{O}$ c. $\overline{P}$ d. $\overline{O \cup P}$

4.1.5. *Set Differences*

a. $E - O$ b. $E - P$ c. $P - E$ d. $O - P$

Exercises 6–8: True or False
Are the following statements true or false? Explain your answer.

4.1.6. Two sets are equal if and only if each one has elements of the other.

4.1.7. $\{1, 2, 3\} = \{3, 2, 1\}$

4.1.8. The complement of the intersection of two sets is the intersection of their complements.

Exercises 9–12: Proper Subset Inclusion
Prove the following properties:

4.1.9. *Proper Containment:* $S \subset T \leftrightarrow S \subseteq T \wedge (\exists x \in T)(x \notin S)$

4.1.10. *Non-Reflexivity:* $S \not\subset S$

4.1.11. *Non-Symmetry:* $S \subset T \rightarrow T \not\subset S$

4.1.12. *Transitivity:* $S \subset T \wedge T \subset R \rightarrow S \subset R$

Exercises 13–16: Properties of the Empty Set
Prove the following:

4.1.13. $S \subseteq \emptyset \rightarrow S = \emptyset$

4.1.14. $\emptyset \cap S = \emptyset = S \cap \emptyset$

4.1.15. $\emptyset \cup S = S = S \cup \emptyset$

4.1.16. $S - \emptyset = S$; $\emptyset - S = \emptyset$

Exercises 17–21: Proofs
Prove the following propositions. First construct a Venn diagram for the proposition, then give an argument for it. Where one exists, the associated PL Replacement Rule should help.

4.1.17. *Intersection*
 a. *Proposition 7a*

4.1.18. *Unions*
 a. *Proposition 6b*
 b. *Proposition 7b*
 c. *Proposition 9c*
 d. *Proposition 9d*

4.1.19. *Intersection and Union*
 a. *Proposition 8a*
 b. *Proposition 8b*

4.1.20. *Complements*
 a. *Proposition 10b*

4.1.21. Rewrite the proof of *Proposition 4*, filling in any steps you think are still needed and giving a reason for each step.

Exercises 22–51: Theorems or Not?
Draw Venn diagrams to decide whether or not each of the following is a theorem of Set Theory. If it is, prove it; if not, give a counterexample. Assume that complements are taken relative to some set U containing R, S, and T.

4.1.22. $R \cup S = R \cup T \rightarrow S = T$

4.1.23. $R \cap S = R \cap T \rightarrow S = T$

4.1.24. $R \subseteq S \rightarrow R \cap T \subseteq S \cap T$

4.1.25. $R \subseteq S \rightarrow R \cup T \subseteq S \cup T$

4.1.26. $S \subseteq T \leftrightarrow S \cup T = T$

4.1.27. $S \subseteq T \leftrightarrow S \cap T = S$

4.1.28. $R \subseteq T \vee S \subseteq T \leftrightarrow R \cap S \subseteq T$

4.1.29. $R \subseteq S \vee R \subseteq T \leftrightarrow R \subseteq S \cup T$

4.1.30. $S \cap T = \emptyset \leftrightarrow S = \emptyset \vee T = \emptyset$ **4.1.39.** $(T - S) - R = (T - R) - S$

4.1.31. $S \cup T = \emptyset \leftrightarrow S = \emptyset \wedge T = \emptyset$ **4.1.40.** $T - S = T - R \rightarrow S = R$

4.1.32. $S \cap T = S \leftrightarrow S \cup T = T$ **4.1.41.** $S - R = T - R \rightarrow S = T$

4.1.33. $S \cap T = \emptyset \leftrightarrow S - T = \emptyset$ **4.1.42.** $S \subseteq T \rightarrow S - R \subseteq T - R$

4.1.34. $S - T \subseteq S$ **4.1.43.** $R \subseteq S \rightarrow T - S \subseteq T - R$

4.1.35. $S - T = S \cap \overline{T}$ **4.1.44.** $S - (S - T) = S \cap T$

4.1.36. $S - T = S - (S \cap T)$ **4.1.45.** $R \cup S = R \cup T \leftrightarrow S - R = T - R$

4.1.37. $S \cup T = (S - T) \cup (T - S)$ **4.1.46.** $(\overline{\overline{S}}) = S$

4.1.38. $(S - T) \cap T = \emptyset$ **4.1.47.** $S - T = \overline{T} - \overline{S}$

4.1.48. $T - (S \cap R) = (T - S) \cap (T - R)$

4.1.49. $T - (S \cup R) = (T - S) \cap (T - R)$

4.1.50. $(T \cup S) - R = (T - R) \cup (S - R)$

4.1.51. $(T - S) - R = (T - S) \cap (T - R)$

Exercises 52–64: Symmetric Difference

*Using the definition $S \oplus T = (S - T) \cup (T - S)$ for the **symmetric difference** of sets S and T, prove the following results:*

4.1.52. $S \oplus S = \emptyset$ **4.1.58.** $S \oplus T = (S \cup T) - (S \cap T)$

4.1.53. $S \oplus \emptyset = S$ **4.1.59.** $\overline{S \oplus T} = (S \cap T) \cup (\overline{S \cup T})$

4.1.54. $S \oplus T = T \oplus S$ **4.1.60.** $S \oplus T = (S \cup T) \oplus (S \cap T)$

4.1.55. $S \oplus T \subseteq S \cup T$ **4.1.61.** $(S \oplus T) \oplus R = S \oplus (T \oplus R)$

4.1.56. $(S \oplus T) \cap (S \cap T) = \emptyset$ **4.1.62.** $S \cap (T \oplus R) = (S \cap T) \oplus (S \cap R)$

4.1.57. $S = T \leftrightarrow S \oplus T = \emptyset$ **4.1.63.** $R \oplus S = R \oplus T \rightarrow S = T$

4.1.64. $S \subseteq S \oplus T \leftrightarrow T \subseteq S \oplus T \leftrightarrow S \cap T = \emptyset \leftrightarrow S \cup T = S \oplus T$

4.2 Collections of Sets and the Power Set

Collections of sets play an important role in many fields of mathematics. We'll reconsider set operations in this broader context, concluding this section by introducing the power set.

4.2.1 *Collecting Sets into Sets*

So far we have two levels in *Set Theory*, elements and sets, which you may think of as very different things. But nothing prohibits us from taking sets as elements of a collection on a still higher level. Being legitimate mathematical entities, sets can be collected to form sets of sets.

Sports can help us understand this new level of sets. Baseball players are members of teams, which are members of leagues. Players are not members of leagues, and teams are not subsets of leagues. Leagues provide a third level of set-theoretic reality in the world of major league baseball. Without being able to form sets of sets, there would be no World Series.

Mathematics also needs such sets, even on an elementary level. In geometry, for example, figures like triangles can be conceptualized as infinite sets of points in a certain configuration. A pair of right triangles, therefore, is a collection of two point sets. If they were merely a conjoined conglomeration of points instead of a set of sets of points, we couldn't say that there were *two* triangles there—it would be an *infinite* collection of points instead.

✤ Example 4.2.1

Discuss the collection of remainder sets that results when the integers are divided by the number 4.

Solution

- Our universe of discourse here is $\mathbb{Z}$. Dividing an integer by 4 leaves a remainder (residue) of 0, 1, 2, or 3.
- This gives rise to four distinct residue classes R_0, R_1, R_2, and R_3, where $n \in R_k$ if and only if $n \div 4$ leaves remainder k. For example, $7 \in R_3$ and $-16 \in R_0$ because these numbers leave remainders of 3 and 0, respectively.
- As we'll see in Section 6.4, these residue classes can be treated much like numbers, yielding *modular arithmetic*. These numbers have important uses in many areas of mathematics and computer science.

Sets of sets occur often in abstract settings. In advanced mathematics, algebraic structures are constructed as *quotient structures*, which are like classes of residue classes. Analysis and topology consider other collections of sets as a basis for defining their central notions. Working with collections of sets is an important skill to learn if you are going further in mathematics.

✤ Example 4.2.2

Let S and T be two distinct sets. Discuss
a) the set whose sole member is a set S;
b) the set whose elements are the sets S and T; and
c) the set whose elements are the sets S and S.

Solution

a) The *singleton* $\{S\}$ contains a single element. Clearly $S \in \{S\}$, but $S \neq \{S\}$.[5] Nor is either set a subset of the other one. For example, if S were the set of all even numbers, S would be infinite, while $\{S\}$ would contain only a single member, the collection of these numbers.

b) The *doubleton* $\{S, T\}$ is formed by pairing up the sets S and T as elements of another set. In general, elements of S and T will not be elements of the doubleton. $\{S, T\}$ thus differs from both S and T. Nor can $\{S, T\}$ be obtained from S and T by taking an intersection, union, or set difference. It lies on a higher set-theoretic level than these sets. This is certainly the case, for example, when S is the set of even numbers and T is the set of odd numbers (see Exercise 3).

The doubleton $\{T, S\}$ is identical with $\{S, T\}$ since they have exactly the same elements.

---

[5] This assumes no set has itself as an element. We'll touch on this briefly in Section 5.3.

c) The *doubleton* $\{S, S\}$ is identical with the *singleton* $\{S\}$, because both contain S as their sole element. Multiplicity is irrelevant to *set identity*. An object either is or is not a member of a set; it cannot be doubly present, even if it is listed twice.[6]

4.2.2 Total Intersections and Unions

Ordinary intersection and union are binary operations. Repeating them yields finite intersections and unions (see Exercises 13–14). However, the *total intersection* and *total union* are needed for taking intersections and unions of an infinite collection of sets.

Definition 4.2.1: Total Intersection of a Collection
If C is a non-empty collection of sets, then $\bigcap_{S \in C} S = \{x : (\forall S \in C)(x \in S)\}$.

Definition 4.2.2: Total Union of a Collection
If C is a non-empty collection of sets, then $\bigcup_{S \in C} S = \{x : (\exists S \in C)(x \in S)\}$.

In words, the intersection of a family of sets consists of all the elements that belong to every set in the collection, and the union of a collection of sets consists of all those elements that belong to at least one set in the collection.

✤**Example 4.2.3**
Determine the total intersection and total union for the indexed collection of concentric closed discs $D_r = \{(x, y) : x^2 + y^2 \leq r\}$ for $\frac{1}{2} < r < 1$, $r \in \mathbb{R}$.

Solution
Each disc is centered about the origin, so the intersection of any two discs is the smaller of the two. The total intersection would thus be the smallest disc of all, if there were one. However, since the disc radius r is always greater than $\frac{1}{2}$ and there is no smallest real number greater than $\frac{1}{2}$, there is no smallest disc in this collection.

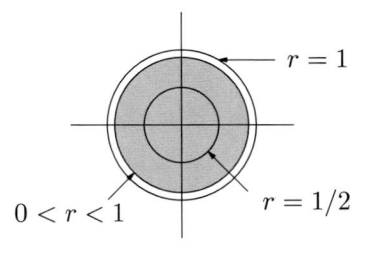

$D_{\frac{1}{2}}$ is certainly contained in all D_r, and given any disc bigger than this, we can always find a smaller disc in the collection by choosing radius r a bit closer to $\frac{1}{2}$. So the total intersection of all discs cannot extend beyond $D_{\frac{1}{2}}$. Therefore, $\bigcap_{\frac{1}{2} < r < 1} D_r = D_{\frac{1}{2}}$.

A similar argument shows that every point strictly inside the unit circle belongs to the union (see Exercise 17). We'll denote the open interior of this unit circle (the disc minus its boundary) by O_1.
Therefore, $\bigcup_{\frac{1}{2} < r < 1} D_r = O_1$.

---

[6] There are mathematical entities, called *multisets* or *bags*, however, in which multiplicity is taken into account. We'll make use of this idea in Section 4.3.

4.2.3 Properties of Total Intersections and Unions

Properties holding for simple intersection and union also hold for arbitrary intersections and unions. We'll state two of these—the generalized *Distributive Laws* and *De Morgan's Laws*. Most of the proofs, along with other results, will be left as exercises.

Proposition 4.2.1: *Distributivity*

a) $R \cap \left(\bigcup_{S \in \mathcal{C}} S \right) = \bigcup_{S \in \mathcal{C}} (R \cap S)$;   b) $R \cup \left(\bigcap_{S \in \mathcal{C}} S \right) = \bigcap_{S \in \mathcal{C}} (R \cup S)$.

Proof:

a) We'll prove the first part and leave the second (see Exercise 18).

$$x \in R \cap \left(\bigcup_{S \in \mathcal{C}} S \right) \leftrightarrow x \in R \wedge (\exists S \in \mathcal{C})(x \in S)$$
$$\leftrightarrow (\exists S \in \mathcal{C})(x \in R \wedge x \in S)$$
$$\leftrightarrow (\exists S \in \mathcal{C})(x \in R \cap S)$$
$$\leftrightarrow x \in \bigcup_{S \in \mathcal{C}} (R \cap S). \quad \blacksquare$$

Proposition 4.2.2: *De Morgan's Laws*

Let U be any set and let $\overline{S} = U - S$ be the complement of S relative to U.

a) $\overline{\bigcap_{S \in \mathcal{C}} S} = \bigcup_{S \in \mathcal{C}} \overline{S}$;   b) $\overline{\bigcup_{S \in \mathcal{C}} S} = \bigcap_{S \in \mathcal{C}} \overline{S}$.

Proof:

See Exercise 19ab.

Proposition 4.2.3: *Intersections, Unions, and Subsets*

a) $\bigcap_{S \in \mathcal{C}} S \subseteq T$ for all $T \in \mathcal{C}$;

b) $R \subseteq S$ for all $S \in \mathcal{C}$ if and only if $R \subseteq \bigcap_{S \in \mathcal{C}} S$;

c) $T \subseteq \bigcup_{S \in \mathcal{C}} S$ for all $T \in \mathcal{C}$;

d) $S \subseteq R$ for all $S \in \mathcal{C}$ if and only if $\bigcup_{S \in \mathcal{C}} S \subseteq R$.

Proof:

b) We'll prove part *b* and leave the rest as exercises (see Exercise 20abc).

· First suppose R is a subset of every set S in the collection $\mathcal{C}$ and let x be any element of R.

Then, since $R \subseteq S$, $x \in S$ for every set $S \in \mathcal{C}$.

This implies that $x \in \bigcap_{S \in \mathcal{C}} S$.

Therefore, $R \subseteq \bigcap_{S \in \mathcal{C}} S$. ✓

· Conversely, suppose $R \subseteq \bigcap_{S \in \mathcal{C}} S$.

Then, for any x in R, x lies in every S belonging to $\mathcal{C}$.

Thus, $R \subseteq S$ for every S in $\mathcal{C}$. $\blacksquare$

Summarizing this last proposition in words: parts a and b say that the total intersection is the largest set contained in each member of the collection, and parts c and d say that the total union is the smallest set containing each member of the collection. As mentioned in the last section, these ideas will come up again in connection with *Boolean Algebra* (see Section 7.2).

4.2.4 *Partitions*

Collections of sets often arise when a set is partitioned into subsets. A *partition* is a collection of pairwise disjoint subsets that together exhaust the given set. Partitions are one of the main reasons mathematicians are interested in collections of sets, something we'll explore later (see Sections 6.3 and 6.4).

Definition 4.2.3: *Pairwise Disjoint Sets*
*A collection of sets is **pairwise disjoint** if and only if $S \cap T = \emptyset$ for any two distinct sets S and T in the collection.*

✤ **Example 4.2.4**
Determine whether the collection of all open intervals of real numbers of the form $(n, n+1)$ is pairwise disjoint:
a) when $n \in \mathbb{Z}$;
b) when $n \in \mathbb{Q}$.

Solution
a) If only integer values of n are allowed, the collection of open intervals is pairwise disjoint. The nearest neighbors in the collection are then of the form $(n-1, n)$ and $(n, n+1)$, and these sets have no points in common.
b) However, if n is permitted to take on rational number values, the collection of intervals is certainly not pairwise disjoint: for example, $(0, 1) \cap (.5, 1.5) = (.5, 1)$.

Pairwise disjoint collections are strongly disjoint—more is required than having an empty total intersection (see Exercises 21–23). Being pairwise disjoint is sometimes necessary in order for a property to hold. For instance, given a finite collection of finite sets, the total number of elements in the union is the sum of the individual numbers if and only if the collection is pairwise disjoint (see Section 4.5).

Definition 4.2.4: *Partition of a Set*
*A **partition** of a set S is a collection $\mathcal{C}$ of subsets of S which is pairwise disjoint and whose total union $\bigcup_{R \in \mathcal{C}} R$ is S.*

✤ **Example 4.2.5**
Does the collection of all open intervals of real numbers $(n, n+1)$ form a partition of $\mathbb{R}$ if $n \in \mathbb{Z}$? if $n \in \mathbb{Q}$? Find a collection of finite intervals that partitions $\mathbb{R}$.

Solution
· Neither collection forms a partition of $\mathbb{R}$.
The first one ($n \in \mathbb{Z}$) doesn't because while it is pairwise disjoint, its union misses all integers and so doesn't yield $\mathbb{R}$.
The second collection also fails to be a partition, for while its union equals $\mathbb{R}$, it is not pairwise disjoint (see Example 4b).
· The collection of half-closed/half-open intervals $[n, n+1)$ for $n \in \mathbb{Z}$ forms a partition of $\mathbb{R}$: it's pairwise disjoint, and its union is all of $\mathbb{R}$.

4.2.5 *Sets of Subsets: The Power Set*

Given any set S, its subsets include at least the extreme possibilities $\emptyset$ (nothing) and S (everything). There will generally be many other subsets besides. The collection of all the subsets of S is called the *power set* of S and is denoted by $\mathcal{P}(S)$.

Definition 4.2.5: *Power Set*
$\mathcal{P}(S) = \{R : R \subseteq S\}$.

The *power set operator* $\mathcal{P}$ is a powerful unary operator, generating large new collections of sets (see Exercises 30–31).

✤**Example 4.2.6**
Determine the power set for the set $S = \{1, 2, 3\}$.

Solution
The following eight subsets of S are the *elements* of $\mathcal{P}(S)$:
$\emptyset$, $\{1\}$, $\{2\}$, $\{3\}$, $\{1, 2\}$, $\{1, 3\}$, $\{2, 3\}$, $\{1, 2, 3\}$

The power set operator is *monotone increasing*, in the following sense: given two sets, one a subset of the other, their power sets have the same subset relationship. This is the content of the next proposition. Related results exploring how the power set operator interacts with intersection and union are left for the exercises (see Exercises 32–37).

Proposition 4.2.4: *Subset Inclusion and Power Sets*
$S \subseteq T \to \mathcal{P}(S) \subseteq \mathcal{P}(T)$

Proof:
The proof here is easy if we begin in the right way. To see what to do, we'll use the *Method of Backward Proof Analysis*.
Assuming $S \subseteq T$ for *Conditional Proof*, we must prove $\mathcal{P}(S) \subseteq \mathcal{P}(T)$, which is a subset claim. We can prove this by taking an element of $\mathcal{P}(S)$ and showing that it also belongs to $\mathcal{P}(T)$.
Suppose, then, that $S \subseteq T$ and let $X \in \mathcal{P}(S)$.
Then $X \subseteq S$.
Since $S \subseteq T$, $X \subseteq T$, too.
Thus, $X \in \mathcal{P}(T)$. ∎

EXERCISE SET 4.2

Exercises 1–3: Pairing Sets
The following problems explore the notions of singleton and doubleton.

4.2.1. Let $S = \{0\}$ and $T = \{0,1\}$. What is $\{S,T\}$? Exhibit this set using only set braces and the numbers 0 and 1.

4.2.2. Let $S = \{0\}$, $T = \{0,1\}$, and $R = T \cup \{T\}$.
a. Exhibit the set R using only set braces and the numbers 0 and 1.
b. Is $S \subseteq R$? Is $S \in R$? Explain.
c. Is $T \subseteq R$? Is $T \in R$? Explain.

4.2.3. Verify the claims made in the solution to Example 2b that the doubleton $\{S,T\}$ differs from the sets S, T, $S \cap T$, $S \cup T$, and $S - T$ in the case where S is the set of even numbers and T is the set of odd numbers.

Exercises 4–5: Finite Collections of Sets
Work the following problems involving finite collections of sets.

4.2.4. A collection $\mathcal{C}$ consists of the sets I_2, I_3, and I_4, where I_n denotes the set of all integers that are multiples of n.
a. List the elements of I_2, I_3, and I_4.
b. Determine $\bigcap_{S \in \mathcal{C}} S$.   c. Determine $\bigcup_{S \in \mathcal{C}} S$.

4.2.5. A collection $\mathcal{C}$ consists of the sets I_2, I_3, I_9, and I_{12}, where I_n denotes the set of all integers that are multiples of n.
a. List the elements of I_2, I_3, I_9, and I_{12}.
b. Determine $\bigcap_{S \in \mathcal{C}} S$.   c. Determine $\bigcup_{S \in \mathcal{C}} S$.

Exercises 6–8: Plenty of Nothing
The following problems focus on the empty set.

4.2.6. Explain why $\{\emptyset\} \neq \emptyset$.

4.2.7. Explain why $\{\{\emptyset\}\}$ is different from both $\{\emptyset\}$ and $\emptyset$.

4.2.8. Explain why $\{\emptyset, \{\emptyset\}\}$ is different from $\emptyset$, from $\{\emptyset\}$, and from $\{\{\emptyset\}\}$.

Exercises 9–10: True or False
Are the following statements true or false? Explain your answer.

4.2.9. $\{5,6,\emptyset\}$ is a subset of $\{5,6,7\}$, since $\{5,6\}$ is a subset of $\{5,6,7\}$ and $\emptyset$ is a subset of everything.

4.2.10. The basic properties holding for intersection and union of two sets also hold for total intersections and total unions of any collection of sets.

Exercises 11–12: Explanations
Explain the following terms/results in your own words.

4.2.11. Explain what a *partition* is and illustrate it with a concrete example.

4.2.12. Explain what the *power set* of a set is. If S is a set of people and subsets of S are committees of these people, what does $\mathcal{P}(S)$ represent?

Exercises 13–14: Extending Set-Theoretic Definitions
The following problems extend binary set operations to finitely many sets.

4.2.13. Give a recursive definition of $\bigcap_{i=1}^{n} S_i$ for finitely many sets S_i.

4.2.14. Give a recursive definition of $\bigcup_{i=1}^{n} S_i$ for finitely many sets S_i.

Exercises 15–16: Infinite Indexed Collections of Sets
The following problems involve indexed collections of sets. The notation used is analogous to that of infinite series.

4.2.15. For each i in $\mathbb{N}^+$, let $Q_i = \left(-\frac{1}{i}, \frac{1}{i}\right)$, an open interval about 0, and $D_i = \left[-\frac{1}{i}, \frac{1}{i}\right]$, the associated closed interval. Determine the following:

a. $\bigcap_{i=1}^{\infty} Q_i$ c. $\bigcup_{i=1}^{\infty} Q_i$

b. $\bigcap_{i=1}^{\infty} D_i$ d. $\bigcup_{i=1}^{\infty} D_i$

4.2.16. For each i in $\mathbb{N}^+$, let $O_i = \left(-\frac{i}{i+1}, \frac{i}{i+1}\right)$, an open interval about 0, and $C_i = \left[-\frac{i}{i+1}, \frac{i}{i+1}\right]$, the associated closed interval. Determine the following:

a. $\bigcap_{i=1}^{\infty} O_i$ c. $\bigcup_{i=1}^{\infty} O_i$

b. $\bigcap_{i=1}^{\infty} C_i$ d. $\bigcup_{i=1}^{\infty} C_i$

4.2.17. *Example 3*
Finish the second part of Example 3 by arguing that $\bigcup_{\frac{1}{2} < r < 1} D_r = O_1$.

Exercises 18–20: Properties of Intersections and Unions
Prove the following propositions.

4.2.18. Prove *Proposition* 1b: $R \cup \left(\bigcap_{S \in \mathcal{C}} S\right) = \bigcap_{S \in \mathcal{C}} (R \cup S)$.

4.2.19. Prove *Proposition* 2, *De Morgan's Laws* for complements.

a. $\overline{\bigcap_{S \in \mathcal{C}} S} = \bigcup_{S \in \mathcal{C}} \overline{S}$ b. $\overline{\bigcup_{S \in \mathcal{C}} S} = \bigcap_{S \in \mathcal{C}} \overline{S}$

4.2.20. Prove the following subset order properties from *Proposition* 3.

a. $\bigcap_{S \in \mathcal{C}} S \subseteq T$ for all $T \in \mathcal{C}$ c. $(\forall S \in \mathcal{C})(S \subseteq R \leftrightarrow \bigcup_{S \in \mathcal{C}} S \subseteq R)$

b. $T \subseteq \bigcup_{S \in \mathcal{C}} S$ for all $T \in \mathcal{C}$

Exercises 21–25: Pairwise Disjoint Sets
The following problems explore notions of disjoint sets.

4.2.21. Prove that if a collection $\mathcal{C}$ of two or more sets is pairwise disjoint, then $\bigcap_{S \in \mathcal{C}} S = \emptyset$.

4.2.22. Is the converse to Exercise 21 true or false? If it is true, prove it. If it is false, give a counterexample.

4.2.23. Is it possible to find a collection $\mathcal{C}$ so that the intersection of every pair of distinct sets in $\mathcal{C}$ is non-empty while the total intersection of the collection is empty? Support your claim.

4.2.24. Given a finite collection of distinct sets S_i for $i = 1, 2, \ldots, n$, show how to generate a new but related collection of sets D_i that has the same union as the original collection but is pairwise disjoint.

4.2.25. Given an infinite collection of distinct sets S_i for $i \in \mathbb{N}$, is it possible to generate a collection of sets D_i that has the same union as the original collection but is pairwise disjoint? Why or why not?

Exercises 26–29: Partitions
The following problems explore the idea of a partition.

4.2.26. Let R_n denote all natural numbers leaving remainder n when divided by 7 for $n = 0, 1, 2, \ldots, 6$. Explain why these R_n form a partition of $\mathbb{N}$.

4.2.27. Let $S_n = \{0, 1, \ldots, n\}$ denote the initial segment of $\mathbb{N}$ from 0 through n. Does this collection of S_n form a partition of $\mathbb{N}$? Why or why not?

4.2.28. Let P_n be the set of all natural numbers that are powers of a prime number n. Does the collection of P_n for all prime numbers n form a partition of $\mathbb{N}$? Why or why not?

4.2.29. Let S_i denote a collection of n finite sets, and let $S = \bigcup_{i=1}^{n} S_i$.
If $C_m = \{x : x \text{ is in exactly } m \text{ sets of the collection}\}$ for $m = 1, 2, \ldots, n$, is the collection $\{C_m\}$ a partition of S or not? Explain.

Exercises 30–31: Numerosity of the Power Set
The following problems concern the size of the power set $\mathcal{P}(S)$ of a set S.

4.2.30. Determine $\mathcal{P}(S)$ for the following sets S:
  a. $S = \{1\}$ b. $S = \{1, 2\}$ c. $S = \{1, 2, 3, 4\}$

4.2.31. *Numerosity of the Power Set*
  a. Generalize Exercise 30 and Example 6: if S has n elements, then $\mathcal{P}(S)$ contains _____ elements. Prove your result using mathematical induction.
  b. How is the result you obtained in part *a* related to the alternative notation that is sometimes used for the power set, namely, 2^S? Why do you think $\mathcal{P}(S)$ is called the *power set* of S?

Exercises 32–37: Properties of Power Sets
Prove the following results on properties of power sets:

4.2.32. $\mathcal{P}(S \cap T) = \mathcal{P}(S) \cap \mathcal{P}(T)$ **4.2.34.** $\mathcal{P}(S \cup T) \supseteq \mathcal{P}(S) \cup \mathcal{P}(T)$

4.2.33. $\mathcal{P}\left(\bigcap_{S \in \mathcal{C}} S\right) = \bigcap_{S \in \mathcal{C}} \mathcal{P}(S)$ **4.2.35.** $\mathcal{P}\left(\bigcup_{S \in \mathcal{C}} S\right) \supseteq \bigcup_{S \in \mathcal{C}} \mathcal{P}(S)$

4.2.36. Can the superset relation in Exercises 34–35 be turned around? If so, prove it; if not, give a counterexample.

4.2.37. Can the conditional in *Proposition* 4 be turned around? If so, prove it; if not, give a counterexample.

4.2.38. *Duality Principle for Set Theory?*
Several propositions in this chapter have exhibited a duality between intersection and union (see the remarks following *Proposition* 4.1.5). Formulate a
Duality Principle for *Set Theory* and then explore the truth of your statement
by verifying it or refuting it using a variety of specific instances.

4.3 Multiplicative Counting Principles

The previous two sections laid the set-theoretic groundwork for the rest of
this chapter on *Combinatorics* and for several themes later in the text. This
section will focus on some multiplicative counting methods, but before we do
so, we'll introduce one more set of ideas to help frame our discussion.

4.3.1 Ordered Pairs and Cartesian Product of Sets

The *Cartesian product* of sets is an operation that enables us to treat relations
and functions as an integral part of *Set Theory*. We'll introduce this notion
using an everyday example and then give the formal definition.

✥**Example 4.3.1**
   A not-very-style-conscious mathematics professor owns eight different shirts
   $S_1, S_2, \ldots, S_8$ and six different pants $P_1, P_2, \ldots, P_6$. If it were up to him,
   he might wear any shirt with any pants. We'll use ordered pairs and a
   Cartesian product to indicate the potential outfits this prof might wear.

Solution
   Each ordered pair (S_i, P_j) represents one outfit that might be worn. The set
   $S \times P$ of all such ordered pairs $\{(S_i, P_j) : 1 \leq i \leq 8, 1 \leq j \leq 6\}$ represents
   his collection of possible outfits. This gives a total of 48 different outfits,
   some of them probably not very well matched.

Definition 4.3.1: *Cartesian Product of Sets*
   $S \times T = \{(x, y) : x \in S \wedge y \in T\}$.

   Our definition of Cartesian product assumes the idea of an ordered pair as
known. This seems reasonable; everyone is familiar with ordered pairs from
graphing points and functions in elementary algebra. In a rigorous development of this topic, however, the Cartesian-product operator can be more
thoroughly grounded in *Set Theory* by defining ordered pairs in terms of sets.
As this introduces a higher order of abstractness into the discussion, though,
we'll leave it for the exercises (see Exercises 26–28).
   Cartesian products can't be pictured with Venn diagrams, but they can
often be graphed, as the following example illustrates.

✥**Example 4.3.2**
   a) If $A = \{1, 2, 3, 5\}$ and $B = \{1, 3, 4\}$, graph the Cartesian product $A \times B$.
   b) If $S = [1, 5]$ and $T = [1, 4]$, graph the Cartesian product $S \times T$.

Solution
  a) $A \times B = \{(1,1), (1,3), (1,4), (2,1), (2,3), (2,4), (3,1), (3,3), (3,4), (5,1),$
     $(5,3), (5,4)\}.$
     This can be plotted as a set of 12 distinct points in a grid, as below.
  b) For S and T being the entire closed intervals of real numbers $[1,5]$ and
     $[1,4]$, respectively, $S \times T$ consists of all possible points on or inside the
     rectangular region $\{(x,y) : 1 \leq x \leq 5; 1 \leq y \leq 4\}$. $S \times T$ is plotted below
     as a shaded region.

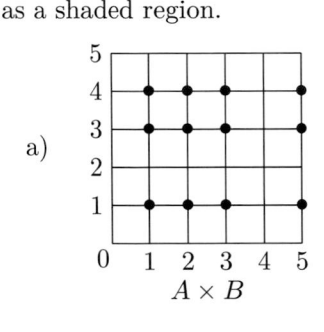

a)

$A \times B$

b)

$[1,5] \times [1,4]$

  One could now explore how Cartesian product interacts with the operations already introduced for sets. A number of these properties are in the exercises (see Exercises 37–47), but we'll have little occasion to use them.

  We can generalize the notions of ordered pairs and Cartesian products to *ordered n-tuples* and *n-fold Cartesian products*. As above, we'll assume the notion of an ordered n-tuple $(x_1, x_2, \ldots, x_n)$ as given (see Exercises 29–31), and we'll use it to define finite Cartesian products. The related notion of a *finite sequence* is also defined using these ideas.

Definition 4.3.2: *Finite Cartesian Products*
  a) $S_1 \times S_2 \times \cdots \times S_n = \{(x_1, x_2, \ldots, x_n) : x_i \in S_i\}.$
  b) $S^n = \{(x_1, x_2, \ldots, x_n) : x_i \in S\}.$

Definition 4.3.3: *Finite Sequences*
  a) A *finite sequence* $(x_1, x_2, \ldots, x_n)$ from a set S is an element of S^n.
  b) A *finite sequence without repetition* from a set S is a finite sequence
     in which no element of S appears more than once: $x_i \neq x_j$ if $i \neq j$.

4.3.2 *Multiplicative Counting Principle*

We'd now like to develop ways to count events compounded out of simple events. To help formulate our results, we'll begin by defining cardinality.

Definition 4.3.4: *Cardinality of a Set*
  The **cardinality** $|S|$ of a set S is the number of elements contained in S.

  This definition defines cardinality in terms of *numerosity*, a synonym. It tells *how many* members a set S has. What sense this makes for infinite sets will be addressed later, but in the rest of this chapter we'll assume a context of finite sets, where the concept is intuitively clear. In formal set theory, cardinality is given a more abstract definition, which we won't pursue.

The following proposition uses $\times$ in two ways: as a symbol for Cartesian product and as ordinary multiplication of numbers. The connection between these products explains why $\times$ is used for the Cartesian product operator.

Proposition 4.3.1: *Cardinality of Cartesian Products*
a) $|S \times T| = |S| \times |T|$
b) $|S_1 \times S_2 \times \cdots \times S_n| = |S_1| \times |S_2| \times \cdots \times |S_n|$
c) $|S^n| = |S|^n$

Proof:
See Exercises 32–33. Part c is an immediate consequence of part b.

We'll call a particular set of outcomes for an experiment or action an *event*. If one event can occur in m ways and for each outcome another event can occur in n ways, then the compound event — first outcome, second outcome — can occur in $m \times n$ ways. We saw this in Example 1: there were $8 \times 6 = 48$ outfits that could be chosen from the given shirts and pants.

Corollary 4.3.1.1: *Multiplicative Counting Principle*
If one event can occur in m ways, and for each of these a second event can occur in n ways, the compound event can occur in $m \times n$ ways.

Proof:
We can model the combined event using ordered pairs (first outcome, second outcome) from two sets F and S. Then the set of joint outcomes is the associated Cartesian product $F \times S$. *Proposition* 1a yields the result. ∎

This result can be generalized to any finite sequence of events.

Corollary 4.3.1.2: *Generalized Multiplicative Counting Principle*
If k events in sequence can occur in n_i ways for $i = 1, 2, \ldots, k$, then the number of ways the compound event can occur is $\displaystyle\prod_{i=1}^{k} n_i = n_1 \cdot n_2 \cdots n_k$.
Proof:
This follows immediately from *Proposition* 1b or by applying mathematical induction to *Corollary* 1 (see Exercise 34). ∎

The *Multiplicative Counting Principle* lies behind several counting techniques. We'll explore some of these here and others in Section 4.4.

✤Example 4.3.3
A computer program contains the lines

```
1 for i := 1 to 30
2   for j := 1 to 12
3     for k := 1 to 7
4       print i + j + k
```

How many times will line 4 be executed?

Solution

An application of the *Multiplicative Counting Principle* yields the answer: line 4 will be executed $30 \cdot 12 \cdot 7 = 2520$ times.

✤ Example 4.3.4

A *byte* is an eight-bit string, such as 01010101, each *bit* (<u>bi</u>nary digi<u>t</u>) being either a 0 or a 1.

a) How many distinct bytes are possible?

b) How many of them begin or end with four 0s?

Solution

a) Since each bit has two possibilities that can be assigned independently of the others, an eight-bit string can be formed in $2^8 = 256$ ways.

b) If a byte begins with four 0s, there are $2^4 = 16$ different ways to finish the byte. This is the total number of such bytes.

   If a byte ends with four 0s, there are 16 ways to begin the byte, and so again there are 16 different bytes.

   The only byte counted as a part of both sets is 00000000.

   Counting this once, we have $16 + 15 = 31$ distinct bytes that start or end with four zeros.

4.3.3 Ordered Outcomes With Repetition

Example 4 has *outcome sequences* of length eight, where each bit comes from the same set $\{0, 1\}$. Each outcome is independent of how previous outcomes occurred—both 0 and 1 can be used repeatedly. This is an example of *Ordered Outcomes with Repetition*.

Proposition 4.3.2: *Counting Ordered Outcomes with Repetition*

Suppose S is a set of n elements. Then the total number of ordered outcomes of size k selected from S, allowing repetition, is n^k.

Proof:

This follows from the *Generalized Multiplicative Counting Principle* with all $n_i = n$. ■

✤ Example 4.3.5

Iowa license plates have three letters followed by three numbers. If any letters and numbers can be used, how many license plates can be made?

Solution

Solving this requires a combination of counting methods.

Counting ordered outcomes with repetition, there are 26^3 three-letter words, and there are 10^3 three-digit numbers (including 000).

By the *Multiplicative Counting Principle*, this gives $26^3 \cdot 10^3 = 17,576,000$ different license plates. As there are less than $4,000,000$ people in Iowa (and pigs, cows, and chickens don't drive), this number is quite adequate, even if every adult owned a few vehicles.

4.3.4 *Permutations: Ordered Outcomes, No Repetition*

Let's now consider the case in which a set S of possibilities is gradually being depleted each time an outcome occurs, so that there is no repetition. Suppose, however, that outcome order remains important—think of outcomes occurring sequentially, as before. We can consider the final result either as an arrangement of S or as a sequence in S^n with no repeated entries. Either way, the number of possible outcomes can be counted using the *Generalized Multiplicative Counting Principle*.

✦**Example 4.3.6**

In a cross-country race, the places of the first five runners of each team to cross the finish line are added together to get the team score. If a team has 12 quality runners entered in a race, in how many different ways might these players contribute to the score of their team?

Solution

The number of different ways runners can potentially finish in the first five places for their team is $12 \cdot 11 \cdot 10 \cdot 9 \cdot 8 = 95,040$.

Note that if this number is both multiplied and divided by $7!$, we will have our answer in the more compact form $12!/7!$.

To calculate the number of outcome sequences without repetition for short sequences (as in the last example), it may be easiest to multiply the k factors together. But the factorial formula that is indicated there is better for long sequences. It also comes in handy when you're using such a result in a computer program or multiplying the numbers with a calculator, since you can then make use of a built-in factorial function.

Let's now look at the theory behind the counting process exhibited in the last example. *Proposition 3* summarizes the result, but we'll first introduce some terminology and notation.

A *permutation* of a set S is *an ordered arrangement* of all its elements. Arrangements arise by selecting elements in succession. Each permutation is thus uniquely associated with an ordered outcome sequence in which all the elements of the set appear exactly once.

If k distinct things from the full set S are arranged in some order without repetition, we have a *k-permutation* of S. Such arrangements are uniquely associated with outcome sequences of size k without repetition. A k-permutation in which $k = |S|$ is just an ordinary permutation of S.

The number of k-permutations from a set of size n is denoted (in print and on calculators) by $P(n, k)$ or $_nP_k$.

Definition 4.3.5: *Permutations*

a) *A **permutation** of S with $|S| = n$ is an ordered n-tuple $(x_1, x_2, \ldots, x_n)$ with $x_i \neq x_j$ for $i \neq j$.*

b) *A **k-permutation** of S is an ordered k-tuple $(x_1, x_2, \ldots, x_k)$ with $x_i \neq x_j$ for $i \neq j$.*

Proposition 4.3.3: *Counting k-Permutations*
Suppose S is a set of n elements. Then the total number of k-permutations
of S is $_nP_k = n \cdot (n-1) \cdots (n-(k-1)) = \dfrac{n!}{(n-k)!}$.

Proof:
From the *Generalized Multiplicative Counting Principle*, the number of
length-k outcome sequences, without repetition, is $n \cdot (n-1) \cdots (n-(k-1))$.
This is thus the number of k-permutations of S.
Multiplying and dividing by $(n-k)!$ gives the factorial form. ∎

Corollary 4.3.3.1: *Counting Permutations*
The total number of distinct permutations of a set S is $n!$, where $n = |S|$.

Proof:
This follows immediately, taking $k = n$. ∎

❖ **Example 4.3.7**
A quiz has 10 *matching* questions on it with 10 possible answers. If it is
answered randomly by someone who forgot to study, how many different
quizzes can be turned in, assuming each answer is matched exactly once.

Solution
There are $10! = 3,628,800$ different ways this quiz can be filled in. Presumably, only one is correct, so pure guessing isn't a high-percentage strategy.

❖ **Example 4.3.8**
In the game of Scrabble, each player has seven tiles with letters on them
for making words. Call any character string formed by letters $a, b, \ldots, g$ a
full scrabble segment.
a) How many full scrabble segments are there?
b) How many full scrabble segments have a and e next to each other?
c) How many full scrabble segments have a and e separated?
d) Do any full scrabble segments make a genuine word (called a *bingo*)?

Solution
a) There are $7! = 5040$ full scrabble segments: $S = \{a, b, c, d, e, f, g\}$ is our
   set, and we are counting its permutations.
b) To work this, think of a and e as forming a vowel-block, and consider
   the other letters as individual consonant-blocks.
   We must have six blocks in succession without repetition. There are
   $6! = 720$ of these block-sequences.
   Since the vowel-block can be either ae or ea, the joint outcome (block-
   sequence, vowel-arrangement) can be done in $720 \cdot 2 = 1440$ ways.
c) If 1440 of the 5040 full scrabble segments have the a and e next to one
   another, the other $5040 - 1440 = 3600$ full scrabble segments do not.
d) Here we leave the realm of mathematics. Evidently, no full scrabble segment forms a real (English-language) word.

EXERCISE SET 4.3

Exercises 1–3: Cartesian Products
Determine the following Cartesian products.

4.3.1. Write out the elements of the Cartesian product $E \times P$, where E is the set of even positive integers less than 10 and P is the set of primes less than 10. Then graph this set in a coordinate grid.

4.3.2. Describe the Cartesian product $\mathbb{Z} \times \mathbb{Z}$, where $\mathbb{Z}$ is the set of integers. What does its graph look like? (This is the set of *integer lattice points*.)

4.3.3. What does the graph of $R \times R$ look like, where R is the set of non–negative real numbers?

Exercises 4–6: True or False
Are the following statements true or false? Explain your answer.

4.3.4. Let S be any set and $T = \{1\}$. Then $S \subseteq S \times T$.

4.3.5. If $S = \{2, 4, 6\}$ and $T = \{3, 5, 7\}$, then $S \times T = \{6, 20, 42\}$.

4.3.6. The number of permutations of a set S is the number of ordered outcomes without repetition that can be formed using all of S's elements.

Exercises 7–13: Counting Everyday Permutations
The following problems concern everyday situations where permutations arise.

4.3.7. A tray of eggs contains five rows of six eggs. Six trays are stacked in a box, and boxes are loaded onto a pallet, five layers of three boxes across and four boxes deep. How many eggs are on each pallet? Explain how your calculation illustrates the *Multiplicative Counting Principle*.

4.3.8. A restaurant dinner menu has five choices for meat; three choices of potato, pasta, or rice; four types of vegetable; and five different desserts. How many different dinners can be served from this menu?

4.3.9. A chain letter is sent out by a crank to five of his friends, asking them to forward copies to five of their friends. If this is done four times by all involved, with no recipient receiving more than one letter, how many letters will have been sent out in all?

4.3.10. Mastermind is a game played by two people on a board with rows of four holes that hold colored pegs. One player chooses four colored pegs from an ample collection of six differently colored pegs and sets them up behind a shield. A second player makes a sequence of guesses to determine which colored pegs are present and in which order.
  a. How many different setups can be made?
  b. How many setups have no color repeated?
  c. How many setups have one color for the outer pegs and another color for the inner pegs?

4.3.11. A cell phone having a four-digit pin code has four prominent smudges on its face above four different numbers. What's the maximum number of distinct tries an attacker needs to gain access to the phone?

4.3.12. An RNA *codon* consists of three *nucleotides* combined in a certain order, chosen from four possibilities, denoted by A, C, G, and U. How many different codons are possible?

4.3.13. A coin is flipped five times. How many different total outcomes are possible? How many have opposite outcomes on the first and last flip? the same on the first and last flip?

Exercises 14–16: Scrabble Segments
The following problems deal with full scrabble segments (see Example 8).

4.3.14. How many full scrabble segments have all five consonants together?

4.3.15. How many full scrabble segments have vowels *a* and *e* separated by *c*?

4.3.16. How many full scrabble segments have vowels *a* and *e* separated by one consonant? by two consonants?

Exercises 17–21: Palindromes
*A **palindrome** is a numeral that reads the same forward or backward, such as 54321012345. Repetition is allowed, but no such numeral has a leading 0.*

4.3.17. How many seven-digit palindromes are there? How many are even?

4.3.18. How many eight-digit palindromes are there? How many are odd?

4.3.19. How many palindromes are there of length $2n + 1$?

4.3.20. How many palindromes are there of length $2n$?

4.3.21. Find a formula for the total number of palindromes of length n.

Exercises 22–24: Counting Divisors
The following problems have to do with the number of divisors of a number. Recall that every number is a unique product of powers of prime numbers.

4.3.22. *Factors of* 60
  a. List and count the number of distinct divisors of 60 (include 1 and 60).
  b. Factor 60 into a product of powers of primes. How do the prime factors of divisors of 60 relate to the prime factors of 60?
  c. Using your result in part *b* and the methods of this section count the total number of distinct divisors of 60.

4.3.23. *Factors of* 72
  a. List and count the number of distinct divisors of 72 (include 1 and 72).
  b. Factor 72 into a product of powers of primes. How do the prime factors of divisors of 72 relate to the prime factors of 72?
  c. Using your result in part *b* and the methods of this section, count the total number of distinct divisors of 72.

4.3.24. *Numbers of Factors and Prime Factorization*
  a. If $n = p \cdot q$, where p and q are distinct prime numbers, how many factors does n have? Explain, using the methods of this section.
  b. If $n = p \cdot (q \cdot r)$, where p, q, and r are distinct prime numbers, how many factors does n have? Explain, using the methods of this section.

c. If $n = p^k \cdot q^m$, where p and q are distinct prime numbers, how many factors does n have? Explain, using the methods of this section.

d. If $n = p_1^{n_1} \cdot p_2^{n_2} \cdot p_3^{n_3}$ where all p_i are distinct prime numbers, how many distinct divisors does n have? Explain, using the methods of this section.

e. Generalize part d, with proof, to any number of distinct prime factors.

4.3.25. *Final Zeros for Factorial Products*

a. How many final 0s does the product 5! have? the product 14!? Explain.

b. How many final 0s does the product 25! have? the product 100!? Explain.

c. Explain how to determine the final number of 0s in the product $n!$.

Exercises 26–28: Wiener-Kuratowski Definition of Ordered Pairs

The following unusual but ingenious set-theoretic definition of ordered pair is due to Norbert Wiener (1914), as simplified by Kazimierz Kuratowski (1921). It treats ordered pairs as special (non-ordered) sets, thus incorporating them into elementary Set Theory as a derived notion.

Definition of Ordered Pair: $(x, y) = \{\{x\}, \{x, y\}\}$.

4.3.26. Write the set-theoretic representation of the ordered pair $(0, 1)$.

4.3.27. What ordered pair does $\{\{2, 3\}, \{3\}\}$ represent?

4.3.28. Using the set-theoretic definition and your intuitions about singletons and doubletons conjecture when $(a, b) = (c, d)$. Prove your result.

Exercises 29–31: Defining Ordered n-Tuples

Ordered n-tuples can be defined recursively using ordered pairs as a basis. The recursive clause is: $(x_1, x_2, \ldots, x_n, x_{n+1}) = ((x_1, x_2, \ldots, x_n), x_{n+1})$.

4.3.29. *Ordered Triples*

a. State what an ordered triple (a, b, c) is in terms of ordered pairs.

b. Determine what an ordered triple (a, b, c) is in its most primitive form using the Wiener-Kuratowski definition of ordered pair (see above).

4.3.30. *Ordered Quadruples*

Write down the definition for a 4-tuple (a, b, c, d) and then work it backward to express it in terms of ordered pairs.

4.3.31. If the above definition of ordered n-tuples is the recursive clause of the definition, what is the base case? Can this definition begin with $n = 1$?

Exercises 32–34: Cardinality of Cartesian Products

Prove the following results about the cardinality of Cartesian products.

4.3.32. Prove *Proposition* 1a: *If* $|S| = m$ *and* $|T| = n$, *then* $|S \times T| = m \times n$. *Hint*: what proof techniques show that a result holds for all natural numbers m and n? Use a combination of both direct approaches.

4.3.33. Use mathematical induction and *Proposition* 1a (Exercise 32) to prove *Proposition* 1b: $|S_1 \times S_2 \times \cdots \times S_n| = |S_1| \times |S_2| \times \cdots \times |S_n|$.

4.3.34. Use mathematical induction to prove *Corollary* 2 to *Proposition* 1: *If* k *events in sequence can occur in* n_i *ways, then the number of ways the compound event can occur is* $n_1 \cdot n_2 \cdots n_k$.

Exercises 35–36: Strings and Finite Sequences
The following problems concern finite strings (see Section 3.3). Recall that strings $a_1 a_2 \cdots a_k$ are finite sequences of length k, each a_i coming from a common alphabet A. Let all strings of length k be denoted by A^k. Then $A^* = \bigcup A^k$ is the set of all finite strings formed from the alphabet.

4.3.35. Let A be the English alphabet $\{a, b, \ldots, z\}$, so that $|A| = 26$.
  a. How many strings are there of size 6? How many of these have no repeated letters?
  b. What does A^* represent in this case?

4.3.36. Let A be the set of digits $0, 1, \ldots, 9$.
  a. How many strings are there of size 7? How many are there if the first entry is non-0?
  b. What does A^* represent in this case?

Exercises 37–47: Theorems About Cartesian Products?
Determine whether or not each of the following results about Cartesian products is a theorem of Set Theory. If it is, illustrate it via an appropriate diagram and then prove it; if it is false, provide a specific counterexample.

4.3.37. $\emptyset \times S = \emptyset = S \times \emptyset$

4.3.38. $S \times T = \emptyset \leftrightarrow S = \emptyset \vee T = \emptyset$

4.3.39. $S \times T = T \times S$

4.3.40. $S_1 \subseteq S_2 \wedge T_1 \subseteq T_2 \leftrightarrow S_1 \times T_1 \subseteq S_2 \times T_2$

4.3.41. $R \times (S \cap T) = (R \times S) \cap (R \times T)$

4.3.42. $R \times (S \cup T) = (R \times S) \cup (R \times T)$

4.3.43. $R \times (T - S) = (R \times T) - (R \times S)$

4.3.44. $(S_1 \times T_1) \cap (S_2 \times T_2) = (S_1 \cap S_2) \times (T_1 \cap T_2)$

4.3.45. $(S_1 \times T_1) \cup (S_2 \times T_2) = (S_1 \cup S_2) \times (T_1 \cup T_2)$

4.3.46. $(S_1 \times T_1) - (S_2 \times T_2) = (S_1 - S_2) \times (T_1 - T_2)$

4.3.47. $\overline{S \times T} = \overline{S} \times \overline{T}$

4.4 Combinations

We can now count how many ways an event can occur, provided order is important. In this section our focus is on *combinations*—choices where order is irrelevant. Here subsets rather than sequences are our focus. We'll develop the theory needed for counting combinations, and we'll also look at several mathematical applications.

4.4.1 *Unordered Outcomes, No Repetition*

Suppose S is a set of n elements. How many different combinations of k elements can be chosen from this set, ignoring the order in which they are chosen and not allowing repetition? In other words, how many subsets of size k does a set of size n have? This number is denoted by $C(n, k)$ or $_nC_k$ or $\binom{n}{k}$, all of which can be read as n *choose* k.

Suppose we first choose a subset of k elements sequentially. As we saw in Section 4.3, there are $_nP_k = n \cdot (n-1) \cdots (n-(k-1)) = \frac{n!}{(n-k)!}$ of these ordered subsets/k-permutations. For each one of these, there are $k!$ permutations of exactly the same elements but in different orders. To count these permutations exactly once, as required, we must divide the total number of k-permutations by this multiplicity. This gives the number of combinations of size k from a set of size n, proving the following proposition.

Proposition 4.4.1: *Counting Combinations without Repetition*
Let S be a set of n elements. The total number of subsets of S of size k is
given by
$$\binom{n}{k} = \frac{n \cdot (n-1) \cdots (n-(k-1))}{k!} = \frac{n!}{k!\,(n-k)!} \,.$$

An easy way to remember these formulas is to note that in the first expression the numerator and denominator have the same number of factors (k of them), one going down from n and the other coming up from 1. In the last expression, the values whose factorials are being multiplied in the denominator add up to the number whose factorial is being taken in the numerator.

✤ **Example 4.4.1**
   a) A math prof has seven whiteboard markers on her desk. If she takes three of them to class, in how many different ways can she do this?
   b) How many subsets of size 3 does a set of size 7 have? How many subsets of size 4?

Solution
   a) The number of combinations of size 3 chosen from a collection of size 7 is $\binom{7}{3} = \frac{7 \cdot 6 \cdot 5}{1 \cdot 2 \cdot 3} = \frac{7!}{3!\,4!} = 35$.
   b) There are 35 distinct subsets of size 3 in a set of size 7. There are also 35 distinct subsets of size 4—every choice of 3 elements from a 7-element set leaves a corresponding subset of 4 elements behind, its complement. Alternatively, the formula for the number of subsets of size k is exactly the same as that for subsets of size $n - k$: $\binom{n}{k} = \binom{n}{n-k}$ (see Exercise 3).

4.4.2 Combinations and Binomial Coefficients

The method of counting combinations can be used to determine the coefficients appearing in the expansion of the binomial $(a+b)^n$ for a positive integer n. For instance, $(a+b)^3 = a^3 + 3a^2b + 3ab^2 + b^3$. Calculating $(a+b)^n$ as an n-fold product gives the theorem below. Newton and others generalized this result to additional kinds of exponents, giving an important tool for working with functions like $\sqrt{1-x} = (1-x)^{\frac{1}{2}}$ or $1/(1+x^2) = (1+x^2)^{-1}$.

Theorem 4.4.1: *Binomial Expansion Theorem*
 Let a and b be real numbers and n be a natural number. Then
$$(a+b)^n = \sum_{k=0}^{n} \binom{n}{k} a^{n-k} b^k.$$

Proof:
- Since $(a+b)^n = (a+b)(a+b)\cdots(a+b)$, the different terms in the expansion arise by choosing one factor from each binomial expression (either a or b) and then multiplying them together to get an n-fold product.
- Let's focus on the ways to form the various powers b^k.
  Choosing no b's gives a^n; there is $\binom{n}{0} = 1$ way to do this, giving a^n.
  Choosing one b and $n-1$ a's gives $\binom{n}{1} a^{n-1} b = n a^{n-1} b$.
  More generally, the coefficient of $a^{n-k} b^k$ for any k is $\binom{n}{k}$.
- Thus, the full binomial expansion is $(a+b)^n = \sum_{k=0}^{n} \binom{n}{k} a^{n-k} b^k$. ∎

Binomial coefficients for nonnegative integer powers of n can be put into a triangular array known as Pascal's triangle.[7] Row n of the triangle gives the sequence of coefficients $\left\{\binom{n}{k}\right\}_{k=0}^{n}$, starting with $n=0$ as the top row.

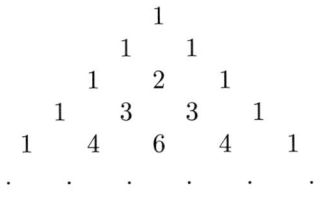

$$\textit{Pascal's Triangle: Binomial Coefficients}$$

Pascal's triangle can be generated recursively from the top two rows. Each later row begins and ends with a 1, and all intermediate numbers arise by adding the two adjacent numbers directly above them. In the last row exhibited, $4 = 1 + 3$, $6 = 3 + 3$, and $4 = 3 + 1$.

Pascal's triangle is a favorite for finding patterns of numbers. Some of these are explored in the exercises (see Exercises 27–34).

✤ Example 4.4.2
 Expand $(1+1)^n$ as a binomial. Then prove that $|\mathcal{P}(S)| = 2^n$ when $|S| = n$.

Solution
 On the one hand, $(1+1)^n = 2^n$.
 On the other hand, $(1+1)^n = \sum_{k=0}^{n} \binom{n}{k} 1^{n-k} 1^k = \sum_{k=0}^{n} \binom{n}{k}$.

---

[7] This is named after the seventeenth-century French mathematician Blaise Pascal, who investigated its properties. It was known several centuries earlier, however, both to Arabic mathematicians and Chinese mathematicians.

Thus, $2^n = \sum\limits_{k=0}^{n} \binom{n}{k}$.

If S is a set of size n, it has $\binom{n}{k}$ subsets of size k.

Therefore, $\sum\limits_{k=0}^{n} \binom{n}{k}$ represents the total number of subsets of S.

Hence, the cardinality of $\mathcal{P}(S)$ is 2^n. $\checkmark$ (see also Exercise 4.2.31)

4.4.3 Permutations of Multisets

In Section 4.3 we saw how to count sequences of elements chosen from a set, with or without allowing repetition in the sequence. Here we'll show how to count sequences formed by choosing elements that are indistinguishable. The construct needed for formulating this is that of *multiset*, a collection where repetition is allowed. We'll first look at an example and then state a proposition covering such situations.

✤**Example 4.4.3**

How many distinct permutations (letter-strings) can be formed from the word *chincherinchee*, the name of a South African star-shaped flower?

Solution

· The 14-letter word *chincherinchee* has six distinct letters, three of them occurring three times (c, h, e), two of them twice (i, n), and one once (r). Switching duplicate letters will yield the same permutation, so we can't count them more than once. The value 14! is thus larger than the number of distinct permutations of the multiset $\{c, h, i, n, c, h, e, r, i, n, c, h, e, e\}$.

· In a 14-letter string, we need three places each for c, h, and e, two places each for i and n, and one place for r.

The number of ways to make these choices is $100,900,800$—calculated as

$$\binom{14}{3}\cdot\binom{11}{3}\cdot\binom{8}{3}\cdot\binom{5}{2}\cdot\binom{3}{2}\cdot\binom{1}{1} = \frac{14!}{3!\,11!}\cdot\frac{11!}{3!\,8!}\cdot\frac{8!}{3!\,5!}\cdot\frac{5!}{2!\,3!}\cdot\frac{3!}{2!\,1!}\cdot\frac{1!}{1!\,0!} = \frac{14!}{3!\,3!\,3!\,2!\,2!\,1!}\,.$$

This is a large number, but two orders of magnitude smaller than 14!.

· Note how the numerator in the final fraction involves the cardinality of the multiset, while the denominator's factors divide by the multiplicities of the repeated letters. Thus, we divide by 3! because that's how many ways the c's can be permuted with no change in the outcome, by 2! for the repeated n's, and so on.

Proposition 4.4.2: *Counting Permutations of Multisets*

The number of distinct permutations of a multiset $\{x_1, x_2, \ldots, x_n\}$ is

$$\binom{n}{m_1}\cdot\binom{n-m_1}{m_2}\cdots\binom{n-m_1-\cdots-m_{k-1}}{m_k} = \frac{n!}{m_1!\,m_2!\cdots m_k!}\,,$$

where m_i, $1 \le i \le k$, indicate the distinct multiplicities for the multiset, $m_1 + m_2 + \cdots + m_k = n$.

Proof:
- If the k distinct elements of the multiset are present with multiplicities m_i, $1 \le i \le k$, we choose m_1 places in an n-sequence for the first distinct element in $\binom{n}{m_1}$ ways, then we choose m_2 places among the remaining $n - m_1$ places in $\binom{n-m_1}{m_2}$ ways, etc. The combined choice can be done in $\binom{n}{m_1} \cdot \binom{n-m_1}{m_2} \cdots \binom{n-m_1-\cdots-m_{k-1}}{m_k} = \frac{n!}{m_1!\, m_2!\cdots m_k!}$ ways. ∎
- The final fraction in the formula results from canceling common factorial terms in the numerator and denominator. This same value will result no matter in which order the distinct elements are positioned.

4.4.4 Discrete Probability

A *sample space* for an experiment is the set of all possible ways it might turn out. In this context, an *event* is a particular set of outcomes for an experiment, a subset of the associated sample space. The probability of an event measures how likely that event is relative to all the possibilities that might occur. Being able to count permutations and combinations, we can calculate the theoretical probability of events. We'll state the classical relative-frequency definition for discrete probability and then look at a few examples.

Definition 4.4.1: *Discrete Probability*
  *The **probability of an event** is the ratio of the number of ways it can occur to the total number of outcomes in the associated sample space.*

This definition presumes that each outcome in a sample space S is equally likely. Under this assumption, the probability of an event E is given by $P(E) = |E|/|S|$. The probability of an event thus satisfies the double inequality $0 \le P(E) \le 1$ since $\emptyset \subseteq E \subseteq S$.

✤ **Example 4.4.4**
  A fair coin is tossed five times.
  a) What is the probability that exactly three heads will occur?
  b) What is the probability that exactly three tosses land the same way?

Solution
  a) The sample space here is all 5-tuples of H's and T's (potential toss outcomes). There are $2^5 = 32$ possible outcomes, all of them equally likely if the coin is fair.
  To get exactly 3 heads, this must happen on 3 specific tosses. The number of ways 3 tosses can be chosen out of 5 is $\binom{5}{3} = 10$ ways.
  Therefore, the probability of exactly 3 heads is $10/32 = 5/16 = .3125$.
  b) If exactly 3 tosses land the same way, these can be either heads or tails. There are 10 ways for each of these events to occur. So the probability of this joint event is $20/32 = 5/8 = .625$. Because these two sub-events are disjoint (3 heads, 3 tails), their individual probabilities add up to the total probability: $5/16 + 5/16 = 5/8$.

✤**Example 4.4.5**

Two dice are rolled. What is the probability of rolling a 7? An 11? A 7 or an 11?

Solution

· Each die has 6 outcomes, so two dice yield 36 distinct pairs of numbers, constituting our sample space.
· To get a 7, the outcomes must be 1 and 6, 2 and 5, or 3 and 4. Each of these can occur in two ways, so there are 6 ways to roll a 7.
  This gives a probability of $6/36 = 1/6 = .1\bar{6}$ for rolling a 7.
· An 11 only comes by rolling a 5 and a 6, which happens in 2 ways.
  The probability of throwing an 11, therefore, is $2/36 = 1/18 = .0\bar{5}$.
· Thus, the probability of rolling either a 7 or an 11 is $8/36 = 2/9 = .\bar{2}$.

✤**Example 4.4.6**

For a standard 52-card deck (four suits with 13 different kinds), which five-card hand has a higher probability: a full house (three cards of one kind, two of another) or a flush (five cards of the same suit)?

Solution

· Our sample space here consists of all possible five-card hands.
  There are $\binom{52}{5} = \frac{52!}{5!\,47!} = 2,598,960$ hands in all.
· The number of ways to get a full house is calculated by multiplying the number of ways to choose one kind times the number of ways to get three of this kind times the number of ways to choose a second kind times the number of ways to get two of this other kind.
  This number is $13 \cdot \binom{4}{3} \cdot 12 \cdot \binom{4}{2} = 156 \cdot 4 \cdot 6 = 3744$.
  (*Note*: the product $\binom{13}{2} \cdot \binom{4}{3} \cdot \binom{4}{2}$ is off by a factor of 2 because it doesn't take into account which values are three of a kind vs. two of a kind.)
· A flush can be generated by first choosing a suit and then choosing 5 cards from that suit.
  The number of ways this can occur is $4 \cdot \binom{13}{5} = 4 \cdot 1287 = 5148$.
· Thus, a flush is slightly more likely than a full house: its probability is $5148/2,598,960 \approx .00198$ vs. $3744/2,598,960 \approx .00144$ for a full house.
  A flush will occur about 54 more times than a full house in 100,000 hands.

4.4.5 *Unordered Outcomes with Repetition*

The most complex combinatorial situation is counting unordered sets when repetition is allowed. Here collections no longer correspond to subsets of a set, because sets don't allow an element to be present multiple times. Furthermore, while we could begin the counting process like we did for ordered sets, there is no constant factor to divide by in order to cancel out the duplication. So we need to involve multisets to systematically count such possibilities. Let's look at a simple example to illustrate how we can proceed.

✤**Example 4.4.7**
Consider a set S with three elements, say $S = \{a, b, c\}$. How many doublets (multiset pairs) from S are there, allowing repetition?

Solution
- The numbers involved in this problem are small enough so we can list all doublets, which we'll denote by $<x, y>$ for $x, y \in S$. Order isn't important, so $<y, x> = <x, y>$, but we don't require that $x \neq y$.
  Here is the full list: $<a, a>$, $<a, b>$, $<a, c>$, $<b, b>$, $<b, c>$, $<c, c>$.
- Now, we could have started out with all nine ordered pairs from $S \times S$, which allows for repeated elements. However, there is no fixed duplication number to divide this total by. The ordered pairs (a, b) and (b, a) both generate the same doublet $<a, b>$, so we'd have to divide this part of the count by 2, but we can't divide by 2 in general, because (a, a) is present only once.
- We'll shift our focus to count these doublets more efficiently. Instead of focusing on the two *spots* that we want to fill with a letter, let's concentrate on the three *letters* that can be chosen. And let's distribute two winning $*$ tags to the letters a, b, and c to indicate how many times, if any, they've been chosen to be in a doublet $<x, y>$.
- Using three separated blanks $\_ \mid \_ \mid \_$ to stand for the three letters in order, we need to count the number of ways to assign two tags. The doublet $<a, b>$ would be represented by $\underline{*} \mid \underline{*} \mid \_$ and the doublet $<c, c>$ by $\_ \mid \_ \mid \underline{* *}$. Blanks can obviously be ignored; the essential thing is the location of the stars relative to the separator bars. So we can represent these two doublets by $* \mid * \mid$ and $\mid \mid * *$, sequences of stars and bars.[8]
- The new question, therefore, is this: how many 4-sequences of stars and bars contain two stars? This is precisely the number of ways to choose two positions in a 4-sequence for the two $*$'s: there are $\binom{4}{2} = 6$ ways.

The following proposition and its proof generalize the last example and give us a formula. Applying the *Stars-and-Bars Method* of the proof, though, is probably more important than memorizing the formula.

Proposition 4.4.3: *Counting Unordered Collections with Repetition*
The total number of unordered collections of size k from a set of size n,
$$\text{allowing repetition, is } \binom{k + (n-1)}{k} = \frac{n \cdot (n+1) \cdots (n + (k-1))}{1 \cdot 2 \cdots k}.$$

Proof:
- Create n compartments (n blanks separated by $n-1$ bars) for the n elements of the set.
- Any distribution of k $*$'s to these n compartments will represent a distinct way of choosing k elements, allowing repetition. This results in sequences with $n - 1$ bars and k stars.

---

[8] This *Stars-and-Bars Method* for counting unordered collections with repetition was popularized by William Feller in his classic 1950 treatise on probability.

· The total number of ways that k stars can be positioned within such a $k+(n-1)$-sequence is $\binom{k+(n-1)}{k}$. Expanding this expression and canceling the common term $(n-1)!$ gives the final fraction. ∎

Note how the various ideas we've been studying come together in this proof. To count the number of unordered collections, allowing repetitions, we first formed ordered sequences to represent the collections. Then, to count the relevant sequences, we counted combinations of positions within the sequence, without concern for the order in which they were chosen.

Note also that the fractional formulas we've developed for counting unordered collections, with or without repetition, have a satisfying symmetry (see Exercise 1). Both types of unordered collections have $k!$ in the denominator. For combinations without repetition, the numerator has k factors, starting with n and counting down. For combinations with repetition, the numerator also has k factors, but this time starting with n and counting up.

❖**Example 4.4.8**

A doting grandmother wants to give 20 five-dollar bills to her four grandchildren. In how many different ways can she distribute the money?

Solution

· Let's model this the way we counted unordered collections with repetition. Create four money compartments, one for each child, and distribute the 20 five-dollar bills among them in some way.
· This can be symbolized by a 23-sequence containing 20 five-dollar symbols and 3 compartment-separator symbols.

There are $\binom{23}{20} = \binom{23}{3} = \frac{23 \cdot 22 \cdot 21}{1 \cdot 2 \cdot 3} = 1771$ ways to do this.
· Of course, to avoid favoritism, she'd better give each $25.

It may not always be clear which counting procedure a given situation requires. At times it may even seem like there are two perfectly good ways to count, though they give different answers! Everyone experiences this dissonance sometime or other. We remarked on this in connection with counting five-card hands in Example 6, but it happens elsewhere as well. To choose a counting method, carefully check that you've correctly conceptualized the variability involved, that you haven't overlooked any restrictions, and that you aren't counting things more or less often than you should. The following example, a variation on Example 1, illustrates a couple of ways in which things might go wrong.

❖ **Example 4.4.9**

A math prof has seven whiteboard markers on her desk, three blue, two green, and two red. If three of them are chosen, how many of these include the following color combinations:
a) exactly one red?
b) at least one red?

Solution

a) Here's a way to count the first case (exactly one red out of three).

- There are two red markers, so if the trio of markers has one red and two others, there are $2 \cdot \binom{5}{2} = \underline{20}$ trios.
- Unfortunately, this answer is wrong. If *marker combinations* were being counted, it would be correct. But *color combinations* are being asked about, so we shouldn't distinguish markers of the same color.
- If we choose one red, the other two must be blue or green: 2 and 0, 1 and 1, or 0 and 2. These $\underline{3}$ options/trios can also be counted using two stars and one bar.

b) For the second case (at least one red out of three), we might reason as follows:

- We can choose one red in 2 ways and two markers from the remaining six in $\binom{6}{2} = 15$ ways. So the total number of trios is $2 \cdot 15 = \underline{30}$.
- After discussing the first case, you probably know this is wrong. But now it's even wrong if *marker combinations* are being counted, since some trios are counted more than once (the 5 marker trios with two reds). The easiest way to count this event is probably to first count its complement: how many marker trios contain no reds? There are $\binom{5}{3} = 10$ of these, while there are $\binom{7}{3} = 35$ marker trios in all, so $35 - 10 = \underline{25}$ trios contain at least one red. This checks with the fact that the first count included 5 duplicates.
- To count *color combinations* with at least one red, we can again count a complement event using stars and bars or *Proposition 3*. We must still be on guard, though, because we have a limited supply of markers: we can't choose three green or three red. But let's first assume counterfactually that we can. Then there would be $\binom{5}{3} = 10$ distinct color trios (three stars, two bars). Subtracting out the 2 cases that can't occur (three green, three red), we end up with 8 total color trios. Similarly, since there are only two greens, there are $\binom{4}{3} - 1 = 3$ color trios with no red. This means $8 - 3 = \underline{5}$ color trios have at least one red. These numbers are small enough to list all the possibilities: BBR, BGR, GGR; BRR, GRR.

✤ Example 4.4.10

An *ordered number partition* of a positive integer n is an *ordered sum* of positive integers, $x_1 + \cdots + x_p = n$. A *number partition* of a positive integer n is an *unordered sum* of positive integers, $x_1 + \cdots + x_p = n$. Number partitions are studied in *Number Theory* and *Combinatorics*.

a) How many ordered number partitions of 6 into two numbers are there? How many number partitions of 6 into two numbers? How many distinct partitions into two non-empty subsets are there for a set S with $|S| = 6$?

b) How many ordered number partitions of a number n into p summands are there? How many number partitions into p numbers?

Solution

a) We can decompose 6 into an ordered sum of two positive integers in the following five ways: $1+5$, $2+4$, $3+3$, $4+2$, and $5+1$.

   · More systematically, we can solve $x_1 + x_2 = 6$ for positive integer pairs (x_1, x_2) using the following reasoning. Since each summand x_i must be positive, we must distribute four additional units to two-summand compartments already containing one unit each.

   · Using the last proposition, there are $\binom{4+1}{4} = 5$ ways to do this.

   · If unordered sums are used, there are only three two-summand partitions: $1+5$, $2+4$, and $3+3$. These can be pictured as follows:

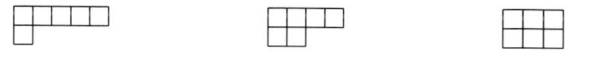

   · If S is a set with six elements, six partitions include singletons, $\binom{6}{2} = 15$ partitions include doubletons, and $\binom{6}{3}/2 = 10$ partitions include tripletons, for a total of 31 different partitions of S into two subsets.

b) Now we'll generalize to any n and p.

   · To decompose n into a sum of p terms, we must distribute n units into p compartments. Each compartment requires at least one unit, so we need to distribute an additional $n - p$ units to the p compartments.

   · Using stars and bars yields sequences of length $(n-p)+(p-1) = n-1$ in which locations for the $n-p$ stars must be chosen. This gives $\binom{n-1}{n-p}$ ordered partitions.

   · For example, if $n = 8$ and $p = 5$, there are $\binom{7}{3} = 35$ distinct ordered sums of 5 positive integers adding up to 8.

   · If order of the summands isn't important, the problem is more difficult. No explicit formula is known for how many ways to partition a number n into p summands, though there are ways to calculate this.

   · For the case where $n = 8$ and $p = 5$, the number of partitions is only 3, pictured below: $1+1+1+1+4$, $1+1+1+2+3$, and $1+1+2+2+2$.

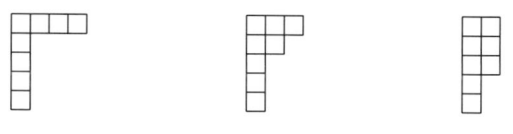

EXERCISE SET 4.4

4.4.1. *Creating a Combinatorics Counting-Formula Chart*
Create a 2×2 chart for the various outcome counting formulas in this section: ordered vs. unordered, repetition vs. no repetition.

Exercises 2–5: True or False
Are the following statements true or false? Explain your answer.

4.4.2. For all natural numbers n and k, $P(n, k) \le C(n, k)$.

4.4.3. For all natural numbers $k \le n$, $\binom{n}{n-k} = \binom{n}{k}$.

4.4.4. The coefficients in the expansion of $(a + b)^n$ can be calculated by counting combinations.

4.4.5. To count combinations, with or without repetition, first count ordered sequences and then divide out by a duplication number, if any.

Exercises 6–17: Counting Events and Possibilities
Use the methods of this and the last section to work the following exercises.

4.4.6. *Diagonals in a Polygon*
a. How many diagonals can be drawn in a convex pentagon (a five-sided figure with no indentations)? How many distinct triangles will this form?
b. How many diagonals can be drawn in a convex polygon of n sides? How many triangles will this form?

4.4.7. *Polite Handshakes*
a. At a party of 15 people, everyone shook hands once with everyone else. How many handshakes took place?
b. Find a formula for the number of handshakes made if n people were present at a party and each person shook hands once with everyone else.

4.4.8. A *Discrete Mathematics* class of 18 students has seven women in it. If three students are picked each period to exhibit their homework solutions in front of class, determine how many different groups contain:
a. Any number of women, from zero to three.
b. Exactly one woman.   d. At least two women.
c. Three women.   e. No more than one woman.

4.4.9. *Circular Arrangements*
A five-member committee is seated in a circle around a table.
a. How many essentially different arrangements of the members are possible if there are five chairs? if there are six chairs and one remains empty? if there are seven chairs and two remain empty?
b. Calculate the number of essentially different arrangements for part *a* if relative location among the members and empty places (rather than which chair each occupies) is all that counts.
c. Rework part *b* if the orientation of the circle is unimportant.

4.4.10. *Paths in a Grid*
a. How many paths can be drawn along an integer grid to go from the origin $(0, 0)$ to the point $(2, 3)$ if one is only allowed to go either up or right at each integer lattice point? Generalize: how many paths are there from the origin to the point (m, n) in the first quadrant? Explain.
b. How many ways are there to pass from triple $(0, 0, 0)$ to triple $(3, 4, 5)$ if you can only increase one coordinate at a time by adding 1 to it? Explain.

4.4.11. *Parity of Sums*
a. How many sums $a + b$ formed by choosing two different numbers from $\{1, 2, \ldots, 2n\}$ are even? How many are odd?
b. How many sums $a + b + c$ formed by choosing three different numbers from $\{1, 2, \ldots, 3n\}$ are divisible by 3?

4.4.12. How many distinct letter-strings can you make from the letters in the word MISSISSIPPI? Explain.

4.4.13. An office mailroom has 15 mailboxes for its employees. In how many different ways can 22 pieces of mail be distributed to the mailboxes? Explain.

4.4.14. *The Daily Donut Run*

Sara makes a donut run for her staff every morning. She always buys a dozen donuts, choosing from five different types of donuts.

  a. If there are no restrictions on what she should buy, how many different donut dozens are possible?

  b. If she always buys at least three chocolate donuts, how many different donut dozens are possible?

  c. If she always gets at least one of each type of donut, how many different donut dozens are possible?

4.4.15. *Raiding a Piggy Bank*

If five coins are taken from a piggy bank with many pennies, nickels, dimes, and quarters, how many types of coin collections are possible?

4.4.16. *Whiteboard Marker Combinations*

A teacher has 12 whiteboard markers—5 blue, 4 green, and 3 red.

  a. How many color combinations are possible if three markers are chosen?

  b. How many color combinations are possible if four markers are chosen?

  c. How many color combinations are possible if five markers are chosen?

4.4.17. *More Partitions of 6*

  a. How many ordered number partitions of 6 into a sum of three numbers are there? How many number partitions of 6 into a sum of three numbers? How many partitions of a six-element set into three subsets are there?

  b. Repeat part *a* for partitions into four numbers and four subsets.

  c. Repeat part *a* for partitions into five numbers and five subsets.

  d. Repeat part *a* for partitions into six numbers and six subsets.

Exercises 18–19: Counting Game Pieces

The following problems ask you to count pieces for board games.

4.4.18. *Dominoes*

A *double-twelve domino* is a flat rectangular piece divided in two on one side, each half having between 0 and 12 dots on it, repetitions allowed. How many different dominoes are there in a full set? Explain.

4.4.19. *Triominoes*

A *triomino* is a flat triangular piece containing a number between 0 and 5 ($0 \leq N \leq 5$) in each corner on one side. Repetitions of numbers are allowed.

  a. How many triominoes are there in a full set if each triomino is oriented clockwise with numbers in nondecreasing size?

  b. How many triominoes can be made if they can be oriented either clockwise or counterclockwise?

  c. How many triominoes can be made using the numbers 0 through 6, oriented clockwise?

Exercises 20–22: Counting and Probability
The following problems involve counting and discrete probability.

4.4.20. Count the five-card hands from a standard 52-card deck that are:
  a. One pair (two cards of the same kind plus three cards of different kinds from the pair and from one another).
  b. Two pairs (two of one kind plus two of another kind plus one card of a third kind).
  c. A straight (a run of five consecutive values, not all the same suit, where cards after the ten are ordered as jack, queen, king, and ace).
  d. Determine the probabilities associated with the above hands and compare them to one another and the probabilities calculated in Example 6.

4.4.21. An urn contains five red balls, eight white balls, and ten blue ones.
  a. How many different sets of three red, three white, and three blue balls can be taken out of the urn?
  b. What is the probability of drawing three of each color if nine balls are drawn from the urn?

4.4.22. Yahtzee is a game played by two or more players, each rolling five dice in turn, with two chances to re-roll some or all of the dice in order to get a score in one of 13 different categories.
  a. What's the probability of rolling a full house (three of one kind, two of another)?
  b. What's the probability of rolling a small straight (four numbers in a row)?
  c. What's the probability of rolling a large straight (five numbers in a row)?
  d. What's the probability of rolling a Yahtzee (all five numbers the same)?
  e. If a player rolls three 5's, a 2, and a 3, what is his probability of getting a Yahtzee on his next two rolls if he decides to keep the 5's and roll the other two dice again?

Exercises 23–26: Binomial Identities
Prove the following binomial identities, either using algebraic formulas or by using a combinatorial interpretation.

4.4.23. $\binom{2n}{2} = 2\binom{n}{2} + n^2$

4.4.25. $\binom{n}{m} \cdot \binom{m}{k} = \binom{n}{k} \cdot \binom{n-k}{m-k}$

4.4.24. $\binom{n}{1} + 6\binom{n}{2} + 6\binom{n}{3} = n^3$

4.4.26. $\binom{n}{m} \cdot \binom{n-m}{k} = \binom{n}{k} \cdot \binom{n-k}{m}$

Exercises 27–34: Pascal's Triangle
The following problems explore some of the patterns in Pascal's triangle. Expand the triangle given in the text to facilitate answering them.

4.4.27. Using the factorial formula for binomial coefficients prove the basic recursion formula on which Pascal's triangle depends: $\binom{n-1}{k-1} + \binom{n-1}{k} = \binom{n}{k}$. Explain why this formula is basic to generating the triangle of coefficients.

4.4.28. Prove that $\binom{n-1}{k-1} + \binom{n-1}{k} = \binom{n}{k}$, this time using the fact that $\binom{n}{k}$ is the number of subsets of size k in a set S of size n. *Hint*: pick some element of S and partition the collection of subsets into two classes, depending on whether the element is in the subset or not.

4.4.29. Prove the *Binomial Expansion Theorem* using mathematical induction. Make use of Exercise 27 where appropriate.

4.4.30. Example 2 establishes that $\sum_{k=0}^{n} \binom{n}{k} = 2^n$. What is $\sum_{k=0}^{n} (-1)^k \binom{n}{k}$? Prove your result using a binomial expansion.

4.4.31. Show that $\binom{2n}{n} = \sum_{k=0}^{n} \binom{n}{k}^2$ when $n = 3$ by direct calculation. Then show via the recursive formula in Exercise 27 how this formula comes about in this case: trace the value of $\binom{6}{3}$ back up Pascal's triangle until you reach the sides of the triangle (row $n = 3$).

4.4.32. Prove the general result stated in Exercise 31.

4.4.33. Identify the pattern of the following result and check it using Pascal's triangle when $n = 6$. Then prove in general that $\sum_{k=1}^{n} \binom{k}{1} = \binom{n+1}{2}$.

4.4.34. *Binomial Expansion Evaluated*

Evaluate the sum $\sum_{k=0}^{n} \binom{n}{k} 3^k$. *Hint*: multiply each term by 1 in the form 1^{n-k}.

4.4.35. *Multinomial Expansion*

Expanding $(x_1 + x_2 + \cdots x_k)^n$, one obtains terms of the form $c x_1^{m_1} x_2^{m_2} \cdots x_k^{m_k}$, where $m_1 + m_2 + \cdots m_k = n$.

a. Explain why the multinomial coefficient c is the number of permutations of the multiset in which each x_i is present m_i times.
b. Use part *a* to calculate the multinomial coefficients and the multinomial expansion of $(x_1 + x_2 + x_3)^4$.

4.5 Additive Counting Principles

The parts of *Combinatorics* we've considered so far are grounded primarily in the *Multiplicative Counting Principle*, which is based on *Proposition* 4.3.1: *the cardinality of a Cartesian product is the product of its cardinalities.*

Paradoxically, now that we can count by multiplying, we'll also be able to count by adding, the focus of this section. The simplest case counts the number of elements in the disjoint union of two sets using the *Additive Counting Principle* (*Proposition* 1). To deal with more complex situations involving unions of several sets, though, we'll need to draw upon our ability to count combinations, yielding the counting method known as the *Principle of Inclusion and Exclusion* (*Theorem* 1).

4.5.1 *Cardinality of Finite Sets and Unions*

The number of elements in two finite sets can be found by counting. If the sets overlap, the common elements should be counted only once. Otherwise, the total can be found by adding the two cardinalities. This is the content of the following axiom and proposition.

Axiom 4.5.1: *Cardinality of Disjoint Unions*
  If S and T are disjoint finite sets, $|S \cup T| = |S| + |T|$.

Proposition 4.5.1: *Additive Counting Principle*
  If an event can occur in one of two mutually exclusive ways, the first in m ways and the second in n ways, then the number of ways the event can occur is $m + n$.

Proof:
  Let disjoint sets represent the event's outcomes and apply the axiom. ∎

  If two sets overlap, then adding their cardinalities counts the common part twice, so it should be subtracted once to compensate.

Proposition 4.5.2: *Cardinality of Unions*
  $|S \cup T| = |S| + |T| - |S \cap T|$

Proof:
  We'll break $S \cup T$ into two disjoint parts and use that to relate the various cardinalities, applying *Axiom 1* twice.

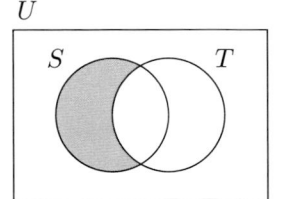

$$S \cup T = (S - T) \cup T$$
  Since these sets are disjoint,
$$|S \cup T| = |S - T| + |T|.$$
  Similarly, $|S| = |S - T| + |S \cap T|$.
  Thus, $|S - T| = |S| - |S \cap T|$.

  Substituting, this yields $|S \cup T| = |S| + |T| - |S \cap T|$. ∎

Corollary 4.5.2.1: *Generalized Additive Counting Principle*
  If an event can occur in one of two ways, the first in m ways and the second in n ways, then the number of ways it can occur is $m + n - b$, where b is the number that can occur in both ways.

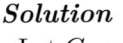**Example 4.5.1**
  A department chair sent out an email to all 19 mathematics majors and 28 computer science majors. Yet the email only went to 43 students in total. How did this happen?

Solution
  Let C represent students with a computer science major and M those majoring in mathematics.
  Then $|C| + |M| = 28 + 19 = 47$.
  Since $|C \cup M| = 43$, $|C \cap M| = 47 - 43 = 4$.
  Thus, four students are double-majoring in mathematics and computer science.

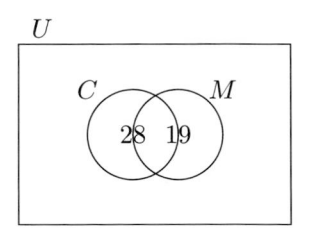

4.5.2 *Principle of Inclusion and Exclusion: Three Sets*

Working with two sets doesn't need any fancy principles: you draw a diagram and use your common sense and some simple arithmetic. But with more sets, things start getting complicated. Let's first focus on three sets, which can still be illustrated and worked using a standard Venn diagram.

If the sets are pairwise disjoint, it's easy to count their union—add the individual cardinalities. This is true for any number of sets, and it forms the basis for counting the union of three sets in concrete instances: add the numbers for distinct components. For more than three sets, though, this gets messy, so we'll need a more efficient procedure for counting larger unions.

✦**Example 4.5.2**

An upper level general education (GE) class has 71 men and women students. Of these, 37 are women; 57 are taking it for GE credit, and the rest are taking it as an elective; 49 are seniors; 7 women are taking it as an elective; 23 women are seniors; 40 seniors are taking it for GE credit; and 21 senior men are taking it for GE credit. How many non-senior men are taking it as an elective?

Solution

· The diagram below shows the final result for the class C; it would be good to start with a blank Venn diagram and fill in the numbers step by step.
· The sets W, G, and S represent, respectively, women in the course, students taking the course for GE credit, and seniors in the course.

· Start with the fact that 21 senior men are taking the course for GE credit: that enumerates a single inner component. Since 40 seniors are taking it for GE credit, 19 must be women. Thus, 4 senior women and 3 non-senior women are taking it as an elective. Therefore, 11 non-senior women are taking it for GE credit. This leaves 6 non-senior men taking it for GE credit.

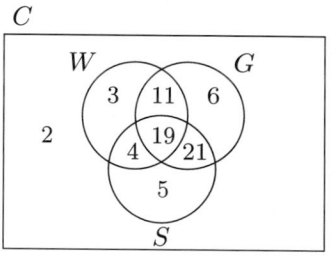

· Since 49 seniors are in the class, 5 senior men are taking it as an elective. In a class of 71 students that leaves 2 non-senior men taking it as an elective.

Now consider any three finite sets S_1, S_2, and S_3. What is the cardinality of the full union $S_1 \cup S_2 \cup S_3$? If we tally up the members in each set, we've counted those in $S_1 \cap S_2$, $S_1 \cap S_3$, and $S_2 \cap S_3$ twice, so we should subtract their cardinalities once from the total. But wait: we've actually counted the members of $S_1 \cap S_2 \cap S_3$ three times, once for each set. So if we subtract off the numbers for each double-intersection, we'll have subtracted off the triple-intersection three times, and it won't have been counted at all. So, then, we'd better add that back in. In symbols we have

$$|S_1 \cup S_2 \cup S_3| = (|S_1| + |S_2| + |S_3|) - (|S_1 \cap S_2| + |S_1 \cap S_3| + |S_2 \cap S_3|) + |S_1 \cap S_2 \cap S_3|$$

In words: *the cardinality of a triple-union is the sum of the cardinalities of the individual sets minus the sum of the cardinalities of the double-intersections plus the cardinality of the triple-intersection.*

✚**Example 4.5.3**

Rework Example 2 using the formula just developed. How efficient is this?

Solution

· The quantity we're interested in is $|\overline{W \cup G \cup S}|$.
  Since $(W \cup G \cup S) \subseteq C$, the universal set, $|\overline{W \cup G \cup S}| = |C| - |W \cup G \cup S|$.
· By the above formula,

$$|W \cup G \cup S| = (|W| + |G| + |S|) - (|W \cap G| + |W \cap S|$$
$$+ |G \cap S|) + |W \cap G \cap S|$$
$$= (37 + 57 + 49) - (|W \cap G| + 23 + 40) + |W \cap G \cap S|$$

  Since $7 = |W - (W \cap G)| = |W| - |W \cap G| = 37 - |W \cap G|$,
  $|W \cap G| = 30$.

And since $|G \cap S - W \cap G \cap S| = 21$, $|W \cap G \cap S| = 19$.

· Substituting these two new values in the above equation, we get

$$|W \cup G \cup S| = (37 + 57 + 49) - (30 + 23 + 40) + 19$$
$$= 143 - 93 + 19 = 69.$$

Thus, $|\overline{W \cup G \cup S}| = 71 - 69 = 2$.

· It's not really faster to use the formula for three sets, unless the sets whose cardinalities we know are the ones in the formula. Otherwise cardinalities need to be determined using other relations that may still be calculated most easily using a diagram, as in Example 2.

· However, the formula does give us a systematic way to work these sorts of problems and one we can generalize to situations where it becomes difficult or impossible to use a diagram for figuring out the numbers.

4.5.3 *Principle of Inclusion and Exclusion in General*

Suppose we have any number of sets whose union we want to count. As mentioned above, if these sets are pairwise disjoint, the total is the sum of the sets' cardinalities. But if they aren't, we must account for the duplications involved, alternatively subtracting and adding values to get the correct count. How to do this in terms of the different intersections involved is the content of the *Principle of Inclusion and Exclusion*. It generalizes the method we just illustrated for three sets. This result is stated largely in words since a fully symbolic formulation is messy. Its proof, however, is rather elegant.

Theorem 4.5.1: *Principle of Inclusion and Exclusion*

Let S_1, S_2, ..., S_n be a collection of n sets. $|S_1 \cup S_2 \cup \cdots \cup S_n|$ is the sum of the cardinalities of all possible odd-fold intersections $S_{i_1} \cap \cdots \cap S_{i_o}$ minus the sum of the cardinalities of all possible even-fold intersections $S_{j_1} \cap \cdots \cap S_{j_e}$.

Proof:
- Our proof strategy is to show that the formula counts each element of the total union *exactly once*.
- Let x be an arbitrary element of the union, and let n denote the number of sets in the collection to which x belongs.
  It will then belong to a k-fold intersection if and only if it is a member of all k sets, i.e., if and only if these k sets are chosen from among the n sets to which x belongs.
- So, x will be in $\binom{n}{k}$ k-fold intersections. These numbers must be added in if k is odd and subtracted off if k is even.
  Thus, x gets counted $\binom{n}{1} - \binom{n}{2} + \cdots + (-1)^{k+1}\binom{n}{k} + \cdots + (-1)^{n+1}\binom{n}{n}$ times by the formula. We need to show that this *alternating series* equals 1.
- This can be seen by using the *Binomial Expansion Theorem* to expand $(1 + {}^-1)^n$ (see also Exercise 4.4.30):
$$0 = (1 + {}^-1)^n = \sum_{k=0}^{n} \binom{n}{k}1^{n-k}(-1)^k = \binom{n}{0} - \binom{n}{1} + \cdots + (-1)^n\binom{n}{n}.$$
  Solving this for $\binom{n}{0}$, we have $\binom{n}{1} - \binom{n}{2} + \cdots + (-1)^{n+1}\binom{n}{n} = \binom{n}{0} = 1$.
- Thus, our formula counts each x exactly once, as claimed. ∎

Corollary 4.5.1.1: *Inclusion and Exclusion, Complementary Form*
If $S_1, S_2, \ldots, S_n \subseteq S$, then $|\overline{S_1 \cup S_2 \cup \cdots \cup S_n}| = |S| - |S_1 \cup S_2 \cup \cdots \cup S_n|$, which is $|S|$ plus the sum of the cardinalities of all possible even-fold intersections $S_{j_1} \cap \cdots \cap S_{j_e}$ minus the sum of the cardinalities of all possible odd-fold intersections $S_{i_1} \cap \cdots \cap S_{i_o}$.

Proof:
This follows from the rule for calculating cardinalities of complements (see Exercise 4a) and the *Principle of Inclusion and Exclusion*. ∎

✚**Example 4.5.4**
How many positive integers $k \leq 360$ are *relatively prime* to 360, i.e., have no factors besides 1 in common with 360?

Solution
- We'll first factor: $360 = 2^3 3^2 5$.
  So we must find how many positive integers have no factor of 2, 3, or 5.
- Let S_2 be the set of all multiples of 2 less than or equal to 360,
    S_3 be the set of all multiples of 3 less than or equal to 360,
    S_5 be the set of all multiples of 5 less than or equal to 360,
  and $S_1 = \{n : n \leq 360\}$.
- We must find $|S_1 - (S_2 \cup S_3 \cup S_5)| = |S_1| - |(S_2 \cup S_3 \cup S_5)|$.
  $|S_1| = 360$, $|S_2| = 360/2 = 180$, $|S_3| = 360/3 = 120$, and
    $|S_5| = 360/5 = 72$.
  Also, $|S_2 \cap S_3| = 360/6 = 60$, $|S_2 \cap S_5| = 360/10 = 36$, and
    $|S_3 \cap S_5| = 360/15 = 24$.
  Finally, $|(S_2 \cap S_3 \cap S_5)| = 360/30 = 12$.

· Using the *Principle of Inclusion and Exclusion*,

$$
\begin{aligned}
|(S_2 \cup S_3 \cup S_5)| &= (|S_2| + |S_3| + |S_5|) - (|S_2 \cap S_3| + |S_2 \cap S_5| \\
&\quad + |S_3 \cap S_5|) + |(S_2 \cap S_3 \cap S_5)| \\
&= (180 + 120 + 72) - (60 + 36 + 24) + 12 \\
&= 372 - 120 + 12 = 264.
\end{aligned}
$$

Thus, $360 - 264 = \underline{96}$ numbers are relatively prime to 360.

EXERCISE SET 4.5

Exercises 1–3: True or False
Are the following statements true or false? Explain your answer.

4.5.1. The cardinality of the union of two sets is the sum of their cardinalities.

4.5.2. If a first action can be done in m ways and for each of these a second action can be done in n ways, the joint action can be done in $m + n$ ways.

4.5.3. The *Principle of Inclusion and Exclusion* counts the number of elements in a finite union of finite sets.

Exercises 4–8: Cardinality and Subsets
Suppose all the sets below are finite sets. Prove the following without using the (more general) Principle of Inclusion and Exclusion or its Corollary.

4.5.4. *Subtraction*
  a. $S \subseteq U \rightarrow |\overline{S}| = |U - S| = |U| - |S|$
  b. Find and prove a general formula for $|S - T|$, regardless of how S and T are related.

4.5.5. $|S \cup T| = |S - T| + |S \cap T| + |T - S|$

4.5.6. $S \subseteq U \rightarrow |S| \leq |U|$

4.5.7. $|S \cap T| \leq |S| \leq |S \cup T|$

4.5.8. *Finite Collections of Sets*
  a. For a finite collection of sets, which number is smaller: the sum of the cardinalities of each set or the cardinality of its union? Formulate a precise answer symbolically and then prove it.
  b. Suppose $\{S_i\}_{i=1}^{n}$ forms a partition of a set S. Prove that $|S| = \sum_{i=1}^{n} |S_i|$.

Exercises 9–14: Counting Sets
Work the following problems, using a counting method from this or an earlier section. Explain your calculations and justify the counting method you chose.

4.5.9. A one-room schoolhouse contains 20 children. Of these students, 14 have brown eyes, 15 have dark hair, 17 weigh more than 80 pounds, and 18 are over four feet tall. Show that at least 4 children have all four features.

4.5.10. There are 150 men and women faculty at a community college: 100 faculty are full time; 60 faculty are women, but only 25 of these are full time; 40 faculty teach a liberal arts course, of which 30 are women and 20 are full time; and 10 full-time women faculty teach a liberal arts course. How many full-time men faculty do not teach a liberal arts course? How many part-time men faculty does the college employ? Explain your reasoning.

4.5.11. A bowl of fruit contains 4 mangoes, 5 kiwis, and 6 bananas.
  a. In how many ways can three pieces of fruit of the same kind be chosen?
  b. In how many ways can three pieces of fruit be chosen if no two fruits are of the same kind?
  c. In how many ways can three pieces of fruit be chosen if at least two fruits are of the same kind?

4.5.12. *Binary Events*
  a. How many different bytes (eight-bit strings; see Example 4.3.4) do not contain five consecutive 1's?
  b. A coin is flipped eight times. What is the probability that it lands heads up four or fewer times in a row? Assume a fair coin that heads and tails are equally likely to occur.

4.5.13. *Collections of Coins*
A loose-change jar contains 62 coins in the following denominations: 31 pennies, 10 nickels, 12 dimes, and 9 quarters. To answer the following questions do not distinguish collections with the same number of coins of the same denominations, and assume that each coin is as likely as the next to be chosen.
  a. How many different collections of seven coins can be chosen from this jar?
  b. How many different collections of seven coins can be chosen from this jar having at least one coin of each type? What is the probability of choosing such a collection?
  c. How many different collections of seven coins have exactly 2 quarters or 2 dimes? exactly 2 quarters, 2 dimes, or 2 nickels?
  d. What are the probabilities associated with the two collections in part *c*?

4.5.14. *Card Hands*
  a. How many five-card hands contain three aces? two kings? three aces and two kings? three aces or two kings?
  b. How many five-card hands have exactly three cards of one kind? exactly two of one kind? a full house (three of one kind and two of another)? either three of a kind or two of a kind?
  c. How many five-card hands are a flush (five nonconsecutive cards in the same suit)? How many hands are a straight flush (five consecutive cards in the same suit, ace being either high or low)? How many hands are a straight (five consecutive cards not all in one suit)? How many hands are either a flush, a straight flush, or a straight?

Exercises 15–28: Exploring Euler's Phi-Function

Use the Principle of Inclusion and Exclusion to answer the following questions about Euler's φ-function, where $\varphi(n)$ equals the number of positive integers k

less than or equal to n that are relatively prime to n (have no factors besides 1 in common with n).

4.5.15. Determine $\varphi(21)$. How is $\varphi(21)$ related to the factors of 21?

4.5.16. Determine a formula for $\varphi(pq)$, where p and q are distinct primes. Put your formula in factored form and prove your result. Will your formula still hold if $p = q$? Will it hold if p and q are only relatively prime?

4.5.17. Determine $\varphi(25)$. Relate this value to the prime factor of 25. Find a formula for $\varphi(p^2)$, where p is a prime. Prove your result.

4.5.18. Determine $\varphi(105)$. Relate this number to the prime factors of 105.

4.5.19. Determine a formula for $\varphi(pqr)$, where p, q, and r are distinct primes. How is this expression related to the factors of pqr? Does your formula hold if the number is p^3 instead of pqr? If it is p^2q? If the numbers p, q, and r are relatively prime to one another? Justify your answers.

4.5.20. Determine $\varphi(27)$. Relate this value to the prime factor of 27. Find a formula for $\varphi(p^3)$, where p is a prime. Prove your result.

4.5.21. Determine $\varphi(546)$ using the *Principle of Inclusion and Exclusion*. How is this value related to the prime factors of 546?

4.5.22. State and prove a formula for $\varphi(pqrs)$, where p, q, r, and s are distinct primes. Use the *Principle of Inclusion and Exclusion*.

4.5.23. Generalize the results of Exercises 16, 19, and 22 to obtain a formula for $\varphi(p_1 \cdots p_n)$ for the product of n distinct prime numbers p_i.

4.5.24. Determine and prove a formula for $\varphi(p^k)$, where p is a prime number and k is a positive integer.

4.5.25. Use the *Principle of Inclusion and Exclusion* to develop a formula for $\varphi(p^m q^n)$ for distinct primes p and q. Take $m = 2$ and $n = 3$ for a case study if this helps. Prove that $\varphi(p^m q^n) = \varphi(p^m)\varphi(q^n) = p^{m-1}q^{n-1}(p-1)(q-1)$. Does this formula still work if $m = 1 = n$?

4.5.26. Generalize the results of Exercises 23–25 to develop a formula for $\varphi(m)$ for any positive integer m whose prime factorization is $m = p_1^{k_1} \cdots p_n^{k_n}$.

4.5.27. Given Exercise 26's formula show that $\varphi(m) = m \cdot \prod_{p|m} (1 - 1/p)$, where p is any prime divisor of m.

4.5.28. Use your work from Exercises 25 and 26 to show that $\varphi(ab) = \varphi(a)\varphi(b)$ if a and b are relatively prime.

Chapter 5
Set Theory and Infinity

Set Theory was first developed in order to handle infinite collections. Mathematicians discovered (with effort) that the realm of infinity is both less paradoxical and more structured than they had thought earlier. In this chapter we'll see how to distinguish infinite sets from finite sets and from one another. We'll also note some foundational uses of *Set Theory*. In the final section, we'll look at the axiomatization of *Set Theory* first put forward to deal with some perplexing results for infinite sets, and we'll look at an application of *Set Theory* to the theory of computation.

5.1 Countably Infinite Sets

We'll begin our study by considering the simplest infinite sets—those that are *countably infinite*, like the set of positive natural numbers $\{1, 2, 3, \dots\}$. We'll first define when sets are equinumerous, so we can compare their sizes, and then we'll show that many familiar sets are countably infinite. But before doing any of this, we'll put *Set Theory* into historical perspective.

5.1.1 *Historical Context of Cantor's Work*

The British mathematicians Boole and De Morgan used sets in their systems of logic in the middle of the nineteenth century, but it was the work of the German mathematicians Richard Dedekind and especially Georg Cantor a quarter-century later that gave rise to a *theory* of sets. Dedekind used sets to provide a real-number foundation for calculus and as the theoretical basis for the natural number system. Cantor, who is pictured in Figure 5.1, employed infinite sets in his research on Fourier series and to settle some open questions in analysis. Cantor subsequently developed *Set Theory* into a branch of mathematics, its centerpiece being his treatment of transfinite (infinite) sets.

Although *Set Theory* is a part of mathematics, for Cantor it was closely connected with philosophy and theology. In fact, Cantor viewed himself as God's prophet of the infinite, revealing necessary truths of mathematics and theology grounded in the Mind of God.[1] These views prompted strong negative reactions from some who believed mathematics had lost its moorings and drifted into the field of

Fig. 5.1 Georg Cantor

---

[1] For the theological and philosophical context of Cantor's work, see Joseph Dauben's article *Georg Cantor and the Battle for Transfinite Set Theory*, available online at http://www.acmsonline.org/journal/2004/Dauben-Cantor.pdf.

© Springer Nature Switzerland AG 2019
C. Jongsma, *Introduction to Discrete Mathematics via Logic and Proof*,
Undergraduate Texts in Mathematics,
https://doi.org/10.1007/978-3-030-25358-5_5

religion or metaphysics. However, since infinity enters nearly every part of mathematics in essential ways, the topic couldn't be avoided. While many considered infinity a quagmire of confusion, Cantor's virtue lay in his persistent belief that it could be consistently treated if one used clear definitions. Cantor exposed the sources of conceptual difficulties in earlier philosophical treatments, and he provided a way to deal with infinity that contemporary philosophers find attractive and definitive.

Research mathematicians and logicians since Cantor have continued to expand the field. David Hilbert's seminal 1900 address to the International Congress of Mathematicians on the 23 most important unsolved problems of the time gave prominent attention to transfinite *Set Theory*.

Once *Set Theory* was available, some mathematicians advocated using it as a foundation for other parts of mathematics, a trend encouraged by major schools of thought in the early twentieth century—by logicism, led by Bertrand Russell, and by formalism, under David Hilbert. This was programmatically pursued in the 1930s and later by a group of prominent mathematicians writing anonymously under the French pseudonym Nicolas Bourbaki.

New Math educators jumped on this bandwagon around 1960. They believed school mathematics could be learned more efficiently (a crucial concern in the U.S. at that time due to the space race with the U.S.S.R.) if children were introduced to the conceptual structure of mathematics from the start. Drawing from earlier developments, they promoted *Set Theory* as the theoretical basis and unifying idiom for all mathematics education.

A strong reaction to this by mathematicians and educators was mounted in the 1970s. Many resisted teaching abstractions to young children, and although mathematics education reform is still ongoing, sets now play only a modest role in most elementary mathematics programs. Where they do come in, they are used more concretely than earlier. Notwithstanding this turnaround, *Set Theory* remains an important part of advanced mathematics, for its unifying concepts and notation, even if not as a grand foundation.

5.1.2 *One-to-One Correspondence and Numerosity*

How can we tell if two finite sets are the same size? One way is simply to count them. This determines whether they're the same size, and also what sizes they are.

A second way is to match their elements one by one. If there are no unmatched elements in either set, they're the same size; otherwise the one with unmatched elements is more numerous.

Matching is less informative than counting, but it's simpler. Children can tell that there are the same number of glasses as plates on the dinner table even if they can't count how many there are. As Cantor discovered, this matching technique is also valuable for developing a theory of cardinality, even when sets are infinite. Cantor made the idea of a one-to-one correspondence (matching) central to his whole approach to numerosity.

Definition 5.1.1: *One-to-One Correspondence Between Sets*
 *S is in **one-to-one correspondence** with T if and only if all the elements of S can be matched with all the elements of T in a one-to-one fashion.*

This definition will become rigorous after we formally introduce functions in Section 6.1. For now, the notion of infinity has enough conceptual complexity without adding more technical complications.

Note that this definition mentions S being in one-to-one correspondence with T (in that order), but *being in one-to-one correspondence* is actually an equivalence relation (see Exercise 1). In particular, if S is in one-to-one correspondence with T, then T is also in one-to-one correspondence with S. This justifies saying something like S *and* T *are in one-to-one correspondence.*

5.1.3 Numerosity Relations Among Sets

Definition 5.1.2: *Equinumerous Sets, Less-Numerous Sets*
 a) *S is **equinumerous to** T, written $S \sim T$, if and only if S and T are in one-to-one correspondence.*
 b) *S is **less numerous than or equinumerous to** T, written $S \preceq T$, if and only if S is equinumerous with some subset of T.*
 c) *S is **less numerous than** T, written $S \prec T$, if and only if $S \preceq T$ but $S \nsim T$.*

We've taken the order relation $\preceq$ as primary because its definition is simpler than that of $\prec$. However, the usual connection between these relations holds: $S \preceq T$ if and only if $S \prec T$ or $S \sim T$ (see Exercise 8). A number of other results also hold, though their proofs aren't always as simple as one might expect, because one-to-one correspondences behave differently for infinite sets than for finite sets (see Exercise 2).

The following theorem is a basic but nontrivial result for $\preceq$. Cantor conjectured this theorem but was unable to prove it. Dedekind gave the first proof, communicating it in a letter to Cantor in 1887. In 1896, Schröder gave a flawed proof of the result; a year later, Bernstein published a valid proof. Today, the theorem is usually called the *Schröder-Bernstein Theorem*, though a more accurate name would probably be the *Cantor-Bernstein Theorem*.

Theorem 5.1.1: *The Schröder-Bernstein Theorem*
 If $S \preceq T$ and $T \preceq S$, then $S \sim T$.

Proof:
 Given the proof's complexity for sets in general, we'll not present it here. A proof can be easily constructed, however, using the following lemma (see Exercise 4a). Of course, this merely concentrates the theorem's difficulty in the lemma, which we also won't prove (see Exercise 4b).

Lemma 5.1.1: *Nested Equinumerous Sets*
 If $S_2 \subseteq S_1 \subseteq S_0$ and $S_2 \sim S_0$, then $S_2 \sim S_1 \sim S_0$.

5.1.4 *Finite and Infinite Sets: Some Distinctions*

What makes one set finite and another infinite? At its most basic level, a set is finite if and only if counting its elements ends after some number of steps. If $|S| = n$ for some $n \in \mathbb{N}$, S is finite; otherwise it's infinite. Thus, $\mathbb{N}$ itself is not finite—there is no largest natural number. The set of all points on a line segment is likewise infinite. Infinite sets are larger than finite sets: they belong to a *transfinite* realm, as Cantor termed it.

Traditionally, *being finite* was assumed to be primary; *being infinite* was a purely negative concept, meaning *being not finite*. Prior to Cantor, most people adhered to Aristotle's viewpoint on infinity, that a quantity is only *potentially infinite*, something that had the potential of being made larger. Magnitudes weren't infinitely extended, only infinitely extendable. Infinite collections weren't completed totalities, they were sets that could be enlarged. *Being infinite* described an *unending process*, not an entity.

Aristotle's ideas on infinity reigned among orthodox mathematicians concerned with logical rigor for more than 2000 years. In support of this view, mathematicians observed that if you considered certain infinite totalities as bona fide mathematical objects, then paradoxical contradictions arose. We'll illustrate this with a historical example.

✤**Example 5.1.1**

Which infinite set of natural numbers is larger, $\{1, 2, 3, \cdots\}$ or $\{1, 4, 9, \cdots\}$?

Solution
- In the seventeenth century, Galileo noted that if infinite collections could be compared with respect to size, paradoxes would arise. Then one could say that the set of perfect squares is less numerous than the set of positive integers, being a proper subset, but that the sets are also equinumerous, because positive integers can be matched one to one with their squares.
- For Leibniz and others, this paradoxical state of affairs seemed to substantiate the notion that completed infinities are contradictory.

The contradiction here only arises, however, if being a proper subset entails being less numerous. *Axiom* 5 in Euclid's *Elements* takes this as self-evident: *the whole is greater than its part*. This agrees with our experience of finite collections and finite magnitudes, such as line lengths. However, common sense fails us in the realm of the infinite, where we face a conscious choice: what criterion should we use to compare the numerosity of infinite sets?

Cantor chose to treat transfinite sets as completed totalities. He rejected Aristotle's view on infinite collections and adopted a more "theological" attitude, as some mathematicians characterized it. Other mathematicians followed his lead, though some balked at importing "metaphysical ideas" into mathematics.

Cantor made one-to-one correspondence the basis of cardinality comparisons, just as we did above. It is still true that the cardinality of a finite whole is greater than the cardinality of any (proper) part. But this property fails dramatically when sets are infinite, as our example from Galileo shows. In Cantor's opinion, this is not because infinite sets shouldn't be compared or taken to be completed totalities but because generalizing from the finite case is invalid. Self-evident properties that hold for finite sets may not simply be taken as true for sets in general. A set can be a proper part of the whole and still be equinumerous with it. In fact, this is true for all infinite sets, as we'll prove below (see *Corollary* 1 to *Theorem* 10). So long as we don't assert Euclid's axiom universally, there is no automatic contradiction.

We may still be left with a sense of unease as we learn how strangely infinite sets behave in comparison with finite ones, but this discomfort should dissipate after working with infinite sets for a while.

5.1.5 *Countably Infinite Sets: Basic Results*

Let's begin exploring the realm of infinity by looking at *countably infinite* sets. These are sets whose elements can be counted off one by one without stopping. Because we can also count off finite sets, we'll include them as well in the class of *countable* sets.

Definition 5.1.3: *Countably Infinite and Countable Sets*
  a) S is ***countably infinite*** if and only if $\mathbb{N}^+ \sim S$, where $\mathbb{N}^+ = \{1, 2, \dots\}$.
  b) S is ***countable*** if and only if it is countably infinite or finite.

To show that a set is countably infinite, we match its elements with the positive natural numbers, which induces a sequential order on the set (first, second, etc.). This leads us to the following definition and proposition.

Definition 5.1.4: *Enumeration of a Set*
  An ***enumeration*** of a set S is a non-repetitious listing of its elements as a finite or infinite sequence. Such a listing ***enumerates*** S.

Proposition 5.1.1: *Countably Infinite Sets and Enumerations*
  S is countably infinite if and only if it can be enumerated by an infinite sequence.

Proof:
  · Suppose S is countably infinite. Then it can be put into one-to-one correspondence with $\mathbb{N}^+$. If x_n is the element of S matched with n, then $x_1, x_2, \dots$ is an enumeration of S. ✓
  · Conversely, suppose S is enumerated by $x_1, x_2, \dots$. This sets up a one-to-one correspondence between S and $\mathbb{N}^+$, so S is countably infinite. ∎

Corollary 5.1.1.1: *Countably Infinite Subsets of* $\mathbb{N}$
  a) $\mathbb{N}^+$ *is countably infinite.*
  b) $\mathbb{N}$ *is countably infinite.*

Proof:
Both parts of this corollary follow from *Proposition 1*.

Proposition 5.1.2: *Equinumerosity of Countably Infinite Sets*
  If T is countably infinite, then S is countably infinite if and only if $S \sim T$.

Proof:
  This uses the fact that $\sim$ is an equivalence relation. See Exercise 6.

Corollary 5.1.2.1: *Countably Infinite, Alternative Characterization*
  S is countably infinite if and only if $\mathbb{N} \sim S$.

Proof:
  This follows immediately from the last two results.

   Whether to compare a set with $\mathbb{N}$ or with $\mathbb{N}^+$ depends upon the circumstances. Sometimes it's convenient to use both characterizations. We also have the option of using *Proposition* 1's enumeration characterization in our proofs.

5.1.6 *Countably Infinite Sets and Their Subsets*

Both $\mathbb{N}^+$ and $\mathbb{N}$ are countably infinite. Are other familiar sets of numbers, such as the integers, the rational numbers, and the real numbers also countably infinite? We'll start investigating this in the next subsection, generalizing what we discover to other sets as well. But let's first make some observations about the size of countably infinite sets in comparison to subsets and to smaller sets.

   Countably infinite sets are infinite and thus larger than finite sets. On the other hand, no infinite sets are smaller than them—strictly smaller sets are finite. This is the intuitive content of the next few results. We'll begin with the result that was earlier taken to be paradoxical.

Theorem 5.1.2: *Countably Infinite Sets and Equinumerous Subsets*
  Countably infinite sets contain equinumerous proper subsets.

Proof:
  This holds for $\mathbb{N}$ according to the *Corollary* to *Proposition* 1: $\mathbb{N} \sim \mathbb{N}^+$. We can use this to show a similar thing in general.
  Suppose S is countably infinite. Then it's equinumerous with $\mathbb{N}$.
  Let $x_0, x_1, x_2, \ldots$ be the associated enumeration of its elements.
  Then $S^* = S - \{x_0\}$ is enumerated by $x_1, x_2, \ldots$, so it is also countably infinite and thus equinumerous with its superset S by *Proposition* 2. ∎

Theorem 5.1.3: *Finite Subsets of Countably Infinite Sets*
  If S is countably infinite, then S contains finite subsets of all sizes.

Proof:
  Let S be countably infinite and let $\{x_1, x_2, \ldots, x_n, \ldots\}$ enumerate S.
  Then $S_n = \{x_1, x_2, \ldots, x_n\}$ is a finite subset of S with n elements. ∎

Corollary 5.1.3.1: *Countably Infinite Sets and Finite Sets*
  If S is countably infinite and F is finite, then $F \prec S$.

Proof:
  See Exercise 16.

Theorem 5.1.4: *Infinite Subsets of Countably Infinite Sets*
  Infinite subsets of countably infinite sets are countably infinite.

Proof:
  Let $T = \{x_1, x_2, x_3, \ldots\}$ be countably infinite, S an infinite subset of T.
  Deleting those elements of T not in S leaves $S = \{x_{k_1}, x_{k_2}, x_{k_3}, \ldots\}$.
  This gives an infinite sequence enumeration because S is infinite.
  Thus, S is countably infinite by *Proposition 1*. ∎

Corollary 5.1.4.1: *Subsets of Countably Infinite Sets*
  If $S \subseteq T$ and T is countably infinite, then S is finite or countably infinite.

Proof:
  See Exercise 23a.

Corollary 5.1.4.2: *Less Numerous Than Countably Infinite*
  If $S \prec T$ and T is countably infinite, then S is finite.

Proof:
  See Exercise 23b.

5.1.7 *Set Operations on Countably Infinite Sets*

Although we began comparing the numerosity of countably infinite sets with that of other sets in the last subsection, we'll temporarily postpone going further, because of a technical complication. Instead, we'll look at how countably infinite sets interact with set-theoretic operations. Let's begin by noting that we can adjoin/remove a finite number of elements to/from such a set and the result will still be countably infinite.

Theorem 5.1.5: *Disjoint Union of Finite and Countably Infinite*
  If S is finite and T is a countably infinite set disjoint from S, then $S \cup T$ is countably infinite.

Proof:
  Enumerate the finite set $S = \{x_1, x_2, \ldots, x_n\}$ prior to $T = \{y_1, y_2, y_3, \ldots\}$.
  $S \cup T = \{x_1, x_2, \ldots, x_n, y_1, y_2, y_3, \ldots\}$, so it is countably infinite. ∎

Corollary 5.1.5.1: *Union of Countably Infinite with Finite*
  If S is finite and T is countably infinite, then $S \cup T$ is countably infinite.

Proof:
  See Exercise 24a.

Proposition 5.1.3: *Deletion of Finite from Countably Infinite*

If S is finite and T is countably infinite, then $T - S$ is countably infinite.

Proof:

See Exercise 24b.

The import of the last result is as follows: you cannot make countably infinite sets finite by lopping off finite parts any (finite) number of times. The reverse is thus also true: countably infinite sets cannot be obtained by successively aggregating finite sets, no matter how large. An immense gulf lies between the finite and the infinite. Talk about large numbers being "nearly infinite" is a picturesque way to describe their size, but that's all it is.

The countably infinite sets encountered so far were all enumerated taking numbers in their natural orders, but this need not be true. In fact, as we'll see, elements will often need to be rearranged in order to enumerate a set.

Proposition 5.1.4: *The Integers are Countably Infinite*

$\mathbb{Z} = \{\ldots, -2, -1, 0, 1, 2, \ldots\}$ is countably infinite.

Proof:

To enumerate $\mathbb{Z}$, we must disrupt its usual order—$\mathbb{Z}$ has no first element and is infinite in two directions. We'll give the standard *zigzag enumeration*. $\mathbb{Z} = \{0, 1, -1, 2, -2, 3, -3, \ldots, n, -n, \ldots\}$, which proves the claim. ∎

This proposition generalizes, as the following result demonstrates.

Theorem 5.1.6: *Union of Countably Infinite Sets*

If S and T are countably infinite, then $S \cup T$ is countably infinite.

Proof:

Enumerate S and T and then merge their lists in a zigzag fashion, omitting any elements already listed. ∎

The last two theorems can be summarized in a single statement: *If S is countable and T is countably infinite, then $S \cup T$ is countably infinite.* We can illustrate this with a cute example of dubious practicality.[2]

♣ **Example 5.1.2**

Hilbert's Hotel is an imaginary spacious inn containing a countable infinity of rooms. Show that even if the inn is full for the night, it can still accommodate one or even a countably infinite number of new arrivals.

Solution

· A single new arrival can be accommodated by having everyone move down one room. But even if there are a countably infinite number of new arrivals, the full group, those present and those arriving, is still only countably infinite, so we can reassign rooms according to the zigzag-merge-enumeration process mentioned in the proof of the last theorem.

---

[2] Hilbert introduced this example without fanfare in a 1925 lecture to illustrate the difference between finite and infinite sets. Later it was popularized in works discussing infinity, and it has even been used in debates about cosmology and theology.

- Of course, reassigning rooms is a nuisance for those already settled in for the night, since infinitely many people will need to change rooms when each new group arrives, even if that's only one person.
- A set-theoretically astute innkeeper would instead house her guests so as to *always* leave a countable infinity of rooms open for further occupancy (see Exercise 25).

The last theorem can be generalized to give the following corollary. *Theorem* 7 goes even further and requires a whole new proof strategy.

Corollary 5.1.6.1: *Finite Union of Countably Infinite Sets*

If $S_1, S_2, \ldots, S_n$ are countably infinite sets, then so is $\bigcup\limits_{i=1}^{n} S_i$.

Proof:
This follows from *Theorem* 6, using induction. See Exercise 26.

Theorem 5.1.7: *Countably Infinite Union of Countably Infinite Sets*

If $\{S_1, S_2, \ldots\}$ is a countably infinite collection of countably infinite sets, then $\bigcup\limits_{i=1}^{\infty} S_i$ is countably infinite.

Proof:
- A simple *diagonal argument* establishes this result.
  We'll zigzag back and forth through all the sets, as it were, not just through two of them, as before.
- Let x_{ij} list the elements of each S_i for $j = 1, 2, 3, \ldots$.
- Make an infinite array of these lists, putting the elements of S_i in row i:

$$x_{11} \ x_{12} \ x_{13} \ x_{14} \ \cdots$$
$$x_{21} \ x_{22} \ x_{23} \ x_{24} \ \cdots$$
$$x_{31} \ x_{32} \ x_{33} \ x_{34} \ \cdots$$
$$\vdots \quad \vdots \quad \vdots \quad \vdots \quad \ddots$$

- Now list all of the elements in the union by moving through the array diagonally, starting at the top-left corner and going down along the minor diagonals: x_{11}; x_{12}, x_{21}; x_{13}, x_{22}, x_{31}; $\cdots$
- If S_i are pairwise disjoint, this enumeration includes every element in the array exactly once. If the sets overlap, elements will repeat, so to avoid duplication, pass over any element already in the list. The resulting enumeration establishes the union as countably infinite. ∎

We've already shown that $\mathbb{N}$ and $\mathbb{Z}$ are countably infinite. What about $\mathbb{Q}$? $\mathbb{Z}$ was not difficult to enumerate, because while it had no first element, at least its elements were all successors. But $\mathbb{Q}$ is not sparsely populated; its elements are densely packed together. Given two rational numbers, a third one always lies between them, so no rational number directly follows or precedes any other. On the face of it, it seems impossible to enumerate the set of rational numbers. However, having proved *Theorem* 7, the job isn't difficult.

Proposition 5.1.5: $\mathbb{Q}$ is Countably Infinite
$\mathbb{Q} = \left\{ \dfrac{m}{n} : m, n \in \mathbb{Z}, \ n \neq 0 \right\}$ is countably infinite.

Proof:

· A diagonal argument like the last one shows that the set of positive rationals is countably infinite: list m/n for $m, n \in \mathbb{N}^+$ as x_{mn} to set up the array. Similarly, the set of negative rational numbers is countably infinite. Thus, as a union of countably infinite sets, the set of all non-zero rational numbers is countably infinite. Putting 0 at the front of such a list shows that $\mathbb{Q}$ is countably infinite. ∎

· For a second, more geometric way of demonstrating this same result, first identify each fraction m/n with the point (m, n) in the plane.

These integral lattice points may then be enumerated in a counterclockwise spiral.

Begin with $(0, 0)$ and proceed to $(1, 0)$, $(1, 1)$, $(0, 1)$, $(-1, 1)$, $(-1, 0)$, $(-1, -1)$ and so on.

To obtain a listing of unique rational-number representatives (m, n), omit all pairs ending in a 0 and all pairs that are integral multiples of earlier ones; for example, omit $(1, 0)$, and omit $(-1, -1)$ because $(1, 1)$ is already in the list. This sequential listing shows $\mathbb{Q}$ is countably infinite. ∎

The Cartesian product of two countably infinite sets is often listed in array form, as above. This gives a way to prove the next theorem.

Theorem 5.1.8: Cartesian Product of Countably Infinite Sets
If S and T are countably infinite, then $S \times T$ is countably infinite.

Proof:
See Exercise 28a.

Corollary 5.1.8.1: Finite Product of Countably Infinite Sets
If S_i is countably infinite for $i = 1, 2, \ldots, n$, then so is $S_1 \times S_2 \times \cdots \times S_n$.

Proof:
Apply induction to the number of sets involved. See Exercise 28b.

On the basis of what we've proved so far, you might be ready to leap to the following conclusion: *the countably infinite Cartesian product of countably infinite sets is countably infinite*. However, generalizing from known results doesn't always work, as Cantor knew well. So if this new result is true, it will need a careful proof. Due to some complications, we'll leave this as a conjecture for now and take it up in Section 5.2. In the meantime let's consider one more standard set of numbers. Before we formulate the next proposition, which Cantor proved in 1874, we need a definition.

Definition 5.1.5: *Algebraic Numbers*
Complex number a is **algebraic** *if and only if* $p(a) = 0$ *for some polynomial* $p(x) = a_n x^n + a_{n-1} x^{n-1} + \cdots + a_1 x + a_0$, *where* a_i, *not all* 0, *are integers.*

Examples of algebraic numbers include rational numbers, as well as radicals, such as $\sqrt{2}$, and even $i = \sqrt{-1}$ (see Exercise 30ab). We'll use $\mathbb{A}$ to denote the set of all algebraic numbers.

Proposition 5.1.6: $\mathbb{A}$ *is Countably Infinite*
$\mathbb{A}$, *the set of algebraic numbers, is countably infinite.*

Proof:
We'll sketch the proof, leaving the details as an exercise (see Exercise 30c).
There are countably many polynomials with integer coefficients of degree n.
Each of these polynomials has at most n distinct zeros or roots.
Thus, there are countably many zeros associated with degree-n polynomials.
Taking the union of these sets of zeros for each positive integer n, we obtain the countably infinite set $\mathbb{A}$. ∎

5.1.8 *Countably Infinite Sets and Infinite Sets*

The theorems proved so far partially determine where countably infinite sets lie in the cardinality hierarchy of sets. Such sets are more numerous than finite sets, and every subset of a countably infinite set is either finite or countably infinite. So, countably infinite sets seem to be the smallest infinite sets. Comparing countably infinite sets to infinite sets in general still needs to be done, though—sometimes the obvious turns out to be false.

We left this comparison until now because it requires the *Axiom of Choice*, which says that we can construct a set by choosing one element from each set in a pairwise disjoint collection of sets, no matter how large. This axiom seems intuitively clear, at least for the kinds of collections we're familiar with, but it is one of the more controversial results of *Set Theory* and leads to some peculiar consequences. This is the case for the *Banach-Tarski Paradox*, which is highly counterintuitive. You can explore this result further by consulting a textbook on *Set Theory* or by looking up *Axiom of Choice* online (see Exercise 33). In Section 5.3 we'll briefly look at a proposition (the *Well-Ordering Theorem*) that's logically equivalent to the *Axiom of Choice*.

Theorem 5.1.9: *Infinite Sets and Countably Infinite Subsets*
a) *If* T *is an infinite set, then* T *contains a countably infinite subset* S.
b) *If* S *is a countably infinite subset of* T, *then* $S \preceq T$.

Proof:
a) Suppose T is an infinite set. We'll construct a countably infinite subset S by selecting and enumerating its elements in stages.

- $T \neq \emptyset$, so it contains some element x_0.
  Let $S_0 = \{x_0\}$ and $T_0 = T - S_0$. Since S_0 is finite, its complement T_0 inside T is infinite and hence non-empty.
  Continuing recursively, suppose that $S_n = \{x_0, x_1, \ldots, x_n\}$ is a set of distinct elements chosen from T. Then since S_n is finite, its complement $T_n = T - S_n$ is non-empty.
  Choose $x_{n+1} \in T_n$ and let $S_{n+1} = S_n \cup \{x_{n+1}\}$.
- Now take $S = \bigcup_{n=0}^{\infty} S_n = \{x_0, x_1, \ldots, x_n, \ldots\}$.
- Since all x_i are distinct, S is a countably infinite subset of T. ∎

b) See Exercise 31.

Given *Theorem* 2 and *Theorem* 9, we can prove the following result.

Theorem 5.1.10: *Infinite Sets Have Equinumerous Proper Subsets*
If T is an infinite set, then T is equinumerous with a proper subset.

Proof:
- Let T be any infinite set, and let $S = \{x_0, x_1, x_2, \ldots\}$ denote a countably infinite subset.
- Following the method of *Theorem* 2, we'll first put S into one-to-one correspondence with a proper subset of *itself* before tackling the full set.
  Match each element x_n in S with its successor x_{n+1} in the list. This places S in one-to-one correspondence with the subset $S^* = S - \{x_0\}$.
- We will next extend this matching to get a one-to-one correspondence between T and $T^* = T - \{x_0\}$. If an element is in S, match it as just indicated. If an element is outside S, match it with itself.
  This matches all elements in T with those in T^* in a one-to-one way.
- Thus, $T \sim T^*$, one of its proper subsets. ∎

Corollary 5.1.10.1: *Infinite Sets (Alternate Characterization)*
T is infinite if and only if T is equinumerous to a proper subset of itself.

Proof:
See Exercise 32a.

Fig. 5.2 Richard Dedekind

This corollary contradicts Euclid's *Axiom* 5. For infinite sets, the whole need not be greater than its parts—they're equinumerous with a proper subset. Since this distinguishes them from finite sets, Dedekind (Figure 5.2) adopted this characterization as the defining property for being infinite. This definition does not require any prior knowledge of the natural numbers, as our earlier definition does. Being finite can then be defined, according to Dedekind, as being not infinite.

Although this approach is legitimate, it is abstract and non-intuitive, which is why we didn't adopt it as our main approach. In addition, the validity of this result depends on a version of the *Axiom of Choice*. Without the *Axiom of Choice*, there can be sets that are infinite in the sense of our original definition that are not infinite in Dedekind's sense. Equivalently, there are sets that are Dedekind-finite that are not finite according to our definition.

Our final corollary is an analogue to earlier results on finite and countably infinite sets.

Corollary 5.1.10.2: *Infinite Sets Absorb Countable Sets*
   a) *If S is countable and T is infinite, then $S \cup T \sim T$.*
   b) *If S is countable and T is uncountably infinite, then $T - S \sim T$.*
Proof:
Modify the proof of *Theorem* 10; see Exercise 32bc.

At this point we don't know that there are uncountably infinite sets, but we do know by what we've already proved that if there are any, they must be larger than countably infinite ones. We'll explore this more in Section 5.2.

EXERCISE SET 5.1

Exercises 1–6: Equinumerosity
Prove the following results concerning equinumerosity.

5.1.1. Show that the relation of equinumerosity is an equivalence relation on sets. (*Note*: do *not* use any supposed properties of $|S|$ here.)
   a. *Reflexive Property*: $S \sim S$.
   b. *Symmetric Property*: If $S \sim T$, then $T \sim S$.
   c. *Transitive Property*: If $R \sim S$ and $S \sim T$, then $R \sim T$.

5.1.2. Show that it is possible for sets S and T to be matched up in a one-to-one fashion by two different matchings so that in the first matching all elements of S are matched up with elements of T, but T has elements with no mates, while in the second matching just the opposite is true (the roles of S and T are reversed). *Hint*: could this occur for finite sets?

5.1.3. Prove that $S \times T \sim T \times S$.

5.1.4. *The Schröder-Bernstein Theorem*
   a. Given the *Lemma on Nested Equinumerous Sets*, prove the *Schröder-Bernstein Theorem*: If $S \preceq T$ and $T \preceq S$, then $S \sim T$.
   b. Prove the *Lemma on Nested Equinumerous Sets*: If $S_2 \subseteq S_1 \subseteq S_0$ and $S_2 \sim S_0$, then $S_2 \sim S_1 \sim S_0$. (This requires some ingenuity to prove.)

5.1.5. True or false? Prove or disprove your claim, using results about $\sim$. If a result is false but can be qualified to make it true, fix it and then prove it.
   a. $R \sim T \wedge S \sim V \to R \cup S \sim T \cup V$
   b. $R \sim T \wedge S \sim V \to R \cap S \sim T \cap V$
   c. $R \sim T \wedge S \sim V \to R \times S \sim T \times V$

5.1.6. *Proposition 2*
If T is countably infinite, then S is countably infinite if and only if $S \sim T$.
Hint: Use logic to design a proof strategy.

Exercises 7–10: Numerosity Order Comparisons

Using the definitions and results in this section about $\preceq$, prove the following.

5.1.7. *Monotonicity Property of $\preceq$*
  a. If $S \subseteq T$, then $S \preceq T$.
  b. Why isn't the following true: if $S \subset T$, then $S \prec T$?
  c. Using part a, argue why $S \preceq S$, $S \cap T \preceq S$, $S \preceq S \cup T$, and $S - T \preceq S$.
5.1.8. $S \preceq T \leftrightarrow (S \prec T \vee S \sim T)$ **5.1.9.** $S \prec T \leftrightarrow (S \preceq T \wedge T \npreceq S)$
5.1.10. $(S \preceq T \wedge T \preceq U) \rightarrow S \preceq U$

Exercises 11–16: Strict Numerosity Order Comparisons

Using the definitions and results in this section about $\prec$, prove the following.

5.1.11. $S \nprec S$ **5.1.13.** $S \prec T \rightarrow T \nprec S$
5.1.12. $S \prec T \rightarrow T \npreceq S$ **5.1.14.** $R \prec S \wedge S \prec T \rightarrow R \prec T$
5.1.15. $S_1 \sim S_2 \wedge T_1 \sim T_2 \rightarrow (S_1 \prec T_1 \leftrightarrow S_2 \prec T_2)$
5.1.16. Prove the *Corollary* to *Theorem* 3: If S is countably infinite and F is finite, then $F \prec S$.

Exercises 17–19: True or False

Are the following statements true or false? Explain your answer.

5.1.17. $S \sim T$ if and only if for each element $x \in S$ there is a unique element $y \in T$ that can be associated with it.

5.1.18. $S \prec T$ if and only if S can be put into one-to-one correspondence with a proper subset of T.

5.1.19. Since $\mathbb{Q}$ is dense (between any two rational numbers there is a third one) while $\mathbb{Z}$ is not, there are more rational numbers than integers.

Exercises 20–30: Countably Infinite Sets

Work the following problems involving countably infinite sets.

5.1.20. *Countable Subsets of $\mathbb{N}$*
  a. Prove that $\{n \in \mathbb{N}: n > 100\}$ is countably infinite by enumerating its elements. Then give an explicit formula for the nth element in your list.
  b. Prove that $\{10^n : n \in \mathbb{N}\}$ is countably infinite.
5.1.21. *The Zigzag Argument*
Determine a formula $f(n)$ for the nth element in the zigzag enumeration used to prove *Proposition 4*. Take $f(0) = 0, f(1) = 1, f(2) = -1$, etc.   *Hint*: use the floor function $\lfloor x \rfloor =$ the largest integer less than or equal to x.
5.1.22. *The Odd Integers are Countably Infinite*
  a. Prove that the set of all odd integers (positive and negative) is countably infinite by listing its elements.
  b. Give a formula $f(n) = x_n$ for the nth element in your listing of part a.

5.1.23. *Theorem 4*
a. Prove *Corollary 1* to *Theorem 4*: If $S \subseteq T$ and T is countably infinite, then S is finite or countably infinite.
b. Prove *Corollary 2* to *Theorem 4*: If $S \prec T$ and T is countably infinite, then S is finite.

5.1.24. *Finite and Countably Infinite Sets*
a. Prove *Corollary 1* to *Theorem 5*: If S is finite and T is countably infinite, then $S \cup T$ is countably infinite.
b. Prove *Proposition 3*: If S is finite and T is countably infinite, then $T - S$ is countably infinite.

5.1.25. *Hilbert's Hotel*
a. Imagine that you're the innkeeper of *Hilbert's Hotel* (see Example 2). Explain how you can house a countably infinite number of guests and still have room for countably many late arrivals.
b. Suppose you've housed a countably infinite number of guests as in part *a*, and then two more such groups arrive. Explain how you will house these in succession without making anyone move to a new room. If still more groups arrive, will you be able to accommodate them in the same way?

5.1.26. *Theorem 6*
Prove the *Corollary* to *Theorem 6*: If $S_1, S_2, \ldots, S_n$ are countably infinite sets, then $\bigcup_{i=1}^{n} S_i$ is countably infinite.

5.1.27. *Theorem 7*
Determine a formula for enumerating the infinite rectangular array given in the proof of *Theorem 7*. This may be easier if you find a formula $f(m,n)$, where x_{mn} is listed in the $f(m,n)^{\text{th}}$ place in the enumeration.

5.1.28. *Theorem 8*
a. Prove *Theorem 8*: If S and T are countably infinite, then $S \times T$ is countably infinite.
b. Prove the *Corollary* to *Theorem 8*: If S_i is countably infinite for $i = 1, 2, \ldots, n$, then $S_1 \times S_2 \times \cdots \times S_n$ is countably infinite.

5.1.29. *Finite Sequences*
a. How many finite sequences of 0's and 1's are there? Prove your answer.
b. How many finite subsets does $\mathbb{N}$ have? *Hint*: model finite subsets using strings of 0's and 1's and then use part *a*.
c. How many finite sequences of natural numbers are there? Prove your answer.

5.1.30. *Algebraic Numbers*
a. Prove that any rational number m/n is an algebraic number.
b. Prove that both $\sqrt{2}$ and $i = \sqrt{-1}$ are algebraic numbers.
c. Flesh out the set-theoretic details for the proof of *Proposition 6*.
d. Prove that the set of all real algebraic numbers $(\mathbb{A} \cap \mathbb{R})$ is countably infinite. What does this mean for the rational numbers? the integers?

Exercises 31–32: Infinite Sets and Countably Infinite Sets
Work the following problems relating infinite sets and countably infinite sets.

5.1.31. *Theorem 9*
Prove *Theorem 9b: If T is an infinite set and S is a countably infinite subset of T, then $S \preceq T$.*

5.1.32. *Theorem 10*
  a. Prove *Corollary 1 to Theorem 10: T is infinite if and only if T is equinumerous to a proper subset of itself.*
  b. Prove *Corollary 2a to Theorem 10: If S is countable and T is infinite, then $S \cup T \sim T$.*
  c. Prove *Corollary 2b to Theorem 10: If S is countable and T is uncountably infinite, then $T - S \sim T$.*

5.1.33. *Axiom of Choice*
Look up *Axiom of Choice* and *Banach-Tarski Paradox* online. Explain why some believe the *Axiom of Choice* is a questionable principle when used in full generality (there are less controversial restricted versions of the axiom).

5.2 Uncountably Infinite Sets

Prior to Cantor, mathematicians and philosophers distinguished finite quantities from infinite ones, but they had little inkling that there were distinct orders of infinity. *Infinite* meant *not finite*. Cantor's discovery that there were different-sized infinities with definite relations among them led him to develop his ideas about sets into a mathematical theory. In this section we'll continue looking at some important ideas unearthed by his investigations.

5.2.1 *Infinity and Countable Infinity*

We've already learned that $\mathbb{N}$, $\mathbb{Z}$, $\mathbb{Q}$, and $\mathbb{A}$ are countably infinite. We also proved that the union and Cartesian product of two (and thus finitely many) countably infinite sets are countably infinite. Even the countably infinite union of countably infinite sets is countably infinite.

Given the evidence amassed thus far, it's fair to wonder whether all the hoopla about countably infinite sets isn't sophisticated hot air, and whether it isn't the case that all infinite sets are countably infinite. A set like $\mathbb{Q}$ didn't look like it could be enumerated, but by creatively rearranging its elements, we found out that it's countable. Maybe this can always be done?

The answer to this, we'll soon see, is a resounding "no." Every infinite set does contain a countably infinite subset (*Theorem 5.1.10*), but unless we can show that an enumeration catches every element of the set, we don't know that it's countably infinite. And, in fact, not all infinite sets are countable, as we'll show in some powerful ways. Some infinite sets are *uncountable*, in the sense that they cannot be listed by an infinite sequence, no matter how ingeniously devised.

5.2.2 *Uncountably Infinite Sets, Geometric Point Sets*

We'll begin with the obvious definition, which only tells us what to call such sets if they exist.

Definition 5.2.1: *Uncountable Sets*
 S is **uncountably infinite** *if and only if S is infinite but not countably infinite.*

The problem with uncountable sets is *not* that they can't be matched up with other sets. *Theorem* 5.1.10 shows that every infinite set is equinumerous with a proper subset of itself. The problem is that they're too big to be matched with $\mathbb{N}^+$, even though that, too, is infinite. The following proposition gives some initial comparison results for uncountably infinite sets.

Proposition 5.2.1: *Numerosity and Uncountably Infinite Sets*
 a) *If S is uncountably infinite and $S \sim T$, then T is uncountably infinite;*
 b) *If S is uncountably infinite and $S \subseteq T$, then T is uncountably infinite;*
 c) *If S is countably infinite and T is uncountably infinite, then $S \prec T$.*

Proof:
 See Exercise 16.

You may have noticed in our discussion of countably infinite sets that we failed to consider the set of real numbers and the set of complex numbers. The reason is that these sets aren't countably infinite. We'll now show this, starting with the real-number continuum. We'll then consider higher dimensional spaces to see what they contribute to the picture. Our train of consequences starts with the fact that the set of real numbers between 0 and 1 (the open unit interval) is uncountable.

Theorem 5.2.1: *Numerosity of* $(0,1)$ (Cantor 1874)
 The open interval $(0,1) = \{x \in \mathbb{R} : 0 < x < 1\}$ is uncountably infinite.

Proof:
· We'll give Cantor's second famous diagonal argument (1891), which involves proof by contradiction.
· We can *uniquely represent* any number in the interval $(0,1)$ by choosing its nonterminating decimal representation $.a_1a_2a_3\cdots$. For example, we can represent .5 by $.4\overline{9}$, 1/7 by $.\overline{142857}$, and $\pi/4$ by $.78539816339\cdots$.
· Suppose now that this set is countably infinite, and let the following sequence list all the elements of $(0,1)$. In our notation, x_{ij} denotes the jth digit in the decimal representation of the ith number in the list.

$$. x_{11}x_{12}x_{13}\cdots$$
$$. x_{21}x_{22}x_{23}\cdots$$
$$. x_{31}x_{32}x_{33}\cdots$$
$$\vdots$$

· Such a list is *inherently incomplete*. We'll construct a number missed by this list by proceeding along the list's main diagonal.
· Construct a nonterminating decimal $d = .d_1 d_2 d_3 \cdots$ as follows:
Let d_n be any definite digit different from x_{nn}, say

$$d_n = \begin{cases} 1: & x_{nn} \neq 1, \\ 2: & x_{nn} = 1. \end{cases}$$

Since $d = .d_1 d_2 d_3 \cdots$ differs in the nth place from the nth number listed, it isn't in the list, contradicting our assumption.
· Thus, the real numbers in $(0, 1)$ can't be enumerated. ■

As an immediate consequence via *Proposition* 1b, we have the following:

Corollary 5.2.1.1: $\mathbb{R}$ *is Uncountably Infinite*
If $S \supseteq (0, 1)$, then S is uncountable. Hence, $\mathbb{R}$ is uncountable, and $\mathbb{N} \prec \mathbb{R}$.

This result still leaves open whether all real-number supersets of $(0, 1)$ are of the same size as $(0, 1)$. On first thought you might suspect the answer is no; after all, $(0, 1)$ is only a short line segment, while $\mathbb{R} = (-\infty, \infty)$ can be thought of as an infinite line. The next propositions, though, answer this question in a surprising way. We'll motivate them by an example.

✦ Example 5.2.1
Show that the following real-number intervals are equinumerous to $(0, 1)$:
a) $(1, 2)$ b) $(0, 2)$ c) $(1, 4)$

Solution
The following formulas define one-to-one correspondences between the unit interval $(0, 1)$ and the given sets and so demonstrate equinumerosity. A 2D geometric interpretation is also described for each of these matchings.
a) $y = x + 1$; this translates points in $(0, 1)$ one unit right.
Alternatively, points (x, y) on the graph of $y = x + 1$ match x-values from $(0, 1)$ in a one-to-one way with y-values in the y-interval $(1, 2)$.
b) $y = 2x$; this stretches the unit interval to twice its length.
The graph of $y = 2x$ establishes a one-to-one correspondence between the x-interval $(0, 1)$ and the y-interval $(0, 2)$, as in part *a*.
c) $y = 3x + 1$; this stretches the unit interval to three times its length and then translates it one unit right.
The graph of $y = 3x+1$ establishes a one-to-one matching between $(0, 1)$ and $(1, 4)$, as in parts *a* and *b*.

Proposition 5.2.2: *All Open Intervals Are Equinumerous*
All finite open intervals are equinumerous and are uncountably infinite:
$(a, b) \sim (c, d)$ for any real numbers $a < b$, $c < d$.

Proof:
The proof for this proposition generalizes the arguments given in the last example. See Exercise 8.

Stretching or shrinking an interval by a finite factor doesn't affect numerosity. All finite open intervals contain the same number of points. This also holds if the interval is closed or half-open and half-closed (see Exercise 11). In particular, $(0, 1]$ and $[0, 1]$ are equinumerous with $(0, 1)$—adjoining one or two points doesn't change the infinite size of a set (see *Corollary* 2a to *Theorem* 5.1.10). The next result shows that magnifying an interval's length by an infinite factor, as it were, also doesn't affect numerosity.

Proposition 5.2.3: $\mathbb{R}$ *is Equinumerous with* $(0, 1)$
  $\mathbb{R} \sim (0, 1)$

Proof:
  The formula $y = \dfrac{x - .5}{x(x - 1)}$ [alternatively, $y = \tan(\pi x - \pi/2)$] sets up a one-to-one correspondence between $(0, 1)$ and $\mathbb{R}$ (see Exercise 9). ∎

Let's look at an important consequence of the fact that while the algebraic numbers are countable, the set of all real numbers is uncountable. First another definition.

Definition 5.2.2: *Transcendental Numbers*
  *A real number is **transcendental** if and only if it is not algebraic.*

Transcendental real numbers are irrational, but the converse is not true. Irrational numbers such as $\sqrt{2}$ are algebraic. Examples of familiar transcendental real numbers are π and e. How many more are there? Finitely many? A countably infinite number? Uncountably many? The next result by Cantor is surprising, given we don't know many particular transcendental numbers. Dedekind saw no practical consequence of $\mathbb{A}$ being countable; Cantor did.

Proposition 5.2.4: *Transcendental Numbers Are Uncountable* (1874)
  There are uncountably many transcendental real numbers.

Proof:
  We'll argue this, again using *Proof by Contradiction*.
  Since the set $\mathbb{A}$ of algebraic numbers is countably infinite, its restriction to the reals, $\mathbb{A} \cap \mathbb{R}$, is countable (see also Exercise 5.1.30d).
  Thus, if the transcendental real numbers were countably infinite, the union of this set with the set of algebraic real numbers, namely, $\mathbb{R}$, would also be countable by *Theorem* 1.1.6, but it's not.
  So, the set of transcendental real numbers is uncountably infinite. ∎

This result gives a slick answer to a difficult problem. In 1844, Liouville had demonstrated how to construct infinitely many transcendental reals, such as $0.11000100000000000000001\cdots$, where 1s occur in the $n!^{\text{th}}$ decimal places. In 1873, Hermite proved that e is transcendental. The following year, Cantor showed that every interval of real numbers contains infinitely many transcendentals, constructively generating such numbers. The above proof, published by Felix Klein in 1894, fails to construct or identify any specific transcendental number, but it nevertheless demonstrates that there are uncountably many of them. In essence, then, real numbers are typically (almost all) transcendental numbers. In 1882, Lindemann showed that π is transcendental. Research done in the twentieth century showed that numbers like $2^{\sqrt{2}}$ are transcendental.[3]

Once we know that there are infinite sets of two different sizes, it's natural to wonder whether all other infinite sets are one of these two types. We've found many sets that are countably infinite and discovered that every infinite set is at least as large as $\mathbb{N}$, because it contains a countably infinite subset. Are all uncountably infinite sets the same size as $\mathbb{R}$? There are two or three subquestions to this query:
1) Are there any other well-known sets equinumerous to $\mathbb{R}$?
2) Are there sets more numerous than $\mathbb{R}$?
3) Are there sets more numerous than $\mathbb{N}$ but less numerous than $\mathbb{R}$?

Let's take up these questions in the order given. We'll begin by looking at several sets that might be more numerous than $\mathbb{R}$ and show that they're the same size as $\mathbb{R}$. We'll go on to show, however, that there are sets strictly larger than $\mathbb{R}$—in fact, there is no largest infinite set! Finally, we'll look briefly at what is known about sets of intermediate size.

5.2.3 *Sets Equinumerous with the Continuum*

We'll first explore what happens when dimensionality is increased. $\mathbb{R}$ quantifies the linear continuum, so let's consider $\mathbb{R}^2 = \mathbb{R} \times \mathbb{R}$, which quantifies two-dimensional space. From there we can go on to consider $\mathbb{R}^3$ and $\mathbb{R}^n$.

We saw that the Cartesian-product operator didn't produce anything new for countably infinite sets—all $\mathbb{N}^n$ are countably infinite. The same thing happens when we move from one-dimensional space to any higher dimension, as we'll see in the next theorem and its corollary. Increasing the dimension does not increase the number of points in space. Cantor became convinced of this, though at first, he, like others, thought higher dimensional space might contain a greater multiplicity of points. His initial proof contained a gap (filled by the *Schröder-Bernstein Theorem*) that he thought he could fix, but when he was unable to do so, he developed a more complex proof, which he asked Dedekind to check. It was in this context that Cantor wrote "I see it, but I don't believe it"—indicating some unease with the correctness of his argument, though not with the result.[4] Our proof below is a simple modification of Cantor's original proof.

---

[3] Problem 7 in Hilbert's famous 1900 list asked whether α^β is transcendental when α is an algebraic base ($\alpha \neq 0, 1$) and β is an irrational algebraic exponent. Russian mathematician Aleksandr Gelfond answered this in the affirmative in 1934.

[4] See Fernando Govêa's *Was Cantor Surprised?* in the March 2011 issue of the ***American Mathematical Monthly*** for clarifying the story about this result and quote.

Theorem 5.2.2: *Planes are Equinumerous with Lines* **(1878)**
$$\mathbb{R}^2 \sim \mathbb{R}$$

Proof:
- Since $\mathbb{R} \sim (0, 1]$, $\mathbb{R}^2 \sim (0, 1] \times (0, 1]$ (see Exercise 5.1.5c).
- To prove our result, it suffices to show $(0, 1] \times (0, 1] \sim (0, 1]$.
  The basic intuition behind matching these two sets is the following:
  Given any point (x, y) in the unit square, use each number's nonterminating decimal representation, writing it as $(. x_1 x_2 \cdots , . y_1 y_2 \cdots)$.
  Match this ordered pair of real numbers to the real number z constructed in zigzag fashion: $z = . x_1 y_1 x_2 y_2 \cdots$.
- This matching very nearly works. It matches different ordered pairs to different numbers, but some numbers will not be matched with any ordered pair. For instance, both $z_1 = .1101010 \cdots$ and $z_2 = .101010 \cdots$ have no ordered pair (x, y) related to them: z_1 would need $x = .100 \cdots$, which is not nonterminating, while z_2 would need $y = .000 \cdots$, which is neither positive nor nonterminating.
- A clever modification of this procedure, however, using *blocks of digits* instead of single digits, avoids these problems: first list a block of x's digits, continuing until a non-zero digit is reached, then do the same for y's digits, then go back to x and list another block of its digits, etc. This procedure yields a one-to-one correspondence between the unit square and the unit interval (see Exercise 13a). ∎

Corollary 5.2.2.1: *Size of the Complex Numbers*
$$\mathbb{C} \sim \mathbb{R}$$

Proof:
$\mathbb{C} \sim \mathbb{R}^2 \sim \mathbb{R}$, so $\mathbb{C} \sim \mathbb{R}$ ∎

Corollary 5.2.2.2: *N-Dimensional Spaces and Lines*
$$\mathbb{R}^n \sim \mathbb{R}$$

Proof:
Use *Proof by Mathematical Induction*. See Exercise 13b.

5.2.4 *Sets Larger than the Continuum: the Power Set*

Finite Cartesian products don't increase the cardinality of $\mathbb{R}$, but do other operations? Intersections and set differences only make sets smaller. Unions don't increase numerosity, either—infinite sets absorb smaller sets without increasing their size (see *Corollary* 2 to *Theorem* 5.1.10 and Exercise 20).

The other operator we can use is the power set operator. We saw earlier that this greatly increased the size of finite sets: if $|S| = n$, then $|\mathcal{P}(S)| = 2^n > n$. By now, your intuitions about cardinality have probably been sufficiently challenged so that you don't trust *any* generalization from the finite to the infinite. That's a good instinct, but in this case, the generalization holds. *Cantor's Theorem* says every set is strictly less numerous than its power set. Let's first look at an example, which is important in its own right.

Proposition 5.2.5: $\mathcal{P}(\mathbb{N})$ *is Equinumerous with* $\mathbb{R}$
   $\mathcal{P}(\mathbb{N}) \sim \mathbb{R}$

Proof:
- Real numbers in the interval $(0, 1]$ can be uniquely represented in *binary notation* by nonterminating sequences of 0's and 1's following a "binary point," where the nth bit has the place value of 2^{-n} instead of 10^{-n}. For example, $1/2 = .0111 \cdots$ because $1/2 = 1/4 + 1/8 + 1/16 + \cdots$.
- Let B denote the associated set of nonterminating binary sequences—just drop the binary point. $B \sim (0, 1]$, and thus $B \sim \mathbb{R}$.
- Now let T denote the set of terminating binary sequences—those that are eventually all 0s. T is essentially the set of finite binary sequences, which is countably infinite (see Exercise 5.1.29a).
- Let S be the set of all binary sequences, i.e., $S = B \cup T$, the disjoint union of an uncountable set with a countable set.
   By *Corollary* 2 to *Theorem* 5.1.10, $S \sim B$. But $B \sim \mathbb{R}$, so $S \sim \mathbb{R}$, too.
- To finish the proof, it suffices to prove $S \sim \mathcal{P}(\mathbb{N})$.
   Each sequence in S determines a subset P of $\mathbb{N}$ as follows: $n \in P$ if and only if there is a 1 in the nth term of the sequence (starting with $n = 0$). This sets up a one-to-one correspondence between S and $\mathcal{P}(\mathbb{N})$.
   Thus, $S \sim \mathcal{P}(\mathbb{N})$, and therefore $\mathcal{P}(\mathbb{N}) \sim \mathbb{R}$. ∎

   Since $\mathbb{R}$ is uncountable, $\mathcal{P}(\mathbb{N})$ is, too. Thus, $\mathbb{N} \prec \mathcal{P}(\mathbb{N})$. $S \prec \mathcal{P}(S)$ also holds for finite sets, so we're ready to generalize to *Cantor's Theorem*. With this theorem, the population of the transfinite realm explodes.

Theorem 5.2.3: *Cantor's Theorem: Cardinality of Power Sets* (**1892**)
   $S \prec \mathcal{P}(S)$

Proof:
- Let S be any set. We'll show that S is strictly less numerous than $\mathcal{P}(S)$ using a type of diagonalization procedure (see Exercise 21).
- Clearly $S \preceq \mathcal{P}(S)$—match each x in S with singleton $\{x\}$ in $\mathcal{P}(S)$.
- We'll show $S \not\sim \mathcal{P}(S)$ by proving that *any* attempted matching M misses some subset of S, i.e., some element of $\mathcal{P}(S)$.
   - Match each $x \in S$ with some $M_x \in \mathcal{P}(S)$, and take $N = \{x \in S : x \notin M_x\}$.
   - *Claim*: $N \subseteq S$, but N is *not* matched with any $x \in S$:
     * For suppose $N = M_a$ for some $a \in S$. Where does a lie relative to N?
     * According to the definition of N, $a \in N \leftrightarrow a \notin M_a$.
     * Thus, $a \in N \leftrightarrow a \notin N$, which is a contradiction.
   - Hence there is no a such that $N = M_a$.
   - Consequently, $M : x \mapsto M_x$ is not a one-to-one correspondence.
   - Since M was perfectly general, no matching exists between S and $\mathcal{P}(S)$.
- Thus, $S \not\sim \mathcal{P}(S)$, and so $S \prec \mathcal{P}(S)$. ∎

   So, this is what we know about infinite sets so far. $\mathbb{N}$ is the smallest type of infinite set, and many other sets are also countably infinite. $\mathbb{R}$ is bigger than $\mathbb{N}$, and all intervals of real numbers as well as all n-dimensional real spaces are the same size as $\mathbb{R}$. *Cantor's Theorem* then greatly expands the domain of infinite sets in one fell swoop. The power set of any set is strictly larger than the set. Thus, $\mathcal{P}(\mathbb{N})$ is bigger than $\mathbb{N}$, though it's the same size as $\mathbb{R}$, $\mathcal{P}(\mathbb{R})$ is strictly larger than $\mathbb{R}$, $\mathcal{P}(\mathcal{P}(\mathbb{R}))$ is bigger yet, and so on—the process never stops. Successive power sets form a sequence of ever-increasing size. Infinite cardinalities go on and on, just like finite ones! In fact, the realm of transfinite sets is far richer and more bizarre than even this suggests. Our discussion only begins to touch on what has been discovered by twentieth-century set theorists. You may start to see why some mathematicians wondered whether this is still mathematics or whether it had morphed into speculative philosophy or theology.

5.2.5 *The Continuum Hypothesis*

One problem Cantor bequeathed to mathematics was determining the exact size of the linear continuum. This made it to the top of Hilbert's famous list of 23 open problems published in 1900. The *Continuum Problem's* two-stage solution during the mid-twentieth century brought fame to those involved.

We know that sets continue to grow in size if we apply the power set operator, but do they grow successively or by leaps and bounds? In the finite realm, $|\mathcal{P}(S)|$ is much larger than $|S|$, with many sets T of intermediate size. Is this also true for transfinite sets? Or is the power set now the next size set, with no intervening size?

This is the *Continuum Problem*. In the case of $\mathbb{R}$, which was Cantor's concern, the question amounts to asking whether there is a subset of $\mathbb{R}$ larger than $\mathbb{N}$ but smaller than $\mathbb{R}$. This question seems unambiguous and easy to understand, but the eventual answer was rather surprising.

We can formulate the *Continuum Problem* and Cantor's conjecture about its solution using some notation and ideas from his *Transfinite Set Theory*.[5] Cantor introduced the symbol $\aleph_0$ (read: *aleph-null* or *aleph-naught*) to stand for the cardinality of countably infinite sets: $|\mathbb{N}| = \aleph_0$. The cardinality of the real numbers is then denoted by $|\mathbb{R}| = |\mathcal{P}(\mathbb{N})| = 2^{\aleph_0}$. The *Continuum Problem* asks, in cardinality terms, whether it's true for some set S that $\aleph_0 < |S| < 2^{\aleph_0}$, or whether $2^{\aleph_0}$ is the next cardinal number after $\aleph_0$.

Cantor's intuition was that there's no such set, that any set S satisfying $\mathbb{N} \subseteq S \subseteq \mathbb{R}$ is either the size of $\mathbb{N}$ or $\mathbb{R}$. This conjecture became known as the *Continuum Hypothesis* (*CH*). Cantor's formulation took different forms at different times, going back to 1874, and more definitively to 1883.

Cantor's Continuum Hypothesis
   There are no sets with cardinality between $|\mathbb{N}| = \aleph_0$ and $|\mathbb{R}| = 2^{\aleph_0}$.

Cantor expended much time and effort trying to prove this result throughout his life. At times he was sure he had succeeded, only to discover a flaw

---

[5] Cantor developed transfinite arithmetics for both cardinal and ordinal numbers.

in his reasoning. In the end, he failed. We now know that there was good reason for his failure, because *it cannot be proved*, given the usual axioms of *Set Theory*. But, surprisingly, *neither can it be disproved*.

When Hilbert put Cantor's *Continuum Problem* at the top of his list of open problems for twentieth-century mathematics, it gave great impetus to attempts to solve it. The first significant step toward its resolution was made by the Austrian Kurt Gödel (Figure 5.3), already one of the leading mathematical logicians of the twentieth century. In 1940, though he believed the *Continuum Hypothesis* to be false, Gödel proved that it was consistent with the rest of *Set Theory* as it had been axiomatized by that time.

Fig. 5.3 Kurt Gödel

Theorem: *Consistency of the Continuum Hypothesis* **(Gödel 1940)**
The Continuum Hypothesis is consistent with Set Theory.

Being consistent with the axioms of *Set Theory* is not the same as being a logical consequence of the axioms, so the *Continuum Hypothesis* still awaited proof. In 1963, however, the American mathematician Paul Cohen, using a new technique called *forcing* that he developed for the task, demonstrated that *CH* could not be proved as a theorem of *Set Theory* because its negation was also consistent with the axioms of *Set Theory*.

Theorem: *Independence of Continuum Hypothesis* **(Cohen 1963)**
The negation of the Continuum Hypothesis is consistent with Set Theory.

These results show that we're free to assume either *CH* or some form of its opposite without contradicting the rest of *Set Theory*. In other words, the truth of the *Continuum Hypothesis* cannot be decided on the basis of the usual axioms of *Set Theory*—it's an *undecidable* result, whose proof or disproof requires adopting a new axiom for *Set Theory*.

Set-theoretical research has explored axioms strong enough to decide this issue since the 1960s. Many set theorists believe the *Continuum Hypothesis* is false, but no one has proposed a widely accepted axiom to decide the matter. Some mathematicians take a more formalistic attitude toward all of this, saying the hypothesis is neither true nor false. Like the parallel postulate in geometry, its truth value depends on what system of mathematics you want to develop. Given the familiar nature of $\mathbb{N}$ and $\mathbb{R}$, however, this outlook seems unsatisfying. Surely there either is or is not a set S between $\mathbb{N}$ and $\mathbb{R}$ of intermediate size. Assuming the *Law of Excluded Middle*, of course.

Delving into the *Continuum Hypothesis* and related matters is usually reserved for graduate-level courses in *Set Theory* or *Mathematical Logic*; it's also a favorite topic of contemporary philosophers of mathematics. We've gone about as deeply into this matter as we can in an introductory course. Our last pass through *Set Theory* in Section 5.3 will focus on its standard axiomatization and some related matters.

EXERCISE SET 5.2

Some of the following exercises are nontrivial, so don't get discouraged if you find them difficult.

Exercises 1–4: True or False
Are the following statements true or false? Explain your answer.

5.2.1. The points in 2D space are more numerous than those in 1D space but less numerous than those in 3D space.

5.2.2. If S and T are uncountably infinite, then $S \sim T$.

5.2.3. $\mathcal{P}(\mathbb{N}) \sim \mathbb{R}$.

5.2.4. For uncountable sets, $S \sim \mathcal{P}(S)$.

Exercises 5–8: Intervals Equinumerous with (0,1)
The following problems deal with finite intervals of real numbers.

5.2.5. *Equinumerous Intervals*
  a. Find a one-to-one correspondence to show that $(0,1) \sim (-1,3)$. Then explain its meaning geometrically.
  b. Find a one-to-one correspondence to show that $(0,1) \sim (-1/2, 2/3)$. Then explain its meaning geometrically.

5.2.6. *Equinumerous Unit Intervals*
Prove that $(0,1] \sim (0,1)$ in the following two ways:
  a. Graph some (discontinuous) function that exhibits a one-to-one correspondence from $(0,1]$ to $\mathbb{R}$ and find a formula for it. Then use the transitivity of equinumerosity to draw your final conclusion.
  b. Show that adjoining a single element to $(0,1)$ doesn't change its cardinality. Either use *Corollary* 2 to *Theorem* 5.1.10 (but only if you've proved it in Exercise 5.1.32b), or else prove *Corollary* 2 for this particular case using an argument similar to the proof of *Theorem* 5.1.10.

5.2.7. What is wrong with the following "proof" that $(0,1)$ is uncountable? *Start with $(0,1)$ and choose its midpoint $1/2$. Then take each of the two subintervals created by $1/2$, and choose their midpoints as the new cut points to list: $1/4$, $3/4$. Continuing, at stage k list the 2^{k-1} new midpoints. This procedure eventually lists all numbers of the form $m/2^n$. However, since $1/3$, among other numbers, is omitted from the list, the set $(0,1)$ is uncountable.*

5.2.8. *Proposition 2*
Prove that $(a,b) \sim (c,d)$ for any real $a < b$, $c < d$ in the following ways:
  a. *Pictorially*: given two intervals of different finite lengths, place the shorter one above and parallel to the other and explain how to match the two intervals point for point to yield a one-to-one correspondence.
  b. *Algebraically*: determine a formula that matches these sets. *Hint*: match x in (a,b) to the y in the same relative position in the interval (c,d).

Exercises 9–14: Sets Equinumerous with $\mathbb{R}$
The following problems deal with sets equinumerous with $\mathbb{R}$.

5.2.9. *Proposition 3*

Show that the formula $y = \frac{x-.5}{x(x-1)}$ given in the proof of *Proposition 3* is a one-to-one correspondence between $(0,1)$ and $\mathbb{R}$. Carefully graph the formula and explain how points in the unit interval are related to real numbers.

5.2.10. *Intervals Equinumerous with* $\mathbb{R}$

a. Show that $\mathbb{R} \sim (-1,1)$ in the following way. Construct a unit circle with center $(0,1)$, and then take the lower semicircle minus its top endpoints. Project the points on this open semicircle onto the x-axis in some way, and explain how this induces a one-to-one correspondence between $(-1,1)$ and $\mathbb{R}$. Why does this mean in turn that $(0,1) \sim \mathbb{R}$?

b. Modify the formula given in Exercise 9 to show directly that $(-1,1) \sim \mathbb{R}$. Explain why your formula is a one-to-one correspondence.

5.2.11. *Equinumerous Finite Intervals*

Show that any two intervals of finite length, whether open, closed, or half-open/half-closed, are equinumerous with one another and with $\mathbb{R}$. Use whatever results are already known about these cases, including earlier exercises.

5.2.12. *Equinumerous Infinite Intervals*

Show that any two intervals of infinite length, whether open or half-closed, are equinumerous with one another and with $\mathbb{R}$. Use whatever results are already known about these cases.

5.2.13. *Theorem 2*

a. Show that the modified correspondence between $(0,1] \times (0,1]$ and $(0,1]$ used in the proof of *Theorem 2* is a genuine one-to-one correspondence, i.e., show that each ordered pair of numbers has a unique number it is associated with, that no numbers are left without a mate, and that given any number z there is a unique ordered pair (x,y) that is related to it.

b. Prove *Corollary 2* to *Theorem 2*: $\mathbb{R}^n \sim \mathbb{R}$.

5.2.14. *Irrational Numbers*

a. Prove that the set of irrational numbers is equinumerous with $\mathbb{R}$.

b. Knowing how rational and irrational numbers are represented by infinite decimal expansions, show that between every two rational numbers there is an irrational number, and between every two irrational numbers there is a rational number. Do you find this paradoxical? Explain.

c. Prove there are uncountably many irrational values of a^b for irrational exponents b when $a > 1$.

5.2.15. Prove that if $S \sim \mathbb{R}$ and $T \sim \mathbb{R}$, then $S \cup T \sim \mathbb{R}$. *Hint:* first prove this result for the case of disjoint sets.

Exercises 16–21: Uncountable Sets

5.2.16. *Prove Proposition 1*

a. If S is uncountably infinite and $S \sim T$, then T is uncountably infinite.

b. If S is uncountably infinite and $S \subseteq T$, then T is uncountably infinite.

c. If S is countably infinite and T is uncountably infinite, then $S \prec T$.

5.2.17. Is the counterpart to *Proposition* 5.1.2, i.e., *if T is uncountably infinite, then S is uncountably infinite if and only if $S \sim T$*, true or false? If it is true, prove it. If it is false, give a counterexample.

5.2.18. Prove that the collection of *co-infinite sets* in $\mathbb{N}$, i.e., *subsets of $\mathbb{N}$ whose complement in $\mathbb{N}$ is infinite,* is uncountably infinite. What size is it? *Hint*: first calculate the cardinality of all *co-finite* sets.

5.2.19. Using a diagonalization argument, prove that the countably infinite Cartesian product of countably infinite sets $\prod_{i \in \mathbb{N}} S_i = \{(x_0, x_1, \dots) \colon x_i \in S_i\}$ is uncountable.

5.2.20. Show that if T is any infinite set with $S \prec T$, then $S \cup T \sim T$.

5.2.21. *Cantor's Theorem and Diagonalization*
Analyze the proof of *Cantor's Theorem* for the case when $S = \mathbb{N}$. Represent each subset M_k of $\mathbb{N}$ by an infinite sequence of 0s and 1s, putting a 1 in the nth place if $n \in M_k$ and otherwise putting in a 0. Explain what N is in this case and why $N \neq M_a$ for any a. Then tell why the proof of *Cantor's Theorem* is considered a (generalized) diagonalization argument.

5.3 Formal Set Theory and the Halting Problem

Cautious mathematicians and philosophers have always thought that if you mess around with infinity long enough, you'll get into trouble. Cantor, however, judged that earlier scruples about infinity were misplaced and confidently developed a theory of transfinite *Set Theory*.

Cantor's theory seemed solid, but he eventually came to believe that besides transfinite quantities there is also an absolute infinity, which is as large as it gets. In the mid-1880s, he speculated that this was connected with God and that it could not be treated mathematically like other infinities.

Mathematicians largely ignored Cantor's theological reflections, but they, too, found it necessary to distinguish between the infinity of everything and the infinities associated with familiar mathematical objects and constructions. Around the turn of the century, mathematicians and logicians discovered that informal *Set Theory* harbored paradoxes (unexpected contradictions) if an absolute infinity was accepted. These paradoxes could be avoided by banning an unrestricted universal set, but a more intellectually honest approach would be to systematically and consciously fence out the paradoxes without using *ad hoc* stratagems. This required axiomatizing *Set Theory* and then proving that the problematic sets don't exist according to the theory and so cannot create theoretical havoc.

5.3.1 *Paradoxes: the Need for Axiomatic Set Theory*

Concern for logical consistency is often treated as the main motivation behind formalizing *Set Theory*. Historically speaking, though, avoiding paradoxes was only one impetus for axiomatizing *Set Theory*, and not even the most

important one.[6] Any mathematical theory with a rich body of results begs
to be deductively organized. Once sufficiently many results are known, mathematicians try to put them into a single coherent system. Complex propositions are demonstrated from simpler ones, starting from a limited number
of fundamental principles chosen as the deductive basis of the whole theory.
Naturally, a well-developed theory with problems that need straightening out
gives an even stronger impulse toward organizing it theoretically.

This is what happened in *Set Theory*. In 1908,
Ernst Zermelo (Figure 5.4) published the first
axiomatization of *Set Theory* in two papers.
Zermelo's immediate goal was to substantiate a
surprising result he had proved four years earlier,
namely, the *Well-Ordering Theorem*, which claimed
that any set can be well-ordered.[7] His initial proof
and the result itself had been vigorously challenged
in the interim, so Zermelo offered a second proof,
identifying in detail the various assumptions that
were needed for its execution. In the process, he

Fig. 5.4 Ernst Zermelo

axiomatized *Set Theory* as a whole, which he considered foundational to
mathematics.

Zermelo's axiomatization also solved the paradoxes of infinity, an explicit
concern of his second paper, by showing that they disappeared under his
axiomatization. At the time, the cause of the paradoxes was still being
debated—some proposed reforming logic to avoid the contradictions, but Zermelo chose to restrict set formation. We'll follow Zermelo's approach on this,
but let's first look more closely at what the problem was.

5.3.2 *Russell's Paradox and Inconsistent Sets*

In 1901, Bertrand Russell (Figure 5.5) was pondering *Cantor's Theorem*
(*Theorem* 5.2.3) and wondered what would happen if the universe $\mathcal{U}$ of all
possible sets was used as the base set. Would $\mathcal{P}(\mathcal{U})$ be larger than $\mathcal{U}$? The
theorem says yes, but if $\mathcal{U}$ is the collection of *all* sets, wouldn't *it* be the
largest set of sets? This baffled Russell, so he investigated it further.

Carefully poring over the proof to discover the flaw either in his own or
Cantor's reasoning, Russell was led to consider the class of all sets that are
not elements of themselves. Such sets are the normal case, though $\mathcal{U}$ would
be an exception. We'll denote this class of normal sets by $\mathcal{N} = \{X : X \notin X\}$.
Russell's next question can be formulated as follows: Is $\mathcal{N}$ itself normal? That
is, does $\mathcal{N}$ contain itself as an element or not? Russell concluded, paradoxically, that if it does, then it doesn't; while if it doesn't, then it does. Today
this is known as *Russell's Paradox*. A more popular version, also offered by
Russell, is the *Barber Paradox* (see Exercise 1).

---

[6] This episode is explored in chapter 3 of Gregory Moore's ***Zermelo's Axiom of
Choice: Its Origins, Development, and Influence***.

[7] S is *well-ordered* by a relation if and only if every non-empty subset has a first element.

Using the above definition of $\mathcal{N}$, the argument goes as follows. First suppose $\mathcal{N} \in \mathcal{N}$. Then $\mathcal{N} \notin \mathcal{N}$, because that's the criterion for being in $\mathcal{N}$. On the other hand, suppose that $\mathcal{N} \notin \mathcal{N}$. Then $\mathcal{N}$ belongs to $\mathcal{N}$, which contains all such sets. Combining these conclusions, we end up with the contradiction $\mathcal{N} \in \mathcal{N} \leftrightarrow \mathcal{N} \notin \mathcal{N}$.

Try as he might, Russell could not explain the paradox away. The reasoning is impeccable, as he was forced to concede, so the contradictory conclusion follows. The problem, then, seems to lie not with the logic of the argument but with the mathematics.[8] *Set Theory* itself must be the source of the contradiction. To purge *Set Theory* of this blight, set formation will somehow need to be restricted so that "inconsistent sets," as Cantor called them, cannot be generated.

Fig. 5.5 Bertrand Russell

Zermelo did this by axiomatizing *Set Theory*. The main axiom that rescues us from the paradoxes, as we'll see shortly, is the *Axiom of Separation*. But before considering this axiom, we'll take up a few preliminary matters.

5.3.3 *Syntax and Semantics of Set Theory*

Set Theory's underlying syntax is that of *First-Order Logic*, augmented by symbols for particular sets, such as $\emptyset$; symbols for operations, such as $\cap$; and symbols for relations, such as $\subseteq$. Many of these can be introduced by definitions, as we've already done, but we'll take the membership relation $\in$ as primitive. The sentence $x \in S$ asserts a relation whose intended meaning is intuitively clear but that cannot be defined in simpler terms.

Informal *Set Theory* often distinguishes between individuals and sets by using lowercase and uppercase letters, as we did earlier. We could continue this practice on a formal level—some axiomatizations of *Set Theory* (including Zermelo's original one) do permit both individuals and sets in their universe of discourse. However, most axiomatizations assume that sets are all there are (for the purpose of developing *Set Theory*, of course), so we will, too. Letters, whether uppercase or lowercase, will represent only one type of object: sets. Their elements, if they have any, will themselves be sets, without exception. This may strike you as highly counterintuitive, but it simplifies the theory by having only one sort of object. It is also adequate for developing versions of familiar theories, such as *Peano Arithmetic*, inside *Set Theory*.

5.3.4 *The Axiom of Extensionality*

First-Order Logic already has a fixed, standard notion of identity governed by rules of inference. $S = T$ logically implies $(\forall x)(x \in S \leftrightarrow x \in T)$—merely

---

[8] Russell, however, concluded that logic, which for him included a theory of classes, was at fault. His theory of logical types was explicitly designed to handle the problem.

apply *Sub* to the last clause of the tautology $\forall x(x \in S \leftrightarrow x \in S)$. Thus, identical sets must contain exactly the same elements. However, we can't turn this claim around and say that sets having the same elements are identical—*set equality for coextensive sets doesn't follow from logic alone*. We'll therefore postulate this as an axiom, using the formal notation of FOL.

Axiom 5.3.1: *Axiom of Extensionality*
$$\forall x(x \in S \leftrightarrow x \in T) \to S = T$$

The criterion for equal sets (see *Definition* 4.1.1) follows immediately.

Proposition 5.3.1: *Equal Sets*
$$S = T \leftrightarrow \forall x(x \in S \leftrightarrow x \in T)$$

5.3.5 The Axiom of Separation and Its Consequences

Let's now revisit *Russell's Paradox*. How can *Set Theory* avoid this contradiction? Not much of *Set Theory* is used in the offending argument! To construct $\mathcal{N}$, we made use of what's called *Cantor's Comprehension Principle*: *elements satisfying a common condition compose a set*. This principle lies at the heart of informal *Set Theory*—could it be illegitimate? Regrettably, it seems it may be, because $\{x : P(x)\}$ isn't always a bona fide set.

But perhaps we don't have to discard this principle completely. We do want to reject very large sets, such as $\mathcal{N}$, the collection of all normal sets, and the full universe of sets $\mathcal{U}$, but maybe the *Comprehension Principle* can be used with smaller sets. It seems reasonable to suspect that *Russell's Paradox* arises not because a defining characteristic is used to pick out the members of a set, but because the size of the universe in which it operates is too large.[9] Given a set U *already known to exist*, we should be able to *separate off within this set* all those elements that satisfy some definite condition to produce a new set.

Zermelo's *Axiom of Separation* codifies this approach. His axiom schema provides the justification/rationale for using restricted set-descriptor notation. Given a set U and a proposition $P(x)$, the collection $S = \{x \in U : P(x)\}$ is also a set. In fact, given its existence, we can show that S is well defined. Uniqueness follows, here and elsewhere, from the *Axiom of Extensionality*, something we'll assume as we proceed without much ado.

Axiom 5.3.2: *Axiom of Separation*
$$\forall U \exists S(S = \{x \in U : P(x)\})$$

A set S exists, therefore, whenever it can be *separated out* from an existing universe of discourse U by means of some criterion. A defining condition $P(x)$ is no longer allowed to operate in an unrestricted universe. Given this axiom, we can (re)prove the following *constructive version* of *Russell's Paradox*, which demonstrates that the class $\mathcal{N}$ of all normal sets does *not* exist.

---

[9] This was Zermelo's intuition, but it also lies behind an alternative axiomatization by von Neumann, Bernays, and Gödel, who distinguished between large classes and sets.

Theorem 5.3.1: *Russell's Paradox* (*Constructive Version*)
$\neg\exists N\,\forall x(x \in N \leftrightarrow x \notin x)$

Proof:
- This is proved by *Contradiction*.
  Suppose $\exists N\,\forall x(x \in N \leftrightarrow x \notin x)$, and let $\mathcal{N}$ denote such a set.
  Thus, $\forall x(x \in \mathcal{N} \leftrightarrow x \notin x)$.
  If $\mathcal{N}$ is a set, we can instantiate this universal sentence to $\mathcal{N}$:
  $\mathcal{N} \in \mathcal{N} \leftrightarrow \mathcal{N} \notin \mathcal{N}$. But this is a contradiction.
- Thus, we can conclude that $\neg\exists N\,\forall x(x \in N \leftrightarrow x \notin x)$.   ■

The *Axiom of Separation* is central to the axiomatic development of *Set Theory*. It or one of its consequences comes into play every time we want a set whose existence is not guaranteed by an axiom. We'll illustrate this by proving that the intersection and difference of two sets exist, i.e., are sets. As mentioned above, these sets are uniquely defined.

Proposition 5.3.2: *Existence of Intersections*
$\exists I(I = \{x \in S : x \in T\})$, *i.e.*, $I = S \cap T$ *is a set.*

Proof:
Substitute S for U and $x \in T$ for $P(x)$ in the *Axiom of Separation*.   ■

Proposition 5.3.3: *Existence of Set Differences*
$\exists D(D = \{x \in S : x \notin T\})$, *i.e.*, $D = S - T$ *is a set.*

Proof:
See Exercise 10c.

5.3.6 *The Empty Set*

As crucial as the *Axiom of Separation* is, it only generates new sets inside old ones. So, to actually produce something, we need an existing superset. This isn't furnished by the *Axiom of Separation* or the *Axiom of Extensionality*. We need an axiom to claim the existence of a set to start everything off.

Once we have a set U, we can prove that its subsets exist. In particular, we can prove that the *empty set* exists. The *Axiom of Separation* proves that $\{x \in U : x \neq x\}$ is a set. This very nearly agrees with our earlier definition (*Definition 4.1.4*), except there we used the now-outlawed unrestricted *Comprehension Principle*. Even with our new approach, however, the *empty set* membership criterion follows: $(\forall x)(x \notin \emptyset)$ (see Exercise 8). The *Axiom of Separation* may generate empty sets inside each set U, but all these sets are identical: the *empty set* is unique (see Exercise 9).

Which initial set should we postulate to exist? Let's be frugal and only claim that the *empty set* itself exists. Remember that we've just argued that if anything exists, then $\emptyset$ must. We don't know whether we can do anything worthwhile if we're this thrifty, but let's see what happens.

We'll therefore take the following as our third axiom. Uniqueness of the empty set follows, which we'll prove to illustrate the process once more and because it involves a slightly unusual argument.

Axiom 5.3.3: *Empty Set Axiom*
$\exists E \forall x(x \notin E)$, *i.e., the empty set $\emptyset$ exists.*

Proposition 5.3.4: *Unique Existence of the Empty Set*
$\exists! E \forall x(x \notin E)$, *i.e., the empty set $\emptyset$ is unique.*

Proof:
· Existence follows from the *Empty Set Axiom*.
· To show uniqueness, let E_1 and E_2 denote such sets.
  For any x, $x \notin E_1$, and $x \notin E_2$.
  But this weakens to $x \notin E_1 \leftrightarrow x \notin E_2$ (consult a truth table, if necessary).
  By the biconditional counterpart to *Contraposition*, $x \in E_1 \leftrightarrow x \in E_2$.
  Thus, $E_1 = E_2$. So the *empty set* postulated by *Axiom 3* is unique. ∎

Because the empty set is well defined, we are permitted to introduce the standard symbol $\emptyset$ to denote it, as we did earlier in *Definition 4.1.4*.

5.3.7 *Finite Unions*

We can now claim that sets exist, though we have only $\emptyset$ to back this up. Given $\emptyset$, we can use it as our reference set U to generate other sets via the *Axiom of Separation* by taking intersections and set differences. Unfortunately, this doesn't give us anything new, because the sole subset of $\emptyset$ is $\emptyset$ itself.

Postulating the existence of $\emptyset$ doesn't seem very fruitful. We could blame this on the meager size of $\emptyset$, but an alternative cause might be the limited construction tools available for *working on* $\emptyset$. Separating subsets, intersecting sets, and taking set differences move us *downward into a given set* instead of *upward and out from that set* into the rest of the universe of sets, whatever that might be.

Here we see the restrictive nature of the *Axiom of Separation*. While it helps us avoid contradictions, such as *Russell's Paradox*, it also puts limitations on what sets are legitimate. To proceed further, we'll have to adopt one or more axioms that move outward into a bigger universe. So we'll address this problem before reassessing the possible need for a bigger initial set.

$S \cup T$ goes beyond its component sets and so requires a new axiom—it can't be separated out of an existing set like intersection and set difference.

Axiom 5.3.4: *Finite Union Axiom*
$\exists U(U = \{x : x \in S \lor x \in T\})$, *i.e., $U = S \cup T$ is a set.*

From this axiom, we can prove the existence of the union of a finite collection of sets via induction (hence the axiom's name). To get arbitrary unions, however, a stronger assertion is needed. We'll turn to this next.

5.3.8 *Total Intersections and Unions*

The axioms adopted to this point validate intersection, set difference, and union. We can obtain finite intersections and unions by repeating ordinary intersection and union. Furthermore, we can show that the total intersection of a non-empty collection of sets (*Definition* 4.2.1) exists, because being inside each member, it exists by the *Axiom of Separation* (see Exercise 12).

Total unions of arbitrary collections of sets can't be proved to exist given our current tools, however, so we'll adopt an axiom asserting their existence. Given this axiom, our earlier definition and notation (*Definition* 4.2.2) are now properly supported. And, using this axiom and others, we'll be able to prove the *Finite Union Axiom* as a theorem (see Exercise 29).

Axiom 5.3.5: *Total Union Axiom*
$\forall \mathcal{C} \, \exists U (U = \{x : \exists S[S \in \mathcal{C} \land x \in S]\})$, *i.e.*, $U = \bigcup_{S \in \mathcal{C}} S$ *is a set.*

5.3.9 *Existence of Sets of Sets: Pairs*

It's time for a brief progress report. The *Finite Union Axiom* and the *Total Union Axiom* don't help us expand the universe of sets if we only have the *empty set* to start with—all they give back is $\emptyset$. So at this point our inventory remains only $\emptyset$. But we have more tools we should be able to use.

The above axioms justify forming subsets, intersections, set differences, and unions, but they don't legitimate sets whose *elements* are given sets. We have nothing that generates *collections of sets*. You might think that a collection of sets exists if all of its elements do, but *Russell's Paradox* should make us cautious—the set of all (existing) sets, for instance, doesn't exist. So, we must separately postulate the existence of any collections of sets. There is no prior guarantee that such sets won't lead to contradictory results, but we'll try to avoid known contradictions.

In Section 4.2, we argued intuitively for the legitimacy of familiar sets of collections. Here, we'll introduce axioms that stipulate the existence of some fairly conservative collections. The first axiom of this type enables us to pair up any two sets to form a new one. How much trouble could this cause?

Axiom 5.3.6: *Pairing Axiom*
$\forall x \forall y \exists P(P = \{z : z = x \lor z = y\})$; *i.e.*, $P = \{x, y\}$ *is a set.*

Uniqueness of pairs follows in the usual way (see Exercise 14). This axiom guarantees doubletons, and singletons are once again defined as special doubletons.

Definition 5.3.1: *Doubletons and Singletons*
 a) $\{x, y\} = \{z : z = x \lor z = y\}$
 b) $\{x\} = \{x, x\}$

Since singletons are doubletons, results for doubletons also apply to them. This takes getting used to, but we encountered this idea earlier (see Example 4.2.2). Going from doubletons to singletons is the standard approach in *Set Theory* (see, however, Exercises 22–23).

Sets with three, four, or any finite number of sets as members can now be defined in a standard fashion (see Exercises 17–21) and given appropriate names. Whether or not collections of these sizes exist, of course, depends on how many distinct sets are available for us to collect together in the first place.

Let's take stock of our supply of sets again. We have $\emptyset$ by fiat, but nothing else. Using the *Pairing Axiom*, though, we can generate $\{\emptyset\} = \{\emptyset, \emptyset\}$. This is new! It's a set with one element, $\emptyset$, while the *empty set* has no elements.[10]

So now we have two sets: $\emptyset$ and $\{\emptyset\}$. With these, we can form the new singleton $\{\{\emptyset\}\}$, and we can pair them to get $\{\emptyset, \{\emptyset\}\}$, which is also new. Continuing in this way, we can generate sets of increasing size and complexity. It's not clear yet whether the universe of sets so obtained is rich enough to do anything worthwhile from a mathematical point of view (it nearly is), but we can already see that the *empty set* coupled with the *Pairing Axiom* is more fruitful than we might have expected.

5.3.10 *The Power Set Axiom*

The *Axiom of Separation* allows us to form subsets, but it does not justify collecting them to form a power set. The *Pairing Axiom* will give us pairs of subsets, but this is insufficient to guarantee the existence of the power set in general. That requires a new axiom, which will then warrant the results we proved earlier about power sets (see Section 4.2).

Axiom 5.3.7: *Power Set Axiom*
$\forall S \exists P (P = \{R : R \subseteq S\})$, *i.e.*, $P = \mathcal{P}(S)$ *is a set.*

Cantor's Theorem tells us that the power set is larger than the original set, so this axiom increases the collection of sets we can prove to exist. We already have $\emptyset, \{\emptyset\}, \{\emptyset, \{\emptyset\}\}$, and so on, obtained from $\emptyset$ by using the pairing operator. The power set operator applied to $\emptyset$ gives $\mathcal{P}(\emptyset) = \{\emptyset\}$, then $\mathcal{P}(\mathcal{P}(\emptyset)) = \{\emptyset, \{\emptyset\}\}$, and so on. Nothing new arises here yet. The pairing set operator in conjunction with taking unions works slower than the power set operator in generating successive power sets, but it can eventually construct them (see Exercise 24). This is definitely not true, however, if we have an infinite base set—the power set operator yields sets that can't be gotten in any other way. We can't construct $\mathcal{P}(\mathbb{N})$ without the *Power Set Axiom*, for example, pairing and unions will only generate a finite portion of $\mathcal{P}(\mathbb{N})$ (see Exercise 25).

---

[10] More generally, a singleton $\{x\}$ is distinct from its element x, but this cannot be proved without an additional rather technical axiom, known as the *Axiom of Foundation*. This was added to Zermelo's original axiomatization by John von Neumann (1925) and reformulated by Zermelo (1930). See Section 5.3.12.

5.3.11 *The Axiom of Infinity*

None of *Set Theory's* axioms to this point produce infinite sets. Applying the power set operator recursively to our initial set $\emptyset$, the best we can do is to get sets of arbitrarily large finite size.

It's possible to model the natural numbers inside set theory using the definitions $0 = \emptyset$ and $\mathcal{S}(n) = n \cup \{n\}$ (see Exercises 32–38), but without another axiom, there is no way to collect these number sets together into an infinite set $\mathbb{N}$. What we have so far, then, is insufficient for mathematical purposes. To get sets like $\mathbb{N}$, we need an *Axiom of Infinity*. That such a result cannot be proved but must be postulated was first recognized by Zermelo.

Asserting the existence of an infinite set can be done in a way that relates to how natural numbers are modeled inside *Set Theory*. This axiom asserts the existence of *inductive sets*, i.e., *sets closed under the successor operator* that contain all natural numbers. Like the *Empty Set Axiom*, this axiom is a simple existential sentence, giving us the unconditional existence of another set, a starting point for generating other infinite sets by means of set operations. Since the axiom's formulation involves successors, we'll first define this unary operation on all sets, using von Neumann's 1923 notion of a successor.

Definition 5.3.2: *Successors*
  $\mathcal{S}(X) = X \cup \{X\}$

Axiom 5.3.8: *Axiom of Infinity; Inductive Sets*
  $\exists I(\emptyset \in I \wedge \forall X(X \in I \rightarrow \mathcal{S}(X) \in I))$, *i.e.,* ***inductive sets*** *exist.*

Inductive sets need not be unique, but based on this axiom we can define $\mathbb{N}$ as the smallest inductive set, containing all and only the successors of $\emptyset$. Since we're not committed to developing *Peano Arithmetic* within *Set Theory*, we'll leave the details to be explored in the exercises (see Exercises 32–52).

5.3.12 *Other Axioms for Set Theory*

We now have eight axioms for *Set Theory*, though not all of them are independent of one another. For example, the *Finite Union Axiom* is redundant, given the *Total Union Axiom* and the *Pairing Axiom* (see Exercise 29). Also, $\emptyset$ exists if any set does—we can separate it out via the property $x \neq x$ (see Exercise 31). Thus, the *Axiom of Infinity* implies the *Empty Set Axiom*. In addition, the *Pairing Axiom* is superfluous given the *Power Set Axiom* and some others (see Exercise 30). It's clear that we could have chosen fewer axioms than we've adopted without losing any deductive power.

On the other hand, some results in *Set Theory* still cannot be proved with these axioms. The proposition that prompted Zermelo to axiomatize *Set Theory*, the *Well-Ordering Theorem*, is a deep result that requires the *Axiom of Choice* for its proof (see Section 5.1.8). Two other more technical axioms that round out the standard axiomatization of *Set Theory* are the

Axiom of Foundation $(\forall y \neq \emptyset)(\exists x \in y)(x \cap y = \emptyset)$ which rules out unusual things like $x \in x$ and circular membership strings like $x \in y \in x$, and the *Axiom of Replacement*, which says that the functional image of a set is also a set. Given this latter axiom, the *Axiom of Separation* can be proved, so it, too, may be omitted from our list of axioms.

Taken together, these axioms form the deductive basis for *Zermelo-Fraenkel Set Theory* (ZF).[11] If the *Axiom of Choice* is included, the theory is known as ZFC. Some researchers have proposed additional axioms having to do with the existence of very large infinite cardinals, but they go beyond what we can consider here. Several excellent texts and monographs go into more depth on *Set Theory* for anyone interested in pursuing this topic further.

5.3.13 *Set Theory and The Halting Problem*

Set Theory is foundational for many mathematical theories. It's also used by *Computer Science*. Some applications use only the basic ideas considered in Chapter 4. But some areas of theoretical computer science, such as the theory of computation, draw upon more advanced parts of *Set Theory*.[12] One significant proof technique used there is that of diagonalization. We looked at this in a couple of guises, first in demonstrating the uncountability of the real numbers and later in proving *Cantor's Theorem*. There it was less clear how the argument was related to diagonalization (see Exercise 5.2.21), since it appears in a form making use of self-reference. It's this latter version of diagonalization that's prominent in foundations of mathematics and computer science. Self-reference is also a standard tactic used by modern philosophers.

So, in this concluding subsection, we'll look at how diagonalization is used in computer science for dealing with the *Halting Problem*. We'll do this informally, since it would take us too far afield to define precisely what a computer program is and describe how it can be numerically encoded so that a computer can implement it. Our approach will be simple enough that someone with only a basic familiarity with computers and programming should be able to follow it. However, the argument given below can be formalized and made perfectly rigorous.

Computer programmers want to make sure that the programs they design behave as they should for all possible inputs. There is always the possibility that a program will work well for most values but fail spectacularly for an outlier. In particular, a program might go into an infinite loop for some input, so that the computer will never halt and produce an output.

It would be wonderful if a super-checker could be designed that, given any program and any input, would be able to determine whether or not

---

[11] Abraham Fraenkel offered some improvements to Zermelo's axiomatization of *Set Theory* in the early 1920s, which included the *Replacement Axiom*.

[12] Several twentieth-century pioneers in *Computer Science*, such as Alan Turing and John von Neumann, were also active researchers in *Mathematical Logic* and *Set Theory*. Others, such as Kurt Gödel, Alonzo Church, and Rózsa Péter focused on computability issues in logic, making indirect but foundational contributions to *Computer Science*.

the program halts/produces an output for that input. Can such a universal program-checker be designed? This question is known as the *Halting Problem*.

This problem was first tackled and answered in 1936 by the British computer scientist Alan Turing (Figure 5.6), probably best known to the public for his pivotal work in decoding Nazi messages for England during World War II. Turing's answer to the *Halting Problem* was that no such super-checker program exists. You no doubt suspected this was true, but Turing actually *proved* that no such program *can* exist using a proof by contradiction diagonalization argument.

Fig. 5.6 Alan Turing

Theorem: *Insolubility of the Halting Problem*
No program checker exists that will determine for any program/input pair whether the program will halt and produce an output for that input.

Proof:
· Suppose to the contrary that such a program checker H exists.
  Suppose H returns a 1 when a program P halts after a finite number of steps and a 0 when it doesn't, i.e., $H(P,x) = 1$ when program P halts for input x, and $H(P,x) = 0$ otherwise.
· Since computer programs are formulated as character strings that, like inputs, get encoded as sequences of bits for the computer, programs may themselves be treated as input data.
· Thus, we can consider what happens when H checks program P for input P: $H(P,P) = 1$ if and only if P halts when fed P as input.
· Let's now design a procedure C that acts on programs in the following contrarian way: if $H(P,P) = 0$, $C(P)$ will halt; while if $H(P,P) = 1$, C will go into an infinite loop and fail to halt.
  In other words, using H to do its checking, C halts for a program P acting on input P if and only if P itself does not halt with that input.
· We'll now employ a self-referential maneuver inspired by Cantor's diagonalization argument/*Russell's Paradox*.
  Question: What does C do when given itself as the program to act upon?
  Answer: C acting on C halts if and only if C acting on C doesn't halt.
  This is a blatant contradiction of the form $Q \leftrightarrow \neg Q$.
· Thus, we must reject the original supposition: no universal program checker H for halting exists.   ∎

The *Halting Theorem* moves us into the field of *Computability Theory*, an area of research with links to other parts of *Computer Science* and *Mathematical Logic*. We won't explore these connections any further here.

EXERCISE SET 5.3

Exercises 1–4: *Russell's Paradox*
The following problems deal with various aspects of Russell's Paradox.

5.3.1. *Russell's Paradox and the Barber Paradox*
 a. Legend has it that a barber in *Russellville* shaves all and only those who do not shave themselves. If this is so, who shaves the barber? Explain.
 b. Compare the *Barber Paradox* in part *a* with *Russell's Paradox*. How are they similar?

5.3.2. *Russell's Paradox vs. Theorem 1*
Why does the set $\mathcal{N}$ formed in connection with *Russell's Paradox* force us to question the consistency of *Set Theory*, while the same construction used to prove *Theorem 1* simply shows that no such set exists?

5.3.3. *No Absolute Universal Set Exists*
Show that adopting $\exists U \forall x (x \in U)$ as an axiom of *Set Theory* leads to *Russell's Paradox*, contradicting *Theorem 1* and making *Set Theory* inconsistent.

5.3.4. *Russell's Paradox and Cantor's Theorem*
Review the proof of *Cantor's Theorem*, and explain what happens if the set used as a base set there is allowed to be the "set" of all sets S. Do you see how Russell might have been led to consider his paradoxical set $\mathcal{N}$?

Exercises 5–7: True or False
Are the following statements true or false? Explain your answer.

5.3.5. The existence of all the elements belonging to a given class guarantees the existence of that class as a set.

5.3.6. *Set Theory* was axiomatized in order to deal with *Russell's Paradox*.

5.3.7. Cantor was the first mathematician to propose an axiomatization of *Set Theory*.

Exercises 8–9: Properties of Interior Empty Sets
Let $\emptyset_U = \{x \in U : x \neq x\}$ for any set U. Then prove the following, without using the Empty Set Axiom.

5.3.8. *Empty Set Membership Criterion*
  $\forall x (x \notin \emptyset_U)$

5.3.9. *Uniqueness of the Empty Set*
  If U and V are any sets, then $\emptyset_U = \emptyset_V$.

Exercises 10–15: Existence and Uniqueness of Sets
Use the Axiom of Separation and the Axiom of Extensionality (along with any given axiom) to prove the following existence and uniqueness results.

5.3.10. *Sets Formed by Separation are Unique*
 a. *Separated Sets*: $\forall U \exists! S (S = \{x \in U : P(x)\})$.
 b. *Intersections*: $\forall S \forall T \exists! I (I = \{x \in S : x \in T\})$.
 c. *Set Differences*: $\forall S \forall T \exists! D (D = \{x \in S : x \notin T\})$.

5.3.11. *Finite Unions*
$$\forall S \forall T \exists! U (U = \{x : x \in S \vee x \in T\})$$

5.3.12. *Total Intersections*
If C is any non-empty collection of sets, its total intersection $\bigcap\limits_{S \in C} S$ exists and is unique.

5.3.13. *Total Unions*
If C is any non-empty collection of sets, its total union $\bigcup\limits_{S \in C} S$ exists and is unique.

5.3.14. *Pairs*
$$\forall x \forall y \exists! P (P = \{z : z = x \vee z = y\})$$

5.3.15. *Power Sets*
$$\forall S \exists! P (P = \{R : R \subseteq S\})$$

5.3.16. *Set Equality from Above*
Prove that $S = T$ if and only if they are members of exactly the same sets U, i.e., prove $S = T \leftrightarrow \forall U (S \in U \leftrightarrow T \in U)$. *Hint*: use what you know about singletons to prove one direction.

Exercises 17–21: Defining Tripletons, Quadrupletons, and N-pletons
The following problems explore ways to define tripletons and other size sets.

5.3.17. Give two distinct ways to define the notion of a *tripleton*, using an appropriate union of doubletons or singletons or both.

5.3.18. Taking your favorite definition from Exercise 17 as the official definition, formulate and then prove the proposition giving the set membership criterion for tripletons: $x \in \{a, b, c\}$ if and only if _____ .

5.3.19. Take the definition from Exercise 17 that you did *not* use in Exercise 18 and show that the set so defined is equal to the one officially defined as the tripleton. You may assume the result of Exercise 18 to prove this identity.

5.3.20. What do you think would be a natural definition for a quadrupleton (a set with four elements)? Would you use two doubletons for this? Or should your definition use tripletons? Explain.

5.3.21. Give a recursive definition of an n-pleton for any positive integer $n \geq 2$ to standardize the process you developed in Exercises 17–20.

Exercises 22–23: Defining Singletons and Doubletons
The following exercises are intended as an alternative way to define singletons and doubletons, so base all your work on Exercise 22a. Do not assume any material from the text unless you can demonstrate its legitimacy here.

5.3.22. *Defining Singletons as Primitive*
  a. Formulate an axiom that affirms the existence of singletons directly.
  b. State and prove a uniqueness proposition for singletons. Then define singletons using the standard notation.
  c. State and prove the set membership criterion for singletons.

5.3.23. *Doubletons from Singletons*

a. Given your definition of singletons from Exercise 22, define doubletons in terms of them and whatever else you need. What guarantees that doubletons are unique?

b. State and prove the set membership criterion for doubletons.

c. Compare the approach taken in the text with what you did in Exercises 22–23ab. What are the relative merits of each approach?

Exercises 24–25: Generating Power Sets Using Pairs and Unions

The following problems explore how much of a power set can be obtained using pairs and unions.

5.3.24. *Power Sets Obtained by Pairing*

Indicate how the pairing operation and union can be used on $\emptyset$ to generate both $\{\emptyset, \{\emptyset\}\}$ and its power set, $\{\emptyset, \{\emptyset\}, \{\{\emptyset\}\}, \{\emptyset, \{\emptyset\}\}\}$.

5.3.25. *The Power Set of* $\mathbb{N}$

a. Let $A_1, A_2, \ldots, A_n$ be subsets of $\mathbb{N}$. Explain how $\{A_1, A_2, \ldots, A_n\}$ can be obtained using the pairing and union operators. Thus, finite portions of $\mathcal{P}(\mathbb{N})$ can be obtained without the power set operator.

b. Tell what's wrong with the following argument, purporting to show that a countably infinite portion of $\mathcal{P}(\mathbb{N})$ can be gotten using pairs and unions. If it were valid, could this be used to obtain all of $\mathcal{P}(\mathbb{N})$? To obtain $\mathbb{N}$?
Let P_k denote $\{A_1, A_2, \ldots, A_k\}$, which can be gotten using pairing and union. Then the infinite collection of subsets $\{A_1, A_2, \ldots A_n, \ldots\} = \bigcup_{k=1}^{\infty} P_k$ results from taking the total union of these sets.

Exercises 26–28: Defining Ordered Pairs and Ordered Triples

The following explore potential definitions for ordered pairs and triples.

5.3.26. *Set-Theoretic Definitions of Ordered Pairs*

a. Show that $(x, y) = \{x, \{y\}\}$ would not be a good definition for ordered pairs. *Hint*: show that two different ordered pairs yield the same unordered pair. Use $\emptyset$, $\{\emptyset\}$, and $\{\{\emptyset\}\}$ in some combination.

b. Show that $(x, y) = \{x, \{x, y\}\}$ could be taken as a definition for ordered pairs, because it, too, leads to the fundamental result about equal ordered pairs: $(a, b) = (c, d)$ if and only if $a = c \wedge b = d$. Explain why you think this definition is not the one settled upon by Wiener or Kuratowski.

c. Wiener's original definition for ordered pairs is $(x, y) = \{\{\{x\}, \emptyset\}, \{\{y\}\}\}$. Show with this definition that the fundamental result about equal ordered pairs still holds: $(a, b) = (c, d)$ if and only if $a = c \wedge b = d$.

d. Compare Wiener's definition with the official one, due to Kuratowski: $(x, y) = \{\{x\}, \{x, y\}\}$. What are the merits of the standard definition?

5.3.27. *Ordered Triples*

One might think that ordered triples can be defined directly, in a way that is similar to the definition for ordered pairs, by $(x, y, z) = \{\{x\}, \{x, y\}, \{x, y, z\}\}$. Determine what is wrong with this definition.

5.3.28. *Existence of Cartesian Product*
Given the official definition of ordered pairs (see Exercise 26d), show how to obtain the Cartesian product $S \times T$ using the power set operator and finite unions. *Hint*: find an appropriate universe for the *Separation Axiom*.

Exercises 29–31: Redundancy of Axioms

The following look at deductive relations among Set Theory's axioms.

5.3.29. *Proving the Finite Union Axiom*
Using the *Pairing Axiom* and the *Total Union Axiom* prove the *Finite Union Axiom*, i.e., using the ability to form pairs and take total unions show how to generate $S \cup T$ from S and T.

5.3.30. *Proving the Pairing Axiom*
Using the *Finite Union Axiom*, the *Power Set Axiom*, the *Axiom of Extensionality*, and the *Axiom of Separation* prove the *Pairing Axiom*.

5.3.31. *Proving the Empty Set Axiom*
Using the *Axiom of Infinity*, the *Axiom of Separation*, and the *Axiom of Extensionality* prove the *Empty Set Axiom*.

Exercises 32–38: Peano Arithmetic Inside Set Theory

The following problems explore modeling natural number arithmetic inside Set Theory. Prove these results in order, which show among other things, that the Peano Postulates are consequences of the axioms of Set Theory. Assume the existence of $\mathbb{N}$, the set whose elements belong to all inductive sets (sets satisfying the Axiom of Infinity), take $0 = \emptyset$, and let $S(n) = n \cup \{n\}$.

5.3.32. Write the natural numbers 1, 2, 3, and 4 in the following two ways.
  a. Using set braces and the numbers 0, 1, 2, and 3.
  b. Using only set braces and the *empty set* $\emptyset$.
  c. Based on the pattern emerging in part a, what do you expect $S(n)$ to be? Can this be written as a set without ellipses? How might you formulate and prove your conjecture using the formal language of *Set Theory*?

5.3.33. $\mathbb{N}$ is the smallest inductive set, i.e., $\mathbb{N}$ satisfies the *Axiom of Infinity* and is a subset of all inductive sets. Consequently, the *Axiom of Induction* holds: $(\forall P \subseteq \mathbb{N})(0 \in P \wedge \forall n(n \in P \to S(n) \in P) \to P = \mathbb{N})$.

5.3.34. $0 \in \mathbb{N}$ **5.3.36.** $(\forall n \in \mathbb{N})(S(n) \neq 0)$
5.3.35. $(\forall n \in \mathbb{N})(S(n) \in \mathbb{N})$ **5.3.37.** $(\forall m, n \in \mathbb{N})(m \in n \to m \subseteq n)$
5.3.38. $(\forall m, n \in \mathbb{N})(S(m) = S(n) \to m = n)$

Exercises 39–47: Properties of $\mathbb{N}$'s Membership Relation

Assuming the definitions of 0 and $S(n)$ preceding Exercise 32 prove the following properties holding for the membership relation on $\mathbb{N}$. Some results can be proved via UG (and may be true of more than just natural numbers); others will require mathematical induction (use the successor form for these).
Some parts involve the exclusive or connective $\underline{\vee}$. The key result here is the trichotomy property of Exercise 45.

5.3.39. $(\forall n \in \mathbb{N})(n \in \mathcal{S}(n))$

5.3.40. $(\forall m, n \in \mathbb{N})(m \in n \rightarrow m \in \mathcal{S}(n))$

5.3.41. $(\forall m, n, p \in \mathbb{N})(m \in n \in p \rightarrow m \in p)$ [Transitivity of $\in$]

5.3.42. $(\forall m, n \in \mathbb{N})(m \in n \rightarrow \mathcal{S}(m) \in \mathcal{S}(n))$

5.3.43. $(\forall m, n \in \mathbb{N})(m \in n \rightarrow n = \mathcal{S}(m) \veebar \mathcal{S}(m) \in n)$

5.3.44. $(\forall m \in \mathbb{N})(m = 0 \veebar 0 \in m)$

5.3.45. $(\forall m, n \in \mathbb{N})(m \in n \veebar m = n \veebar n \in m)$ [Trichotomy Law]

5.3.46. $(\forall m, n \in \mathbb{N})(m \in n \leftrightarrow m \subset n)$ [Note the *proper inclusion* here.]

5.3.47. $(\forall m, n \in \mathbb{N})(m \in n \rightarrow m \in \mathbb{N})$

[*Note*: this result shows that $\mathbb{N}$ is an extension of the sequence of natural numbers, a "number" of sorts following all the natural numbers.]

Exercises 48–52: Set-Theoretic Definitions of $<$ and $\leq$ for $\mathbb{N}$

The order relation $<$ on $\mathbb{N}$ is defined by $m < n \leftrightarrow m \in n$.
The order relation $\leq$ on $\mathbb{N}$ is defined by $m \leq n \leftrightarrow m < n \vee m = n$.

5.3.48. Given the definitions of $<$ and $\leq$ translate the results of Exercises 39–45 into the language of numerical inequality.

5.3.49. Given the definitions of $<$ and $\leq$ show using the results in Exercises 45–46 that $m \leq n \leftrightarrow m \subseteq n$.

5.3.50. Given the above definitions and the results of Exercises 45–46 prove that $m \leq n \wedge n \leq m \rightarrow m = n$ for any natural numbers m and n.

5.3.51. Show that $\mathbb{N}$ is *well-ordered*, i.e., show that any non-empty set S of natural numbers has a least element m in the sense that $m \leq x$ for every element $x \in S$.

Hint: note that the intersection of two natural numbers is always the smaller one, and use the relevant results of Exercises 39–47 and the above problems.

5.3.52. Discuss the set-theoretic approach to *Peano Arithmetic* outlined in Exercises 33–51. What do you think about how the natural numbers and $\mathbb{N}$ are defined? What aspects of these definitions or the subsequent theoretical developments, if any, bother you? What value might such an approach have?

Exercises 53–54: The Halting Problem

The following problems deal with the Halting Problem.

5.3.53. Speedy Gonzalez claims that given today's advanced technology any computer program that will produce an output for a given input will do so in less than 10,000 hours. Comment on his claim based on what you know about the *Halting Problem*.

5.3.54. Ellen Touring claims to have solved the *Halting Problem* using the following procedure S: *Given a program P and an input x, let P run with input x. If it produces an output, let $S(P, x) = 1$; otherwise let $S(P, x) = 0$.* Explain what's wrong with her solution.

Chapter 6
Functions and Equivalence Relations

6.1 Functions and Their Properties

Modern science and contemporary culture are unthinkable without mathematics. Quantitative thinking, mathematical ideas, algorithmic techniques, and symbolic reasoning permeate the ways we interact with the world around us. After numbers and shapes, *functions* are the most pervasive mathematical tool used in scientific theories and applications. Functions capture causal connections between variable scientific measures, and they are used to compare algebraic structures in advanced mathematics.

Our interest in functions is different from that of elementary algebra and calculus. We won't be graphing functions or calculating extreme values or finding areas under a curve. We'll investigate some algebraic features pertinent to discrete mathematics applications and more advanced topics in mathematics. Some discrete mathematics texts also explore algorithmic properties of functions, such as the complexity of calculations needed to compute function values, but we'll not pursue that here.

After two sections on functions, we'll introduce the idea of a *relation*. The key notions here are *equivalence relation* and *quotient structure*, which connect to some earlier ideas and provide a theoretical basis for later material. The final section applies these ideas to constructing number systems.

6.1.1 *Historical Background: What's a Function?*

The concept of a function first emerged near the end of the seventeenth century. Leibniz coined the term *function*, but others also worked with the idea. Earlier scientists had made use of direct and inverse proportionality, whose roots go back to ancient times, but the function concept expanded this idea. A quantity was considered a function of other quantities if it depended upon them in a way that could be expressed by a computational formula. Algebra, geometry, and especially calculus teased out connections among an equation's variable quantities.

In the eighteenth century, function formulas could also involve infinite sums and products, even though convergence criteria were initially lacking. This expanded the realm of functions beyond polynomials and rational functions. Trigonometric functions, for example, could be represented using infinite power series.

Mathematicians have long associated functions with graphs. A function was considered continuous if its graph was a connected curve. But did every continuous graph represent a function? Disputes over the nature of functions

© Springer Nature Switzerland AG 2019
C. Jongsma, *Introduction to Discrete Mathematics via Logic and Proof*,
Undergraduate Texts in Mathematics,
https://doi.org/10.1007/978-3-030-25358-5_6

broke out in the mid-eighteenth century in connection with solving differential equations for wave phenomena in physics. Even stranger functions arose in Fourier's early nineteenth-century work on heat. By mid-century, however, a fairly general function concept had emerged. This is what we'll explore below.

6.1.2 Toward a Definition of Function

One way to think about a function is as a machine that, given an appropriate *input*, produces a uniquely determined *output*. If x stands for the input and y for the output, we write $y = f(x)$ to indicate that output y is produced by function f from input x.

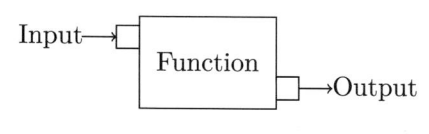

People occasionally get lazy and treat $f(x)$ as if *it* were the function, but the function f should be distinguished from its output $f(x)$. As the directed connection between inputs and outputs, a function can be considered a mathematical entity in its own right. How outputs are calculated is irrelevant from an abstract mathematical perspective (though that might be the main interest in some settings). What's important about a function is that, and how, the function assigns output values to the inputs. We thus have the following informal definition of a function, illustrated by a function diagram.

Definition 6.1.1: *Function (Informal Version)*
  a) *f is a **function** from a set D to a set C if and only if f is a correspondence assigning a unique value $f(x)$ in C to each value x in D; in symbols, $f : D \to C$ if and only if $(\forall x \in D)(\exists! y \in C)(y = f(x))$.*
  b) *The set D of all inputs is called the **domain** of the function. The set C of all potential outputs is called the **codomain** of the function; the set of all actual outputs inside C is called the **range** of the function.*
  c) *Functions are also called **maps** or **transformations**. If $y = f(x)$, y is called the **value** of f at x or the **image** of x under f, and x is called the **preimage** of y. The functional element assignment can also be indicated by the notation $f : x \mapsto y$.*

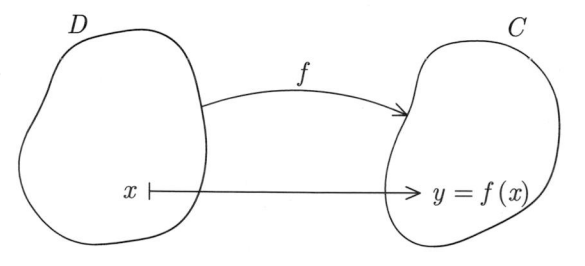

Functions are correspondence relations in which every input value in the domain corresponds to some definite output value in the codomain. They may have other important features—a causal relationship between the variables, a single formula, a continuous graph—but what defines them as functions is their being the right sort of association between sets of values.

✜ Example 6.1.1

Explain why the following define functions.

a) The angle formula for converting degrees into radians: $r = \dfrac{\pi}{180} d$.

b) $P(x) =$ the refund/income tax payment calculated for the current year owed to/by the person whose social security number is x.

Solution

a) The conversion formula $r = \dfrac{\pi}{180} d$ stipulates r as a function of d. The opposite conversion formula $d = \dfrac{180}{\pi} r$ specifies the same relationship, but this time from the other direction, so it defines a different function. In both cases, the domain and codomain can be taken to be $\mathbb{R}$.

b) Because a definite amount of money is either owed to or by each person, P is a function from the set of social security numbers to the set of two-decimal-place numbers. P is a function even though there is no formula for calculating $P(x)$. In fact, such an equation would be useless as a formula, because it would only summarize a discrete set of data, not cover any new cases that might arise.

✜ Example 6.1.2

Determine whether the following define functions. Unless specified, assume that all variables range over real numbers.

a) $x - 3y = 7$

b) $s = 16t^2$

c) $x^2 + y^2 = 1$

d) $\chi_S(x) = \begin{cases} 0 : x \notin S \\ 1 : x \in S \end{cases}$ for any $S \subseteq \mathbb{R}$

e) $M(S)$ is the minimum value of S, $\emptyset \neq S \subseteq \mathbb{N}$.

Solution

a) The formula $x - 3y = 7$ defines y as a function of x. The formula $y = (x - 7)/3$ gives y's unique value explicitly for each preimage x.

b) The formula $s = 16t^2$ also defines a function from $\mathbb{R}$ to $\mathbb{R}$. Given any real value t, a unique real number s is associated with it by s.

c) $x^2 + y^2 = 1$ may or may not define a functional relation.
  · First of all, it does not define a function from $\mathbb{R}$ to $\mathbb{R}$ because not all real numbers can be inputs. If $|x| > 1$, $x^2 + y^2 \neq 1$. We can fix this by restricting x to the closed interval $[-1, 1]$.
  · However, $x^2 + y^2 = 1$ still does not define a function from $[-1, 1]$ to $\mathbb{R}$. The problem now is that some x are associated with more than one y. For example, when $x = 0$, $y = \pm 1$ both satisfy the equation. We can remedy this by ruling out negative y-values.

- The formula $x^2 + y^2 = 1$ now defines y as function of x from the closed interval $[-1, 1]$ into the nonnegative real numbers $\mathbb{R}^+ \cup \{0\}$.
- As this example shows, whether an equation defines a function depends on what the domain and codomain are.

d) $\chi_S : \mathbb{R} \to \mathbb{R}$ defines the *characteristic function of S*, assigning 0 (no) to numbers outside S and 1 (yes) to those inside S. It's defined piecewise, but that doesn't make it any less a function.

- $\chi_{\mathbb{Q}}$ is a well-known function due to Dirichlet, the nineteenth-century mathematician who first proposed thinking of functions as correspondences. Dirichlet's function is often discussed when Riemann's theory of integration (used in introductory calculus) is being generalized.

e) This description gives a function $M : \mathcal{P}(\mathbb{N}) - \{\emptyset\} \to \mathbb{N}$. Each non-empty subset $S \subseteq \mathbb{N}$ has a unique least element $M(S)$.

To graph a real-valued function of a real variable, we plot all (x, y) in the usual way for x in D and $y = f(x)$ in C. Outputs of functions are unique, so the following *vertical-line criterion* tells when a curve is the graph of a function: *a graph is the graph of a function if and only if for each input x, a vertical line drawn at x passes through exactly one point of the graph.*

This criterion gives us an informal way to determine whether something is a function, but it's no substitute for a rigorous argument. For one thing, graphs are never made by plotting outputs for all possible inputs but are drawn by extrapolating from a finite amount of information.

The notion of a graph suggests another way to think about functions, however. A function is represented by graphing ordered pairs, so the set of these ordered pairs captures the action of the function without bringing in the idea of a correspondence. This leads into the following more formal definition of a function. We'll not use this definition much in our work (it takes time to get used to thinking about functions as sets of ordered pairs), but it illustrates a tendency we identified earlier, treating everything in terms of *Set Theory*.[1]

Definition 6.1.2: *Function (Formal Set-Theoretic Version)*
*F is a **function** from D to C if and only if $F \subseteq D \times C$ such that every x in D is the first element of some unique pair (x, y) belonging to F; in symbols, $F : D \to C$ if and only if $F \subseteq D \times C$ and $(\forall x \in D)(\exists! y \in C)((x, y) \in F)$.*

6.1.3 Definition of Function, Equality of Functions

There are two ways to approach function equality. A minimalist view considers a function as an input-output assignment. Then functions are equal if and only if the same inputs yield the same outputs. In this approach, the domains and ranges of equal functions are identical, while codomains are ignored.

This is not completely satisfactory. To define function composition in Section 6.2, we only need the codomain of the first function to agree with the

---

[1] *Category Theory*, a recent branch of *Abstract Algebra*, takes a completely opposite approach. There functions are taken as undefined, and they are used to help define sets.

domain of the second. A function's range is less important here than its codomain. Furthermore, a function's range may be difficult to determine, though we can easily specify a codomain. Finally, unless we distinguish range and codomain, all functions will automatically be onto functions (see below).

We'll insist on including the codomain of a function in conceptualizing a function. Our approach, favored in algebraic circles, is given in the following definition and proposition.

Definition 6.1.3: *Function* (*Final Version*)
*A **function** consists of a **domain** D, a **codomain** C, and an **assignment** $f : D \to C$ such that each x in D corresponds to a **unique** $y = f(x)$ in C.*

Proposition 6.1.1: *Equality of Functions*
*Functions $f_1 : D_1 \to C_1$ and $f_2 : D_2 \to C_2$ are **equal**, denoted by $f_1 = f_2$, if and only if $D_1 = D_2$, $C_1 = C_2$, and $(\forall x \in D_1)(f_1(x) = f_2(x))$.*

Proof:
This is an immediate consequence of the equality of ordered triples. ∎

✤ **Example 6.1.3**
Show that the following real-valued functions of a real variable are not equal.
a) $2^x \neq x^n$ for any $n \in \mathbb{N}$.
b) $\lfloor x + 1 \rfloor \neq \lceil x \rceil$.

Solution
The domains and codomains agree, so we must show that the actions differ.
a) Let $f(x) = 2^x$ and $g_n = x^n$.
   Since $f(0) = 1 \neq 0 = g_n(0)$ and $f(1) = 2 \neq 1 = g_n(1)$, $f \neq g_n$. ✓
   · But f and g_n disagree even if we exclude these inputs from our domain. For $f(2) = 2^2$ and $g_n(2) = 2^n$. If these are equal, $n = 2$. However, $2^x \neq x^2$ when $x = 1$.
   Thus, $2^x \neq x^n$; exponential functions are not power functions. ✓
b) $\lfloor \; \rfloor$ is the *floor function*, returning the greatest integer less than or equal to its input, while $\lceil \; \rceil$ is the *ceiling function*, yielding the least integer greater than or equal to its input.
   · Letting $x = 0$, we get $1 = \lfloor 0 + 1 \rfloor \neq \lceil 0 \rceil = 0$. In fact, these two functions disagree on all integer values, though they agree everywhere else. ✓

One rarely needs to show that two particular functions are equal. Showing they are unequal is more common, usually because the actions differ.

However, to develop a mathematical theory for functions, we'll often need to show that functions formed in certain ways are equal (see Section 6.2; also Exercises 26–27 on decomposing real-valued functions of a real variable).

6.1.4 One-to-One and Onto Functions

Functions have two very important algebraic properties: being *one-to-one* and being *onto*. While we'll define and illustrate these properties here, their significance will become more apparent in Section 6.2.

Definition 6.1.4: *One-to-One Functions*

*A function f is **one-to-one** (an **injection**) if and only if distinct inputs produce distinct outputs, i.e., if and only if $x_1 \neq x_2 \rightarrow f(x_1) \neq f(x_2)$; equivalently, via Contraposition, if and only if $f(x_1) = f(x_2) \rightarrow x_1 = x_2$.*

According to this definition, a function is one-to-one if and only if every element of the codomain has *at most one* preimage in the domain. A codomain element need not be related to any input, but if there is one, it must be unique.

On the other hand, all functions have unique outputs, whether or not they are one-to-one. This is part of the definition, so *be careful not to confuse* the uniqueness requirement for being a function (having *unique images*/being *many-to-one*) with the *additional uniqueness requirement* of being *one-to-one* (having *unique preimages*).

If a function is given by an equation, you can sometimes determine if it's one-to-one by solving the equation for x in terms of y and seeing how many solutions there are. As we'll see in Section 6.2, this strategy is also valuable for calculating a function's inverse, if it has one.

The idea of being one-to-one is captured by the *horizontal-line criterion: A function is one-to-one if and only if a horizontal line through any y-value of its codomain meets the graph in at most one point*. Again, however, this criterion *is not a rigorous method* for proving a function is one-to-one.

✤ **Example 6.1.4**

Determine whether the following functions are one-to-one (injective).
a) $f \colon \mathbb{R} \to \mathbb{R}$, defined by $y = f(x)$, where $x - 3y = 7$.
b) $s \colon \mathbb{R} \to \mathbb{R}$, defined by $s(t) = 16t^2$.
c) $M \colon \mathcal{P}(\mathbb{N}) \to \mathbb{N}$, defined by $M(S) =$ the minimum value of $S \subseteq \mathbb{N}$.
d) $\lfloor \ \rfloor \colon \mathbb{R} \to \mathbb{Z}$, defined by $\lfloor x \rfloor =$ the greatest integer $n \leq x$.

Solution

a) The function f is one-to-one.
   - First note that $f(x) = \dfrac{x - 7}{3}$, and suppose $f(x_1) = f(x_2)$.
     Then $\dfrac{x_1 - 7}{3} = \dfrac{x_2 - 7}{3}$.
     Multiplying this equation by 3 and then adding 7 gives $x_1 = x_2$.
     Hence, f is one-to-one. ✓
   - This can also be argued by solving $x - 3y = 7$ for x.
     Let y be any element in the range of f.
     Then there is exactly one x associated with it: $x = 7 + 3y$.
     So f is one-to-one. ✓
b) The position function s is not one-to-one.
   This is because two inputs give the same output: $f(-1) = 16 = f(1)$. ✓
   - In fact, s is nearly a two-to-one function: only 0 has a unique preimage.
   - However, if the domain of s were all nonnegative real numbers, then s would be one-to-one. The preimage of a given real number y would either be nonexistent (for $y < 0$) or it would be given by $t = \sqrt{y}/4$.

c) The least number function M is not one-to-one.
  · Let $S_1 = \{0\}$ and $S_2 = \{0, 1\}$. The minimum element of both sets is 0, so different preimages have the same image. ✓
  · In fact, each image (minimum value) has infinitely many preimages, so the function is infinitely-many-to-one.

d) The greatest integer function $\lfloor \ \ \rfloor$ is also not one-to-one. For example, all real numbers x in the interval $[0, 1)$ return $\lfloor x \rfloor = 0$ as an output. ✓

Definition 6.1.5: *Onto Functions*

*A function $f : D \to C$ is **onto** (a **surjection**) if and only if every element of C has a preimage, i.e., if and only if $(\forall y \in C)(\exists x \in D)(f(x) = y))$.*

Thus, f is *onto* if and only if the range of f is the same as its codomain, for every element y of the codomain has *at least one* preimage x in the domain.

The *Method of Analysis* can often be used to show that f is onto: assume that an arbitrary y in the codomain is related to some x and then find out what x must be, checking that x belongs to the domain and yields y.

There is a *horizontal-line criterion* for being an onto function: *a horizontal line through any y-value of f's codomain must meet its graph in at least one point.* Again, this criterion is not a rigorous method of proof.

✦ **Example 6.1.5**
Determine whether the following functions are onto (surjective).
a) $f : \mathbb{R} \to \mathbb{R}$ defined by $y = f(x)$, where $x - 3y = 7$.
b) $s : \mathbb{R} \to \mathbb{R}$ defined by $s(t) = 16t^2$.
c) $M : \mathcal{P}(\mathbb{N}) \to \mathbb{N}$, defined by $M(S) =$ the minimum value of $S \subseteq \mathbb{N}$.
d) $\lfloor \ \ \rfloor : \mathbb{R} \to \mathbb{Z}$, defined by $\lfloor x \rfloor =$ the greatest integer $n \leq x$.

Solution
a) The function f defined by $f(x) = y = \dfrac{x - 7}{3}$ is an onto function.
  Let y be any real number.
  Since $x = 7 + 3y$ gives x in terms of y and since x is real when y is, x lies in the domain of the function.
  This x is the preimage of y: $f(7 + 3y) = \dfrac{(7 + 3y) - 7}{3} = y$.
  So f is onto. ✓

b) The function s is not onto.
  This is because no negative number has a preimage in $\mathbb{R}$. ✓
  · However, if the codomain were restricted to nonnegative real numbers, s would be onto. Whether a function is onto, then, just like whether a function is one-to-one, depends upon the domain and codomain.

c) The function M is surjective.
  Every $n \in \mathbb{N}$ has a preimage: $M(\{n\}) = n$. ✓

d) The floor function is also onto. For any integer y, $\lfloor y \rfloor = y$. ✓

Definition 6.1.6: *One-to-One-and-Onto Functions*

*A function $f : D \to C$ is **one-to-one-and-onto** (a **bijection**) if and only if f is a one-to-one function from D onto C (an injection and a surjection).*

We've already worked with one-to-one-and-onto functions. We called them *one-to-one correspondences* in our work with numerosity in Chapter 5. Here's a reformulation of two definitions from there.

$S \sim T$ if and only if there is a bijective function $f : S \to T$.

$S \preceq T$ if and only if there is an injective function $f : S \to T$.

If a function f is one-to-one-and-onto, then for any y in f's codomain there is *exactly one* x in f's domain that corresponds to it. If f is presented via an equation, we may be able to show that it's one-to-one-and-onto by solving the equation for x in terms of y and arguing that each x is unique.

The graphic criteria given above can be combined to give a *horizontal-line criterion* for bijective functions: *A function is one-to-one-and-onto if and only if a horizontal line passing through any y-value of the codomain meets the function's graph in exactly one point.* This criterion helps us identify bijective functions informally, but a rigorous argument requires using the definition.

✜ **Example 6.1.6**

Determine which of the following functions are one-to-one-and-onto.

a) $f : \mathbb{R} \to \mathbb{R}$ defined by $y = f(x)$, where $x - 3y = 7$.

b) $s : \mathbb{R} \to \mathbb{R}$ defined by $s(t) = 16t^2$.

c) $M : \mathcal{P}(\mathbb{N}) \to \mathbb{N}$, defined by $M(S) =$ the minimum value of $S \subseteq \mathbb{N}$.

d) $\lfloor \ \rfloor : \mathbb{R} \to \mathbb{Z}$, defined by $\lfloor x \rfloor =$ the greatest integer $n \le x$.

Solution

a) The function f is both one-to-one and onto. See Examples 4 and 5.

b) Examples 4 and 5 show that s is neither one-to-one nor onto.

   If the domain and codomain were both restricted to nonnegative reals, though, s would be one-to-one and onto.

c) The function M is onto but not one-to-one.

d) The greatest integer function $\lfloor \ \rfloor$ is also onto but not one-to-one.

✜ **Example 6.1.7**

Show that $f(x) = \dfrac{x}{x + 3}$ is one-to-one onto its range; then find the range.

Solution

· No domain is given; we'll assume that it's as large as possible: $\mathbb{R} - \{-3\}$.

· To prove that f is one-to-one, suppose $\dfrac{x_1}{x_1 + 3} = \dfrac{x_2}{x_2 + 3}$.

  Cross-multiplying and simplifying yields $x_1 = x_2$.

  Thus, f is one-to-one. ✓

· f is obviously onto its range by definition.

· To determine the range, suppose $y = \dfrac{x}{x + 3}$ for $x \ne -3$.

  All such y are real numbers, but not all real numbers can be put into this form. To see which can, we'll solve this equation for x.

  Cross-multiplying, $xy + 3y = x$, so $x(1 - y) = 3y$.

Hence, $x = \dfrac{3y}{1-y}$, which is defined for all y except 1, which would give

$1 = \dfrac{x}{x+3}$, leading to $3 = 0$, which is false.

For any real number $y \neq 1$, the x-value we found will produce it:

$$\dfrac{\dfrac{3y}{1-y}}{\dfrac{3y}{1-y}+3} = \dfrac{3y}{3y+(3-3y)} = \dfrac{3y}{3} = y.$$

Thus, the only y-value that must be excluded from $\mathbb{R}$ is 1.

Hence, the range of f is $\mathbb{R} - \{1\}$. ✓

Note that $y = 1$ is the equation of the asymptote for the function f.

6.1.5 Images, Preimages, and Set Theory

The following results explore how functions interact with subsets and set operations. Additional results are in the exercise set (see Exercises 17–25).

Definition 6.1.7: *Image of a Set*
  Let $f : D \to C$ be a function from D into C with $S \subseteq D$. Then **the image of S under f** is the set of all images $f[S] = \{f(x) : x \in S\}$.

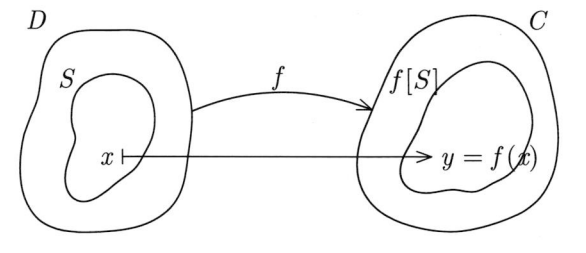

Definition 6.1.8: *Preimage of a Set*
  Let $f : D \to C$ be a function from D into C with $V \subseteq C$. Then **the preimage of V under f** is the set of all preimages $f^*[V] = \{x : f(x) \in V\}$.

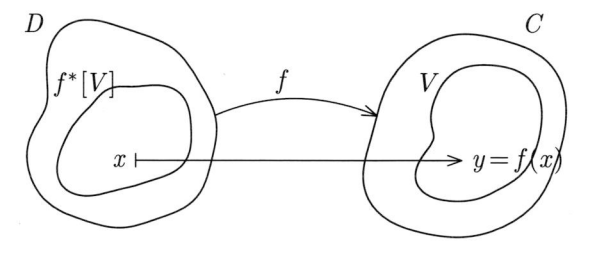

Proposition 6.1.2: *Images of Subsets*
  If $f : D \to C$ is any function with $R \subseteq S \subseteq D$, then $f[R] \subseteq f[S]$.

Proof:
  Suppose $R \subseteq S \subseteq D$, and let
  y be any element of $f[R]$.
  Then $y = f(x)$ for $x \in R$.
  But since $R \subseteq S$, $x \in S$, too.
  Hence, $y = f(x)$ is in $f[S]$.
  Therefore $f[R] \subseteq f[S]$. ∎

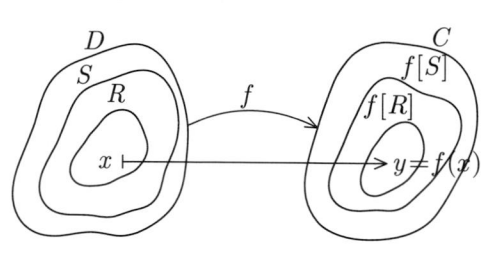

Proposition 6.1.3: *Preimages of Subsets*
  If $f : D \to C$ is any function with $U \subseteq V \subseteq C$, then $f^*[U] \subseteq f^*[V]$.

Proof:
  See Exercise 17.

Proposition 6.1.4: *Images of Unions, Intersections, Differences*
  Let $f : D \to C$ be any function with $R \subseteq D$ and $S \subseteq D$. Then:
  a) $f[R \cup S] = f[R] \cup f[S]$
  b) $f[R \cap S] \subseteq f[R] \cap f[S]$
  c) $f[R - S] \supseteq f[R] - f[S]$

Proof:
  See Exercise 19. The inclusions for intersection and set difference are as
  good as we can get; see Exercise 20.

Proposition 6.1.5: *Preimages of Unions, Intersections, Differences*
  Let $f : D \to C$ be any function with $U \subseteq C$ and $V \subseteq C$. Then:
  a) $f^*[U \cup V] = f^*[U] \cup f^*[V]$
  b) $f^*[U \cap V] = f^*[U] \cap f^*[V]$
  c) $f^*[U - V] = f^*[U] - f^*[V]$

Proof:
  See Exercise 22. Note that preimages behave differently than images.

  Besides seeing how the images and preimages interact with subsets and
set operations, we can explore how they interact with each other. Two results
hold, each of which be sharpened for certain kinds of functions (see Exercise 25).

Proposition 6.1.6: *Preimages of Images, Images of Preimages*
  Let $f : D \to C$ be any function with $S \subseteq D$, $V \subseteq C$. Then:
  a) $f^*[f[S]] \supseteq S$
  b) $f[f^*[V]] \subseteq V$

Proof:
  See Exercise 24.

EXERCISE SET 6.1

Exercises 1–4: Some Simple Functions
Determine whether the following are functions and whether they are one-to-one or onto. Justify your answers using the relevant definitions.

6.1.1. Let $S(x) = x + 1$ be the successor function on $\mathbb{N}$.
a. Explain why S is a function from $\mathbb{N}$ to $\mathbb{N}$.
b. Is S one-to-one? onto? one-to-one-and-onto?

6.1.2. Let $B = \{0, 1\}$ and $B^2 = B \times B$. Define $f : B^2 \to B$ by $f(x, y) = x \cdot y$.
a. Explain why f defines a function from B^2 to B.
b. Is f one-to-one? onto? one-to-one-and-onto?

6.1.3. Let $f(x) = 2x + 4$.
a. Explain why f is a function from $\mathbb{R}$ to $\mathbb{R}$.
b. Is f one-to-one? onto? one-to-one-and-onto?

6.1.4. Let $A(x) = |x|$.
a. Explain why A is a function from $\mathbb{R}$ to $\mathbb{R}$. Give a piecewise-defined formula for $A(x)$, taking the cases $x \geq 0$ and $x < 0$.
b. Is A an injection? a surjection? a bijection?

Exercises 5–8: True or False
Are the following statements true or false? Explain your answer.

6.1.5. $\lfloor x + y \rfloor = \lfloor x \rfloor + \lfloor y \rfloor$
6.1.6. $\lceil xy \rceil = \lceil x \rceil \cdot \lceil y \rceil$
6.1.7. Functions are equal if and only if domains, ranges, and actions agree.
6.1.8. A function is one-to-one if and only if each input has a unique output.

Exercises 9–13: More Complex Functions
Work the following problems on more complex functions. Justify your answers using the relevant definitions.

6.1.9. *Quadratic Polynomials*
Let $p(x) = x^2 - 3x + 2$.
a. Explain why $p(x)$ defines a function $p : \mathbb{R} \to \mathbb{R}$. Is p one-to-one? onto? one-to-one-and-onto?
b. Explain why $p(x)$ defines a function $p : \mathbb{R} \to [-1/4, +\infty)$. Is p one-to-one? onto? one-to-one-and-onto?   *Hint:* complete the square.
c. Explain why $p(x)$ defines a function $p : [3/2, +\infty) \to [-1/4, +\infty)$. Is p one-to-one? onto? one-to-one-and-onto?   *Hint:* complete the square.
d. What do parts *a–c* say about one-to-one and onto functions?

6.1.10. *Cubic Polynomial*
a. Explain why $c(x) = 2x^3 - 1$ defines a function $c : \mathbb{R} \to \mathbb{R}$.
b. Is c one-to-one? onto? one-to-one-and-onto?

6.1.11. Determine functions $f : \mathbb{N} \to \mathbb{N}$ that satisfy the following conditions:
a. Neither one-to-one nor onto.
b. One-to-one but not onto.

c. Onto but not one-to-one.

d. One-to-one and onto.

6.1.12. *Functions on Finite Sets*

Let $f : S \to S$ be a function from a finite set S to itself. Answer the following, giving an example or an argument.

a. Can f be one-to-one and not onto?

b. Can f be onto and not one-to-one?

c. Answer parts a and b if the domain and codomain are distinct finite sets.

6.1.13. *Residues modulo n*

Let $\mathbb{Z}$ be the set of integers and $\mathbb{Z}_n = \{0, 1, \ldots, n-1\}$, $n \geq 1$.

Define $r : \mathbb{Z} \to \mathbb{Z}_n$ by $r(m) =$ the remainder when m is divided by n.

a. Explain why r is a function from $\mathbb{Z}$ to $\mathbb{Z}_n$.

b. Is r a one-to-one function? onto? one-to-one-and-onto?

Exercises 14–15: Numbers of Functions

Count the number of functions of the types indicated. Explain your answers.

6.1.14. Consider the set of all functions from $S = \{1, 2, 3\}$ to $T = \{1, 2\}$.

a. List all such functions in ordered-pair format and name them f_1, f_2, etc. How many functions are there from S into T?

b. Classify each function in part a as one-to-one, onto, both, or neither.

6.1.15. Let $F = \{f \mid f : S \to T\}$, where $S = \{1, 2, \ldots, m\}$, $T = \{1, 2, \ldots, n\}$.

a. How many functions are there from S into T? Explain.

b. F is often denoted by T^S. Is this a good notation? Why?

c. How many of the functions in part a are one-to-one? onto? one-to-one-and-onto? *Hint*: use Chapter 4's counting formulas.

6.1.16. *Function Ranges*

a. Prove that if D is countably infinite and $f : D \to C$ is one-to-one, then its range $f[D]$ is countably infinite.

b. Is the conclusion of part a true for onto functions?

c. If f is an injection, how is the cardinality of the range $f[D]$ related to that of D (any size set)? Explain.

Exercises 17–23: Images and Preimages

The following problems explore properties of images and preimages.

6.1.17. *Proposition 3*

Prove that *If $f : D \to C$ is a function with $U \subseteq V \subseteq C$, then $f^*[U] \subseteq f^*[V]$.*

6.1.18. *Sufficient Conditions for a Strengthened Proposition 3*

Can equality be asserted in *Proposition 3*, yielding $f^*[U] = f^*[V]$, if f is one-to-one? onto? If so, prove it; if not, give a counterexample.

6.1.19. *Proposition 4*

Let $f : D \to C$ be any function with $R \subseteq D$ and $S \subseteq D$. Prove the following:

a. $f[R \cup S] = f[R] \cup f[S]$

b. $f[R \cap S] \subseteq f[R] \cap f[S]$

c. $f[R - S] \supseteq f[R] - f[S]$

6.1.20. *Strengthened Versions of Proposition* 4

a. Show that set equality fails for *Proposition* 4b: find sets R and S so that $f[R] \cap f[S] \not\subseteq f[R \cap S]$. Does equality hold if f is one-to-one? onto? If so, prove it; if not, give a counterexample.

b. Show that equality fails for *Proposition* 4c: find sets R and S so that $f[R - S] \not\subseteq f[R] - f[S]$. Does equality hold if f is one-to-one? onto? If so, prove it; if not, give a counterexample.

6.1.21. *Images of Disjoint Sets*

Prove or disprove the following. If a result is false, give a counterexample and tell why it's false. If requiring f to be one-to-one or onto will convert a false statement into a true one, add that condition and then prove the result.

a. Let $f : D \to C$, $R \subseteq D$, and $S \subseteq D$. Then $f[R] \cap f[S] = \emptyset$ if $R \cap S = \emptyset$.

b. Let $f : D \to C$, $R \subseteq D$, and $S \subseteq D$. Then $R \cap S = \emptyset$ if $f[R] \cap f[S] = \emptyset$.

6.1.22. *Proposition* 5

Let $f : D \to C$ be any function with $U \subseteq C$ and $V \subseteq C$. Prove the following:

a. $f^*[U \cup V] = f^*[U] \cup f^*[V]$

b. $f^*[U \cap V] = f^*[U] \cap f^*[V]$

c. $f^*[U - V] = f^*[U] - f^*[V]$

6.1.23. *Preimages of Disjoint Sets*

Prove or disprove the following. If a result is false, give a counterexample and tell why it's false. If requiring f to be one-to-one or onto will convert a false statement into a true one, add that condition and then prove the result.

a. If $f : D \to C$, $U \subseteq C$, and $V \subseteq C$, $U \cap V = \emptyset$ only if $f^*[U] \cap f^*[V] = \emptyset$.

b. If $f : D \to C$, $U \subseteq C$, and $V \subseteq C$, $U \cap V = \emptyset$ if $f^*[U] \cap f^*[V] = \emptyset$.

Exercises 24–25: Images and Preimages of One Another

The following problems relate set images and set preimages.

6.1.24. *Proposition* 6

Let $f : D \to C$ be a function with $S \subseteq D$ and $V \subseteq C$. Prove the following:

a. $f^*[f[S]] \supseteq S$

b. $f[f^*[V]] \subseteq V$

6.1.25. *Strengthened Versions of Proposition* 6

a. What additional condition on f allows equality to be asserted in *Proposition* 6a, yielding $f^*[f[S]] = S$? Prove your claim.

b. What additional condition on f allows equality to be asserted in *Proposition* 6b, yielding $f[f^*[V]] = V$? Prove your claim.

Exercises 26–27: Decomposing Functions

The following problems look at ways to decompose functions.

6.1.26. *Positive and Negative Parts of a Function*

Let $f : \mathbb{R} \to \mathbb{R}$ be a real-valued function of a real variable and let

$$f_+(x) = \begin{cases} f(x) : & f(x) \geq 0 \\ 0 : & f(x) < 0 \end{cases} \qquad f_-(x) = \begin{cases} f(x) : & f(x) \leq 0 \\ 0 : & f(x) > 0 \end{cases}$$

Prove that $f_s : \mathbb{R} \to \mathbb{R}$ defined by $f_s(x) = f_-(x) + f_+(x)$ is equal to f.

6.1.27. *Even and Odd Functions*

Definitions: $f: \mathbb{R} \to \mathbb{R}$ *is* **even** *if and only if* $f(-x) = f(x)$ *for all* x.
$f: \mathbb{R} \to \mathbb{R}$ *is* **odd** *if and only if* $f(-x) = -f(x)$ *for all* x.

a. Prove that if $f: \mathbb{R} \to \mathbb{R}$ is a polynomial function having all even powers, then f is an even function.

b. Prove that if $f: \mathbb{R} \to \mathbb{R}$ is a polynomial function having all odd powers, then f is an odd function.

c. If $f: \mathbb{R} \to \mathbb{R}$ is a polynomial function having both even and odd powers, is f even, odd, both, or neither? Explain.

d. Let $f: \mathbb{R} \to \mathbb{R}$ be any function. Determine two functions f_e and f_o such that f_e is even, f_o is odd, and $f_e(x) + f_o(x) = f(x)$. *Hint*: first suppose such functions exist (*Method of Analysis*). Then determine what they are.

6.2 Composite Functions and Inverse Functions

Functions can be combined in various ways. If the domains and codomains of two functions agree and if the codomain supports arithmetic, function operations can be defined using pointwise computations on the images. For example, if $f: \mathbb{R} \to \mathbb{R}$ is given by $f(x) = x^2$ and $g: \mathbb{R} \to \mathbb{R}$ by $g(x) = x - 1$, the function $f + g: \mathbb{R} \to \mathbb{R}$ is defined by $(f + g)(x) = x^2 + x - 1$ and the function $f \cdot g$ by $(f \cdot g)(x) = x^2 \cdot (x - 1) = x^3 - x^2$.

Another operation on functions is more set-theoretic in nature. *Function composition*, which can be performed even when no calculations can be made with function images, is the most fundamental operation on functions.

6.2.1 *Composition of Functions*

Given two functions that hook up properly, we can connect them in series, as it were, to create a composite function. Linking them up like this requires the outputs of the first function to be inputs for the second. This happens if the codomain of the first function agrees with the domain of the second.

Definition 6.2.1: *Composite Functions*

If $f: D \to C$ *and* $g: C \to B$, *then the* **composite function** f *followed by* g *is the function* $g \circ f: D \to B$ *whose action is given by* $(g \circ f)(x) = g(f(x))$.

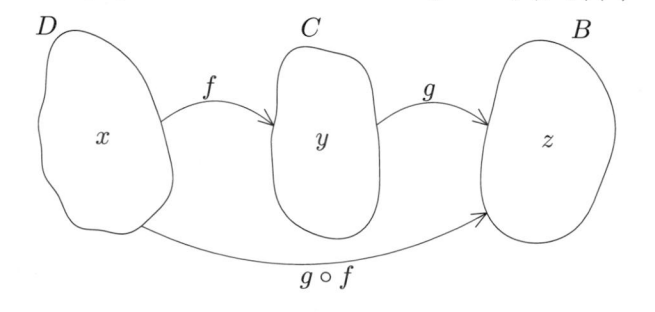

If $f: D \to C$ and $g: C \to B$ are functions, the assignment $h(x) = g(f(x))$ does define a function $h: D \to B$. For, given any $x \in D$, there is a unique $y \in C$ such that $y = f(x)$, since f is a function. Similarly, since g is a function, $g(f(x))$ is a unique output in B. Thus, each input x from D yields a unique output $z = g(f(x))$ in B

Note that the action of $g \circ f$ consists of f's action followed by that of g. This looks backward but is due to the fact that we write input variables to the right of function symbols. Since f acts first on x, we put it closest to x, and we put g to the left of f to show that it acts on $f(x)$.

Composition can't always be performed because f and g may not link up right. However, we can explore what properties composition has when they do. How does composition relate to functions being one-to-one and onto? Is there an identity element for composition? If so, are there also inverses? These are the sorts of questions we'll answer in the rest of this section.

6.2.2 Basic Properties of Composition

Let's begin with a negative result: the order of functions in a composition makes a difference. This idea should be intuitively clear from everyday life. For example, the composite action of putting on your underwear and your outer wear needs to be performed in a certain order to achieve the desired effect. The same is true for functions. Composition is not generally commutative, though some functions do commute (see Exercises 1–3).

Proposition 6.2.1: *Composition is not Commutative*
 $\neg \forall f \, \forall g (g \circ f = f \circ g)$

Proof:
  The functions mentioned above, both from $\mathbb{R}$ to $\mathbb{R}$, are a counterexample:
  if $f(x) = x^2$ and $g(x) = x - 1$, then $(g \circ f)(x) = x^2 - 1$,
  while $(f \circ g)(x) = (x - 1)^2 = x^2 - 2x + 1$.
  These differ on most inputs, e.g., $(g \circ f)(0) = -1 \neq 1 = (f \circ g)(0)$. ∎

  While commutativity fails for composition, associativity holds.

Proposition 6.2.2: *Composition is Associative*
  Suppose $f: D \to C$, $g: C \to B$, and $h: B \to A$ are functions.
  Then $(h \circ g) \circ f = h \circ (g \circ f)$.

Proof:
  The domains and codomains of the two composite functions agree: the common domain is D, and the common codomain is A.
  The actions also yield the same outputs, given the same inputs:
$$((h \circ g) \circ f)(x) = (h \circ g)(f(x)) = h(g(f(x))), \quad \text{and}$$
$$(h \circ (g \circ f))(x) = h((g \circ f)(x)) = h(g(f(x))).$$
  Thus, $((h \circ g) \circ f)(x) = (h \circ (g \circ f))(x)$.
  Therefore, the two functions are equal. ∎

6.2.3 *Composition of Special Functions*

In Section 6.1 we discussed what it means for functions to be one-to-one and onto. Let's now look at how composition relates to these properties. This will help us investigate inverse functions below.

Proposition 6.2.3: *Composition of One-to-One and Onto Functions*
  Let $f : D \to C$ and $g : C \to B$, so that $g \circ f : D \to B$.
  a) *If f and g are both one-to-one functions (injections), then so is $g \circ f$.*
  b) *If f and g are both onto functions (surjections), then so is $g \circ f$.*
  c) *If f and g are both one-to-one-and-onto functions (bijections), then so is $g \circ f$.*

Proof:
  a) To show that $g \circ f$ is one-to-one, we'll work back from B to D through
     C. Draw a function diagram to help you visualize the proof.
   · Suppose, then, that $(g \circ f)(x_1) = (g \circ f)(x_2)$ in B.
     Then $g(f(x_1)) = g(f(x_2))$ by the definition of composition.
     Since g is one-to-one, $f(x_1) = f(x_2)$ in C.
     Since f is also one-to-one, $x_1 = x_2$ in D.
     Thus, $g \circ f$ is one-to-one, too. ✓
  b) This part proceeds similarly. See Exercise 6.
  c) This is an immediate consequence of the last two parts. ∎

The converses of *Proposition* 3 are false, but partial converses do hold (see Exercise 7).

6.2.4 *Identity Functions*

An identity element for a binary operation is one that leaves all elements unaffected when it operates upon them. For example, 0 is the additive identity for the integers, and 1 is the multiplicative identity. Let's explore the notion of an identity for composition of functions.

For composition to be a total binary operation, it must be defined for all pairs of functions in its universe of discourse $\mathcal{F}$. Thus, it's necessary to restrict functions in $\mathcal{F}$ to ones whose domain and codomain are some common set S. This condition is also sufficient, provided $\mathcal{F}$ is *closed* under composition, i.e., provided that the composition of any two functions in $\mathcal{F}$ is also in $\mathcal{F}$.

If we fix a set S and take $\mathcal{F}$ to be the collection of all functions $f : S \to S$, then composition is a binary operation on $\mathcal{F}$. Here it makes sense to ask whether an identity element exists in $\mathcal{F}$ and what properties it might have. Before showing an identity exists, though, we'll define it more precisely.

Definition 6.2.2: *Identity Element for Composition*
  *A function i is an **identity element** for composition defined on a collection $\mathcal{F}$ of functions if and only if $i \circ f = f = f \circ i$ for all f in $\mathcal{F}$.*

If such a function exists, it is unique. This is proved in the usual way: suppose i_1 and i_2 denote identities, and prove $i_1 = i_2$ (see Example 2.4.10).

Assuming that a composition identity i exists (*Method of Analysis*), then $i(f(x)) = (i \circ f)(x) = f(x)$, so i leaves all images $f(x)$ unchanged. That is, i acts as the do-nothing function on these elements. We'll show that such a candidate is indeed the identity element for composition.

Definition 6.2.3: *Identity Function for a Set*
The ***identity function*** *for a set S is the function $I_S : S \to S$ defined by $I_S(x) = x$ for all x.*

I_S is clearly a one-to-one-and-onto function (see Exercise 11). It's also the unique identity element for composition of functions from S into S.

Proposition 6.2.4: *Identity Functions are Composition Identities*
The identity function I_S is the identity element for composition in the collection of all functions $f : S \to S$.

Proof:
Use function equality to show I_S satisfies *Definition 2*. See Exercise 12.

6.2.5 Inverse Functions

For operations with identities, we can ask whether inverses also exist. Two objects are inverses relative to an operation if and only if together they yield the identity. For example, 3 and -3 are additive inverses in $\mathbb{Z}$ because they add up to 0, the additive identity; 3 has no multiplicative inverse in $\mathbb{Z}$, but $1/3$ is its multiplicative inverse in $\mathbb{Q}$ because $3 \cdot 1/3 = 1$, the multiplicative identity.

As we explore this concept of inverses for functions under composition, we'll see that some functions have inverses and some don't.

Definition 6.2.4: *Inverse Functions*
*If $f : D \to C$ and $g : C \to D$, f and g are **inverse functions** of one another relative to composition if and only if $g \circ f = I_D$ and $f \circ g = I_C$.*

The final output after composing inverses is thus the same as the original input; *inverse functions undo one another*. The following proposition is a slight reformulation of this definition in terms of inputs and outputs.

Proposition 6.2.5: *Inverse Functions Cancel One Another*
If $f : D \to C$ and $g : C \to D$, f and g are inverse functions of each other if and only if $g(f(x)) = x$ for all x in D and $f(g(y)) = y$ for all y in C.

Proof:
Use *Definition 3* and *Definition 4*. See Exercise 20.

❖**Example 6.2.1**
Show that the function $g : \mathbb{R} \to \mathbb{R}$ defined by $g(y) = \dfrac{y - 7}{3}$ is an inverse of the function $f : \mathbb{R} \to \mathbb{R}$ defined by $f(x) = 3x + 7$.

Solution

Proposition 5 shows this.

$$g(f(x)) = g(3x + 7) = \frac{(3x + 7) - 7}{3} = x, \text{ and}$$

$$f(g(y)) = f\left(\frac{y - 7}{3}\right) = 3 \cdot \left(\frac{y - 7}{3}\right) + 7 = y.$$

Thus, g is f's inverse. ✓

We'll now state a biconditional variant of *Proposition* 5. The backward direction will help us show that functions are inverses.

Proposition 6.2.6: *Inverse Functions Undo One Another*

If $f : D \to C$ and $g : C \to D$ are functions, then f and g are inverses of each other if and only if $\forall x \forall y (y = f(x) \leftrightarrow x = g(y))$.

Proof:

· Before proving this theorem, let's analyze what it claims, using logic.
· Suppose that f and g are each functions whose domain matches the other's codomain. Then f and g are inverses, according to this theorem, if and only if a universal undoing *biconditional* is satisfied: $y = f(x) \leftrightarrow x = g(y)$.
· Formulating this with PL symbolism gives $P \wedge Q \to (R \leftrightarrow (S \leftrightarrow T))$. While this is a slightly complex logical form, we can use a *Backward-Forward Proof Analysis* to choose an overall proof strategy.
· Begin by supposing $P \wedge Q$ for the purpose of *CP*. Then use *BI*: first supposing R, prove $S \leftrightarrow T$, which itself needs *BI*. Next, supposing $S \leftrightarrow T$ and using *BE*, prove R. This will finish off the consequent of our conditional; the theorem will follow via *CP*.

We'll indicate different proof levels in the argument by indenting them.

· Suppose, then, that f and g are two functions, as given.
  ○ Suppose in the first place that f and g are inverses.
    * Suppose, moreover, that $y = f(x)$.
      [We'll show that $g(y) = x$ by applying g to $y = f(x)$.]
      Since g is a function, $g(y) = g(f(x))$;
        but $g(f(x)) = x$, since g is f's inverse.
      Thus, $g(y) = x$. ✓
    * Suppose now that $x = g(y)$.
      A proof that $y = f(x)$ proceeds in precisely the same way:
      apply f to both sides of the given equation and use the fact that f is a function and is g's inverse.
  ○ On the other hand, suppose that $y = f(x) \leftrightarrow x = g(y)$.
    [We show f and g are inverses by showing that they undo each other.]
    * Let $y = f(x)$ be any image of f.
      Applying the function g, we get $g(y) = g(f(x))$.
      But $g(y) = x$ (by *BE*), so $g(f(x)) = x$. ✓
    * Similarly, if $x = g(y)$, it follows that $f(g(y)) = y$.
  ○ Therefore, by *Proposition* 5, f and g are inverses. ∎

At this point, we still haven't determined when inverses exist, but we can show that when they do, they are unique.

Proposition 6.2.7: *Uniqueness of Inverses*
  If $f : D \to C$ has an inverse $g : C \to D$, then it's unique.

Proof:
  Suppose that f has inverses $g_1 : C \to D$ and $g_2 : C \to D$.
  To show $g_1 = g_2$, we only need to show that they have the same action; their domains (C) and codomains (D) agree.
  Let y be any element of C.
  By *Proposition 5*, since g_1 is an inverse for f, $y = f(g_1(y))$.
  Since g_2 is an inverse for f, applying g_2 yields $g_2(y) = g_2(f(g_1(y))) = g_1(y)$.
  Thus, $g_1 = g_2$. ∎

Because there is only one inverse for a function f when it has one, we can speak of *the* inverse of f and denote it by the standard notation f^{-1}.

  Caution: $f^{-1} \neq 1/f$. Here f^{-1} denotes the *inverse for f relative to composition*, while $1/f$ denotes a *multiplicative inverse*. We could use another notation for the composition inverse, but exponents are useful for symbolizing repeated compositions and their inverses just as they are for repeated multiplication and their inverses (see Exercises 13–18).

6.2.6 *Calculating Inverse Functions*

We now need a way to compute inverse functions. Example 1 showed that two given functions were inverses, but how can we find them in the first place?

  There are two main approaches to calculating formulas for inverse functions, both based on the fact that inverse functions f and g reverse one another's actions. One way decomposes the action of $y = f(x)$ into a sequence of reversible consecutive steps, when that's possible. Then the formula $x = g(y)$ can be computed by a *Simple Reversal Method* (called the *Method of Inversion* when it was first introduced in medieval India): *calculate an inverse function by performing the inverse operations in the reverse order on a given input.*

❖**Example 6.2.2**
  Find the inverse function f^{-1} for the function $y = f(x) = (2x - 5)/3$ by the *Simple Reversal Method*.

  Solution
  The function value $f(x) = (2x - 5)/3$ can be computed by the following series of steps: *input x*; *double*; *subtract 5*; *divide by 3*; *output* $y = f(x)$. The inverse action inverts both the order and the actions and so gives an algorithm for the inverse: *input y*; *multiply by 3*; *add 5*; *halve*; *output x*. Thus, $f^{-1}(y) = (3y+5)/2$ is the inverse function's formula (see Exercise 19).

Perhaps you noticed that the *Simple Reversal Method* matches the process used to solve $y = f(x)$ for x in terms of y. According to *Proposition 6*, this solution yields $x = f^{-1}(y)$ as the inverse for f, if it exists. We'll develop this idea further in a moment, but let's first point out a limitation of the *Simple Reversal Method*. If f's action does not readily decompose into a linear sequence of consecutive operations, each applied to the output of the preceding step, you won't be able to invert the process to determine f^{-1}, even though f may have an inverse. The next example illustrates this.

✤Example 6.2.3

Find the inverse function for $f : [1, +\infty) \to [-2, +\infty)$, where $y = f(x) = x^2 - 2x - 1$, using the *Simple Reversal Method*, if possible.

Solution

· The domain and codomain intervals have been chosen so that f is invertible, but ignore this for the moment.

· To calculate $f(x)$, we start by squaring x to obtain x^2, then we recall x and double it, next we subtract the result of step two from step one, and finally we subtract 1. Because of the branching involved (recalling x and doubling before subtracting), the inverse function can't be found merely by reversing the steps used to calculate y. Try it: you'll get stuck.

· In this case we can get around the problem by recalculating the given function in a more linear way. For instance, if we rewrite the formula for $f(x)$ by completing the square, we'll get a more complicated equivalent expression that can be calculated in serial fashion and thus reversed. (This will also show why the domain and codomain were chosen as they are.) Now the *Simple Reversal Method* will succeed. Showing all this will be left as an exercise (see Exercise 22).

The difficulty in this example points out the need for a more systematic approach to calculating inverse functions. An *Algebraic Solution Method* that works in many instances also draws upon *Proposition 6*: *to find the inverse g for a function f, take the equation $y = f(x)$ and solve it for x; the expression $x = g(y)$ that results gives the action of the inverse function, provided two things check out*:

1) $x = g(y)$ *defines a function with the appropriate domain and codomain.*
   The formula may not define a function, because it may not give a unique output, or it may not be defined for all y-values. For example, solving $y = x^2$ for x yields $x = \pm\sqrt{y}$. Unless the domain of x-values is restricted in some way, say, to nonnegative numbers, two outputs will result from a single input (or, no output will result if $y < 0$).

2) *the two functions undo one another: $y = f(x) \leftrightarrow x = g(y)$.*
   The forward direction here automatically follows from the solution process: passing from $y = f(x)$ to $x = g(y)$ gives $y = f(x) \to x = g(y)$. If the solution process is reversible, the other direction also holds. Otherwise, you need to check, given $x = g(y)$, that $y = f(x)$, i.e., that $y = f(g(y))$ holds.

✤**Example 6.2.4**

Determine the inverse function for $f(x) = \dfrac{x}{x+3}$. Assume f is defined on as large a set of real numbers as possible and that f's codomain is its range.

Solution

· The equation $y = \dfrac{x}{x+3}$ defines a function f on $\mathbb{R} - \{-3\}$.

· We'll stipulate its codomain after we determine what values y can be. Solving $y = \dfrac{x}{x+3}$ for x, we get the following:

$$xy + 3y = x$$

$$xy - x = -3y$$

Thus, $x = \dfrac{-3y}{y-1} = \dfrac{3y}{1-y} = g(y).$ ✓

· Since there are x-values for all y except $y = 1$, our domain for g and our codomain for f must be taken to be $\mathbb{R} - \{1\}$.

For these x- and y-values, the above solution process is reversible. ✓

· The inverse function is therefore given by $f^{-1}(y) = \dfrac{3y}{1-y}.$ ✓

There may be times when an inverse function exists but we're unable to solve $y = f(x)$ for x to find a formula for f^{-1}. When this happens, we may have to invent a new notation to denote the inverse function and leave it at that, at least temporarily. This may sound like cheating, but it is exactly what we do with a number of familiar functions. We'll list three; you may be able to think of others.

1) $f(x) = x^3;$ $f^{-1}(y) = \sqrt[3]{y}.$
2) $g(x) = 2^x;$ $g^{-1}(y) = \log_2(y).$
3) $h(x) = \sin x;$ $h^{-1}(y) = \arcsin(y).$

6.2.7 *Existence of Inverse Functions*

Let's now investigate when a function has an inverse. In doing so, we'll be arguing for a proposition we have yet to state. While not standard textbook practice, this process is typical of how theorems are discovered and proved.

We'll begin by determining a necessary condition for having an inverse—what must hold if f has an inverse (*Method of Analysis* again). We'll then show that this condition is also sufficient.

· Suppose, then, that $g\colon C \to D$ is the inverse function of $f\colon D \to C$.
  ◦ Since g is a function, it must be defined for every y in C. But since g is an inverse of f, $x = g(y) \leftrightarrow f(x) = y$; so f must be an onto function.
  ◦ Furthermore, since g is a function, the images $x = g(y)$ are unique. But since g undoes f, this means there is only one x associated with a given y by f: f must be a one-to-one function.
  ◦ Conjoining these two conclusions, we have the following necessary condition: *If f has an inverse function g, then f is one-to-one-and-onto.* ✓

· Conversely: *If f is one-to-one-and-onto, then f has an inverse function.*
  ◦ Suppose f is one-to-one-and-onto, with $y = f(x)$.
  ◦ Let $g : C \to D$ denote a correspondence such that $x = g(y) \leftrightarrow y = f(x)$.
  ◦ This defines a function: since f is onto, $g(y)$ is defined for all y;
    and since f is one-to-one, the output x is unique for each y.
  ◦ This being the case, *Proposition* 6 implies that g is the inverse of f. ∎

We've thus proved the following result.

Theorem 6.2.1: *Existence of Invertible Functions*
A function $f : D \to C$ is invertible if and only if f is one-to-one-and-onto.

Corollary 6.2.1.1: *Inverse Functions are Bijections*
f is invertible if and only if f^{-1} is one-to-one-and-onto.

Proof:
  This corollary claims that the necessary and sufficient condition holding for
  an invertible function f also holds for its inverse. See Exercise 25a.

Corollary 6.2.1.2: *Bijections Have Bijective Inverses*
f is one-to-one-and-onto if and only if f^{-1} is one-to-one-and-onto.

Proof:
  This follows immediately from the theorem and the last corollary. See Exercise 25b.

Corollary 6.2.1.3: *Composition of Inverses*
*$g \circ f$ is invertible if f and g are. Moreover, $(g \circ f)^{-1} = f^{-1} \circ g^{-1}$, the
composition of the inverses in reverse order.*

Proof:
  See Exercise 25c.

EXERCISE SET 6.2

Exercises 1–3: Commuting Linear Functions under Composition
The following problems explore whether some simple functions commute.
Let $\mathcal{M}$ denote the set of magnifications $f : \mathbb{R} \to \mathbb{R}$ of the form $f(x) = mx$,
* $\mathcal{S}$ the set of shifts $f : \mathbb{R} \to \mathbb{R}$ of the form $f(x) = x + b$, and*
* $\mathcal{A}$ the set of affine functions $f : \mathbb{R} \to \mathbb{R}$ of the form $f(x) = mx + b$.*

6.2.1. *Magnifications*
  a. Show that all functions in $\mathcal{M}$ commute with one another.
  b. Which functions in $\mathcal{S}$ commute with all functions in $\mathcal{M}$? Explain.
  c. Which functions in $\mathcal{A}$ commute with all functions in $\mathcal{M}$? Explain.

6.2.2. *Shifts*

a. Show that all functions in $\mathcal{S}$ commute with one another.

b. Which functions in $\mathcal{M}$ commute with all functions in $\mathcal{S}$? Explain.

c. Which functions in $\mathcal{A}$ commute with all functions in $\mathcal{S}$? Explain.

6.2.3. *Affine Functions*

a. How are the functions in $\mathcal{A}$ related to those in $\mathcal{M}$ and $\mathcal{S}$? Explain.

b. Which functions in $\mathcal{A}$ commute with all functions in $\mathcal{A}$? Explain.

Exercises 4–5: Properties of Composition

Work the following problems related to algebraic properties of composition.

6.2.4. Choose two different nonlinear real-valued functions of a real variable and determine whether or not they commute.

6.2.5. Verify that composition is associative in the case where $f(x) = 2x - 1$, $g(x) = 3 - x$, and $h(x) = x^2 + 1$: calculate $((h \circ g) \circ f)(x)$ and $(h \circ (g \circ f))(x)$ in stages and compare the final results.

Exercises 6–7: Composition of One-to-One and Onto Functions

The following problems explore composition and function properties.

6.2.6. *Proposition 3*

Prove *Proposition 3b: If f and g are onto functions, then so is $g \circ f$.*

6.2.7. *Converses and Partial Converses to Proposition 3*

a. Disprove the following: If $g \circ f$ is one-to-one, then so are f and g.

b. Prove the following: If $g \circ f$ is one-to-one, then so is f.

c. Disprove the following: If $g \circ f$ is onto, then so are f and g.

d. Prove the following: If $g \circ f$ is onto, then so is g.

e. Disprove the following: If $g \circ f$ is a one-to-one-and-onto function, then so are f and g.

f. Formulate and prove a partial converse to *Proposition* 3c (see parts b and d).

g. True or false: f is a one-to-one/onto function if and only if $f \circ f$ is a one-to-one/onto function. Explain, using *Proposition* 3 along with the results of parts b and d. Assume $f \circ f$ is defined.

Exercises 8–10: True or False

Are the following statements true or false? Explain your answer.

6.2.8. The *Algebraic Solution Method* always gives the inverse of an invertible function.

6.2.9. For invertible functions f and g, $(f \circ g)^{-1} = f^{-1} \circ g^{-1}$.

6.2.10. All one-to-one functions have inverses.

Exercises 11–12: Identities

The following problems deal with identity functions and their properties.

6.2.11. *Identity Functions*

Show that the identity function $I_S : S \to S$ is one-to-one-and-onto.

6.2.12. *Proposition 4*

Prove *Proposition 4: The identity function I_S is the identity element for composition in the collection S^S of all functions $f : S \to S$.*

Exercises 13–18: Function Exponents

Function exponents are defined recursively for all natural numbers and then extended to the integers as follows:

Definition: *Let f be a function from a set S into itself.*

$$f^0 = I_S; \qquad f^{n+1} = f^n \circ f \quad \text{for all } n \text{ in } \mathbb{N}.$$

$$f^{-n} = \left(f^{-1}\right)^n \quad \text{for all } n \text{ in } \mathbb{N}^+, \text{ provided } f \text{ is invertible.}$$

6.2.13. Prove that $f^m \circ f^n = f^{m+n}$ for all m and n in $\mathbb{N}$.

6.2.14. Prove that $(f^m)^n = f^{mn}$ for all m and n in $\mathbb{N}$.

6.2.15. Prove that $(f \circ g)^n = f^n \circ g^n$ for all n in $\mathbb{N}$ if and only if $f \circ g = g \circ f$.

6.2.16. Prove that $f^{-n} = (f^n)^{-1}$ for all n in $\mathbb{N}$.

6.2.17. Prove $f^m \circ f^n = f^{m+n}$ for integer exponents m and n.

6.2.18. Prove $(f^m)^n = f^{mn}$ for integer exponents m and n.

Exercises 19–25: Inverse Functions

The following problems deal with inverse functions and their properties.

6.2.19. *Demonstrating Inverses*

Using the definition of inverse functions, show that $g(y) = (3y + 5)/2$ in Example 2 gives the inverse of $f(x) = (2x - 5)/3$.

6.2.20. *Proposition 5*

Prove *Proposition 5: If $f : D \to C$ and $g : C \to D$, then f and g are inverses if and only if $g(f(x)) = x$ for all x in D and $f(g(y)) = y$ for all y in C.*

6.2.21. *Inverses of Inverses*

Prove that if f is invertible, so is its inverse f^{-1}, with $(f^{-1})^{-1} = f$.

6.2.22. *Example 3*

a. To flesh out Example 3, rewrite $f : [1, +\infty) \to [-2, +\infty)$ defined by $f(x) = x^2 - 2x - 1$ by completing the square. Then analyze the computation process this new formula exhibits, and reverse the steps to get the inverse. Show that the function you get is the inverse function f^{-1} using *Proposition 5*. In the process, explain why the domain and codomain are chosen the way they are.

b. Calculate the inverse for $f : [1, +\infty) \to [-2, +\infty)$ defined by $y = f(x) = x^2 - 2x - 1$ by solving for $x = g(y)$ using the *Quadratic Formula*. Compare your result with part *a*.

Caution: you are to solve for x, given y, not find a zero for the function.

6.2.23. *Inverses for Simple Polynomials*

a. Use the *Simple Reversal Method* to determine the inverse function for $y = f(x) = (2 - x)/3$. Check your result using *Proposition 5*.

b. Show that $f : \mathbb{R} \to \mathbb{R}$ defined by $f(x) = 3x - 5$ is one-to-one-and-onto and thus is invertible. Find its inverse $x = f^{-1}(y)$ and justify your choice.

c. Show that $f : [1, +\infty) \to [0, +\infty)$ defined by $f(x) = x^2 - 2x + 1$ is invertible and find its inverse.

d. Show that the function $f : \mathbb{R} \to \mathbb{R}$ defined by $f(x) = x^3 + 2$ is invertible and find its inverse.

6.2.24. *Inverses for Rational Functions*

a. Show that the function f whose action is given by $f(x) = \dfrac{2 - x}{x + 1}$ and whose domain is $\mathbb{R} - \{-1\}$ is a one-to-one function onto its range. Then calculate its inverse $x = f^{-1}(y)$ and specify both its domain and range.

b. Show that $f(x) = x/(1 - x^2)$ is an invertible function from the open interval $(-1, +1)$ into $\mathbb{R}$ by showing that it's one-to-one-and-onto. Then find its inverse $x = f^{-1}(y)$. [Hint: make use of the *Quadratic Formula*.]

6.2.25. *Corollaries to Theorem 1*

Prove the following three corollaries to *Theorem 1*:

a. f is invertible if and only if f^{-1} is one-to-one-and-onto.

b. f is one-to-one-and-onto if and only if f^{-1} is one-to-one-and-onto.

c. $g \circ f$ is invertible if f and g are. Moreover, $(g \circ f)^{-1} = f^{-1} \circ g^{-1}$, the composition of the inverses in reverse order.

Exercises 26–28: Exploring Identities and Inverses

The following problems explore one-sided identities and inverses.

6.2.26. *Exploring One-Sided and Two-Sided Identities*

Let $\mathcal{F}$ denote a collection of functions $f : S \to S$. A function $i : S \to S$ is a *left identity* for $\mathcal{F}$ if and only if $i \circ f = f$ for all $f \in \mathcal{F}$, a *right identity* if and only if $f \circ i = f$ for all f, and a *two-sided identity* if and only if it is both a left and a right identity.

a. Show that I_S is a left identity and a right identity for $\mathcal{F}$.

b. Must left (right) identities be unique? Or could two different functions act like left (right) identities? Support your answer.

c. Must two-sided identities be unique? Explain.

6.2.27. *Exploring One-Sided and Two-Sided Inverses*

a. Using Exercise 26 as your model, define left and right and two-sided inverses for functions.

b. Must left (right) inverses for a function be unique? Can a function have a left inverse without having a right inverse, or conversely?

c. Can a function have both a left and a right inverse, but no two-sided inverse? If a function has a two-sided inverse, must it be unique?

6.2.28. *Inverses and Composition Cancellation*

Let $f : D \to C$ and $g_i : C \to D$ for $i = 1, 2$; $D \neq \emptyset \neq C$. Prove the following:

a. f is one-to-one if and only if it has a left inverse g, i.e., such that $g \circ f = I_D$.

b. If f is one-to-one, then $f \circ g_1 = f \circ g_2 \to g_1 = g_2$.

c. g is onto if and only if it has a right inverse g, i.e., such that $f \circ g = I_C$.

d. If f is onto, then $g_1 \circ f = g_2 \circ f \to g_1 = g_2$.

e. f is one-to-one-and-onto if and only if $f \circ g_1 = f \circ g_2 \to g_1 = g_2$ and $g_1 \circ f = g_2 \circ f \to g_1 = g_2$.

6.3 Equivalence Relations and Partitions

Set Theory gives precision to the theory of functions and relations. In the final analysis this is due to the Wiener-Kuratowski definition of ordered pair (see Exercises 4.3.26–28), but we'll work with a more informal notion.

Functions are specialized binary relations, as are *equivalence relations* and various *order relations*. In this section we'll focus on binary relations in general and equivalence relations in particular. We'll explore an equivalence relation of importance to mathematics and computer science (*congruence mod n*) in Section 6.4. Order relations will be treated in Chapter 7.

6.3.1 *Binary Relations and Relations in General*

Binary relations match elements in one set with those in another. Directionality is important. For example, the relation *is the father of* can't be turned around—the child is not the father of the man. *Being the father of* is a relationship between many pairs of individuals, between your father and you as well as my father and me, and me and my children.

By focusing on the pairs of things being related, we can give an *extensional* representation of that relation. A binary relation can be thought of as the set of all ordered pairs of objects having that relationship.

Definition 6.3.1: *Binary Relation*   (*set-theoretical version*)
 a) *R is a **binary relation between sets S and T** (or **from S to T**) if and only if $R \subseteq S \times T$.*
 b) *R is a **relation on S** if and only if R is a relation from S to S, i.e., if and only if $R \subseteq S \times S$.*
 c) *R is a **binary relation** if and only if R is a binary relation between some S and T.*

✚**Example 6.3.1**
 Here are two binary relations, from everyday life and from mathematics.

Solution
 a) The relation _____ *is a student taking* _____ is a binary relation linking students and courses. According to the definition, we can think of this relation as the set of (x, y) in which x is a student taking course y.
 b) The relation _____ *is less than* _____ is a relation on $\mathbb{R}$. *Definition 1* would indicate $3 < \pi$ is by writing $(3, \pi) \in <$. Frankly, this extensional notation looks weird.[2] We'll mostly use the familiar *infix notation* $3 < \pi$, but occasionally set-theoretic notation will be useful.

Thinking of relations as sets of ordered pairs is admittedly a bit awkward. Mathematicians who are not set theorists usually think of a binary relation as a connection between pairs of things, not as a set of ordered pairs. We see

---

[2] Moreover, the relation of *set membership* is itself necessarily still treated intensionally; how would one write that $((3, \pi), <)$ is in $\in$ without entering a vicious circle?

this in standard mathematical notation. We write $\triangle ABC \cong \triangle A'B'C'$, not $(\triangle ABC, \triangle A'B'C') \in \cong$. If x has a relation R to y, we write $x\,R\,y$. To make use of this notation while formally adhering to a set-theoretical basis, we can posit the following definition.

Definition 6.3.2: *Infix Notation for Binary Relations*
  Let R be a binary relation. Then $x\,R\,y \leftrightarrow (x, y) \in R$.

Strange as the extensional definition of a binary relation may seem, its value lies in what we can do with it inside *Set Theory*. Given *Definition 1*, various general theorems about relations can be rigorously proved. This perhaps justifies its introduction, just as hypotheses in any science are justified by considering *their* consequences.

Associated with each binary relation $R \subseteq S \times T$ are the *domain* and the *range* of the relation. These are, respectively, the set of all first elements of the relation and the set of all second elements. S and T may be the domain and range, but nothing requires them to be so. All we can say for sure is that the domain of R is a subset of S and the range of R is a subset of T.

Definition 6.3.3: *Domain and Range of a Binary Relation*
  Suppose R is a binary relation, with $R \subseteq S \times T$. Then
  a) *The **domain** of R is $\mathrm{Dom}(R) = \{x \in S : x\,R\,y \text{ for some } y \in T\}$.*
  b) *The **range** of R is $\mathrm{Rng}(R) = \{y \in T : x\,R\,y \text{ for some } x \in S\}$.*

✤**Example 6.3.2**
  Show that functions and operations are relations, and specify their domains and ranges.

Solution
  a) Let $F : S \to T$ be a function. F relates inputs to outputs and can be thought of as a subset of ordered pairs (x, y) in $S \times T$ where $y = F(x)$. The notions of domain and range for F as a function agree with those for treating it as a relation.
  b) Consider a binary operation $*$ on some set S. Given elements a and b in S, there exists a unique c in S such that $a * b = c$. This defines a function $* : S \times S \to S$, with $*(a, b) = c$. By part a, $*$ is a binary relation between ordered pairs in $S \times S$ and outputs in S.
  Alternatively, $*$ can be considered a ternary relation comprised of the ordered triples (a, b, c) in S^3, where $a * b = c$.

We're focusing on binary relations, but as the last example indicates, relations may connect more than two things at a time. In fact, computer scientists who work with data structures are dealing with n-ary relations holding among n things. These provide relational models for representing and manipulating information in a database. Any table with n columns of data can be thought of as an n-ary relation. Such relations can be represented using ordered n-tuples and n-fold Cartesian products.

6.3.2 Types of Binary Relations and Their Properties

A binary relation R may be from a set S to a different set T or from a set S to itself. Relations of the first type are important for handling functions and operations. But relations of the second type are also important, because they allow us to treat order and relational structure within a single set.

For the rest of this section we'll focus on relations of this second type, beginning with a look at properties such relations may have. To avoid confusion, note that *it's the relation as a whole* that has or doesn't have these properties. In addition to specifying key properties of relations, we'll define some associated types of relations that help us characterize these properties.

Definition 6.3.4: *Reflexive Relations on* S
  R is *reflexive* if and only if $(\forall x \in S)(x \mathrel{R} x)$.

Definition 6.3.5: *Identity Relation on* S
  The *identity relation* D on a set S is defined by $x \mathrel{D} y$ if and only if $x = y$.

For a relation R to be reflexive, the pair (x, x) must be in R for all x in S. This means that the main diagonal D of $S \times S$ is in (the graph of) the relation R. Consequently, R *is reflexive if and only if* $D \subseteq R$, *where D is the identity relation on* S. Other pairs/points may also be in R, but they need not be in order for R to be reflexive, nor do they affect whether R is reflexive.

Definition 6.3.6: *Symmetric Relations on* S
  R is *symmetric* if and only if $(\forall x, y \in S)(x \mathrel{R} y \to y \mathrel{R} x)$.

In terms of being a subset of $S \times S$, a symmetric relation R has the following property: whenever (x, y) is in (the graph of) R, its mirror image (y, x) across the diagonal $y = x$ is also in (the graph of) R.

Definition 6.3.7: *Converse Relation*
  If R is a relation from S to T, the *converse relation* of R is the relation from T to S defined by $\widehat{R} = \{(x, y) : (y, x) \in R\}$.

Converse relations are the "same" relationship viewed from opposite directions, one the reflection of the other over the diagonal. For example, the order relation $<$ on $\mathbb{R}$ is the converse of $>$: $a < b$ if and only if $b > a$.

Converse relations help us characterize symmetric relations. If elements of S are related by a symmetric relation, they are related in both directions. Thus, R *is symmetric if and only if it's equal to its converse:* $\widehat{R} = R$.

Definition 6.3.8: *Transitive Relations on* S
  R is *transitive* if and only if $(\forall x, y, z \in S)(x \mathrel{R} y \wedge y \mathrel{R} z \to x \mathrel{R} z)$.

Transitive relations have a certain degree of connectivity. If x is related to some y, which in turn is related to some z, then x must be related to z, too. Ordinary $<$ on $\mathbb{R}$ is an example of a transitive relation.

Definition 6.3.9: *Composite Relation*

If R_1 is a relation from S to T and R_2 is a relation from T to U, then $R = \{(x,z)\colon x\,R_1\,y \text{ and } y\,R_2\,z \text{ for some } y \in T\}$ is the **composite relation** $R_1 \circ R_2$ from S to U.

Note the order in which composite relations are symbolized. Unfortunately, this is the reverse of how we write composite functions, though they're also composite relations. Our infix notation is responsible for this—the first relation is put next to the element of the first set: $(x,z) \in R_1 \circ R_2$ if and only if $x\,(R_1 \circ R_2)\,z$.

Composite relations are composed of two simpler relations and usually lead to a new relation, even if the two given relations are the same. For example, the relation *is the father of* composed with itself gives the composite relation *is the grandfather of*.

However, there are times when the composite of a relation with itself gives the original relation. This is roughly what characterizes a transitive relation, a fact that needs arguing: *a relation R on a set S is transitive if and only if $R^2 \subseteq R$, where $R^2 = R \circ R$* (see Exercise 56).

✦ **Example 6.3.3**

Let's look at several transitive relations from earlier in the book and determine whether they're reflexive or symmetric.

Solution

   a) The most basic transitive relation is *equals*. As the FOL inference rules for identity affirm, $=$ is reflexive and symmetric as well as transitive.

   b) We first saw a transitive relation in Section 1.3 when we discussed logical equivalence: if $P \dashv\vdash Q$ and $Q \dashv\vdash R$, then $P \dashv\vdash R$. This relation is also reflexive and symmetric. Thus, true to its name, logical equivalence is an *equivalence relation* (see *Definition* 10).

   c) Interderivability ($\dashv\vdash$) is also reflexive, symmetric, and transitive.

   d) The relations of logical implication ($\vDash$) and derivability ($\vdash$) are transitive and reflexive but not symmetric.

   e) The relation $<$ is a transitive relation on the set of natural numbers $\mathbb{N}$, but it's neither reflexive nor symmetric. The relation $\leq$ is reflexive as well as transitive, but it's still not symmetric.

   f) The relation *divides* on the set of integers $\mathbb{Z}$ is transitive, as we saw in Section 3.5. It's also reflexive but not symmetric. The converse of *divides* is *is a multiple of*, a wholly different relation.

6.3.3 *Equivalence Relations and Partitions*

We've encountered a number of equivalence relations to this point. For *Logic*, we've looked at identity, logical equivalence, and interderivability. *Set Theory* has the equivalence relations of set equality and equinumerosity. Equivalence relations generalize identity by relating objects that are identical with respect to some property (cardinality, truth-value, shape, etc.).

Definition 6.3.10: *Equivalence Relation*
  *R is an **equivalence relation** on S if and only if R is reflexive, symmetric, and transitive.*

❖ **Example 6.3.4**
  Let's look at a few more equivalence relations from mathematics.

Solution
  a) *Congruence* and *similarity* in *Geometry* are equivalence relations—every figure is congruent/similar to itself; if two figures are congruent/similar, they are so in either order; and if one figure is congruent/similar to a second one, which is congruent/similar to a third, the first figure is congruent/similar to the third.
  b) The *Number Theory* relation defined by $x \equiv_2 y \leftrightarrow 2 \mid (x - y)$, is an equivalence relation on the integers. For if $x \equiv_2 y$, then either x and y are both even or they are both odd (see Exercise 47), i.e., x and y have the *same parity*. Thus, $\equiv_2$ is reflexive, symmetric, and transitive.
  c) In *Calculus*, the relation *has the same derivative as* is an equivalence relation among differentiable functions. Every differentiable function has a derivative, which is unique, so the relation is reflexive; if two functions have the same derivative, they do so regardless of which derivative is taken first; and if one function has the same derivative as a second, which has the same derivative as the third, then the first and third functions have the same derivative.
  · The relation *differs by a constant from* agrees with the relation *has the same derivative as* on the set of differentiable functions. This fact is proved in *Calculus* and then used to introduce the operation of antidifferentiation/indefinite integration.

  Given an arbitrary equivalence relation $\sim$ defined on a set S, we know that every element $x \in S$ is related to something because $x \sim x$. This may be the only element related to it, or it may be related to others. If you collect together all the elements related to one another, you'll see a pattern emerge. The following examples illustrate this. But let's first introduce some terminology and notation.

Definition 6.3.11: *Equivalence Class*
  *Let $\sim$ be an equivalence relation on a set S and let $x \in S$. The **equivalence class** of x is the set $[\,x\,] = \{y \in S : x \sim y\}$.*

❖ **Example 6.3.5**
  Let's look at the equivalence class structure of some sets having equivalence relations defined on them.

Solution
  a) Define the relation $\sim$ on $\mathbb{Z}$ by $n \sim m$ if and only if n and m have the same (nonnegative) remainder when divided by 3.
  · Each integer has a unique remainder, and since there are only three

permissible remainders, 0, 1, and 2, each integer belongs to one and only one equivalence class: $[0]$, $[1]$, or $[2]$.
- For example, $8 \in [2]$ because $8 = 3(2) + \underline{2}$; $-6 \in [0]$ because $-6 = 3(-2) + \underline{0}$; and $-2 \in [1]$ because $-2 = 3(-1) + \underline{1}$.

b) Take S to be the set of all equilateral triangles in a plane and let $\sim$ denote the relation *has the same area as*, which is an equivalence relation.
- The equivalence classes here are sets of congruent equilateral triangles, one for each size triangle. We could therefore index this collection of equivalence classes by positive real numbers (the lengths of their sides).
- These classes jointly contain all equilateral triangles without any membership overlap among the classes.

c) Our third example comes from algebra. Let S be the set of all polynomials with real coefficients, and take $\sim$ to be the relation *is a non-zero real multiple of*.
- This relation is an equivalence relation (see Exercise 54), and these equivalence classes contain polynomials of the same degree, related to one another as non-zero scalar multiples. Every polynomial, including the zero polynomial, is in some equivalence class, and none of them belongs to two different classes. All equivalence classes here are infinite except for $[0]$.

d) Finally, suppose $f : S \to T$, and define $a \sim b \leftrightarrow f(a) = f(b)$. Then $\sim$ is an equivalence relation on S (see Exercise 57a). The equivalence classes are preimage subsets of the domain whose elements share a common image: $[a] = \{b \in S : f(b) = f(a)\} = f^*[\{f(a)\}]$.

As we see from these examples, an equivalence relation divides up a set S into mutually exclusive equivalence classes. In technical terms, the equivalence classes form a *partition* of S. We defined this concept in Section 4.2 but will repeat it here for easy reference.

Definition 6.3.12: *Partition of a Set; Cell of a Partition*
*A collection $\mathcal{P}$ of non-empty subsets T of S is a **partition** if and only if $\mathcal{P}$ is pairwise disjoint and $\bigcup_{T \in \mathcal{P}} T = S$. Each $T \in \mathcal{P}$ is a **cell** of the partition.*

✦ **Example 6.3.6**
Determine which of the following collections $\mathcal{P}$ are partitions of S.
a) $S = \mathbb{Z}$; $\mathcal{P}$ is the collection of the three sets $R_0 = \{n \in \mathbb{Z} : n = 3k\}$, $R_1 = \{n \in \mathbb{Z} : n = 3k + 1\}$, and $R_2 = \{n \in \mathbb{Z} : n = 3k + 2\}$.
b) $S = \mathbb{N}$; $\mathcal{P} = \{P, C\}$, where P is the set of prime numbers $\{2, 3, 5, 7, \cdots\}$, and C is the set of composite numbers, $\{4, 6, 8, 9, \cdots\}$.
c) $S = \mathbb{Z}$; $\mathcal{P}$ is the collection $k\mathbb{Z} = \{n \in \mathbb{Z} : n = mk \text{ for some } m \in \mathbb{Z}\}$ for $k \in \mathbb{N}$, for example, $3\mathbb{Z} = \{0, \pm 3, \pm 6, \pm 9, \cdots\}$.

Solution
a) $\mathcal{P}$ is a partition of $\mathbb{Z}$ because its sets are pairwise disjoint and together exhaust $\mathbb{Z}$. By the *Division Algorithm*, remainders are unique, so each integer belongs to exactly one cell of the collection.

· This partition consists of the equivalence classes given in Example 5a. We could thus use this partition, if we wanted, to define that relation on $\mathbb{Z}$: $x \sim y$ if and only if x and y belong to the same cell of $\mathcal{P}$.

b) $\mathcal{P}$ is not a partition of $\mathbb{N}$ in this case, for while P and C are disjoint, they do not exhaust $\mathbb{N}$; neither 0 nor 1 are included in $P \cup C$.

c) This $\mathcal{P}$ is also not a partition, because while the sets $k\mathbb{Z}$ exhaust $\mathbb{Z}$, they are not pairwise disjoint; in particular, every set contains 0.

Generalizing from Examples 5 and 6, an equivalence relation induces a partition of the set, and conversely. This is the content of the next theorem.

Theorem 6.3.1: *Fundamental Theorem of Equivalence Relations*

a) *If $\sim$ is an equivalence relation on S, then its equivalence classes form a partition of S.*

b) *If $\mathcal{P}$ is a partition of S, then $\mathcal{P}$ generates an equivalence relation on S; moreover, the equivalence classes so created are the cells of $\mathcal{P}$.*

Proof:

a) Suppose $\sim$ is an equivalence relation on S, and let $\mathcal{P}_\sim = \{[x] : x \in S\}$.
$[x] \subseteq S$ for each x, so $\bigcup_{[x] \in \mathcal{P}_\sim} [x] \subseteq S$.

Conversely, each x in S is in $[x]$ because $x \sim x$, so $S \subseteq \bigcup_{[x] \in \mathcal{P}_\sim} [x]$.
Thus, $S = \bigcup_{[x] \in \mathcal{P}_\sim} [x]$. ✓

· Now let $[a]$ and $[b]$ be any equivalence classes.
Suppose $[a] \cap [b] \neq \emptyset$. We'll show that then $[a] = [b]$.
First suppose $x \in [a]$. Then $a \sim x$.
Since $[a] \cap [b] \neq \emptyset$, let $c \in [a] \cap [b]$.
Then $a \sim c$ and $b \sim c$.
By symmetry, $c \sim a$.
Thus, since $b \sim c \sim a \sim x$, transitivity yields $b \sim x$, and so $x \in [b]$.
This shows that $x \in [a] \rightarrow x \in [b]$, which gives $[a] \subseteq [b]$.
$[b] \subseteq [a]$ follows similarly (see Exercise 55a), so $[a] = [b]$. ✓
· This proves that $\mathcal{P}_\sim$ is a partition of S.

b) Suppose $\mathcal{P}$ is any partition of S.
Define $\sim_\mathcal{P}$ by $x \sim_\mathcal{P} y$ if and only if x and y are in the same cell of $\mathcal{P}$.
Then $\sim_\mathcal{P}$ is an equivalence relation (see Exercise 55c). ✓
· Now let $[a]$ be any equivalence class for $\sim_\mathcal{P}$.
Given any $x \in [a]$, $a \sim_\mathcal{P} x$.
Thus, a and x are in some cell C of the partition, making $[a] \subseteq C$.
Given any $z \in C$, z is in the same cell of $\mathcal{P}$ as a, too, so $a \sim_\mathcal{P} z$.
Thus, $z \in [a]$, and so $C \subseteq [a]$.
Hence, $[a] = C$: each equivalence class is a cell of the partition. ✓
· Now let C denote an arbitrary cell of the partition.
Let $x \in C$ (all cells are non-empty).
If $z \in C$, then $x \sim_\mathcal{P} z$, and so $z \in [x]$.
If $y \in [x]$, then $x \sim_\mathcal{P} y$.

But then x and y belong to the same cell of $\mathcal{P}$, which must be C, that is, $y \in C$.

Thus, $C = [x]. \checkmark$

· Therefore, the partition of equivalence classes induced by the equivalence relation $\sim_{\mathcal{P}}$ is the same as the original partition $\mathcal{P}$. ∎

The most important half of this theorem is the forward direction—*equivalence classes form a partition*. To say it differently, *equivalence relations induce partitions* on sets. Equivalence classes form a new collection of objects, a mathematical structure known as a *quotient structure*, which can be worked with further. In many instances, this new collection (the partition) has a similar structure to the original set. We'll see how this idea functions in two different contexts in the next section as we construct both the integers and the integers modulo n as quotient structures.

EXERCISE SET 6.3

Exercises 1–6: Binary Relations
Tell, with reasons, which of the following sets represent binary relations. For those that do, identify the relation's domain and range and graph the relation.

6.3.1. $\{(0,0), (0,1), (1,0), (1,1)\}$

6.3.2. $\{0, (0,0), (0,1), (1,0), (1,1), 1\}$

6.3.3. $\{(\emptyset, \emptyset), (\emptyset, \{\emptyset\}), (\{\emptyset\}, \emptyset), (\{\emptyset\}, \{\emptyset\})\}$

6.3.4. $S \times T$, where $S = \{0,1,2\}$, $T = \{2,3,4,5\}$

6.3.5. $R = \{(x,y) \in \mathbb{R}^2 : y^2 = x^2\}$

6.3.6. $\mathbb{N}$

6.3.7. *Counting Binary Relations*
Suppose S is a set with $|S| = n$.
Using the set-theoretic definition of a relation, answer the following:
  a. How many binary relations are there on S?
  b. How many binary relations on S are reflexive? symmetric? both?

Exercises 8–10: True or False
Are the following statements true or false? Explain your answer.

6.3.8. All equivalence relations are transitive relations.

6.3.9. Equivalence classes $[x]$ and $[y]$ are identical if and only if $x = y$.

6.3.10. A binary relation R can be represented as $\mathrm{Dom}(R) \times \mathrm{Rng}(R)$.

Exercises 11–17: Properties of Binary Relations
Determine whether the following relations are reflexive, symmetric, or transitive. Argue your claims.

6.3.11. $D = \{(x,x) : x \in S\}$, the diagonal of $S \times S$, where S is any set.

6.3.12. $\sim$, where $P \sim Q$ if and only if $P \wedge Q$ is a tautology, P and Q being sentences of *Propositional Logic*.

6.3.13. U, where $x \, U \, y \leftrightarrow x^2 + y^2 = 1$; $\quad x, y \in \mathbb{R}$.

6.3.14. C, where $x \, C \, y \leftrightarrow |x - y| < 1$; $\quad x, y \in \mathbb{R}$.

6.3.15. I, where $x \, I \, y \leftrightarrow x \subseteq y$; $\quad x, y \in \mathcal{P}(S)$ for some set S.

6.3.16. S, where $x \, S \, y$ if and only if polygons x and y have the same area.

6.3.17. P, where $x \, P \, y$ if and only if x and y are distinct parallel lines.

Exercises 18–30: Relational Properties and Set Theory

Prove or disprove the following results regarding the interaction of properties of binary relations with the relations and operations of Set Theory.

6.3.18. If R is a reflexive relation on S, then so is any subset of R.

6.3.19. If R is a symmetric relation on S, then so is any subset of R.

6.3.20. If R is a transitive relation on S, then so is any subset of R.

6.3.21. If $\sim$ is an equivalence relation on S and $T \subseteq S$, then $\sim$ restricted to T is an equivalence relation on T.

6.3.22. If R_1 and R_2 are reflexive relations on S_1 and S_2 respectively, then $R_1 \cap R_2$ is a reflexive relation on $S_1 \cap S_2$.

6.3.23. If R_1 and R_2 are symmetric relations on S_1 and S_2 respectively, then $R_1 \cap R_2$ is a symmetric relation on $S_1 \cap S_2$.

6.3.24. If R_1 and R_2 are transitive relations on S_1 and S_2 respectively, then $R_1 \cap R_2$ is a transitive relation on $S_1 \cap S_2$.

6.3.25. If R_1 and R_2 are reflexive relations on S_1 and S_2 respectively, then $R_1 - R_2$ is a reflexive relation on $S_1 - S_2$.

6.3.26. If R_1 and R_2 are symmetric relations on S_1 and S_2 respectively, then $R_1 - R_2$ is a symmetric relation on $S_1 - S_2$.

6.3.27. If R_1 and R_2 are transitive relations on S_1 and S_2 respectively, then $R_1 - R_2$ is a transitive relation on $S_1 - S_2$.

6.3.28. If R_1 and R_2 are reflexive relations on S_1 and S_2 respectively, then $R_1 \cup R_2$ is a reflexive relation on $S_1 \cup S_2$.

6.3.29. If R_1 and R_2 are symmetric relations on S_1 and S_2 respectively, then $R_1 \cup R_2$ is a symmetric relation on $S_1 \cup S_2$.

6.3.30. If R_1 and R_2 are transitive relations on S_1 and S_2 respectively, then $R_1 \cup R_2$ is a transitive relation on $S_1 \cup S_2$.

Exercises 31–40: Independence of Basic Relation Properties

Exhibit relations on sets of your choice having the following properties and show that they have them. You may use either familiar relations or construct them to suit the task (specific relations on $S = \{1, 2, 3\}$ will suffice).

6.3.31. Not reflexive, not symmetric, and not transitive.

6.3.32. Not reflexive, not symmetric, and transitive.

6.3.33. Not reflexive, symmetric, and not transitive.

6.3.34. Not reflexive, symmetric, and transitive.

6.3.35. Reflexive, not symmetric, and not transitive.

6.3.36. Reflexive, not symmetric, and transitive.

6.3.37. Reflexive, symmetric, and not transitive.

6.3.38. Reflexive, symmetric, and transitive.

6.3.39. *Euclidean Relations on a Set and Equivalence Relation Properties*
A relation R is *Euclidean* if and only if $(\forall a, b, c \in S)(a\,R\,b \wedge a\,R\,c \to b\,R\,c)$
(this is Euclid's *Common Notion* 1). *Argue the following*:

  a. A transitive relation need not be Euclidean, nor conversely.

  b. A symmetric, Euclidean relation is transitive, and a symmetric, transitive
     relation is Euclidean. Must a transitive, Euclidean relation be symmetric?

  c. A reflexive, Euclidean relation is an equivalence relation. Must a reflexive,
     transitive relation be Euclidean? Must it be an equivalence relation?

6.3.40. Criticize the following "proof," which argues that every symmetric,
transitive relation is reflexive.

   Suppose R is both symmetric and transitive, and let x and y be elements
     of S such that $x\,R\,y$.
   Then by the symmetric property, $y\,R\,x$ too.
   By the transitive property, then, $x\,R\,x$.
   Thus, R is reflexive. ∎

Exercises 41–46: Properties of Converse Relations
Symbolically formulate and then prove the following results pertaining to con-
verse relations (Definition 7) and set-theoretic operations on relations.

6.3.41. The converse of the converse is the original relation.

6.3.42. The converse of the intersection is the intersection of their converses.

6.3.43. The converse of the difference is the difference of their converses.

6.3.44. The converse of the union is the union of their converses.

6.3.45. A relation is symmetric if and only if it is identical with its converse.

6.3.46. The converse of an equivalence relation is an equivalence relation.

Exercises 47–54: Equivalence Classes and Partitions
Describe and picture in some way, if possible, the partitions induced by the
following equivalence relations.

6.3.47. $m \sim n \leftrightarrow 2 \mid (m - n);\quad m, n \in \mathbb{N}$.

6.3.48. $x \sim y \leftrightarrow |y^2 - x^2| = 0;\quad x, y \in \mathbb{Z}$.

6.3.49. $(x_1, y_1) \sim (x_2, y_2) \leftrightarrow x_1^2 + y_1^2 = x_2^2 + y_2^2;\quad x_i, y_i \in \mathbb{R}$ for $i = 1, 2$.

6.3.50. $x \sim y \leftrightarrow \lfloor x \rfloor = \lfloor y \rfloor;\quad x, y \in \mathbb{R}$.

6.3.51. $l \sim m$ if and only if l and m are parallel (or the same) straight lines.

6.3.52. $T_1 \sim T_2$ if and only if T_1 and T_2 are congruent triangles.

6.3.53. $p \sim q$ if and only if p and q are polynomials of the same degree.

6.3.54. $p \sim q$ if and only if $p = c \cdot q$ for p and q polynomials, $c \neq 0$.

6.3.55. *Fundamental Theorem of Equivalence Relations*
a. Fill the gap in *Theorem* 1a's proof, by showing that $x \in [b] \to x \in [a]$.
b. For *Theorem* 1b, show that the common partition cell C to which two related elements x and y belong must be unique.
c. Show that the relation $\sim_P$, defined in the proof of *Theorem* 1b, is an equivalence relation. Point out where both parts of the definition of a partition are used in your argument.

Exercises 56–57: Composition Results
The following results pertain to composite relations.

6.3.56. *Transitivity and Composition*
Prove that R is a transitive relation on a set S if and only if $R^2 \subseteq R$, where $R^2 = R \circ R$. Give a counterexample to show that R^2 need not equal R if R is transitive.

6.3.57. *Decomposition of Functions*
Let $f: S \to T$ be any function. Show f can be factored into a composition $f = g \circ p$ where p is onto and g is one-to-one by working the following parts.
a. Take $x_1 \sim x_2$ if and only if $f(x_1) = f(x_2)$. Show that $\sim$ is an equivalence relation, and identify the equivalence classes.
b. Let $S_\sim$ denote the set of equivalence classes of $\sim$ in S. Prove that the association $p: x \mapsto [x]$ is an onto function $p: S \to S_\sim$.
c. Prove that $g: [x] \mapsto f(x)$ is a one-to-one function $g: S_\sim \to T$.
d. Prove that $f = g \circ p$.

Exercises 58–64: Transitive Extensions and Closures
Let R be a binary relation on a set S.
Definition: $\overline{R} = R \cup \{(x, z) : (x, y) \in R \wedge (y, z) \in R$ *for some* $y \in S\}$ *is the* **transitive extension** *of* R.
Definition: $R^* = \bigcup_{i=0}^{\infty} R_i$, *where* $R_0 = R$ *and* $R_{n+1} = \overline{R}_n$, *is the* **transitive closure** *of* R.

6.3.58. Let $S = \{a, b, c, d\}$, $R = \{(a, b), (b, c), (c, b), (d, a)\}$. What is $\overline{R}$ and R^*?

6.3.59. Is the transitive extension of a binary relation transitive? Prove it in general, or give a counterexample.

6.3.60. Is the transitive closure of a binary relation transitive? Prove it in general, or give a counterexample.

6.3.61. Prove that $R^* \subseteq T$, where T is any transitive relation containing R.

6.3.62. Prove that the intersection of all transitive relations containing R is a transitive relation containing R. Conclude from this that R^* is the intersection of all transitive relations containing R.

6.3.63. Prove that the transitive closure of a symmetric relation is symmetric.

6.3.64. Is it ever necessary to take infinitely many unions to get R^*, or will finitely many suffice (i.e., eventually $R_{n+1} = R_n$)? Prove that finitely many are enough, or find a counterexample to show that infinitely many are needed.

6.4 The Integers and Modular Arithmetic

In Chapter 3, we explored the natural number system as an inductive structure and saw how the *Peano Postulates* and recursive definitions for addition and multiplication provide a deductive basis for elementary arithmetic. We also discussed divisibility, and we expanded our computational universe to the integers, but without considering its theoretical foundation. We'll do so now. We'll also introduce modular arithmetic, a topic that's important for both mathematics and computer science. Our approach to these topics will use ideas about equivalence relations and partitions from the last section.

6.4.1 *Historical Background on Negative Numbers*

The natural numbers form the oldest and most easily understood number system—we use it all the time for counting. Common fractions also evolved relatively early because cultures needed to divide quantities into equal parts to share them fairly. Irrational quantities like $\sqrt{2}$ arose in order to deal with lengths of line segments compared to some unit, though the Greeks conceived of them as ratios inexpressible in terms of whole numbers. Different cultures developed their own ways to think about and calculate with all of these quantities.

Negative numbers were the most troublesome and were often rejected. How could anything be less than nothing? Subtracting a smaller number from a larger made sense, but taking a larger number away from a smaller was meaningless.

Nevertheless, some cultures learned to calculate with negative numbers. The ancient Chinese incorporated a germinal arithmetic of negative numbers into their method of solving linear problems. Using a standardized process similar to Gaussian elimination on a rectangular system of coefficients that represented the problem's data, they sometimes needed to compute with negative numbers to get their answer. Medieval Indian mathematicians had a more developed arithmetic of negative numbers, which they initially associated with debts. This idea was only rarely accepted in medieval Europe. In the early modern period, negative numbers were adopted by European mathematicians to round out algebraic problem solving, even though they had no good interpretation for negative numbers in most settings.

As late as the early nineteenth century, the meaning and significance of negative numbers were still being debated. Around 1840, British mathematicians put forward a couple of rationales to justify their use. One was to consider them signs that one could calculate with in familiar ways. William Rowan Hamilton found this formalistic approach unsatisfying and tried to explain negative numbers philosophically, using ideas from Kant. We won't follow all the steps in this approach, but some of his techniques came to be accepted as an appropriate mathematical foundation. We'll look at a modern version of his ideas next.

6.4.2 *Theoretical Construction of the Integers*

The arithmetic of negative numbers began when mathematicians calculated with subtracted quantities—those computations suggested the rules that should hold. For example, $7 - (3 - 1) = 5 = 7 - 3 + 1$, so subtracting a negative must be the same as adding its positive. Similarly, $(8 - 5) \cdot (4 - 2) = 6 = 8 \cdot 4 - 8 \cdot 2 - 5 \cdot 4 + 5 \cdot 2$, so multiplying a negative by a positive gives a negative, and multiplying a negative by a negative must give a positive.

Subtracted quantities also gave a way to think about negative numbers—as differences where a larger number is subtracted from a smaller: $^-3 = 0 - 3 = 5 - 8$, etc. Similarly, positive numbers are differences when a smaller is subtracted from a larger: $^+2 = 2 - 0 = 6 - 4$, etc. In this way, all integers are treated as differences of pairs of natural numbers. Two differences $a - b$ and $a' - b'$ represent the same integer when $a - b = a' - b'$. Put in terms of natural number arithmetic, $a - b = a' - b'$ if and only if $a + b' = a' + b$. The following construction develops these ideas rigorously in set-theoretic terms.

❖**Example 6.4.1**

Define an equivalence relation on $\mathbb{N} \times \mathbb{N}$ to capture the notion of equality of integers, and choose representatives for the associated equivalence classes.

Solution

· *Define* $\equiv$ *on* $\mathbb{N} \times \mathbb{N}$ *by* $(a, b) \equiv (a', b') \leftrightarrow a + b' = a' + b$.
Then $\equiv$ is an equivalence relation on $\mathbb{N} \times \mathbb{N}$ (see Exercise 4).
Consequently, this relation partitions $\mathbb{N} \times \mathbb{N}$ into equivalence classes $[(m, n)]$ of pairs of natural numbers.
· Ordered pairs $(m, 0)$ and $(0, n)$ can be used as standard representatives for these classes, for each equivalence class contains exactly one or the other of these ordered pairs (see Exercise 5f). This gives us a way to define positive and negative integers.

Definition 6.4.1: *Set-Theoretic Model of the Integers*

a) *Let* $\equiv$ *be defined on* $\mathbb{N} \times \mathbb{N}$ *by* $(a, b) \equiv (a', b') \leftrightarrow a + b' = a' + b$. *Then*
b) $\mathbb{Z} = \{[(m, n)]\}$ *is the set of equivalence classes induced by* $\equiv$, *with*
c) $^+m = [(m, 0)]$ *and* $^-n = [(0, n)]$ *for* $m \neq 0 \neq n$, *and* $0 = [(0, 0)]$.

A copy of the natural numbers $\mathbb{N}$ lies inside this model of the integers, but strictly speaking, natural numbers are not nonnegative integers. This awkward peculiarity is usually resolved by identifying $\mathbb{N}$ with its *isomorphic image* $\{0\} \cup \{^+m : m \in \mathbb{N}^+\}$ in $\mathbb{Z}$. There is a benefit to this set-theoretic approach, however. Since the integers are defined using only natural number arithmetic and constructs from set theory, the consistency of integer arithmetic is guaranteed by those supposedly more basic portions of mathematics.

Regardless of how $\mathbb{Z}$ is theoretically constructed, nothing hinders us from connecting integers to a variety of meaningful real-world applications, thinking of them there as gains and losses or increases and decreases or moving forward and backward or temperatures above and below 0, etc.

6.4.3 *Addition and Subtraction of Integers*

Addition of integers $x+y$ is first explained to children by considering different cases, depending on the signs and magnitudes of x and y. For us, addition is defined in a more straightforward way as the sum of equivalence classes, based on knowing that $(a - b) + (c - d) = (a + c) - (b + d)$. To make sure that this operation is well defined, we must show that our definition does not depend on the particular ordered-pair representatives used.

Proposition 6.4.1: *Addition is Well Defined*
  If $[(a, b)] = [(a', b')]$ *and* $[(c, d)] = [(c', d')]$, *then*
$$[(a + c, b + d)] = [(a' + c', b' + d')].$$
Proof:
· Suppose $[(a, b)] = [(a', b')]$ and $[(c, d)] = [(c', d')]$.
  Then $(a, b) \equiv (a', b')$ and $(c, d) \equiv (c', d')$,
    so $a + b' = a' + b$ and $c + d' = c' + d$.
· Adding these equations (and using natural number properties to rearrange terms and ignore parentheses) gives $a + c + b' + d' = a' + c' + b + d$.
  But this shows $(a + c, b + d) \equiv (a' + c', b' + d')$.
· Thus, $[(a + c, b + d)] = [(a' + c', b' + d')]$.   ∎

Definition 6.4.2: *Addition of Integers*
  $[(a, b)] + [(c, d)] = [(a + c, b + d)]$

  Addition of integers satisfies the usual algebraic properties. Their proofs depend upon the definition of integer addition and on the associated laws for natural number addition occurring in each coordinate. Universal quantifiers have been omitted in the formulations for the sake of readability.

Proposition 6.4.2: *Algebraic Properties of Integer Addition*
  a) **Associative Law**: $(x + y) + z = x + (y + z)$
  b) **Commutative Law**: $x + y = y + x$
  c) **Additive Identity**: $x + 0 = x = 0 + x$
  d) **Additive Inverses**: *each integer x has a unique additive inverse ^-x.*
  e) **Cancellation**: $x + y = x + z \rightarrow y = z$; $x + z = y + z \rightarrow x = y$

Proof:
  a) Suppose $x = [(x_1, x_2)]$, $y = [(y_1, y_2)]$, and $z = [(z_1, z_2)]$.
    Then $(x + y) + z = [(x_1 + y_1, x_2 + y_2)] + [(z_1, z_2)]$
$$= [((x_1 + y_1) + z_1, (x_2 + y_2) + z_2)]$$
$$= [(x_1 + (y_1 + z_1), x_2 + (y_2 + z_2))]$$
$$= [(x_1, x_2)] + [(y_1 + z_1, y_2 + z_2)]$$
$$= x + (y + z).   ∎$$
  b) See Exercise 7.
  c) This is immediate from the definition of 0. See Exercise 8.
  d) Uniqueness is proved in the usual way. To prove existence, find an integer ^-x such that $x + {}^-x = 0 = {}^-x + x$. See Exercise 9.
  e) This can be proved using additive inverses. See Exercise 10.

Additive inverses can be used to define subtraction on $\mathbb{Z}$, which is well defined because addition is and because inverses are unique. Note that here ^-x denotes the inverse of integer x, not a negative integer. Representing negative integers $[(0,n)]$ by ^-n (*Definition* 1b) involves a similar notation, but there the negative sign is prefixed to a natural number.

Definition 6.4.3: *Subtraction of Integers*
$$x - y = x + {}^-y$$

6.4.4 *Multiplication of Integers*

Multiplication can be defined for integers so that the rules of signs hold as well as all of the usual algebraic properties. In fact, as motivation for how to define integer multiplication, let's review how subtracted quantities are multiplied: $(a - b) \cdot (c - d) = a \cdot c - a \cdot d - b \cdot c + b \cdot d = (ac + bd) - (ad + bc)$. This suggests the formula to use because we're replacing $m - n$ with $[(m,n)]$. Again, before making the definition, we'll make sure that it is well defined. This is more complicated than it was for addition.

Proposition 6.4.3: *Multiplication is Well Defined*
If $[(a,b)] = [(a',b')]$ and $[(c,d)] = [(c',d')]$, then
$$[(ac + bd, ad + bc)] = [(a'c' + b'd', a'd' + b'c')].$$

Proof:
· Suppose $[(a,b)] = [(a',b')]$ and $[(c,d)] = [(c',d')]$.
  Then, due to ordered-pair equivalence, $a + b' = a' + b$ and $c + d' = c' + d$.
· To prove $[(ac + bd, ad + bc)] = [(a'c' + b'd', a'd' + b'c')]$, we'll scale the above equivalence equations and add them together to reproduce the terms involved in the consequent's ordered pairs.
· We'll scale the first equivalence equation by c and d and then add:
  $(a + b')c + (a' + b)d = (a' + b)c + (a + b')d$
  $(ac + b'c) + (a'd + bd) = (a'c + bc) + (ad + b'd)$
  $(ac + bd) + (a'd + b'c) = (ad + bc) + (a'c + b'd)$
  So, $[(ac + bd, ad + bc)] = [(a'c + b'd, a'd + b'c)]$. ✓
· Similarly, we'll scale the second equivalence equation by a' and b' and add:
  $a'(c + d') + b'(c' + d) = a'(c' + d) + b'(c + d')$
  $(a'c + a'd') + (b'c' + b'd) = (a'c' + a'd) + (b'c + b'd')$
  $(a'c + b'd) + (a'd' + b'c') = (a'd + b'c) + (a'c' + b'd')$
  So, $[(a'c + b'd, a'd + b'c)] = [(a'c' + b'd', a'd' + b'c')]$. ✓
· Thus, $[(ac + bd, ad + bc)] = [(a'c' + b'd', a'd' + b'c')]$. ∎

Definition 6.4.4: *Multiplication of Integers*
$$[(a,b)] \cdot [(c,d)] = [(ac + bd, ad + bc)]$$

Multiplication of integers is associative, commutative, has a unique identity, satisfies cancellation, and enters into a distributive law. We'll restate these but leave their proofs for the exercises.

Proposition 6.4.4: *Algebraic Properties of Multiplication*
 a) ***Associative Law***: $(x \cdot y) \cdot z = x \cdot (y \cdot z)$
 b) ***Commutative Law***: $x \cdot y = y \cdot x$
 c) ***Multiplicative Identity***: *the unique multiplicative identity is* $1 = {^+}1$.
 d) ***Cancellation***: $z \neq 0 \to (x \cdot z = y \cdot z \to x = y) \wedge (z \cdot x = z \cdot y \to x = y)$
 e) ***Distributive Laws***: $x \cdot (y + z) = x \cdot y + x \cdot z$; $\quad (x + y) \cdot z = x \cdot z + y \cdot z$

Proof:
 a) This proof is tedious but straightforward. See Exercise 13.
 b) This follows from natural number commutativity. See Exercise 14.
 c) See Exercise 15.
 d) This follows from cancellation for the natural numbers. See Exercise 16.
 e) This uses distributivity for the natural numbers. See Exercise 18.

6.4.5 Division of Integers

Addition and multiplication were total operations on the natural numbers. Expanding $\mathbb{N}$ to $\mathbb{Z}$ gives us a system in which we can do unrestricted subtraction. Expanding $\mathbb{Z}$ to $\mathbb{Q}$, where division becomes a (near) total operation, requires the same sort of construction—representing fractions a/b by ordered pairs of integers, introducing an equivalence relation on them to capture equality of fractions, and then defining operations on equivalence classes. We'll leave this to be explored in an exercise (see Exercise 33).

We can divide integers, though, if we allow remainders. We already considered this possibility when we looked at divisibility in Section 3.5. This gives us the background needed for modular arithmetic. To keep matters simple, we'll ignore the fact that integers were just constructed as ordered pairs and treat them more informally. A rigorous development would also require results about the order relations $<$ and $\leq$ on $\mathbb{Z}$ (see Exercises 19–24).

Theorem 6.4.1: *Division Algorithm*
 If $d > 0$, there exists unique q and r satisfying $n = qd + r$ with $0 \leq r < d$.

Proof:
· Though this is called the *Division Algorithm*, it's a unique existence *theorem* about quotients and remainders, something requiring proof.
· Let n be any integer and d be a positive integer.
 Consider the collection of half-closed, half-open intervals of the form $[md, (m+1)d)$ for all integers m.
 These intervals form a partition of $\mathbb{Z}$, covering it without any overlap.
 Thus, n lies in exactly one of these intervals, say, in $[qd, (q+1)d)$.
· Let $r = n - qd$. Then $n = qd + r$ and $0 \leq r < d$. ✓
· To show uniqueness, suppose $q_1 d + r_1 = n = q_2 d + r_2$, $0 \leq r_1, r_2 < d$.
 Then $r_2 - r_1 = (q_1 - q_2)d$, some integer multiple of d.
 Since $0 \leq r_1, r_2 < d$, $0 \leq |r_2 - r_1| < d$; so $r_2 - r_1 = 0d = 0$.
· Thus, $r_2 - r_1 = 0$, and $q_1 - q_2 = 0$, which means $r_2 = r_1$ and $q_1 = q_2$. ∎

Our main interest here, unlike what it was when we first learned long division, is in the *remainders* that result from division. The following result gives us an equivalence that's useful for developing *Modular Arithmetic*.

Proposition 6.4.5: *Same Remainders Criterion*

Integers m, n divided by d yield the same remainder if and only if $d \mid (m-n)$.

Proof:

· Suppose m and n yield the same remainder r when divided by d.
Then $m = q_1 d + r$ and $n = q_2 d + r$.
Thus, $m - n = q_1 d - q_2 d = (q_1 - q_2)d$, which implies $d \mid (m - n)$. ✓

· Now suppose $d \mid (m - n)$, and let $m = q_1 d + r_1$ and $n = q_2 d + r_2$.
Then $d \mid ((q_1 - q_2)d + (r_1 - r_2))$, which means $d \mid (r_1 - r_2)$, too.
But $0 \leq r_1, r_2 < d$, so $0 \leq |r_1 - r_2| < d$.
Thus, $r_1 - r_2 = 0$, i.e., $r_1 = r_2$. ∎

6.4.6 *Modular Arithmetic*

There are two approaches to *Modular Arithmetic*. One treats it like clock arithmetic, the numbers being hours and the operations causing time changes. For example, in *mod* 12 arithmetic, the numbers are $0, 1, \ldots, 11$, and these are added and subtracted as we would hours on a clock, with 12 being 0: $8 + 5 = 1$, $3 + 9 = 0$, and $2 - 7 = 7$.

A more rigorous approach introduces an equivalence relation on the integers and then develops arithmetic for the resulting equivalence classes using more natural operations. The first approach gives an elementary way to work with modular arithmetic; the second provides it with a theoretical foundation. We'll follow the second approach here, expanding our number concept to a new system using equivalence relations and equivalence classes.

Definition 6.4.5: *Congruence Mod n*

$m \equiv_n r \leftrightarrow n \mid (m - r)$

This is easily shown to be an equivalence relation (see Exercise 27a). Consequently, $\equiv_n$ induces a partition of equivalence classes on $\mathbb{Z}$. Each class contains a unique remainder r relative to division by n (see Exercise 27b), so those remainders can be used as representatives for the *integers mod n*.

Definition 6.4.6: *Integers Mod n*

$\mathbb{Z}_n = \{[0], [1], \ldots, [n-1]\}$, *the set of equivalence classes induced on $\mathbb{Z}$ by $\equiv_n$.*

Proposition 6.4.6: *Addition Mod n is Well Defined*

If $a \equiv_n a'$ and $b \equiv_n b'$, then $a + b \equiv_n a' + b'$.

Proof:

Suppose $a \equiv_n a'$ and $b \equiv_n b'$.
Then $n \mid (a - a')$ and $n \mid (b - b')$, so $n \mid ((a - a') + (b - b'))$.
Thus, $n \mid ((a + b) - (a' + b'))$.
Hence, $a + b \equiv_n a' + b'$. ∎

Definition 6.4.7: *Addition Mod n*
$[a] +_n [b] = [a + b]$

Given this definition, we can prove the *Associative Law* and the *Commutative Law* for *addition mod n*, that $[0]$ is the *Additive Identity*, and that $[r]$ has *Additive Inverse* $^-[r] = [-r]$. Using additive inverses, *subtraction mod n* can be defined. These are explored in the exercises (see Exercises 28 and 30).

Multiplication mod n can also be defined in terms of class representatives, as we'll see next.

Proposition 6.4.7: *Multiplication Mod n is Well Defined*
If $a \equiv_n a'$ and $b \equiv_n b'$, then $ab \equiv_n a'b'$.

Proof:
· Suppose $a \equiv_n a'$ and $b \equiv_n b'$. Then $n \mid (a - a')$ and $n \mid (b - b')$.
· We'd like $n \mid (ab - a'b')$ to draw our conclusion. Scaling the two differences to involve these terms and then adding will give what's needed.
   $n \mid (a - a')b$ and $n \mid (b - b')a'$, so $n \mid (ab - a'b)$ and $n \mid (a'b - a'b')$.
   Thus, $n \mid ((ab - a'b') + (a'b - a'b))$, i.e., $n \mid (ab - a'b')$.
· Hence, $ab \equiv_n a'b'$. ∎

Definition 6.4.8: *Multiplication Mod n*
$[a] \cdot_n [b] = [ab]$

Using this definition and analogous properties for $\mathbb{Z}$, we can prove the *Associative Law* and the *Commutative Law* for multiplication *mod n*, the *Distributive Laws* governing addition and multiplication *mod n*, and that $[1]$ is the *Multiplicative Identity* (see Exercise 29). Like the integers, the integers mod n need not have multiplicative inverses, and so division mod n may not be possible. We will also leave these results for the exercises (see Exercise 31).

We could continue to use equivalence class notation for *mod n* integers, $+_n$ for addition, and $\cdot_n$ for multiplication, but it is customary to drop the brackets and the subscripts in *Modular Arithmetic*, understanding that ordinary-looking symbols for numbers, addition, and multiplication refer to modular entities and operations. This occasionally gives some surprising looking identities—in *mod 12* arithmetic, we have results like $4 \cdot 6 = 0$ and $5 \cdot 3 = 3$.

EXERCISE SET 6.4

Exercises 1–3: True or False
Are the following statements true or false? Explain your answer.

6.4.1. Integer arithmetic is logically consistent if *Peano Arithmetic* and *Set Theory* are.

6.4.2. The *Division Algorithm* is a procedure for doing long division.

6.4.3. If $m \leq n$, then $\mathbb{Z}_m \subseteq \mathbb{Z}_n \subseteq \mathbb{Z}$.

Exercises 4–5: Construction of the Integers
The following problems flesh out the construction of the integers.

6.4.4. Show that the relation $\equiv$ on $\mathbb{N} \times \mathbb{N}$ defined by $(a, b) \equiv (a', b') \leftrightarrow a + b' = a' + b$ (*Definition 1*) is an equivalence relation, as follows:
a. $(a, b) \equiv (a, b)$
b. $(a, b) \equiv (a', b') \rightarrow (a', b') \equiv (a, b)$
c. $(a, b) \equiv (a', b') \wedge (a', b') \equiv (a'', b'')$
$\rightarrow (a, b) \equiv (a'', b'')$

6.4.5. Given the equivalence relation $\equiv$ of Exercise 4, show the following:
a. $[(m, 0)] = [(a, b)] \leftrightarrow m = a - b$
b. $[(0, n)] = [(a, b)] \leftrightarrow n = b - a$
c. $(m_1, 0) \equiv (m_2, 0) \leftrightarrow m_1 = m_2$
d. $(0, n_1) \equiv (0, n_2) \leftrightarrow n_1 = n_2$
e. $(m, 0) \equiv (0, n) \leftrightarrow m = 0 = n$
f. Explain why the above results imply that each equivalence class induced by $\equiv$ on $\mathbb{N} \times \mathbb{N}$ contains exactly one pair of the form $(m, 0)$ or $(0, n)$.

Exercises 6–11: Addition and Subtraction of Integers
Prove the following propositions for addition and subtraction of integers (Definitions 2 and 3). [The symbol $+$ indicates addition of integers and addition of natural numbers as well as being a positive integer. The symbol $-$ indicates negative integers, integer inverses, and subtraction of integers.]

6.4.6. *Addition of Positives and Negatives*
a. $^{+}m_1 + {^{+}m_2} = {^{+}(m_1 + m_2)}$
b. $^{+}m + 0 = {^{+}m} = 0 + {^{+}m}$
c. $^{-}n + 0 = {^{-}n} = 0 + {^{-}n}$
d. $^{-}n_1 + {^{-}n_2} = {^{-}(n_1 + n_2)}$

6.4.7. *Commutativity of Integer Addition:* $x + y = y + x$

6.4.8. *Additive Identity for Integers*
a. $x + 0 = x = 0 + x$
b. $x + x = 0 \leftrightarrow x = 0$

6.4.9. *Additive Inverses for Integers*
a. *Every integer x has a unique additive inverse ^{-}x.*
b. $^{-}(^{-}x) = x$
c. $^{-}(x + y) = {^{-}x} + {^{-}y}$

6.4.10. *Cancellation for Integer Addition*
a. $x + y = x + z \leftrightarrow y = z$
b. $x + z = y + z \leftrightarrow x = y$

6.4.11. *Subtraction Properties*
a. $0 - x = {^{-}x}$
b. $x - (y + z) = x - y - z$

Exercises 12–18: Multiplication of Integers
Prove the following propositions for multiplication of integers (Definition 4).

6.4.12. *Rule of Signs for Multiplication*
a. $^{+}m \cdot {^{+}n} = {^{+}mn}$
b. $^{+}m \cdot {^{-}n} = {^{-}mn}$
c. $^{-}m \cdot {^{+}n} - {^{-}mn}$
d. $^{-}m \cdot {^{-}n} = {^{+}mn}$

6.4.13. *Associativity of Integer Multiplication:* $(x \cdot y) \cdot z = x \cdot (y \cdot z)$

6.4.14. *Commutativity of Integer Multiplication:* $x \cdot y = y \cdot x$

6.4.15. *Multiplicative Identity:* $1 = {^{+}1}$ *is the unique multiplicative identity.*

6.4.16. *Cancellation Laws for Integer Multiplication*
a. $z \neq 0 \rightarrow (x \cdot z = y \cdot z \rightarrow x = y)$
b. $z \neq 0 \rightarrow (z \cdot x = z \cdot y \rightarrow x = y)$

6.4.17. *Zero Products and Zero Divisors in $\mathbb{Z}$:* $x \cdot y = 0 \leftrightarrow x = 0 \vee y = 0$

6.4.18. *Distributive Laws for Integer Arithmetic*
a. $x \cdot (y + z) = x \cdot y + x \cdot z$ b. $(x + y) \cdot z = x \cdot z + y \cdot z$

Exercises 19–24: Strict Order in $\mathbb{Z}$
The strict order relation $<$ for $\mathbb{Z}$ is defined by $x < y \leftrightarrow \exists^+p(x + {}^+p = y)$
(compare how the order relation $\leq$ for $\mathbb{N}$ was defined in Exercise Set 3.4.)
Prove the following results about this order relation.

6.4.19. *Positive Integers*
a. $0 < {}^+p$ for all non-zero natural numbers p.
b. $0 < x \rightarrow \exists^+p(x = {}^+p)$
c. Taking $>$ as the converse of $<$, prove that $x < y \leftrightarrow (\exists d > 0)(x+d = y)$.
d. $0 < x \cdot x$

6.4.20. *Trichotomy Laws for Integer Ordering*
a. Exactly one of the following is true: $x < 0$, $x = 0$, or $0 < x$.
b. Exactly one of the following is true: $x < y$, $x = y$, or $y < x$.

6.4.21. *Non-Reflexivity and Non-Symmetry of $<$*
a. $\neg(x < x)$ b. $\neg(x < y \rightarrow y < x)$

6.4.22. *Transitivity of $<$:* $x < y \land y < z \rightarrow x < z$

6.4.23. *Additive and Multiplicative Properties of $<$*
a. $x < y \rightarrow x + z < y + z$ c. $x < y \land 0 < c \rightarrow c \cdot x < c \cdot y$
b. $x < y \rightarrow x - z < y - z$ d. $x < y \land c < 0 \rightarrow c \cdot y < c \cdot x$

6.4.24. *Properties of $\leq$*
The partial order $\leq$ can be defined on $\mathbb{Z}$ by $x \leq y \leftrightarrow x < y \lor x = y$.
a. Which of the properties listed above for $<$ also hold for $\leq$?
b. Prove the reflexive property of $\leq$: $x \leq x$
c. Prove the antisymmetric property of $\leq$: $x \leq y \land y \leq x \rightarrow x = y$

Exercises 25–26: Division Algorithm
The following problems explore the Division Algorithm.

6.4.25. *Illustrating the Division Algorithm*
Find the quotient q and remainder r for the following:
a. $31 \div 7 = q$ rem. r c. $425 \div 17 = q$ rem. r
b. $16 \div 5 = q$ rem. r d. $-978 \div 12 = q$ rem. r

6.4.26. *Proving the Division Algorithm*
Given n and d as in the *Division Algorithm*, consider the set of all positive
differences $n - qd$. Apply the *Well-Ordering Principle* to this set and use the
result to develop an alternative proof of the *Division Algorithm*.

Exercises 27–32: Modular Arithmetic
The following problems explore aspects of Modular Arithmetic.

6.4.27. *Congruence Mod n*
a. Prove that $\equiv_n$ is an equivalence relation on the integers.
b. Prove that *every equivalence class contains a unique remainder relative
   to division by n.*

6.4.28. *Addition Mod n*

Prove the following propositions for addition *mod n* on $\mathbb{Z}_n$:

   a. *Associative Law*: $([x] +_n [y]) +_n [z] = [x] +_n ([y] +_n [z])$

   b. *Commutative Law*: $[x] +_n [y] = [y] +_n [x]$

   c. *Additive Identity*: $[0]$ is the unique additive identity.

   d. *Additive Inverses*: $[-x]$ is the unique additive inverse for $[x]$.

   e. *Cancellation Laws*: $[x] +_n [z] = [y] +_n [z] \rightarrow [x] = [y]$
$$[z] +_n [x] = [z] +_n [y] \rightarrow [x] = [y]$$

   f. Reformulate the results in parts *a–e* without brackets or subscripts. Do they match what you know about the integers?

6.4.29. *Multiplication Mod n*

Prove the following propositions for multiplication *mod n* on $\mathbb{Z}_n$:

   a. *Associative Law*: $([x] \cdot_n [y]) \cdot_n [z] = [x] \cdot_n ([y] \cdot_n [z])$

   b. *Commutative Law*: $[x] \cdot_n [y] = [y] \cdot_n [x]$

   c. *Multiplicative Identity*: $[1]$ is the unique multiplicative identity.

   d. *Distributive Laws*: $[a] \cdot_n ([b] +_n [c]) = [a] \cdot_n [b] +_n [a] \cdot_n [c]$
$$([a] +_n [b]) \cdot_n [c] = [a] \cdot_n [c] +_n [b] \cdot_n [c]$$

   e. Reformulate the results in parts *a–d* without brackets or subscripts. Do they match what you know about the integers?

6.4.30. *Subtraction Mod n*

   a. Define *subtraction mod n* for $\mathbb{Z}_n$: $[a] -_n [b] = $ _____ .

   b. Using your definition, prove that $[a] -_n ([b] -_n [c]) = [a] -_n [b] +_n [c]$.

   c. Using your definition, prove that $[a] \cdot_n ([b] -_n [c]) = [a] \cdot_n [b] -_n [a] \cdot_n [c]$.

6.4.31. *Multiplicative Inverses Mod n*

Definition: $[b]$ *is a multiplicative inverse of* $[a]$ *if and only if* $[a] \cdot_n [b] = [1] = [b] \cdot_n [a]$.

   a. If an integer mod n has a multiplicative inverse, must it be unique? Why?

   b. Which numbers *mod* 11 have multiplicative inverses? Which numbers *mod* 12 have multiplicative inverses?

   c. *Exploration*: For which $\mathbb{Z}_n$ do all $[m] \neq [0]$ have multiplicative inverses?

6.4.32. *Order Relations on $\mathbb{Z}_n$*

Can a useful order relation $\leq_n$ be defined on $\mathbb{Z}_n$? Why or why not?

6.4.33. *Exploring the Classical Construction of $\mathbb{Q}$*

The following steps outline the classical construction of $\mathbb{Q}$ from $\mathbb{Z}$ using *Set Theory*.

   a. Construct $\mathbb{Q}$ as equivalence classes of ordered pairs of integers. Recall that denominators can't be 0.

   b. Define addition of rational numbers, first proving it is well defined.

   c. Prove that addition is associative and commutative. What is the additive identity for $\mathbb{Q}$? How are additive inverses defined?

   d. Define multiplication of rational numbers, first proving it is well defined.

   e. Prove that multiplication is associative and commutative. What is the multiplicative identity for $\mathbb{Q}$? How are multiplicative inverses defined?

   f. Prove the distributive laws for $\mathbb{Q}$.

Chapter 7
Posets, Lattices, and Boolean Algebra

7.1 Partially Ordered Sets

Elementary mathematics has traditionally focused disproportionately on computation. Students start with arithmetic, and after mastering that, move on to algebra, which is presented as a symbolic arithmetic with a collection of problem-solving techniques. Elementary algebra calculates with letter expressions or applies ready-made formulas to find unknown values, and it manipulates equations to show how variables are interrelated.

Such an operational emphasis is understandable, but there's another side to algebra—the study of relations. Algebra students occasionally solve inequalities, but this can seem like an intrusion disconnected from the main goal of solving equations. Calculus also deals with order relations when defining limits or determining intervals where functions increase and decrease, but it primarily calculates quantities like slopes, rates of change, areas under curves, centers of mass, and so on.

Relational ideas have become more mainstream, though, with the advent of computer science and the rise of discrete mathematics. Many contemporary mathematical applications involve binary or n-ary relations in addition to computations. We began looking at this topic at the end of the last chapter with equivalence relations. Here we'll explore other kinds of *binary relations*, particularly ones that impose order on a set. We'll investigate poset and lattice order-structures and use them to introduce *Boolean Algebra*. Exploring these matters will tie together some earlier ideas in logic and set theory and lead us into areas important to computer science and electrical engineering.

7.1.1 *Partial and Total Orders on a Set*

The prototypical partial order is the subset relation. Let's begin with this example.

✤**Example 7.1.1**

Let S be any set and $\mathcal{A}$ some collection of its subsets. Analyze the properties holding for the subset relation $\subseteq$ on $\mathcal{A}$.

Solution
- The subset relation is *reflexive* and *transitive*, but generally *not symmetric*. For if $X \in \mathcal{A}$, then $X \subseteq X$; and whenever $X \subseteq Y$ and $Y \subseteq Z$, $X \subseteq Z$. ✓
- Furthermore, $X \subseteq Y$ does not force $Y \subseteq X$. In particular, it fails if $\mathcal{A}$ is the full power set $\mathcal{P}(S)$ for a non-empty S. ✓
- What is true, however, is that if $X \subseteq Y$ and $Y \subseteq X$, then $X = Y$. ✓
  This is the *antisymmetric property* of $\subseteq$.

© Springer Nature Switzerland AG 2019
C. Jongsma, *Introduction to Discrete Mathematics via Logic and Proof*,
Undergraduate Texts in Mathematics,
https://doi.org/10.1007/978-3-030-25358-5_7

Definition 7.1.1: *Antisymmetric Relations*
A binary relation R on a set $\mathcal{A}$ is **antisymmetric** *if and only if for all elements x and y of $\mathcal{A}$, whenever $x\,R\,y$ and $y\,R\,x$, then $x = y$.*

Definition 7.1.2: *Partial Order on a Set; Poset*
 a) *A binary relation R on a set $\mathcal{A}$ is a* **partial order** *on $\mathcal{A}$ if and only if R is reflexive, antisymmetric, and transitive.*
 b) *A set $\mathcal{A}$ together with a partial order R on $\mathcal{A}$ is a* **partially ordered set (poset)**.

A poset is a set $\mathcal{A}$ together with a partial order R. When the order is understood, though, we'll simply refer to $\mathcal{A}$ as the poset.

Binary relations link elements in a certain way, but why is a reflexive, antisymmetric, transitive relation called a *partial order*? Why an *order*? And is there such a thing as a *total* order? There is. Defining it and looking at another example might help clarify the terminology.

Definition 7.1.3: *Connected Relations*
A binary relation R on a set $\mathcal{A}$ is **connected** *if and only if for all elements x and y of $\mathcal{A}$, either $x\,R\,y$ or $y\,R\,x$.*

Definition 7.1.4: *Total Order/Linear Order on a Set*
A binary relation R on a set $\mathcal{A}$ is a **total order/linear order** *on $\mathcal{A}$ if and only if R is a connected partial order on $\mathcal{A}$.*

❖**Example 7.1.2**
Show that the relation $\leq$ is a total order on the set of real numbers $\mathbb{R}$.

Solution
The number-relation $\leq$ is reflexive, antisymmetric, transitive, and connected, so $\leq$ is a total or linear order on $\mathbb{R}$. This relation orders real numbers linearly, something that's obvious when we place them on a number line. The ordering is total because any two real numbers can be compared.

Every total order is like this. We can put the elements of $\mathcal{A}$ on a vertical line to graph R by placing x below y if and only if $x\,R\,y$. Since total orders act like $\leq$ for numbers, this symbol is frequently used for any total order. In fact, this practice is extended to partial orders more generally. In an *abstract setting*, $\leq$ is often used to denote a partial order relation (and the terms *lesser* and *greater* are used in an abstract way, too), even when numerical order isn't intended and the relation isn't total. In concrete settings, though, the conventional symbol for a relation is used, if one exists.

A partial order is a total order minus the property of having to be connected. Not every two elements in a poset $\mathcal{A}$ need to be comparable, but if a subset of $\mathcal{A}$ is connected, that part will be ordered like a line. A poset $\mathcal{A}$, therefore, may contain a number of these line-segment paths, possibly branching in some places.

✤ Example 7.1.3

Show that the following are partial orders on $\mathcal{A}$. Is either a total order?

a) $\mathcal{A} = \mathbb{N}$, $m \mid n$ if and only if m divides n.

b) $\mathcal{A} = \{d \in \mathbb{N} : n \mid 36\}$, $m \mid n$ if and only if m divides n.

Solution

a) Since $n \mid n$ for all $n \in \mathbb{N}$, the divides-relation is reflexive.

It's also antisymmetric: if $m \mid n$ and $n \mid m$, then $m = n$.

And it's transitive: if $k \mid m$ and $m \mid n$, then $k \mid n$.

It's not connected, however. For example, neither 2 nor 3 divides the other. Thus, the relation is not a total order.

b) Here the divisibility relation is on $\mathcal{A} = \{1, 2, 3, 4, 6, 9, 12, 18, 36\}$.

The arguments of part a hold here, too: divisibility is reflexive, antisymmetric, and transitive, so it's a partial order. But it's still not connected, so it's not a total order.

Finite posets can be graphed by a diagram in which elements are connected by edges if they're related, the lesser below the greater. Since order relations are transitive, any relation between elements that can be deduced by following the edges upward is left implicit instead of cluttering up the diagram with more edges. These graphs are called *Hasse diagrams* after the twentieth-century German number theorist Helmut Hasse.

✤ Example 7.1.4

Diagram the following posets:

a) The poset of Example 3b: the divisors of 36 ordered by $m \mid n$.

b) The poset $\mathcal{P}(S)$ for $S = \{0, 1, 2\}$, ordered by subset inclusion $\subseteq$.

Solution

These are graphed by the following Hasse diagrams.

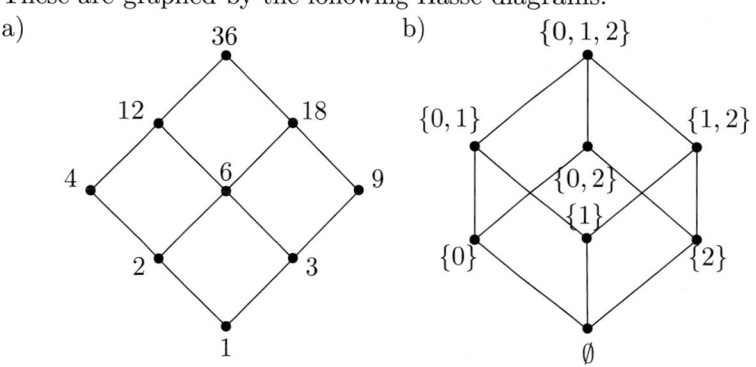

7.1.2 *Strict and Partial Orders*

Given a partial order $\leq$ on a set, we can define the related strict order $<$, which is useful in a variety of settings. Let's first define the notion of strict order in general and then relate it to partial order. It helps at times to use the set-theoretic definition of a binary relation as a set of ordered pairs.

Definition 7.1.5: *Irreflexive and Asymmetric Relations*
  a) *A relation R on $\mathcal{A}$ is **irreflexive** if and only if $(x, x) \notin R$ for all $x \in \mathcal{A}$.*
  b) *A relation R on $\mathcal{A}$ is **asymmetric** if and only if $(x, y) \in R \to (y, x) \notin R$.*

Irreflexive and asymmetric relations are strongly nonreflexive and nonsymmetric. Ordinary $<$ on $\mathbb{R}$ is both irreflexive and asymmetric.

Definition 7.1.6: *Strict Order*
  *A relation R is a **strict order** if and only if R is irreflexive and transitive.*

We can show that *any irreflexive transitive relation is asymmetric*; i.e., *every strict order is asymmetric* (see Exercise 22a), so asymmetry comes for free. Similarly, every strict order is antisymmetric (see Exercise 22b).

With these definitions, we can explore the connection between partial orders and strict orders. Based on what we know of ordinary $\leq$ and $<$, we expect the following correspondence results to be true—which they are.

Proposition 7.1.1: *Strict and Partial Orders*
  a) *If $\leq$ is a partial order on $\mathcal{A}$, then the relation $<$ defined by*
    $x < y \leftrightarrow x \leq y \land x \neq y$ *is a strict order on $\mathcal{A}$.*
  b) *If $<$ is a strict order on $\mathcal{A}$, then the relation $\leq$ defined by*
    $x \leq y \leftrightarrow x < y \lor x = y$ *is a partial order on $\mathcal{A}$.*

Proof:
  a) Let $\leq$ be a partial order with $<$ defined by $x < y \leftrightarrow x \leq y \land x \neq y$.
     The relation $<$ is irreflexive by definition: if $x < y$, then $x \neq y$.
     Now suppose $x < y$ and $y < z$. Then by the transitivity of $\leq$, $x \leq z$.
     But $x \neq z$, for if $x = z$, the antisymmetry of $\leq$ would imply that $x = y$,
     which contradicts $x < y$.
     Thus $x < z$, which means that $<$ is transitive.
     Thus $<$ is a strict order.   ∎
  b) See Exercise 24a.

Set-theoretically, we go from a partial order to its associated strict order by deleting all pairs (x, x) from the partial order (see Exercise 24b). Conversely, starting with a strict order, adjoining all pairs (x, x) generates the associated partial order, because the *reflexive closure* of a strict order remains transitive and is both reflexive and antisymmetric (see Exercise 24c).

Proposition 1 tells us that the same relations hold between $\leq$ and $<$ in general as they do for the numerical orders denoted by these symbols. This provides some justification for using these symbols in a more abstract setting, though this may be confusing at first. The meaning of $<$ in any concrete setting is dependent on the meaning of $\leq$, and conversely.

✤ **Example 7.1.5**
  Discuss the meaning of the strict orders $<$ associated with the following partial orders $\leq$:
  a) $\leq$ is the divisibility partial order on the set of natural numbers.
  b) $\leq$ is the subset ordering on a collection of sets.

Solution
a) If $\leq$ denotes *is a divisor of*, then $<$ indicates proper divisibility: $m < n$ means m *is a proper divisor of* n.
b) If $\leq$ denotes *is a subset of*, then $<$ indicates proper inclusion: $S < T$ represents the relation $S \subset T$.

7.1.3 Partial Orders on Cartesian Products and Strings

Suppose $\mathcal{A}_1$ and $\mathcal{A}_2$ are two posets ordered by $\leq_1$ and $\leq_2$, respectively (we'll drop the subscripts when it's clear which order relation is intended). The Cartesian product $\mathcal{A}_1 \times \mathcal{A}_2$ becomes a poset, according to *Proposition 2*, under the following *dictionary (lexicographical) order relation*.

Definition 7.1.7: *Dictionary Order Relation*
*If $\langle \mathcal{A}_1, \leq_1 \rangle$ and $\langle \mathcal{A}_2, \leq_2 \rangle$ are posets, the **dictionary order** $\leq$ on $\mathcal{A}_1 \times \mathcal{A}_2$ is given by $(x_1, x_2) \leq (y_1, y_2)$ if and only if $x_1 <_1 y_1 \vee (x_1 = y_1 \wedge x_2 \leq_2 y_2)$.*

Proposition 7.1.2: *Partially Ordered Cartesian Products*
The Cartesian product $\mathcal{A}_1 \times \mathcal{A}_2$ of two posets $\mathcal{A}_1$ and $\mathcal{A}_2$ forms a poset under the dictionary order $\leq$.

Proof:
We'll indicate what needs to be argued (without the order subscripts).
a) $\leq$ *is reflexive*: $(x_1, x_2) \leq (x_1, x_2)$ because $x_1 = x_1$ and $x_2 \leq x_2$.
b) $\leq$ *is antisymmetric*: if $(x_1, x_2) \leq (y_1, y_2)$ and $(y_1, y_2) \leq (x_1, x_2)$, then $x_1 = y_1$ and $x_2 = y_2$ (see Exercise 28a), and so $(x_1, x_2) = (y_1, y_2)$.
c) $\leq$ *is transitive*: if $(x_1, x_2) \leq (y_1, y_2)$ and $(y_1, y_2) \leq (z_1, z_2)$, then $(x_1, x_2) \leq (z_1, z_2)$ (see Exercise 28b). ∎

We can generalize this dictionary order to any number of sets. Let's start with $\mathcal{A}_1 \times \mathcal{A}_2 \times \mathcal{A}_3$. Considering ordered triples as nested ordered pairs [see Exercise 4.3.29: $(x_1, x_2, x_3) = ((x_1, x_2), x_3)$], the ordering is immediate. We have a partial order on $\mathcal{A}_1 \times \mathcal{A}_2$, and that partial order joined with the one on $\mathcal{A}_3$ gives the order on $(\mathcal{A}_1 \times \mathcal{A}_2) \times \mathcal{A}_3$. This yields the following partial order on $\mathcal{A}_1 \times \mathcal{A}_2 \times \mathcal{A}_3$: $(x_1, x_2, x_3) \leq (y_1, y_2, y_3)$ if and only if $x_1 < y_1$, or $x_1 = y_1 \wedge x_2 < y_2$, or $x_1 = y_1 \wedge x_2 = y_2 \wedge x_3 \leq y_3$ (see Exercise 31). We can extend this to the Cartesian product of n sets, though writing out the order criterion is messy.

❖**Example 7.1.6**
Exhibit the dictionary order on all finite length words from a partially ordered alphabet $\mathcal{A}$.

Solution
· Let $\mathcal{A}^k$ denote all strings of length k and $\mathcal{A}^*$ denote all strings of any finite length formed from alphabet $\mathcal{A}$ (see Section 3.3).
· Recall that two strings are equal if and only if they have exactly the same number of characters and the same characters in each place.

- We'll specify the dictionary order on $\mathcal{A}^*$ by giving the associated strict order relation $<$. To order strings $x_1 \cdots x_k$ and $y_1 \cdots y_n$, we compare successive initial segments of the two words.
- Suppose $x_1 \cdots x_m$ and $y_1 \cdots y_m$ are the longest initial segments where the words completely agree (m might be 0). Then $x_1 \cdots x_k < y_1 \cdots y_n$ if and only if $m = k$ and $m < n$, or $m < k$, $m < n$, and $x_{m+1} < y_{m+1}$.
- If the alphabet $\mathcal{A}$ is totally ordered (as ordinary alphabets are), then $\mathcal{A}^*$ will be totally ordered by this relation (see Exercise 33).

7.1.4 Extreme Elements in Posets

In a totally ordered set, any two elements can be compared, which gives the set a simple structure—it's linear. Posets are more varied and complex. Elements in a poset may or may not be comparable. But even if they aren't, they might be similarly related to some other element in the poset—they might be *less than* or *greater than* some common element. We'll look at this situation in more detail in the next section, but we need to introduce a few ideas and terms here for what we'll discuss next.

Definition 7.1.8: *Extremal Elements*

Suppose $\langle \mathcal{A}, \leq \rangle$ is a poset, $m \in \mathcal{A}$, $M \in \mathcal{A}$, and $S \subseteq \mathcal{A}$.

a) *m is a **minimal element of S** if and only if $m \in S$ and for no x in S is $x < m$;*

   *m is a **minimum of S** if and only if $m \in S$ and $m \leq x$ for all x in S.*

b) *M is a **maximal element of S** if and only if $M \in S$ and for no x in S is $M < x$;*

   *M is a **maximum of S** if and only if $M \in S$ and $x \leq M$ for all x in S.*

Note that these extremal elements belong to the set of elements they bound: that's part of their definition. Extrema (minimum, maximum) may or may not exist for a given subset, but if they do, they will be unique (see Exercise 26b). Further connections are explored in Exercises 25–27.

✤ **Example 7.1.7**

Identify extreme elements in the following posets:

a) The divisors of 60, ordered by divisibility.

b) The set $\{a, b, c, d, e, f, g, h\}$, ordered like the subsets of $\{0, 1, 2\}$ (see Example 4b).

Solution

- The Hasse diagrams of these posets are given in Figure 7.1. Note how the "dimensionality" of the first diagram corresponds to the number of prime factors in the factorization of 60 (see also Exercises 1–6).
- All non-empty subsets S of both posets have maximal and minimal elements (nothing above them or below them, respectively), though they need not have either a maximum or a minimum. The set $S = \{b, d\}$ from the second poset is such an example—both elements are maximal as well as minimal for S, but S has no maximum or minimum.

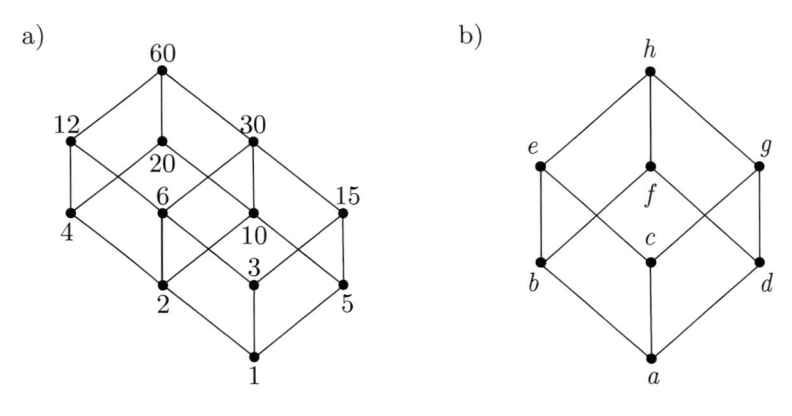

Fig. 7.1 The Hasse diagrams of the posets in Example 7.1.7

7.1.5 *Constructing Total Orders from Partial Orders*

In some situations, we may have a partial order on a set but want a total order that extends it. This can happen when a composite task contains different subactivities, each with its own linear order of prerequisites. When constructing a building, for example, there are structural components, electrical components, plumbing components, etc. Some of these may require that certain activities happen before others. Putting the building up may require you to sequentialize all the actions while respecting the various order priorities. In other words, the partial order gets embedded within a total order. Such a total order always exists for finite posets, though it will not be unique unless the original order is already a total order.

Proposition 7.1.3: *Embedding a Partial Order in a Total Order*
Finite posets can be embedded in totally ordered sets whose order extends the original order.

Proof:
Let $\langle \mathcal{A}, \leq \rangle$ be a finite poset.
· $\mathcal{A}$ contains a minimal element (see Exercise 25b): choose one and call it a_1. The set $\mathcal{A} - \{a_1\}$ remains a poset. If it's non-empty, it too has a minimal element: choose one and call it a_2.
Choose $a_3, \ldots, a_n$ similarly until all the elements of $\mathcal{A}$ are ordered.
· Ordering the elements by how they were chosen— $a_i \leq a_j$ if and only if $i \leq j$—creates a total order on $\mathcal{A}$.
· Moreover, if $a_k \leq a_m$ in the original order, it remains so in the new total order: a_m is chosen only after everything that precedes it, including a_k, has been placed into the sequence of choices, because at each stage a minimal element is chosen.
· Thus a total order exists that respects the original partial order. ∎

✤**Example 7.1.8**
Convert the poset of divisors of 36 into a totally ordered set in two ways.

Solution
- One total order is created by choosing lowest-level elements, moving left to right within the level: $1; 2, 3; 4, 6, 9; 12, 18; 36$.
- Another total order results from choosing the left-most minimal elements each time: $1, 2, 4;\ 3, 6, 12;\ 9, 18, 36$.
- Both of these total orders respect the original partial order. They each stretch and squeeze the branching partial order together, as it were, to convert it into a single line of elements.

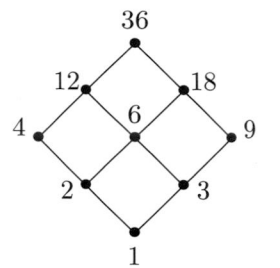

EXERCISE SET 7.1

Exercises 1–7: Divisor Posets

The following explore Hasse diagrams' relation to prime factorization.

7.1.1. *Divisors of* 39
a. Draw and describe the Hasse diagram for the divisors of 39.
b. Relate the shape of the diagram to the prime factorization of 39.

7.1.2. *Divisors of* 343
a. Draw and describe the Hasse diagram for the divisors of 343.
b. Relate the shape of the diagram to the prime factorization of 343.

7.1.3. *Divisors of* 153
a. Draw and describe the Hasse diagram for the divisors of 153.
b. Relate the shape of the diagram to the prime factorization of 153.

7.1.4. *Divisors of* 385
a. Draw and describe the Hasse diagram for the divisors of 385.
b. Relate the shape of the diagram to the prime factorization of 385.
c. If the minimum and maximum elements are deleted from this poset, leaving the proper divisors of 385, is the resulting set still a poset? Explain.

7.1.5. *Divisors of* 84
a. Draw and describe the Hasse diagram for the divisors of 84.
b. Relate the shape of the diagram to the prime factorization of 84.
c. If the minimum and maximum elements are deleted from this poset, leaving the proper divisors of 84, is the resulting set still a poset? Explain.

7.1.6. *Hasse Diagrams and Prime Factorization*
a. Based on Exercises 1–5, conjecture what the Hasse diagram looks like for the divisors of 180. Draw the diagram to verify your conjecture.
b. What sort of Hasse diagram would the divisors of 420 have? How could this be envisioned and described? You do not need to draw the diagram.

7.1.7. Show that the divisibility relation is not a partial order on the set of integers $\mathbb{Z}$. Which property is lacking?

Exercises 8–11: Hasse Diagrams of Small Posets
The following explore Hasse diagrams for some small posets. Elements should be pictured as isolated points if they can't be compared with other points.

7.1.8. Draw a baseball diamond: connect home plate (4) to first base (1) to second (2) to third (3) and back to home; extend the foul lines out to right field (R) and left field (L), and connect these along the warning track through straightaway center field (C). Is this a Hasse diagram for a poset? Explain.

7.1.9. Draw all possible distinct Hasse diagrams for a two-element poset.

7.1.10. Draw all possible distinct Hasse diagrams for a three-element poset.

7.1.11. Draw all possible distinct Hasse diagrams for a four-element poset.

Exercises 12–14: Identifying Extremal Elements
Identify extremal elements in the following posets.

7.1.12. Let D be the divisor poset for 60 (see Example 7a).
a. Explain why $P = D - \{1, 60\}$, the set of proper divisors of 60, is a poset. Does P have a maximum? minimum? Identify all extremal elements.
b. Let $U = \{x \in D : 4 \le x, \ 10 \le x\}$, the set of upper bounds for $\{4, 10\}$. List the elements of U. What is its minimum?
c. Let $L = \{x \in D : x \le 4, \ x \le 10\}$, the set of lower bounds for $\{4, 10\}$. List the elements of L. What is its maximum?

7.1.13. Let $S = \{0, 1, 2, 3\}$ and $\mathcal{P}(S)$ be its power set.
a. Let $U = \{R \subseteq S : \{0, 1\} \subseteq R, \{0, 2\} \subseteq R\}$, the set of upper bounds for $\{\{0, 1\}, \{0, 2\}\}$. List U and identify its minimal elements. Does U have a minimum? How do your answers relate to operations on these sets?
b. Let $L = \{R \subseteq S : R \subseteq \{0, 1\}, R \subseteq \{0, 2\}\}$, the set of lower bounds for $\{\{0, 1\}, \{0, 2\}\}$. List L and identify its maximal elements. Does U have a maximum? How do your answers relate to operations on these sets?

7.1.14. Let S be any set, $\mathcal{P}(S)$ its power set, and $\mathcal{C}$ a collection of S's subsets.
a. Explain why $\mathcal{P}(S)$ is a poset under the subset relation $\subseteq$.
b. Must $\mathcal{C}$ contain maximal elements? minimal elements? Prove it or provide a counterexample.
c. Let $U = \{X \subseteq S : (\forall C \in \mathcal{C})(C \subseteq X)\}$, the set of upper bounds for $\mathcal{C}$. Does U have a minimum, a least upper bound? What is it?
d. Let $L = \{X \subseteq S : (\forall C \in \mathcal{C})(X \subseteq C)\}$, the set of lower bounds for $\mathcal{C}$. Does L have a maximum, a greatest lower bound? What is it?

Exercises 15–18: True or False
Are the following statements true or false? Explain your answer.

7.1.15. Any elements x, y in a poset can be compared: either $x \le y$ or $y \le x$.

7.1.16. Every non-empty subset of a poset is also a poset.

7.1.17. If M is the maximum of a set S, then M is a maximal element.

7.1.18. Every total order is an equivalence relation.

Exercises 19–21: Properties of Relations and Posets
The following consider logical connections between relation properties.

7.1.19. Prove or disprove: *Every relation that's symmetric and antisymmetric is reflexive.*

7.1.20. Determine a necessary and sufficient condition for an equivalence relation to be antisymmetric. When are equivalence relations partial orders?

7.1.21. *Poset Duals*
Prove that if $\mathcal{A}$ is a poset under $\leq$, it also forms a poset under its converse relation: $a \overset{\frown}{\leq} b$ if and only if $b \leq a$. This converse relation is symbolized by $\geq$.

Exercises 22–24: Strict Ordering Relations
Interpret $\leq$ and $<$ below as correlated abstract partial order and strict order relations on some poset P. Don't assume anything else about them.

7.1.22. *Properties of Strict Orders*
 a. Prove that if $<$ is a strict order, then it's *asymmetric*: if $x < y$, then $y \not< x$. Also prove the stronger condition: if $x < y$, then $y \not\leq x$.
 b. Prove that if $<$ is a strict order, then it's *antisymmetric*: if $x < y$ and $y < x$, then $x = y$. *Hint*: when does the antecedent occur?

7.1.23. *Transitive Properties of Strict and Partial Orders*
 a. Prove that if $x < y \leq z$ or $x \leq y < z$, then $x < z$.
 b. Prove or disprove: it's possible to have $x < y < z$ and $z \leq x$.

7.1.24. *Proposition 1*
 a. Prove *Proposition* 1b: *If $<$ is a strict order on $\mathcal{A}$, then $\leq$ defined by $x \leq y \leftrightarrow x < y \lor x = y$ is a partial order.*
 b. Prove the set-theoretic version of *Proposition* 1a: *If $\leq$ is a partial order on $\mathcal{A}$, deleting all pairs (x, x) for $x \in \mathcal{A}$ yields a strict order $<$ on $\mathcal{A}$. Furthermore, $x < y$ if and only if $x \leq y$ and $x \neq y$.*
 c. Prove the set-theoretic version of *Proposition* 1b: *If $<$ is a strict order on $\mathcal{A}$, adjoining all pairs (x, x) for $x \in \mathcal{A}$ yields a partial order $\leq$ on $\mathcal{A}$. Furthermore, $x \leq y$ if and only if $x < y$ or $x = y$.*

Exercises 25–27: Theory of Extremal Elements
Let $\langle \mathcal{A}, \leq \rangle$ be a poset and S a non-empty subset of $\mathcal{A}$.

7.1.25. *Maximal and Minimal Elements*
 a. Show that maximal and minimal elements of S need not be unique.
 b. Prove that finite non-empty posets have maximal and minimal elements.

7.1.26. *Maximum and Minimum Elements*
 a. Show that S need not have either a maximum or a minimum.
 b. Prove that a maximum and a minimum must be unique, if they exist.

7.1.27. *Extremal Elements in Totally Ordered Sets*
Suppose $\langle \mathcal{A}, \leq \rangle$ is totally ordered.
 a. Must S have maximal or minimal elements?
 b. Must S have a maximum or a minimum?
 c. What are the answers to parts a and b if S is finite?

Exercises 28–34: Cartesian Products and Strings
The following problems deal with ordered Cartesian products and strings.

7.1.28. *Proposition 2*
a. Prove that the dictionary ordering $\leq$ of $\mathcal{A}_1 \times \mathcal{A}_2$ is antisymmetric.
b. Prove that the dictionary ordering $\leq$ of $\mathcal{A} = \mathcal{A}_1 \times \mathcal{A}_2$ is transitive.
c. Prove or disprove: If $\mathcal{A}_1$ and $\mathcal{A}_2$ are totally ordered, so is $\mathcal{A}_1 \times \mathcal{A}_2$ under the dictionary order.

7.1.29. *Product Order vs. Dictionary Order*
Dixie Nary thinks the dictionary order placed on $\mathcal{A}_1 \times \mathcal{A}_2$ is unnecessarily complicated, so she offers the following *product ordering* for consideration: $(x_1, x_2) \leq (y_1, y_2)$ if and only if $x_1 \leq_1 y_1$ and $x_2 \leq_2 y_2$.
a. Is the product order a partial order? Justify your answer.
b. Does the product order agree with the dictionary order? Explain.
c. If both component orders are total orders, is the product order a total order? Explain.

7.1.30. *Strict Dictionary Order on Ordered Pairs*
Let $\mathcal{A} = \mathcal{A}_1 \times \mathcal{A}_2$ be the Cartesian product of two posets.
a. If $\leq$ is the dictionary order on $\mathcal{A}$, state and prove a condition for the associated strict order: $(x_1, x_2) < (y_1, y_2)$ if and only if _____ .
b. Prove that if $<$ is defined the way you did it in part a, then $<$ is a strict order, and the associated partial order $\leq$ is the dictionary order; i.e., the associated $\leq$ satisfies the biconditional stated in *Definition 7*.

7.1.31. *Dictionary Order on Triples*
a. Given that dictionary order on pairs of elements is a partial order and that triples are defined in terms of pairs, show that the order extended to the Cartesian product $\mathcal{A}_1 \times \mathcal{A}_2 \times \mathcal{A}_3$ (see the discussion after *Proposition 2*) is a partial order relation. (A lengthy argument isn't required here.)
b. Prove that this induced partial order satisfies $(x_1, x_2, x_3) \leq (y_1, y_2, y_3)$ if and only if $x_1 < y_1$ or $x_1 = y_1 \wedge x_2 < y_2$ or $x_1 = y_1 \wedge x_2 = y_2 \wedge x_3 \leq y_3$.

7.1.32. *Dictionary Order on Strings*
Let $\mathcal{A}$ be the English alphabet, with letters ordered in the usual way, and let $\mathcal{A}^*$ denote the set of all finite strings/words formed using this alphabet. Use the strict dictionary order on $\mathcal{A}^*$ from Example 6 to answer the following:
a. Why does the string *earlier* appear earlier in the dictionary than *later*?
b. Explain why *short, shorter, shortest* are listed in dictionary order.

7.1.33. *Totally Ordered Strings*
Prove that the order relation defined on $\mathcal{A}^*$, the set of all finite strings formed from a totally ordered alphabet $\mathcal{A}$, is a total ordering.

7.1.34. *Binary Strings*
Let $\mathcal{B}^n$ denote the set of all length-n strings of 0's and 1's, and let $\leq$ denote the dictionary order on $\mathcal{B}^n$.
a. Explain why $\mathcal{B}^n$ is a totally ordered set.
b. If the elements of $\mathcal{B}^n$ are considered as numbers in binary notation, how does $\leq$ compare with ordinary numerical order on $\{0, 1, \ldots, 2^{n-1}\}$.

Exercises 35–37: Embedding Posets in Totally Ordered Sets
The following deal with extending a partial order to make it a total order.

7.1.35. Extend the partial order for $\mathcal{P}(\{0,1,2\})$ (see Example 4b) in two different ways to make it into a total order.

7.1.36. Extend the partial order for the poset of divisors of 84 (see Exercise 5) to make it into a total order.

7.1.37. If a poset is infinite, can it be embedded in a totally ordered set? (This goes beyond the material of this section and our treatment of *Set Theory*.)

Exercises 38–42: Well-Ordered Sets
The following problems explore the notion of a well-ordered set.
Definition: *A totally ordered set $\langle A, \leq \rangle$ is **well ordered** if and only if every non-empty subset S of A contains a least element (a minimum).*

7.1.38. Is $\mathbb{N}$ under the usual $\leq$ ordering a well-ordered set? Is $\mathbb{Z}$? Explain.

7.1.39. Is $(0,1)$, the set of all real numbers between 0 and 1, well ordered under the usual ordering? Explain.

7.1.40. Is every totally ordered set with a least element well ordered?

7.1.41. Let $\mathbb{N}$ be ordered by the ordinary $\leq$ order. Show that $\mathbb{N} \times \mathbb{N}$ is well ordered under the induced dictionary order.

7.1.42. Prove that the following *Principle of Induction* holds for any well-ordered set $\langle A, \leq \rangle$: If T is a non-empty subset of A, and if T contains an element whenever it contains all the predecessors of that element, then $T = A$. (*Note:* x is a *predecessor* of y if and only if $x < y$.)

Exercises 43–44: Subset Posets
The following problems involve posets ordered by subset inclusion.

7.1.43. Let $\mathcal{A}$ be the set of all *initial segments* of $\mathbb{N}$, i.e., the collection of subsets of $\mathbb{N}$ of the form $[0,n] = \{0,1,\ldots,n\}$ for some n.
  a. Tell why $\mathcal{A}$ forms a poset under the subset order relation.
  b. Does $\mathcal{A}$ form a totally ordered set? Prove your claim.
  c. Suppose S is a finite subset of $\mathcal{A}$. Must S have maximal or minimal elements? A maximum or a minimum? Prove your answers.
  d. If S is an infinite subset of $\mathcal{A}$, must S have maximal or minimal elements? A maximum or a minimum? Prove your answers.

7.1.44. Let $\mathcal{A}$ be the set of all *cofinite* subsets of $\mathbb{N}$, i.e., subsets whose complement in $\mathbb{N}$ is *finite*.
  a. Tell why $\mathcal{A}$ forms a poset under the subset order relation.
  b. Suppose S is a finite decreasing chain in $\mathcal{A}$, i.e., $S = \{A_1, A_2, \ldots, A_n\}$, where $A_1 \supseteq A_2 \supseteq \cdots \supseteq A_n$. Must S have a least upper bound or a greatest lower bound in $\mathcal{A}$? If it does, must these belong to S?
  c. If S is an infinite decreasing chain $A_1 \supseteq A_2 \supseteq \cdots \supseteq A_n \supseteq \cdots$ of cofinite subsets, must S have a least upper bound or a greatest lower bound in $\mathcal{A}$? If so, must these belong to S?

7.2 Lattices

We'll now turn our attention to the type of poset known as a *lattice*. Lattices provide a good setting in which to introduce *Boolean Algebra*, a field of prime importance for computer science.

To arrive at the type of lattice needed for *Boolean Algebra*, we'll have to define quite a number of new properties for relations. Perhaps the best way to become familiar with these terms is to return to their definitions as you work through material that involves them.

7.2.1 Definition of a Lattice

Lattices are posets with meets and joins, so let's first define those terms. Meets and joins are unique when they exist (see Exercise 1), so we'll assume this in our definition.

Definition 7.2.1: *Meet, Join*
Let $\langle \mathcal{A}, \leq \rangle$ be a poset, and let x and y be any elements of $\mathcal{A}$.
a) *The **meet** of x and y, denoted by $x \wedge y$, is the greatest lower bound for $\{x, y\}$, i.e., $x \wedge y = \max\{w \in \mathcal{A} : w \leq x, w \leq y\}$.*
b) *The **join** of x and y, denoted by $x \vee y$, is the least upper bound for $\{x, y\}$, i.e., $x \vee y = \min\{z \in \mathcal{A} : x \leq z, y \leq z\}$.*

Definition 7.2.2: *Lattice*
*A poset $\langle \mathcal{A}, \leq \rangle$ is a **lattice** if and only if every x and y in $\mathcal{A}$ have a meet and a join.*

Since each pair of distinct elements in a lattice has something above and below it, no lattice (besides the one-point lattice) can have isolated points. The Hasse diagram of a lattice is thus a connected (single component) graph—paths of edges link any two points in the diagram (see Exercise 8). In fact, its crisscross structure looks like what people would call ... a *lattice*.

You'll notice that we're reusing the logical symbols $\wedge$ (*and*) and $\vee$ (*or*) to represent meets and joins in a lattice. In Section 7.3, we'll see that there's a good reason for this. At this point, however, we'll take them as recycled abstract symbols standing for meet and join, defined in terms of the poset's partial order $\leq$. Later, we'll use $\cdot$ and $+$ for meet and join, also interpreted abstractly. All of this notational repetition and flexibility can be confusing as you start learning about lattices. Unfortunately, it's unavoidable, because different books and different areas of thought use different symbols to denote the same concept and use the same symbol to denote different concepts.

❖ **Example 7.2.1**
Show that the following are lattices, and interpret their meets and joins.
a) The poset of the divisors of 60 (see Example 7.1.7a).
b) The poset of the subsets of $\{0, 1, 2\}$ (see Example 7.1.4b).

Solution

We'll repeat these posets' Hasse diagrams for easy reference.

a)
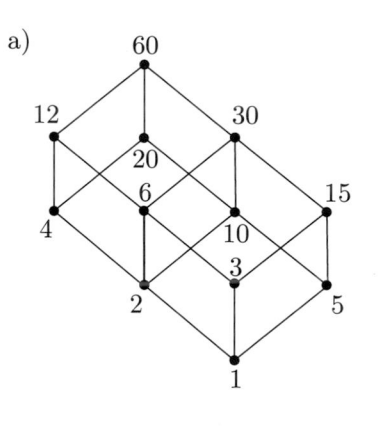

b)
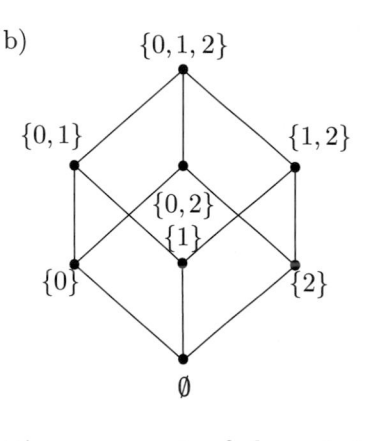

a) The poset of the divisors of 60 is a lattice—every pair of elements has both a meet and a join.
  · The meet of two divisors is their *greatest common divisor*; for example,
    $6 \wedge 20 = \gcd(6, 20) = 2$.
  · The join of two divisors is their *least common multiple*;
    for example, $6 \vee 20 = \text{lcm}(6, 20) = 60$.
b) The poset of the subsets of $\{0, 1, 2\}$ is also a lattice.
  · The meet of two subsets is their *intersection*; for example,
    $\{0, 1\} \wedge \{1, 2\} = \{0, 1\} \cap \{1, 2\} = \{1\}$.
  · The join of two subsets is their *union*; for example,
    $\{0, 1\} \vee \{1, 2\} = \{0, 1\} \cup \{1, 2\} = \{0, 1, 2\}$.

Both examples generalize. The set of all divisors of a positive integer n forms a lattice in which meets are greatest common divisors and joins are least common multiples (see Exercise 4a).

The power set $\mathcal{P}(S)$ of any set S also forms a lattice (see Exercise 13a). Given two subsets R and T of S, $R \wedge T = R \cap T$ and $R \vee T = R \cup T$. This example helps explain the terms *meet* and *join*—the meet is the set in which the two sets meet, and the join is the set formed by joining the two together.

[*Aside*: note that it now makes sense to write, say, $R \wedge T$, whereas it didn't before when $\wedge$ meant *and*, because then it was supposed to connect sentences, not sets. Note also that formal statements involving meet and join will be somewhat confusing if they also involve the propositional connectives *and* and *or*, so we'll now mostly use words in place of logical symbols.]

7.2.2 Basic Properties of Meet and Join

Meet and join satisfy a number of properties as the greatest lower bound and least upper bound of pairs of elements in a poset. These are summarized in the following two propositions. The first proposition contains some basic relational properties that are helpful for establishing the more important operational properties of the second proposition.

Proposition 7.2.1: *Basic Order Properties of Meet and Join*
Let $\langle \mathcal{A}, \leq \rangle$ be a lattice. Then the following hold:
a) $x \wedge y \leq \{x, y\} \leq x \vee y$.
b) $x \leq y$ *if and only if* $x \wedge y = x$.
c) $x \leq y$ *if and only if* $x \vee y = y$.
d) *if* $x \leq y$, *then* $x \wedge z \leq y \wedge z$ *and* $x \vee z \leq y \vee z$.
e) *if* $x \leq y$ *and* $z \leq w$, *then* $x \wedge z \leq y \wedge w$ *and* $x \vee z \leq y \vee w$.

Proof:
a) This holds because $x \wedge y$ is a lower bound for x and y, while $x \vee y$ is an upper bound. In fact, we know that no other element of the lattice lies between these extremal elements and the pair of elements. ∎
b) This follows from the definition of meet. See Exercise 9a.
c) This follows from the definition of join. See Exercise 9b.
d) This follows from the definition of meet and join. See Exercise 9c.
e) In words this says that the meet and join of the smaller elements is, respectively, less than or equal to the meet and join of the larger elements. This can be proved using part d. See Exercise 9d.

Proposition 7.2.2: *Basic Operational Properties of Meet and Join*
Let $\langle \mathcal{A}, \leq \rangle$ be a lattice. Then the following hold:
a) *Commutativity*: $x \wedge y = y \wedge x$; $x \vee y = y \vee x$.
b) *Associativity*: $(x \wedge y) \wedge z = x \wedge (y \wedge z)$; $(x \vee y) \vee z = x \vee (y \vee z)$.
c) *Idempotence*: $x \wedge x = x$; $x \vee x = x$.
d) *Absorption*: $x \wedge (x \vee y) = x$; $x \vee (x \wedge y) = x$.

Proof:
a) These two results are immediate. The meet and join of a pair of elements don't depend on the order in which x and y are listed. ∎
b) We'll prove the first and leave the second as an exercise (Exercise 10b).
 · Let's show this equality using the antisymmetric property of $\leq$. This exactly parallels how we show set equality in the poset of sets.
$(x \wedge y) \wedge z \leq x \wedge y \leq x$ by *Proposition* 1a (twice).
So $(x \wedge y) \wedge z \leq x$ due to transitivity of $\leq$.
Also by *Proposition* 1a, $x \wedge y \leq y$.
By *Proposition* 1d, $(x \wedge y) \wedge z \leq y \wedge z$.
Thus $(x \wedge y) \wedge z$ is below both x and $y \wedge z$.
 · Since the meet of these two elements is their *greatest* lower bound, we have $(x \wedge y) \wedge z \leq x \wedge (y \wedge z)$. ✓
 · Similarly, $x \wedge (y \wedge z) \leq (x \wedge y) \wedge z$ (see Exercise 10a). ∎
c) See Exercise 10c.
d) See Exercise 10d.

7.2.3 Distributive Lattices

If you paid attention to which properties we proved in the last proposition, you may have wondered why we didn't include the distributive property. There's a good reason. Not all lattices are *distributive*, though lattices do satisfy a semi-distributive law (see Exercise 24).

Definition 7.2.3: *Distributive Lattice*
  *A lattice $\mathcal{A}$ is **distributive** if and only if the following hold for any x, y, and z: $x \wedge (y \vee z) = (x \wedge y) \vee (x \wedge z)$, and $x \vee (y \wedge z) = (x \vee y) \wedge (x \vee z)$.*

We already know that meet and join are commutative, so if a lattice satisfies the above distributive laws, it also satisfies the other distributive laws (see Exercise 25): $(x \wedge y) \vee z = (x \vee z) \wedge (y \vee z)$ and $(x \vee y) \wedge z = (x \wedge z) \vee (y \wedge z)$.

❖**Example 7.2.2**
  Show that the following simple lattices are not distributive. In fact, it can be proved that *every non-distributive lattice contains one of these lattices*.
  a)

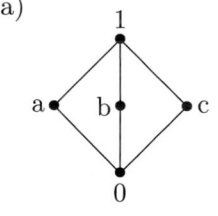

  b)

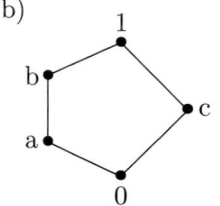

Solution
  a) Using the middle elements of the divided diamond lattice, we have:
     $a \wedge (b \vee c) = a \wedge 1 = a$, but $(a \wedge b) \vee (a \wedge c) = 0 \vee 0 = 0$, and $a \neq 0$. ✓
     The other distributive law likewise fails for these elements (see Exercise 22a).
  b) The pentagon lattice is also not distributive (see Exercise 22b).

  Working by hand, it's usually easier to show that a lattice is *not* distributive than that it is. For example, if a lattice has five elements, like the ones in Example 2, you would need to check the validity of $5 \cdot 15 = 75$ essentially different equations for each distributive law equation in *Definition 3* (see Exercise 23; order is irrelevant for the final pair, due to commutativity). Presumably, if a lattice isn't distributive, you'll stumble upon a counterexample before reaching the final equation. Programming a computer to do the work, though, levels the playing field for checking all these identities to see if a finite lattice is distributive.

❖**Example 7.2.3**
  Explain why the power set lattice $\mathcal{P}(U)$ is a distributive lattice for any $\mathcal{U}$.

Solution
  Since meet is intersection and join is union, and since each set-theoretic operation distributes over the other one (see *Proposition 4.1.8*), $\mathcal{P}(U)$ is a distributive lattice under subset inclusion. ✓

7.2.4 *Bounded and Complemented Lattices*

A second omission from *Proposition 2* is the lack of *De Morgan's Laws* for meet and join. We had such laws for *Propositional Logic* and *Set Theory*; might a version hold here as well?

At this point, because we don't yet have a counterpart for lattices to negation or set complementation, we can't say one way or the other. However, since power sets are lattices under the subset relation, let's see how we can generalize set-theoretic complementation to arbitrary lattices. Once that's done, we can investigate whether *De Morgan's Laws* hold.

Complements are taken with respect to some universal set U. To identify the place of $\overline{S} = U - S$ within the lattice $\mathcal{P}(U)$, we'll need to recast complements in terms of intersection and union, *Set Theory's* meet and join. Doing this, we can characterize set complements as follows (see Exercise 12): $\overline{S}$ is *the set such that* $\overline{S} \cap S = \emptyset$ *and* $\overline{S} \cup S = U$. This gives us a further question to ponder: what are $\emptyset$ and U in a lattice? That's easy to answer: they're the minimum and maximum elements of the lattice. So it seems we need to be in a *bounded lattice* before we can define complements.

Definition 7.2.4: *Bounded Lattices*
  *A lattice $\langle \mathcal{A}, \leq \rangle$ is **bounded** if and only if it has a minimum element and a maximum element. These are denoted by 0 and 1, respectively.*

The extreme elements of bounded lattices interact with other elements of the lattice in obvious ways, as captured by the next proposition.

Proposition 7.2.3: *Extreme Elements in a Bounded Lattice*
  Suppose $\langle \mathcal{A}, \leq \rangle$ is a bounded lattice having minimum 0 and maximum 1, and let x be any element in $\mathcal{A}$. Then
  a) $0 \vee x = x = x \vee 0$; $1 \wedge x = x = x \wedge 1$.
  b) $0 \wedge x = 0 = x \wedge 0$; $1 \vee x = 1 = x \vee 1$.

Proof:
  These hold because $0 \leq x \leq 1$; apply *Proposition* 1b (see Exercise 16).

Definition 7.2.5: *Complements in a Bounded Lattice*
  *Suppose $\langle \mathcal{A}, \leq \rangle$ is a bounded lattice with minimum 0 and maximum 1. Then z is a **complement** of x if and only if $x \wedge z = 0$ and $x \vee z = 1$.*

Note that complements only make sense in a bounded lattice. But even in bounded lattices, complements need not exist. Totally ordered posets are all lattices—distributive lattices, in fact (see Exercise 19), but those that are bounded are rarely complemented (see Exercise 18). Thus, we need still another category of lattice for taking complements.

Definition 7.2.6: *Complemented Lattices*
  *A lattice $\langle \mathcal{A}, \leq \rangle$ is **complemented** if and only if it is bounded and every element has a complement.*

Complemented lattices are, by definition, lattices with complements for all their elements, but the existence of complements still does not guarantee uniqueness (see Exercises 14–15). Nevertheless, if the lattice is also distributive, this leeway disappears, as the next proposition demonstrates. The corollary that follows is immediate.

Proposition 7.2.4: *Unique Complements in Distributive Lattices*
If $\langle \mathcal{A}, \leq \rangle$ is a bounded distributive lattice with minimum 0 and maximum 1, then complements are unique, provided they exist.

Proof:
Suppose that $\overline{x}$ and z are complements of x in $\mathcal{A}$. We'll show that $\overline{x} = z$. The following argument isn't as easy as it looks. To appreciate it, try to prove $\overline{x} = z$ on your own before continuing.
By the definition of being a complement,
$$\overline{x} \wedge x = 0 = x \wedge z \quad \text{and} \quad \overline{x} \vee x = 1 = x \vee z$$

$$\begin{aligned}
\text{Thus,} \quad \overline{x} &= \overline{x} \wedge 1 \\
&= \overline{x} \wedge (x \vee z) \\
&= (\overline{x} \wedge x) \vee (\overline{x} \wedge z) \\
&= 0 \vee (\overline{x} \wedge z) \\
&= (x \wedge z) \vee (\overline{x} \wedge z) \\
&= (x \vee \overline{x}) \wedge z \\
&= 1 \wedge z \\
&= z \quad \blacksquare
\end{aligned}$$

Corollary 7.2.4.1: *Existence of Unique Complements*
Every element in a complemented distributive lattice has a unique complement. The complement of x will be denoted by $\overline{x}$.

Our abstract notation for complements is the same as for sets. Not surprisingly, given our initial motivation for defining complements, power sets are good examples of complemented distributive lattices (see Exercise 13b).

7.2.5 Boolean Lattices: Definition and Properties

Complemented distributive lattices are an important type of lattice. They're called *Boolean lattices*.

Definition 7.2.7: *Boolean Lattices*
*A lattice $\langle \mathcal{A}, \leq, {}^{-}, 0, 1 \rangle$ is a **Boolean lattice** if and only if it's a complemented distributive lattice.*

We finally have the right structure for proving what we wanted. A Boolean lattice $\langle \mathcal{A}, \leq, {}^{-}, 0, 1 \rangle$ satisfies *De Morgan's Laws*.

Proposition 7.2.5: *De Morgan's Laws*
If $\langle \mathcal{A}, \leq, {}^{-}, 0, 1 \rangle$ *is a Boolean lattice, then the following hold:*
a) $\overline{x \wedge y} = \overline{x} \vee \overline{y}$; b) $\overline{x \vee y} = \overline{x} \wedge \overline{y}$.

Proof:
   a) Let x and y be elements of $\mathcal{A}$. To show $\overline{x \wedge y} = \overline{x} \vee \overline{y}$, we'll prove that $\overline{x} \vee \overline{y}$ is a complement of $x \wedge y$, which is unique.

$$
\begin{aligned}
(x \wedge y) \wedge (\overline{x} \vee \overline{y}) &= (x \wedge y \wedge \overline{x}) \vee (x \wedge y \wedge \overline{y}) && \text{Distrib, Assoc} \\
&= 0 \vee 0 && \text{Comm, Compl} \\
&= 0. \checkmark && \text{Extrm Elts} \\
(x \wedge y) \vee (\overline{x} \vee \overline{y}) &= ((x \wedge y) \vee \overline{x}) \vee \overline{y} && \text{Assoc} \\
&= ((x \vee \overline{x}) \wedge (y \vee \overline{x})) \vee \overline{y} && \text{Distrib} \\
&= (1 \wedge (y \vee \overline{x})) \vee \overline{y} && \text{Compl} \\
&= (y \vee \overline{x}) \vee \overline{y} && \text{Extrm Elts} \\
&= \overline{x} \vee (y \vee \overline{y}) && \text{Comm, Assoc} \\
&= \overline{x} \vee 1 && \text{Compl} \\
&= 1. \; \blacksquare && \text{Extrm Elts}
\end{aligned}
$$

   b) See Exercise 26.

We now know several properties about Boolean lattices. But what do Boolean lattices look like? Can we say anything definitive about their size? What shape do their Hasse diagrams have? Do the properties of Boolean lattices constrain their structure in predictable ways? We'll look at a few examples and then continue exploring this topic in the next section.

✤**Example 7.2.4**
   Investigate all Boolean lattices with cardinalities 1–5.

Solution
   · There is only one lattice structure of size 1: a single point. This forms a Boolean lattice, but not an interesting one.
   · There is also only one lattice structure of size 2: two points in a line. This lattice is also a Boolean lattice, but as a totally ordered set it, too, is not very interesting.
   · There are no lattice structures of size 3 except the totally ordered one of three points in a line. However, this structure has no complement for the middle point, so it cannot be a Boolean lattice. In fact, no totally ordered set with more than two elements is a Boolean lattice (see Exercise 18).
   · There are two different lattice structures on four elements. There must be a top point and a bottom point, and the other two must lie in the middle. These can either be side-by-side or in-line. This gives two lattice structures—a four-point line and a diamond. The first isn't a Boolean lattice, as we just noted, but the latter is.

· Finally, given five points, one must be the top and another the bottom. The other three must lie between these two in some order. No such lattice structure is a Boolean lattice (see Exercise 29a).
· Thus, the only posets of cardinality 5 or less that are Boolean lattices are ones with 1, 2, or 4 elements. That's an interesting sequence! More on this soon (see also Exercise 29).
· The Boolean lattice structures found so far, then, are these:

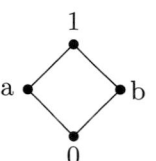

EXERCISE SET 7.2

7.2.1. Show that in a poset $\langle \mathcal{A}, \leq \rangle$, $x \wedge y$ and $x \vee y$ are unique, if they exist.

Exercises 2–4: Divisor Lattices
The following problems have to do with lattices of divisors.

7.2.2. Let $\langle \mathcal{D}_{12}, \, | \, \rangle$ denote the poset of all divisors of 12.
  a. Show that $\mathcal{D}_{12}$ is a lattice by drawing its Hasse diagram and then verifying that each pair of divisors has a meet and a join. What are the meets and joins in terms of divisibility?
  b. Is $\mathcal{D}_{12}$ a complemented lattice? Explain.
  c. Is $\mathcal{D}_{12}$ a distributive lattice? To verify that $x \wedge (y \vee z) = (x \wedge y) \vee (x \wedge z)$, how many different equations must be checked? Explain.
  d. If the bottom point 1 and the top point 12 are deleted from $\mathcal{D}_{12}$, is the result still a lattice? Explain.
  e. Is $\mathcal{D}_{12}$ a Boolean lattice? Explain.

7.2.3. Let $\langle \mathcal{D}_{30}, \, | \, \rangle$ denote the poset of all divisors of 30.
  a. Show that $\mathcal{D}_{30}$ is a lattice. Explain.
  b. Is $\mathcal{D}_{30}$ a complemented lattice? Explain.
  c. Is $\mathcal{D}_{30}$ a distributive lattice? Explain.
  d. Is $\mathcal{D}_{30}$ a Boolean lattice? Explain.

7.2.4. Let $\langle \mathcal{D}_n, \, | \, \rangle$ denote the poset of all divisors of a positive integer n.
  a. Prove that $\mathcal{D}_n$ is a lattice. Carefully explain why $x \wedge y = \gcd(x, y)$ and $x \vee y = \operatorname{lcm}(x, y)$.
  b. When is $\mathcal{D}_n$ a complemented lattice? Explain.
  c. When is $\mathcal{D}_n$ a distributive lattice? Explain.
     Hint: to check $x \wedge (y \vee z) = (x \wedge y) \vee (x \wedge z)$, express x, y, and z via their prime factorizations and relate meet and join to how many factors of each prime (possibly 0) are used (see Exercise 3.5.23a). You may assume that $\langle \mathbb{N}, \leq \rangle$ is a distributive lattice (see Exercise 19).
  d. For which n is $\mathcal{D}_n$ a Boolean lattice?

Exercises 5–6: True or False
Are the following statements true or false? Explain your answer.

7.2.5. Complements are unique in a complemented lattice.

7.2.6. There are no totally ordered Boolean lattices with eight elements.

Exercises 7–8: Lattice Diagrams
The following problems explore Hasse diagrams for lattices.

7.2.7. *Small Lattices*
Draw Hasse diagrams for all lattice structures on the following numbers of points (see Example 4). Explain why you've considered all possibilities.
  a. Lattices with one element.   c. Lattices with three elements.
  b. Lattices with two elements.   d. Lattices with four elements.
  e. Lattices with five elements.
  f. Which of the lattices in parts *a–e* are complemented lattices? Explain.
  g. Which of the lattices in parts *a–e* are distributive lattices? Explain.

7.2.8. *Lattice Paths Between Points*
  a. Prove or disprove: *A Hasse diagram for a lattice is one in which given any two elements a and b it's possible to pass from a to b by first passing through a sequence of connected points in the upward direction and then passing through a sequence of connected points in the downward direction.*
  b. Is your answer to part *a* different if the terms *upward* and *downward* are interchanged? Explain.

Exercises 9–10: Basic Properties of Lattices
Suppose $\langle \mathcal{A}, \leq \rangle$ is an arbitrary lattice. Prove the following. Note: only assume properties that belong to all such lattice partial order relations.

7.2.9. *Proposition 1: Order Properties*
  a. $x \leq y$ if and only if $x \wedge y = x$.
  b. $x \leq y$ if and only if $x \vee y = y$.
  c. If $x \leq y$, then $x \wedge z \leq y \wedge z$ and $x \vee z \leq y \vee z$.
  d. If $x \leq y$ and $z \leq w$, then $x \wedge z \leq y \wedge w$ and $x \vee z \leq y \vee w$.

7.2.10. *Proposition 2: Operational Properties*
  a. Complete the proof of the first part of *Proposition* 2b:
     $x \wedge (y \wedge z) \leq (x \wedge y) \wedge z$.
  b. Prove the second part of *Proposition* 2b: $(x \vee y) \vee z = x \vee (y \vee z)$.
  c. Prove *Proposition* 2c: $x \wedge x = x = x \vee x$.
  d. Prove *Proposition* 2d: $x \wedge (x \vee y) = x = x \vee (x \wedge y)$.

7.2.11. *Dual Lattices*
Let $\langle \mathcal{A}, \leq \rangle$ be a lattice, and let $\geq$ denote the converse relation of $\leq$.
  a. If $w = x \wedge y$ in $\mathcal{A}$ under the order relation $\leq$, how is w related to x and y under the relation $\geq$? Explain.
  b. If $z = x \vee y$ in $\mathcal{A}$ under the order relation $\leq$, how is z related to x and y under the relation $\geq$? Explain.
  c. Explain why $\langle \mathcal{A}, \geq \rangle$ is a lattice, the dual lattice for $\langle \mathcal{A}, \leq \rangle$.

Exercises 12–13: Power Set Lattices
The following problems relate to power set lattices.

7.2.12. *Lattice Characterization of Set Complements*
Show for any $S \subseteq U$ whose complement $\overline{S}$ is taken inside U, $C = \overline{S}$ if and
only if $C \cap S = \emptyset$ and $C \cup S = U$.

7.2.13. *Power Sets Are Boolean Lattices*
   a. Prove that all power set posets $\mathcal{P}(U)$ are lattices. Carefully explain why
      the meet and join for these posets are set intersection and set union.
   b. Given part *a*, explain why all $\mathcal{P}(U)$ are complemented and distributive
      lattices and hence are Boolean lattices.
   c. If the bottom point $\emptyset$ and the top point U are deleted from the power set
      lattice $\mathcal{P}(U)$, is the resulting set a lattice?
   d. If a single subset S is deleted from the power set lattice $\mathcal{P}(U)$, will the
      result still be a lattice? Explain.

Exercises 14–16: Bounded and Complemented Lattices
*Suppose $\langle \mathcal{A}, \leq, 0, 1 \rangle$ is an arbitrary bounded lattice with least element 0 and
greatest element 1. Note: only assume properties that hold for all such rela-
tions and extremes.*

7.2.14. *Complements for Example 2*
   a. Find all complements, when they exist, for the elements in the lattice of
      Example 2a.
   b. Find all complements, when they exist, for the elements in the lattice of
      Example 2b.

7.2.15. *Complements for Exercise* 21.
   a. Find all complements, when they exist, for the elements in the lattice of
      Exercise 21a. Is the lattice there a complemented lattice?
   b. Repeat part *a* for Exercise 21b.
   c. Repeat part *a* for Exercise 21c.

7.2.16. *Proposition 3: Extreme Element Laws*
   a. Prove *Proposition* 3a: $0 \vee x = x = x \vee 0$; $1 \wedge x = x = x \wedge 1$.
   b. Prove *Proposition* 3b: $0 \wedge x = 0 = x \wedge 0$; $1 \vee x = 1 = x \vee 1$.

Exercises 17–20: Totally Ordered Sets and Lattices
Let $\langle \mathcal{A}, \leq \rangle$ be a totally ordered set.

7.2.17. Prove the following.
   a. $\mathcal{A}$ is a lattice.
   b. What is the meet and join of any two elements a and b in $\mathcal{A}$? Explain.

7.2.18. Prove that if $\mathcal{A}$ has more than two elements, then it's not a complemented lattice, even if it has a minimum and a maximum.

7.2.19. Prove that the distributive laws hold when the three elements x, y,
and z chosen all lie along a common line (are collinear) within a lattice $\mathcal{A}$.
Hence, totally ordered sets are distributive lattices.

7.2.20. Explain, based on Exercise 19, why in determining whether a lattice $\mathcal{A}$ is distributive, it suffices to check distributive law equations for distinct elements not all collinear.

Exercises 21–25: Distributive Lattices
The following problems relate to distributive lattices.

7.2.21. Are the following lattices distributive or not? Explain.

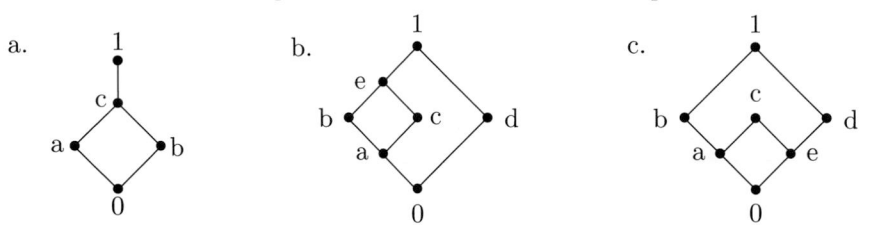

7.2.22. *Distributive Laws for Example 2*
  a. Show that the second distributive law $x \vee (y \wedge z) = (x \vee y) \wedge (x \vee z)$ fails for the lattice of Example 2a.
  b. Show that the lattice of Example 2b is not distributive.

7.2.23. Explain why the calculation given in the text right after Example 2 for the number of distributive law equations that need to be checked for a lattice with 5 elements is correct.

7.2.24. Show that the following semi-distributive laws hold in any lattice.
  a. $x \vee (y \wedge z) \leq (x \vee y) \wedge (x \vee z)$ b. $x \wedge (y \vee z) \geq (x \wedge y) \vee (x \wedge z)$

7.2.25. Show that the following distributive laws hold in a distributive lattice.
  a. $(x \wedge y) \vee z = (x \vee z) \wedge (y \vee z)$ b. $(x \vee y) \wedge z = (x \wedge z) \vee (y \wedge z)$

Exercises 26–29: Boolean Lattices
Suppose that $\langle \mathcal{A}, \leq, {}^{-}, 0, 1 \rangle$ is an arbitrary Boolean lattice. Prove the following results. Note: only assume properties that belong to all such relations, extremes, and complements.

7.2.26. *Proposition 5b: De Morgan's Law*
  a. If $\langle \mathcal{A}, \leq, {}^{-}, 0, 1 \rangle$ is a Boolean lattice, then $\overline{x \vee y} = \overline{x} \wedge \overline{y}$.

7.2.27. *Equivalent Formulations for Boolean Partial Order*
Prove that the following are logically equivalent to $x \leq y$ in a Boolean lattice.
  a. $x \wedge \overline{y} = 0$ b. $\overline{x} \vee y = 1$

7.2.28. *Redundancy Laws*
  a. $x \wedge (\overline{x} \vee y) = x \wedge y$ b. $x \vee (\overline{x} \wedge y) = x \vee y$

7.2.29. *Finite Boolean Lattices*
  a. Show that there are no Boolean lattices with five elements.
  b. Show that there are no Boolean lattices with six elements.
  c. Show that there are no Boolean lattices with seven elements.
  d. Show that there is a Boolean lattice with eight elements.

7.3 From Boolean Lattices to Boolean Algebra

After focusing on equivalence relations in Sections 6.3–6.4 and partial order relations in Sections 7.1–7.2, we're now going to shift back to the operational side of algebra. Boolean lattices provide us with a good lead-in to the computational theory of *Boolean Algebra*.

Let's begin by systematically exploring what Boolean lattices can look like and then show how to convert a Boolean lattice into an operational structure. This will give us a good background for presenting the theory of *Boolean Algebra* axiomatically. Finally, we'll see how to recover a Boolean lattice structure from a Boolean algebra. All of this will provide a solid foundation for working with Boolean functions in later sections.

7.3.1 *Finite Boolean Lattices*

Recall that a Boolean lattice $\langle \mathcal{A}, \leq, {}^-, 0, 1 \rangle$ is a complemented distributive lattice. It has a meet $\wedge$ and a join $\vee$, which satisfy some basic relational and operational properties (*Propositions* 7.2.1 and 7.2.2). $\mathcal{A}$'s minimum 0 and maximum 1 satisfy laws for extreme elements (*Proposition* 7.2.3). And, because $\mathcal{A}$ satisfies the distributive laws and has complements, it satisfies some additional laws, such as *De Morgan Laws* (*Proposition* 7.1.5), which further constrain what such a structure can be.

Let's look at a Boolean lattice of size eight to get a better sense of what finite Boolean lattices are like (see Exercise 7.2.29).

❖**Example 7.3.1**

Investigate the structure of the Boolean lattice $\mathcal{A}$ of size 8 given below.

Solution

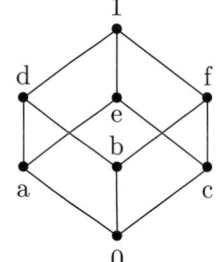

- $\mathcal{A}$ is a finite Boolean lattice with distinct levels. Call the bottom level Level 0, the next level up (the elements right above 0) Level 1, and so on. Here $\mathcal{A}$ has 3 levels above 0.

- Level 1 is the *atomic level* of the lattice—it contains the *atoms* of the lattice, which in this case are a, b, and c. These atoms generate the lattice above it via the join operation $\vee$ (hence the name for these elements).

- Level 2 of lattice $\mathcal{A}$ consists of elements d, e, and f. Note that these $\binom{3}{2} = 3$ elements are all the joins of pairs of atoms: $a \vee b$, $a \vee c$, and $b \vee c$.

- Finally, the top level, Level 3, consists only of the maximum element 1. It lies directly above all the elements on Level 2. Just as importantly, it can be thought of as the join of the atoms taken all together, as $a \vee b \vee c$. There is only $\binom{3}{3} = 1$ element here.

- An insightful way to denote the elements here uses triplet binary notation for indicating the eight elements, thinking of them, say, as representing the elements of the power set of $S = \{a, b, c\}$. The bottom element would be denoted by 000 (no elements present); the next level by 100 (only a present), 010 (only b present), and 001 (only c present); the level above this by 110 (a and b present), and so on. The *product order* on the poset B^3 (see Exercises 34–35), where $B = \{0, 1\}$ ordered as usual by $0 \leq 1$, yields the partial order $(x_1, x_2, x_3) \leq (y_1, y_2, y_3)$ if and only if $x_1 \leq y_1$, $x_2 \leq y_2$, and $x_3 \leq y_3$, which is the order we have here.

The example just given generalizes to any finite Boolean lattice. The properties of a Boolean lattice severely restrict its structure. First a definition, then the surprising result.

Definition 7.3.1: *Atom of a Boolean Lattice*

An **atomic element** (**atom**) of a Boolean lattice $\langle \mathcal{A}, \leq, {}^-, 0, 1 \rangle$ is a minimal element of $\mathcal{A} - \{0\}$.

Theorem 7.3.1: *Stone Representation Theorem* (1936)

Suppose $\langle \mathcal{A}, \leq, {}^-, 0, 1 \rangle$ is a finite Boolean lattice. Then $|\mathcal{A}| = 2^n$ for some n, and the lattice structure of $\mathcal{A}$ is the same as that of the power set lattice $\langle \mathcal{P}(S), \subseteq \rangle$ for $S = \{1, 2, \ldots, n\}$.

Proof:

We'll outline the proof and leave the technical details as exercises.

- First note that $\mathcal{A}$ does have atomic elements (non-0 minimal elements) when $|\mathcal{A}| > 1$: their existence is guaranteed by the fact that $\mathcal{A}$ is finite (see Exercise 7). We'll let $a_1, a_2, \ldots, a_n$ denote the atoms of $\mathcal{A}$.
- Secondly, all atomic elements lie directly above 0 (see Exercise 8).
- Further, every element in $\mathcal{A}$ lies above one or more atoms (see Exercise 9).

Given these observations, we can prove two major claims.

1. *Each non-zero element of $\mathcal{A}$ is the join of all the atoms that lie below it.*

   Suppose element x lies above atoms $a_{i_1}, a_{i_2}, \ldots, a_{i_k}$.

   This claim is proved by showing that $x \leq a_{i_1} \vee a_{i_2} \vee \cdots \vee a_{i_k}$ and $a_{i_1} \vee a_{i_2} \vee \cdots \vee a_{i_k} \leq x$ (see Exercise 11). One direction uses the fact that if $x \wedge \overline{y} = 0$, then $x \leq y$ (see Exercise 7.2.27.a or Exercise 10).

2. *Atomic join representations of elements are unique.*

   This says that two atomic join-combinations $a_{i_1} \vee a_{i_2} \vee \cdots \vee a_{i_k}$ represent the same element if and only if they join exactly the same atoms. This can be proved by taking the meet of the various atoms with both expressions, using the fact that the meet of distinct atoms is 0 (see Exercise 12).

These two claims establish a one-to-one-and-onto correspondence between the elements of $\mathcal{A}$ and all possible join representations. Furthermore, join-combinations of the atoms correspond in an obvious way to distinct subsets of $S = \{a_1, a_2, \ldots, a_n\}$.

Thus, $\mathcal{A}$ perfectly matches up with $\mathcal{P}(S)$. Since this matching respects the ordering of the two Boolean lattices, we've shown that $\mathcal{A}$ *is isomorphic to* (has the same lattice structure as) $\mathcal{P}(S)$. Therefore, $|\mathcal{A}| = 2^n$, too. ∎

The *Stone Representation Theorem* is a deeply satisfying result. We've been exploring posets and lattices as abstract structures, without specifying what their elements are or what order relation connects them. Yet this theorem says that when we have a finite Boolean lattice, it's always a full power set lattice. That may still seem abstract, since it involves the power set, but it's quite specific. We now know exactly what finite Boolean lattices are. Studying finite power set lattices $\mathcal{P}(S)$, therefore, amounts to exploring *all* finite Boolean lattices. Infinite Boolean lattices need not be full power set lattices, but they, too, are essentially substructures of power set lattices.

7.3.2 *Boolean Lattices Are Boolean Algebras*

A Boolean lattice $\langle \mathcal{A}, \leq, \bar{\ }, 0, 1 \rangle$ has meets, joins, and complements, all defined in terms of the order relation $\leq$ and the extreme elements 0 and 1. Meet and join are actually *binary operations* on $\mathcal{A}$. Given any elements x and y, $x \wedge y$ and $x \vee y$ are uniquely defined elements of the lattice. Furthermore, complementation is a unary operation on $\mathcal{A}$: $\bar{x}$ is uniquely defined for any x in $\mathcal{A}$. So while $\mathcal{A}$ may begin as a relational algebraic structure, it has a second life as an operational algebraic structure.

We've already summarized the basic operational properties of meet and join in Section 7.2. They satisfy *Commutative Laws*, *Associative Laws*, *Idempotent Laws*, *Absorption Laws*, *Distributive Laws*, and more. The elements 0 and 1 satisfy various *Extreme Element Laws*. And complementation satisfies *De Morgan's Laws*.

We have here a theory that can be developed independently of its relational origin. This theory is called *Boolean Algebra*, and any associated algebraic structure (a model of the theory) is called a *Boolean algebra*.[1] Historically speaking, *Boolean Algebra* arose before Boolean lattices were studied. Its roots lie in the work of George Boole, who used algebra to investigate logic, a connection we'll explore shortly.

In order to treat $\langle \mathcal{A}, \wedge, \vee, \bar{\ }, 0, 1 \rangle$ as an operational algebra, divorced from its relational origins, we'll need another way to think about meet, join, and the extreme elements. What are these apart from the partial order used to define them? Modern *Boolean Algebra* takes an abstract axiomatic approach to this, treating them as uninterpreted binary operations and distinguished elements that satisfy certain conditions.

It's possible to assume all the laws already proved for these operations, but this is more than we need. As we saw in Section 7.2, some of these properties can be used to prove others—not all are equally basic. If we want to select some properties as axioms, as automatically holding for a Boolean algebra, which ones are most foundational? The following list gives a common axiomatization of *Boolean Algebra*, though it still contains some slight redundancy (see Exercise 39a).

---

[1] Mathematical parlance uses *algebra* in this dual sense, both as a *theory* and a *model*. We'll write <u>*Algebra*</u> to denote the theory and <u>*algebra*</u> for the system of elements.

7.3.3 Boolean Algebra as an Axiomatic Theory

A Boolean algebra $\langle \mathcal{A}, +, \cdot, {}^-, 0, 1 \rangle$ is a set $\mathcal{A}$ together with two binary operations $+$ and $\cdot$, a unary operation ${}^-$, and two distinguished elements 0 and 1 satisfying the following ten axioms:

1. **Commutative Laws**
   a) $x + y = y + x$ b) $x \cdot y = y \cdot x$
2. **Associative Laws**
   a) $(x + y) + z = x + (y + z)$ b) $(x \cdot y) \cdot z = x \cdot (y \cdot z)$
3. **Distributive Laws**
   a) $x \cdot (y + z) = x \cdot y + x \cdot z$ b) $x + (y \cdot z) = (x + y) \cdot (x + z)$
4. **Identity Laws**
   a) $x + 0 = x = 0 + x$ b) $x \cdot 1 = x = 1 \cdot x$
5. **Complementation Laws**
   a) $x + \overline{x} = 1 = \overline{x} + x$ b) $x \cdot \overline{x} = 0 = \overline{x} \cdot x$

We've changed our binary operation symbols from $\vee$ and $\wedge$ to $+$ and $\cdot$. Although we could keep the old symbols, this new choice will help us think about $\mathcal{A}$ apart from any possible lattice structure. It also better aligns with standard algebraic notation. Of course, this is *not* ordinary addition or multiplication. In fact, number systems like $\mathbb{Z}$ or $\mathbb{R}$ do *not* form Boolean algebras under their addition and multiplication. We know this because the second distributive law fails there: $7 = 1 + 2 \cdot 3 \neq (1 + 2) \cdot (1 + 3) = 12$. Many Boolean laws are the same as those in ordinary algebra, but some are not. Despite these differences, most treatments of *Boolean Algebra* use $+$ and $\cdot$ for the binary operations. The same conventions are used as in elementary algebra: $\cdot$ has higher priority than $+$, and xy is typically written for $x \cdot y$.

If $+$ and $\cdot$ are not ordinary addition and multiplication, what are they? They're simply binary operations satisfying the ten axioms listed above. Similarly, ${}^-$ is characterized solely by the *Complementation Laws* and 0 and 1 by the laws that involve them. *Boolean Algebra* forms an abstract template, as it were, which may be given many different interpretations. These ten axioms act like an implicit definition—they help us identify structures we'll be calling Boolean algebras.

Although *Boolean Algebra* is an abstract theory, its models are concrete structures, some of which we already know from our work in Section 7.2.

❖ **Example 7.3.2**

Given a set U, $\mathcal{P}(U)$ forms a Boolean lattice under the partial order relation of $\subseteq$. Interpret $\mathcal{P}(U)$ as a Boolean algebra.

Solution

The operation of $+$ in this case is $\cup$, that of $\cdot$ is $\cap$, and ${}^-$ is set complementation. The elements 0 and 1 are $\emptyset$ and U, respectively.

All the axioms hold for this interpretation, as we know from our study of *Set Theory*. ✓

This provides us with infinitely many models of *Boolean Algebra*.

❖**Example 7.3.3**

Let $B = \{0, 1\}$ be strictly ordered in the usual way by $0 < 1$.
Let $x + y = \max\{x, y\}$, $x \cdot y = \min\{x, y\}$, and $\overline{x} = 1 - x$.
Show that B forms a Boolean algebra under these operations.

Solution

It's routine to check that all five pairs of axioms hold for this interpretation.
The *Commutative Laws* hold, for example, because $\max\{x, y\} = \max\{y, x\}$
and $\min\{x, y\} = \min\{y, x\}$. The other laws also hold (see Exercises 2–6).

An abstract approach is valuable because it provides a unified theory for
treating the essential algebraic features of varied concrete structures. For
example, the divisors of 60 have nothing to do with words in a dictionary,
but both sets of things form lattices under their respective partial orders.
The same is true for Boolean algebras. Very different structures can satisfy
the same set of axioms. *Boolean Algebra* has applications in advanced mathematics and also in cognate areas such as computer science and engineering.
We'll look at the latter in Sections 7.4–7.6.

If we know that a structure satisfies the basic laws of *Boolean Algebra*,
what else do we know about it? The answer is that any result proved from
the axioms will hold for all Boolean algebras, regardless of what their specific
elements and operations are.

Following are some key results we can derive from the axioms of *Boolean
Algebra*. For all of these propositions, we're assuming an abstract Boolean
algebra $\langle \mathcal{A}, +, \cdot, \bar{}, 0, 1 \rangle$ as the universe of discourse, and universal quantifiers
are understood for identities with variables. These propositions are ordered
so that each one can be proved using only the axioms and previous propositions, but other orders are also possible. Proving them in the exercises (see
Exercises 18–26) will give you some practice at constructing proofs in an
abstract setting. In your work, be careful not to use a result unless it has
already been proved.

Proposition 7.3.1: *Uniqueness of Identities and Complements*
  a) If $x + z = x$ for all x, then $z = 0$.
  b) If $x \cdot u = x$ for all x, then $u = 1$.
  c) If $xz = 0$ and $x + z = 1$, then $z = \overline{x}$.

Proof:
See Exercise 18.

Proposition 7.3.2: *Complements of Elements Laws*
  a) $\overline{0} = 1$ c) $\overline{\overline{x}} = x$
  b) $\overline{1} = 0$

Proof:
See Exercise 19.

Proposition 7.3.3: *Idempotence Laws*
  a) $x \cdot x = x$ b) $x + x = x$

Proof:
  See Exercise 20.

Proposition 7.3.4: *Annihilation and Absorption Laws*
  a) $x \cdot 0 = 0$ c) $x(x + y) = x$
  b) $x + 1 = 1$ d) $x + xy = x$

Proof:
  See Exercise 21.

Proposition 7.3.5: *De Morgan's Laws*
  a) $\overline{xy} = \overline{x} + \overline{y}$ b) $\overline{x + y} = \overline{x}\,\overline{y}$

Proof:
  See Exercise 22.

Proposition 7.3.6: *Operation Linkage Laws*
  The following equations are logically equivalent:
  a) $xy = x$ c) $x\overline{y} = 0$
  b) $x + y = y$ d) $\overline{x} + y = 1$

Proof:
  Remark: Note the difference between this proposition and the others.
  What is being asserted here, for example, is a universal biconditional
  expressing an equivalence: $\forall x \forall y (xy = x \leftrightarrow x + y = y)$.
  The individual equations of parts a) through d) are *not* being universally
  asserted as was the case in the other propositions. See Exercise 23.

Proposition 7.3.7: *Redundancy Laws*
  a) $x(\overline{x} + y) = xy$ b) $x + \overline{x}y = x + y$

Proof:
  See Exercise 24.

Proposition 7.3.8: *Consensus Laws*
  a) $xy + \overline{x}z + yz = xy + \overline{x}z$
  b) $(x + y)(\overline{x} + z)(y + z) = (x + y)(\overline{x} + z)$

Proof:
  See Exercise 25.

Proposition 7.3.9: *Cancellation Law*
  If $xy = xz$ and $x + y = x + z$, then $y = z$.

Proof:
  Both of the equations conjoined in the antecedent are needed in order for
  cancellation to be legitimate. See Exercise 26ab.

7.3.4 *Boolean Algebra and Logic*

Boolean Algebra originated in the mid-nineteenth century with the work of George Boole, for whom the theory is named. Boole's interest in logic was piqued by a dispute between De Morgan and a Scottish philosopher over their extensions of *Aristotelian Logic*. This led Boole to the novel idea of using algebra to develop logic (see Section 1.1).

Boole first transformed *Aristotelian Logic*, which considered statements about how classes were related. He used 1 to stand for the universe of discourse, 0 for the class with nothing in it, and letters for particular classes or categories of objects. A product of two classes would be the class of things common to them both, and a sum would be their union, provided they were disjoint.[2] The identity of two classes would be asserted by an equation, which could then be manipulated to determine the consequences of a set of premises. Deductive argumentation, Boole thought, was nothing more than a species of algebraic computation, thus making logic a branch of algebra.

Here's an example that uses his system to demonstrate the most important syllogistic form in *Aristotelian Logic*, traditionally known as *Barbara*.

✦ **Example 7.3.4**

Use Boole's algebraic approach to show the validity of the following:

All X's are Y's

All Y's are Z's

All X's are Z's

Solution

The sentence-form *All A's are B's* is formulated by the equation $A = AB$. We can then deduce the argument's conclusion from its premises as follows:

$$
\begin{array}{ll}
X = XY & \text{Prem} \\
 = X(YZ) & \text{Sub, using the second premise } Y = YZ \\
 = (XY)Z & \text{Assoc Law} \\
 = XZ & \text{Sub, using } X = XY \text{ again}
\end{array}
$$

Though there are some differences, Boole's system is similar to *Set Theory*. His full system had some complications and idiosyncrasies (not our concern here), but it brought mathematics and logic closer together, and it stimulated other mathematicians to explore these connections further. It was also one of the first systems to introduce an algebra whose laws were not identical to those of ordinary algebra, a novel idea at the time. The power law $X^2 = X$ (idempotence for multiplication) was one such law. Boole knew that these anomalies would lead many of his contemporaries to reject his system as not being genuine algebra, but he forestalled them by noting that if one restricted a variable's possible values to 0 and 1, the solutions to $X^2 = X$, then all the algebraic laws of logic would hold.

---

[2] Later logicians changed this to ordinary union, based on *nonexclusive or*, which made it a full binary operation.

Boole also used his algebraic logic to develop a version of *Propositional Logic*. In it, variables represented sentences, 1 and 0 stood for true and false, respectively, addition and multiplication for disjunction and conjunction, and complements for negations.

✤ **Example 7.3.5**

Derive the *Law of Non-Contradiction* from the *Law of Excluded Middle*.

Solution

Here's an abbreviated premise-conclusion argument based on the results listed above. Again, details are left for the exercises (see Exercise 29a).

$$X + \overline{X} = 1 \qquad \text{LEM}$$
$$X \cdot \overline{X} = 0 \qquad \text{Compln, DeM, Compl Elts, Comm}$$

Today we view *Propositional Logic*'s relation to *Boolean Algebra* differently than Boole did, and we don't symbolize sentences being true or false using the syntax of logic, but aspects of our approach are rooted in his seminal work. This is spelled out further in the next example.

✤ **Example 7.3.6**

Interpret *Propositional Logic* in terms of *Boolean Algebra*.

Solution

· Let the variables of *Boolean Algebra* denote sentences of PL, 0 denote a logical falsehood, and 1 a logical truth (pick any favorites, say, $P \wedge \neg P$ and $P \vee \neg P$).
· Interpret $P \cdot Q$ as the conjunction $P \wedge Q$, $P + Q$ as the disjunction $P \vee Q$, and $\overline{P}$ as the negation $\neg P$.
· Then, if we interpret $=$ as $\models\!\dashv$,[3] all axioms of *Boolean Algebra* will be true. Many of them are familiar, being counterparts of the *Replacement Rules* we adopted for making PL deductions.
· For example, the *Distributive Law* becomes $P \wedge (Q \vee R) \models\!\dashv (P \wedge Q) \vee (P \wedge R)$. The *Identity Laws* involving 0 and 1 are a bit unusual, considering the natural deduction approach we took in studying *Propositional Logic*, but they're also true (see Exercise 31).

More can be done to relate *Boolean Algebra* and logic, but we'll save that for the next section, when we look at some later developments.

7.3.5 *Boolean Algebras Generate Boolean Lattices*

Our entry point for discussing *Boolean Algebra* was Boolean lattices. We'll now finish this topic by reversing the process, generating a Boolean lattice from a Boolean algebra. Combining these two procedures will bring us full

---

[3] We're cheating here to keep matters simpler. We're reinterpreting equality to avoid introducing $\models\!\dashv$ as an equivalence relation on the set of PL sentences and then taking its equivalence classes as the elements of the Boolean algebra of logic.

circle: if we start with a Boolean lattice, pass to its Boolean algebra, and then generate the associated Boolean lattice, we end up where we began (see Exercise 37a). The same is true if we begin and end with a Boolean algebra (see Exercise 37b).

Theorem 7.3.2: *Boolean Algebras are Boolean Lattices*

Suppose $\langle \mathcal{A}, +, \cdot, ^-, 0, 1 \rangle$ is a Boolean algebra, and let $x \leq y \leftrightarrow x \cdot y = x$. The resulting structure $\langle \mathcal{A}, \leq, ^-, 0, 1 \rangle$ is a Boolean lattice.

Proof:

We'll sketch the proof's main outline and leave the details for the exercises. First recall that in a Boolean algebra, $x \cdot y = x \leftrightarrow x + y = y$ (*Proposition* 6ab), so we could have defined the order relation by $x \leq y \leftrightarrow x + y = y$. We can therefore use this equivalence, where needed, to argue the following:

1. $\leq$ *is a partial order on $\mathcal{A}$.*
   Showing that $\leq$ is a partial order is easy (see Exercise 36a).

2. $x \cdot y$ *is $x \wedge y$, and $x + y$ is $x \vee y$ relative to the partial order $\leq$.*
   Showing that $x \cdot y$ is a lower bound and $x + y$ is an upper bound for x and y is immediate from the definition and *Idempotence Laws*; proving that they're the greatest lower bound and the least upper bound is almost as easy (see Exercise 36bc).

3. $\overline{x}$ *is the complement of x in the lattice $\mathcal{A}$.*
   This is guaranteed by the *Complementation Laws*.

4. $\mathcal{A}$ *is a distributive lattice.*
   This follows from the definitions of $\wedge$ and $\vee$ and the *Distributive Laws*.

5. Thus, $\langle \mathcal{A}, \leq, ^-, 0, 1 \rangle$ forms a Boolean lattice. ∎

EXERCISE SET 7.3

Note: *In the problems below, treat $+$, $\cdot$, $^-$, 0, and 1 as abstract operations and entities in a Boolean algebra, not as ordinary addition, multiplication, complementation, 0, or 1, unless otherwise instructed. All you can assume about them in this context are properties that hold in Boolean Algebra.*

Exercises 1–6: Boolean algebras of 0 and 1

Let $\mathcal{B} = \{0, 1\}$ denote the Boolean algebra of Example 3.

7.3.1. Write out the addition and multiplication tables for $\mathcal{B}$, putting first elements x along the side and second elements y across the top. Your calculation results $x + y$ and $x \cdot y$ go inside their respective tables' cells.

7.3.2. Verify the *Commutative Laws* for $\mathcal{B}$'s addition and multiplication.

7.3.3. Verify the *Associative Laws* for $\mathcal{B}$'s addition and multiplication.

7.3.4. Verify the *Distributive Laws* for $\mathcal{B}$'s addition and multiplication.

7.3.5. Verify the *Identity Laws* for $\mathcal{B}$'s addition and multiplication.

7.3.6. Verify the *Complementation Laws* for $\mathcal{B}$'s addition and multiplication.

Exercises 7–13: Proving the Stone Representation Theorem

Prove the following results used in the proof of the Stone Representation Theorem, in order.

7.3.7. Every finite Boolean lattice with two or more elements has atoms.

7.3.8. All atoms in a Boolean lattice lie directly above 0.

7.3.9. Every non-zero element in a finite Boolean lattice is above some atom.

7.3.10. In a Boolean lattice, if $b \wedge \overline{c} = 0$, then $b \leq c$. *Hint*: join this equation's expression with an appropriate element.

7.3.11. Each non-zero element of a finite Boolean lattice can be expressed as the join of the distinct atoms below it. *Hint*: make use of Exercise 10.

7.3.12. Atomic join representations of non-zero elements in a finite Boolean lattice are unique. *Hint*: take the meet of these expressions with the atoms of the lattice.

7.3.13. Using the results of the last two problems, explain why a Boolean lattice with n atoms has 2^n distinct elements and is essentially a power set lattice for a finite set.

Exercises 14–17: True or False

Are the following statements true or false? Explain your answer.

7.3.14. No lattice with 10 elements is a complemented distributive lattice.

7.3.15. Finite Boolean lattices have the same lattice structure as the power set lattice formed from the set of its atoms.

7.3.16. There is exactly one Boolean algebra having 12 elements.

7.3.17. While algebra can be used to represent propositions, equation solution procedures don't capture ordinary deductive reasoning very well.

Exercise 18–28: Propositions of Boolean Algebra

Prove the following propositions, using the axioms of Boolean Algebra or any proposition that precedes it. Carefully justify the steps in your arguments.

7.3.18. *Proposition 1: Uniqueness Laws for Identities and Complements*
   a. If $x + z = x$ for all x, then $z = 0$.
   b. If $x \cdot u = x$ for all x, then $u = 1$.
   c. If $xz = 0$ and $x + z = 1$, then $z = \overline{x}$.

7.3.19. *Proposition 2: Complements of Elements Laws*
   a. $\overline{0} = 1$ c. $\overline{\overline{x}} = x$
   b. $\overline{1} = 0$

7.3.20. *Proposition 3: Idempotence Laws*
   a. $x \cdot x = x$ b. $x + x = x$

7.3.21. *Proposition 4: Annihilation and Absorption Laws*
   a. $x \cdot 0 = 0$ c. $x(x + y) = x$
   b. $x + 1 = 1$ d. $x + xy = x$

7.3.22. *Proposition 5: De Morgan's Laws*
   a. $\overline{xy} = \overline{x} + \overline{y}$ b. $\overline{x + y} = \overline{x}\,\overline{y}$

7.3.23. *Proposition* 6: *Operation Linkage Laws*
Prove that the following *Boolean Algebra* statements are logically equivalent.

a. $xy = x$ c. $x\overline{y} = 0$

b. $x + y = y$ d. $\overline{x} + y = 1$

7.3.24. *Proposition* 7: *Redundancy Laws*

a. $x(\overline{x} + y) = xy$ b. $x + \overline{x}y = x + y$

7.3.25. *Proposition* 8: *Consensus Laws*

a. $xy + \overline{x}z + yz = xy + \overline{x}z$ b. $(x+y)(\overline{x}+z)(y+z)=(x+y)(\overline{x}+z)$

7.3.26. *Proposition* 9: *Cancellation Law*

a. If $xy = xz$ and $x + y = x + z$, then $y = z$.

b. Is either equation in the antecedent condition of *Proposition* 9 (part *a*) sufficient by itself for the conclusion? Why or why not?

7.3.27. *Other Boolean Cancellation Laws*
Prove or disprove the following *Cancellation Laws*.

a. If $x + y = x + z$ and $\overline{x} + y = \overline{x} + z$, then $y = z$.

b. If $xy = xz$ and $\overline{x}y = \overline{x}z$, then $y = z$.

c. If $xy = xz$ and $x\overline{y} = x\overline{z}$, then $y = z$.

7.3.28. *Duality Principles for Boolean Algebra*

a. *True or False*: Every Boolean lattice is changed into an (inverted) lattice by changing $\leq$ into $\geq$ and interchanging 0 and 1 and $\wedge$ and $\vee$. Explain.

b. *True or False*: Every theorem in *Boolean Algebra* has a dual theorem that results when $+$ and $\cdot$ as well as 0 and 1 are exchanged. Explain.

Exercises 29–30: Boole's Algebra of Logic

The following problems relate to Boole's algebraic approach to logic.

7.3.29. *Deducing LNC from LEM Algebraically*

a. Fill in the details of Example 5's argument to show that the *Law of Non-Contradiction* follows from the *Law of Excluded Middle*. Begin by explaining why these laws are algebraically formulated the way they are.

b. Deduce LNC from LEM using inference rules from *Propositional Logic*. Compare your proof with that of Example 5 (see part *a*). Which one better captures the process of deductive reasoning?

7.3.30. *Deducing Aristotelian Syllogisms and Conditional Arguments*

a. Formulate the universal negative statement *No A's are B's* as an equation. *Hint*: start with a set-theoretic statement for classes A and B and then translate it into Boolean notation.

b. Use Boole's approach to logic to formulate and deduce the argument-form *Celarent* from Aristotelian Logic: *No X's are Y's, All Z's are X's; therefore, No Z's are Y's*. Use only basic algebra to make your argument.

c. Use Boole's approach to logic to formulate and deduce *Modus Ponens* $(P \rightarrow Q, P \vdash Q)$ and *Modus Tollens* $(P \rightarrow Q, \neg Q \vdash \neg P)$. *Hint*: what's the only case in which $P \rightarrow Q$ is false? Say this in an equation, using an equivalent for $P \rightarrow Q$.

Exercises 31–33: Propositional Logic as Boolean Algebra
The following problems relate to the interpretation of Propositional Logic as a Boolean algebra.

7.3.31. *Identity Laws*
Interpret the *Identity Laws* of *Boolean Algebra* for *Propositional Logic* and explain why they're true, using what you know about PL.

a. $x + 0 = x = 0 + x$ b. $x \cdot 1 = x = 1 \cdot x$

7.3.32. *Complementation Laws*
Interpret the *Complementation Laws* of *Boolean Algebra* for *Propositional Logic* and explain why they're true, using what you know about PL.

a. $x + \overline{x} = 1 = \overline{x} + x$ b. $x \cdot \overline{x} = 0 = \overline{x} \cdot x$

7.3.33. *Uniqueness Laws*
Interpret the *Uniqueness Laws* of *Boolean Algebra* for *Propositional Logic* and explain why they're true, using what you know about PL.

a. If $x + z = x$ for all x, then $z = 0$.   c. If $xz = 0$ and $x + z = 1$, then
b. If $x \cdot u = x$ for all x, then $u = 1$. $z = \overline{x}$.

Exercises 34–37: Boolean Algebras and Boolean Lattices
The following results explore the interconnections between Boolean algebras and Boolean lattices.

7.3.34. *Alternative Characterization of Finite Boolean Lattices*
Let $\mathcal{B}^n$ denote the n-fold Cartesian product of $\mathcal{B} = \{0,1\}$ with itself. The *product order* $\leq$ on $\mathcal{B}^n$ is defined by $(x_1, x_2, \ldots, x_n) \leq (y_1, y_2, \ldots, y_n)$ if and only if $x_i \leq y_i$ for all i.

a. Prove that $\leq$ is a partial order on $\mathcal{B}^n$. (This generalizes the order considered in Exercise 7.1.29).
b. Prove that $\langle \mathcal{B}^n, \leq \rangle$ forms a Boolean lattice.
c. Prove that the n-tuples $x = (x_1, x_2, \ldots, x_n)$ in $\mathcal{B}^n$ can be placed in natural one-to-one correspondence with $\mathcal{P}(S)$ for $S = \{1, 2, \ldots, n\}$. *Hint*: take $i \in R_x \subseteq S$ if and only if $x_i = 1$.
d. Prove that $\mathcal{B}^n$ and $\mathcal{P}(S)$ are *isomorphic* as lattices; i.e., prove that $(x_1, x_2, \ldots, x_n) \leq (y_1, y_2, \ldots, y_n)$ if and only if $R_x \subseteq R_y$, where R_x is the set of indices associated with $x = (x_1, x_2, \ldots, x_n)$ as in part c. Thus, by the *Stone Representation Theorem*, all finite Boolean lattices are given by $\langle \mathcal{B}^n, \leq \rangle$ for some n.

7.3.35. *The Product Order and Finite Boolean Algebras $\mathcal{B}^n$*
a. Explain why when the Boolean lattice $\mathcal{B}^n$, ordered by the product order (see Exercise 34), is converted into a Boolean algebra in the normal way, the resulting operations are performed coordinatewise: that is, $\overline{(x_1, x_2, \ldots, x_n)} = (\overline{x_1}, \overline{x_2}, \ldots, \overline{x_n})$, $(x_1, x_2, \ldots, x_n) + (y_1, y_2, \ldots, y_n) = (x_1 + y_1, x_2 + y_2, \ldots, x_n + y_n)$, and $(x_1, x_2, \ldots, x_n) \cdot (y_1, y_2, \ldots, y_n) = (x_1 \cdot y_1, x_2 \cdot y_2, \ldots, x_n \cdot y_n)$, where the component operations are the usual ones of the basic Boolean algebra $\mathcal{B} = \{0, 1\}$.
b. Explain from part a, Exercise 34, and Exercise 37b why all finite Boolean algebras are essentially $\mathcal{B}^n$ under standard coordinatewise operations.

7.3.36. *Theorem 2*

Suppose $\langle \mathcal{A}, +, \cdot, ^-, 0, 1 \rangle$ is a Boolean algebra, and $\leq$ is defined by $x \leq y$ if and only if $x \cdot y = x$. Prove the following claims for the proof of *Theorem 2*.

a. $\leq$ *is a partial order.*   c. $x + y$ *is the join of* x *and* y.

b. $x \cdot y$ *is the meet of* x *and* y.

7.3.37. *Boolean Conversions Undo One Another*

a. Suppose $\langle \mathcal{A}, \leq, ^-, 0, 1 \rangle$ is a Boolean lattice. Then it forms a Boolean algebra, interpreting $+$ as $\vee$ and $\cdot$ as $\wedge$. Show that if this Boolean algebra is reconverted into a Boolean lattice via the definition $x \leq_1 y$ if and only if $x \cdot y = x$, the resulting order is exactly the same as the original partial order; i.e., $x \leq_1 y$ if and only if $x \leq y$.

b. Suppose $\langle \mathcal{A}, +, \cdot, ^-, 0, 1 \rangle$ is a Boolean algebra. Show that if this is converted into a Boolean lattice by the definition $x \leq y$ if and only if $x \cdot y = x$, the result is a lattice whose join and meet agree with the original $+$ and $\cdot$, i.e., $x \vee y = x + y$ and $x \wedge y = x \cdot y$. Thus, reconverting this lattice into a Boolean algebra will give the original structure.

7.3.38. Let U be an infinite set.

a. Show that the collection $\mathcal{A}$ of all finite subsets of U is not a Boolean algebra under the set operations of $\cup$ and $\cap$. *Hint*: what's missing?

b. Show that the collection $\mathcal{A}^*$ of all finite and cofinite subsets (subsets with finite complements) of U is the smallest Boolean algebra extending $\mathcal{A}$.

c. Explain why part *b* implies that not all Boolean algebras are power set structures.

7.3.39. *Independence Exploration*

a. Show that the *Associative Law* can be proved from the other axioms and their consequences. Thus, it may be dropped from our list of axioms.

b. While the associative laws can be proved from the other axioms (see part *a*), the rest of the axioms of *Boolean Algebra* are independent of one another.[4] Show this by constructing models that satisfy all axioms except each one.

7.4 Boolean Functions and Logic Circuits

We now know that a Boolean system is simultaneously a lattice ordered by $\leq$ and an algebra with operations $^-$, $\wedge$ ($\cdot$), and $\vee$ ($+$). At times it will be more helpful to adopt a relational perspective, at others an operational one. Both viewpoints may help you recall the laws holding for a Boolean structure, both can be used to test an identity's validity, and both are equally legitimate.

We started to explore the relation between *Boolean Algebra* and *Propositional Logic*, and we'll continue that in this section, showing that this connection is foundational to computational circuitry. In preparation for this, let's look at some functions defined on two-valued Boolean algebras.

---

[4] This was shown by American postulate theorist Edward Huntington in a 1904 paper on *Boolean Algebra*.

7.4.1 *Two-Valued Boolean Algebras*

The simplest nontrivial Boolean algebra is $\mathcal{B} = \{0, 1\}$. Its addition and multiplication tables are dictated by the *Identity*, *Annihilation*, and *Absorption Laws* of *Boolean Algebra*.

+	0	1
0	0	1
1	1	1

·	0	1
0	0	0
1	0	1

The *Stone Representation Theorem* tells us that all finite Boolean algebras are of size 2^n, structured like the power set of $\{1, 2, \ldots, n\}$ under the subset relation and the operations of union, intersection, and complementation. We can make this representation less abstract using $\mathcal{B}$.

✤**Example 7.4.1**

Show how all finite Boolean algebras can be represented by $\mathcal{B}^n$.

Solution

· Example 7.3.1 showed how to represent any finite power set as an n-fold Cartesian product $\mathcal{B}^n$: each $S \subseteq \{1, 2, \ldots, n\}$ corresponds to an ordered n-tuple of 0's and 1's, where entry $x_i = 1$ if and only if $i \in S$. For example, $(1, 0, 1)$ represents the subset $S = \{1, 3\}$ in $\mathcal{B}^3$.

· The associated order induced on $\mathcal{B}^n$ by this correspondence is the *product order*: $(x_1, x_2, \ldots, x_n) \leq (y_1, y_2, \ldots, y_n)$ if and only if $x_i \leq y_i$ for all i (see Exercise 7.3.34). The Boolean operations produced are defined coordinatewise: $(x_1, x_2, \ldots, x_n) + (y_1, y_2, \ldots, y_n) = (x_1 + y_1, x_2 + y_2, \ldots, x_n + y_n)$ and $(x_1, x_2, \ldots, x_n) \cdot (y_1, y_2, \ldots, y_n) = (x_1 y_1, x_2 y_2, \ldots, x_n y_n)$ (see Exercise 7.3.35).

· For example, $(1, 0, 0) \leq (1, 0, 1)$ because $1 \leq 1$, $0 \leq 0$, and $0 \leq 1$. Using the above addition and multiplication tables for working in each coordinate, $(1, 0, 0) + (1, 0, 1) = (1, 0, 1)$ and $(1, 0, 0) \cdot (1, 0, 1) = (1, 0, 0)$.

· These are the Boolean algebra representations we'll be working with below.

7.4.2 *Boolean Functions*

Definition 7.4.1: *Boolean Function*

A Boolean function is a function $f : \mathcal{B}^n \to \mathcal{B}$, a Boolean-valued function of several Boolean variables.

✤**Example 7.4.2**

Exhibit Boolean addition and multiplication as Boolean functions.

Solution

The operations of $+$ and $\cdot$ can be considered functions from $\mathcal{B}^2$ to $\mathcal{B}$. The following tables represent these operations. Notice their similarity to the truth tables for $\vee$ and $\wedge$ in PL if 0 is interpreted as F and 1 as T.

x_1	x_2	$+(x_1, x_2)$
0	0	0
0	1	1
1	0	1
1	1	1

x_1	x_2	$\cdot(x_1, x_2)$
0	0	0
0	1	0
1	0	0
1	1	1

❖**Example 7.4.3**

Write out the *ternary majority function's* table, where $f(x_1, x_2, x_3) = m$, the value that appears among $\{x_1, x_2, x_3\}$ the majority of the time.

Solution

The following table gives an output for each input sequence, listed in dictionary (binary numerical) order. This is the exact reverse of how we organized truth tables for *Propositional Logic*.

x_1	x_2	x_3	$f(x_1, x_2, x_3)$
0	0	0	0
0	0	1	0
0	1	0	0
0	1	1	1
1	0	0	0
1	0	1	1
1	1	0	1
1	1	1	1

7.4.3 Boolean Algebra and Propositional Logic

Let's now look at some simple examples of Boolean functions and use them to connect *Boolean Algebra* more closely with *Propositional Logic*.

❖**Example 7.4.4**

Discuss the relation of the following Boolean functions to *Propositional Logic* connectives:

a)

x	$N(x)$
0	1
1	0

b)

x_1	x_2	$C(x_1, x_2)$
0	0	1
0	1	1
1	0	0
1	1	1

c)

x_1	x_2	$D(x_1, x_2)$
0	0	0
0	1	1
1	0	1
1	1	1

Solution

a) The Boolean function N can be thought of as assigning truth values to negations $\neg x$, taking 0 to be F and 1 to be T.

b) Similarly, C assigns truth values to conditional sentences $x_1 \rightarrow x_2$.

c) D corresponds to assigning truth values to disjunctions $x_1 \vee x_2$.

These examples illustrate that every Boolean function has a propositional connective associated with it (possibly unfamiliar, defined simply by its truth table) and that every propositional connective generates a Boolean function via its truth table. Whatever we can say about truth-functional connectives can be translated into language about Boolean functions, and conversely.

7.4.4 *Historical Aside: From Boole to Shannon*

As noted in Section 7.3, Boole introduced ideas about an algebra of 0 and 1 in order to justify his approach to logic and reasoning. He never adopted an abstract approach to *Boolean Algebra*, though, nor did he consider its possible application to other fields of thought.

The American mathematician Claude Shannon was one of the first to recognize the broader potential of an algebraic approach to logic (see Section 1.1). His pathbreaking 1938 M.A. thesis, *A Symbolic Analysis of Relay and Switching Circuits*, showed how *Boolean Algebra* could be physically implemented. Conversely, Shannon showed how to use *Boolean Algebra* to analyze, simplify, and design electrical circuits, establishing it as an essential tool for contemporary digital electronics and basic computation.

Shannon's later work continued to exploit the possibilities in the algebra of 0 and 1. His most important publication, *A Mathematical Theory of Communication* (1948), laid the groundwork for the information revolution of the following decades by treating information in terms of bits (0's and 1's). He also showed that information noise and data corruption could be overcome by transmitting messages with built-in redundancy. And his ideas were instrumental in encrypting messages. The Data Encryption Standard (DES), for example, which saw widespread use for several decades, arose from the work Shannon did during the 1940s.

7.4.5 *Boolean Algebra and Switching Circuits*

The main link between *Boolean Algebra* and computer logic involves logic gates, to which we'll turn shortly, but we'll start by looking at the simpler case of switching circuits. This was Shannon's original idea for how *Boolean Algebra* could be applied to telecommunication.

Simple switches have two states: open (no current through) and closed (current through). We'll represent an open-switch state with 0 and a closed-switch state with 1—the now-standard assignment, though Shannon's original choice reversed these values as well as the operators used for circuit types.

Switches can be connected either in series or in parallel, as shown in Figure 7.2.

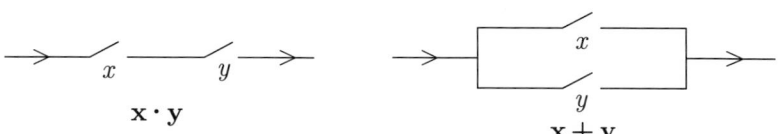

Fig. 7.2 Series and parallel circuits

Switches in series are represented by Boolean products since current flows through the circuit (yields output 1) if and only if it flows through both switches. Switches in parallel are denoted by Boolean sums.

Complex circuits are constructed out of these basic components. If two circuit switches are controlled so that they're always in the same state, the same letter is used for both switches. If one switch is always in the opposite state of another one, a complement is used to indicate this negative relationship. So, every circuit is associated with a Boolean expression involving switch-labels. Switching circuits provide a concrete representation for Boolean functions.

❖**Example 7.4.5**

Design a switching circuit diagram and a Boolean expression for the ternary majority function of Example 3.

Solution

Since we want output 1 (current flowing through) when two or more switches are closed (value 1), the circuit shown models the ternary majority function. The Boolean expression for this function/circuit is given below the diagram.

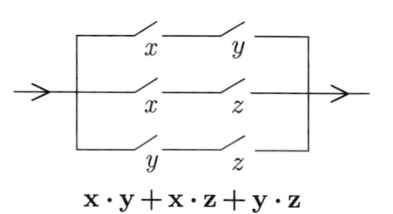

$$\mathbf{x \cdot y + x \cdot z + y \cdot z}$$

❖**Example 7.4.6**

Design a switching circuit for two switches controlling a single light in the usual way (confusingly called *three-way switches* in the U.S.).

Solution

· Let's first determine an appropriate Boolean function F.

· Taking output 1 to represent the light being on and letting x and y stand for the two switches, we choose $F(1,1) = 1$. Changing the state of a single switch will turn the light off, so $F(0,1) = 0$ and $F(1,0) = 0$. Changing the state of both switches should turn the light back on, so $F(0,0) = 1$, giving the following Boolean function table:

x_1	x_2	$F(x_1, x_2)$
0	0	1
0	1	0
1	0	0
1	1	1

- This resembles the truth table for $x_1 \leftrightarrow x_2$, which is logically equivalent to $(x \wedge y) \vee (\neg x \wedge \neg y)$—represented in *Boolean Algebra* by $x \cdot y + \overline{x} \cdot \overline{y}$. This expression can be developed into a circuit as shown below.
- Although this isn't yet a wiring diagram, it can be converted into one, as indicated by the three-way switch diagram.

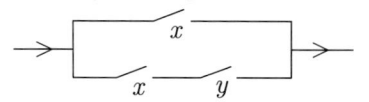

$$\mathbf{x} \cdot \mathbf{y} + \overline{\mathbf{x}} \cdot \overline{\mathbf{y}} \qquad \text{three-way switch}$$

✤**Example 7.4.7**

Determine the Boolean expression representing the following switching circuit. Then design a simpler circuit to accomplish the same thing.

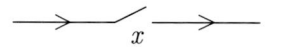

Solution
- The Boolean expression for this circuit is $x + xy$.
- By the *Absorption Law* for *Boolean Algebra*, this equals x, so the circuit can be simplified in this case by dropping off the lower portion of the circuit.
- A simplified circuit is therefore the following:

As this example demonstrates, any compound circuit can be represented by a Boolean expression and then simplified. A simpler, cheaper switching circuit can then be designed that acts exactly like the original. *Boolean Algebra*, abstract as it is, simplifies circuit design. Because such mathematical work leads to greater profitability, Bell Labs for years had mathematicians such as Shannon on its Research and Development payroll.

7.4.6 *Boolean Algebra and Circuits from Logic Gates*

Computers use integrated electronic circuits built out of *logic gates*. Our concern here will be with *combinational circuits*, circuits whose outputs are solely determined by their inputs and not also on the present state/memory of the machine. There are ways of treating this extended situation, such as with finite-state machines or Turing machines, but these go beyond the scope of this text. Already with combinational circuits, however, we'll be able to perform basic numerical computations, as we'll show shortly.

Let's start by considering the logic gates that correspond to standard binary and unary logical connectives. These logic gates are symbolized as shown in Table 7.1. The inputs are placed left of each logic-gate symbol, the outputs on the right. Note how negating or inverting another operator is indicated by including an inverter node before the output.

Strictly speaking, each of the basic logic gates except for the inverter NOT-gate is a double-input gate. However, given the associative and commutative properties for the Boolean operators AND and OR we can allow their respective logic gates (as well as those for NAND and NOR) to take on more than two input lines without introducing any ambiguity.

A value of 1 for input or output represents high voltage; 0 indicates low voltage. If output 1 is the desired state of a logic gate, we can specify which string-inputs xy the gate *accepts* to put it into that state. We'll use this same terminology in speaking about Boolean functions.

Table 7.1 Logic Gates

Name	Symbol	Accepts Strings	Boolean Expression
AND	x y z	$xy = 11$	$z = x \cdot y$
OR	x y z	$xy = 01$, 10, or 11	$z = x + y$
NOT	x z	$x = 0$	$z = \overline{x}$
XOR	x y z	$xy = 01$ or 10	$z = x \oplus y$ $= x \cdot \overline{y} + \overline{x} \cdot y$
NAND	x y z	$xy = 00$, 01, or 10	$z = \overline{x \cdot y}$ $= \overline{x} + \overline{y}$
NOR	x y z	$xy = 00$	$z = \overline{x + y}$ $= \overline{x} \cdot \overline{y}$
XNOR	x y z	$xy = 00$ or 11	$z = \overline{x \oplus y}$ $= x \cdot y + \overline{x} \cdot \overline{y}$

Logic gates can be represented by Boolean expressions (the last column of Table 7.1), just as for switching circuits. The choice of expression, like the name of the gate itself, depends on the functional output of the circuit. For example, the gate that outputs high voltage if and only if it has two high voltage inputs (the AND gate) corresponds to multiplication because that matches the way we get 1 as a product in *Boolean Algebra*.

Logic-gate outputs can be used as inputs for other gates in order to make complex circuits. The most basic logic gates from a logical and algebraic point of view are the AND, OR, and NOT gates. These can be combined to duplicate the action of NAND and NOR gates, but they can also generate any desired output. Together they form a complete set of connectives, something we'll prove in Section 7.5. The NAND gate is the one most often used in integrated circuits, however, because any Boolean expression can be modeled solely with NAND gates, appropriately connected (see Exercise 40).

Different compound circuits (Boolean expressions) may yield the same outputs for the same inputs. The laws of *Boolean Algebra* can help demonstrate this, and so can be used to simplify logic-gate circuit expressions to design simpler equivalent circuits. We'll explore this topic in depth in the next two sections, but here's one example.

✦ Example 7.4.8

Determine the Boolean expression representing the following logic-gate circuit; then design a simpler circuit that accomplishes the same thing.

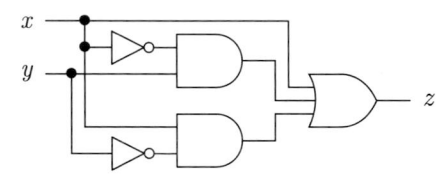

Solution

This circuit realizes the Boolean expression $x + \overline{x}y + x\overline{y}$. Using the *Absorption* and *Redundancy Laws* of *Boolean Algebra*, we can simplify this expression to $x + \overline{x}y + x\overline{y} = x + \overline{x}y = x + y$.

A simpler logic-gate circuit, therefore, is given by the following diagram.

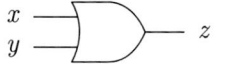

At this point we're using Boolean identities to simplify our expressions, giving us valuable practice working with *Boolean Algebra*. In Section 7.6, we'll learn some easier, more systematic techniques for doing such simplifications.

7.4.7 Computation Performed Via Logic Circuits

Boole used numerical and algebraic calculations to model a version of logic. In the twentieth century, logic-gate circuits were used to return the favor, providing resources for doing numerical calculations. This remarkable development required some ingenious thinking about the computational potential of logical operations in their *Boolean Algebra* format. Significantly, it provided the foundation for modern electronic computation. In its most elementary form, computation is performed by means of *half adders* and *full adders*, using binary (base-two) representation of numbers for doing ordinary addition. This idea was developed toward the end of Shannon's 1938 work as one example to illustrate what could be done with switching circuits and relays.

❖ **Example 7.4.9**

   a) Design a *half adder* to implement the Boolean functions $z_0 = f_0(x_0, y_0)$ and $z_1 = f_1(x_0, y_0)$ for adding two bits: $x_0 + y_0 = z_1 z_0$.

   b) Draw a logic circuit for the half adder.

Solution

   a) A half adder adds two one-place binary digits (0 or 1) to produce their ordinary sum (*not* their Boolean sum). The sum $1+1 = 10$ requires a second column, so the combined table for adding bits x_0 and y_0, outputting the binary numeral $z_1 z_0$, is as follows:

	x_0	y_0	z_1	z_0
x_0	0	0	0	0
$+\ y_0$	0	1	0	1
$\overline{z_1\ z_0}$	1	0	0	1
	1	1	1	0

   b) Knowing the Boolean expressions for basic logic gates, we see that $z_1 = (x_0 \text{ AND } y_0)$ while $z_0 = (x_0 \text{ XOR } y_0)$. The half adder thus has the following circuit diagram. We'll leave making a diagram using only AND, OR, and NOT gates as an exercise (see Exercise 42a).

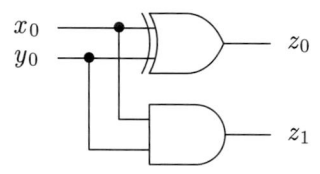

   The half adder adds two one-place numbers, but in adding two multi-place numbers, such as $111 + 11$, there will often be a third bit to add from the previous place's carry. To deal with this, we'll use a *full adder*, a circuit with three inputs and two outputs.

❖ **Example 7.4.10**

   Design a *full adder* for adding three binary digits.

Solution

   The table for the associated Boolean function is given below; a diagram using logic gates and half adders is left as an exercise (see Exercise 43).

	w_0	x_0	y_0	z_1	z_0
	0	0	0	0	0
w_0	0	0	1	0	1
x_0	0	1	0	0	1
$+\ y_0$	0	1	1	1	0
$\overline{z_1\ z_0}$	1	0	0	0	1
	1	0	1	1	0
	1	1	0	1	0
	1	1	1	1	1

EXERCISE SET 7.4

Exercises 1–6: Boolean Functions and Propositional Logic

Write out Boolean function tables to represent the following connectives.

7.4.1. $\text{AND}(x_1, x_2) = (x_1 \wedge x_2)$ **7.4.3.** $\text{NOR}(x_1, x_2) = \neg(x_1 \vee x_2)$

7.4.2. $\text{NAND}(x_1, x_2) = \neg(x_1 \wedge x_2)$ **7.4.4.** $\text{FALSE}(x) = x \wedge \neg x$.

7.4.5. $\text{XOR}(x_1, x_2) = (x_1 \vee x_2) \wedge \neg(x_1 \wedge x_2) = (x_1 \wedge \neg x_2) \vee (\neg x_1 \wedge x_2)$

7.4.6. $\text{XNOR}(x_1, x_2) = (x_1 \wedge x_2) \vee (\neg x_1 \wedge \neg x_2)$. What other name does this connective go under?

Exercises 7–10: True or False

Are the following statements true or false? Explain your answer.

7.4.7. Every PL sentence has a corresponding Boolean function.

7.4.8. Distinct Boolean expressions define distinct Boolean functions.

7.4.9. If an XOR gate rejects a string of inputs, a NOR gate accepts it.

7.4.10. Boole was the first to see how to use logic to perform computations.

Exercises 11–13: Boolean Expressions for Switching Circuits

For each switching circuit below, do the following:

  a. *Determine the Boolean expression corresponding to the switching circuit.*

  b. *Simplify the Boolean expression as much as possible, using the laws of Boolean Algebra.*

  c. *Draw the equivalent circuit for your simplified expression.*

7.4.11.

7.4.12.

7.4.13.

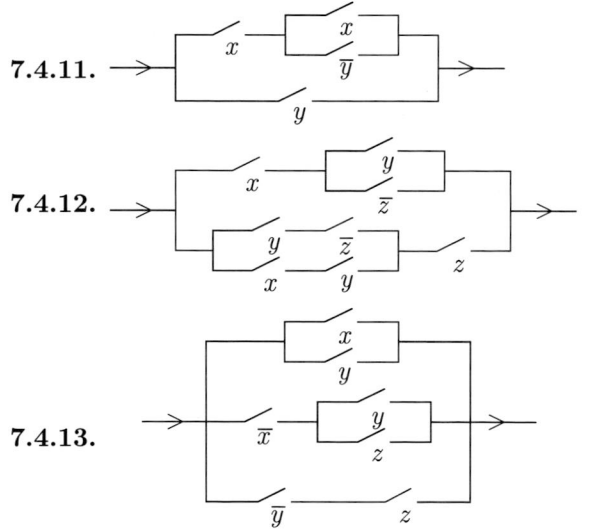

Exercises 14–17: Switching Circuits for Boolean Expressions

For each Boolean expression given below, do the following:

  a. *Draw a switching circuit diagram to realize the given Boolean expression.*

  b. *Simplify the Boolean expression as much as possible, using the laws of Boolean Algebra.*

  c. *Draw the equivalent circuit for your simplified expression.*

7.4.14. $(x + \overline{y})\, xy$

7.4.15. $xy + x\overline{y} + \overline{x}y$

7.4.16. $xyz + \overline{y}z + \overline{x}yz$

7.4.17. $(x + y)(x + z)(y + z)$

Exercises 18–19: Circuit Applications

Use a switching circuit or a logic-gate circuit to model the situation given in each problem.

7.4.18. Develop a Boolean function table, with rationale, and a switching circuit to model three switches controlling a light.

7.4.19. Four naval officers in a submarine have sufficient rank to authorize firing a nuclear warhead. Because of the seriousness of this action, at least three of the four must agree to fire before taking action. Give a Boolean expression and then design a logic circuit that realizes this protocol. You may use generalized logic gates having more than two inputs.

Exercises 20–23: Boolean Expressions for Logic Circuits

For each logic circuit given below, do the following:
  a. *Determine the Boolean expression corresponding to the circuit.*
  b. *Simplify the Boolean expression as much as possible, using the laws of Boolean Algebra.*
  c. *Draw the logic circuit corresponding to your simplified expression.*

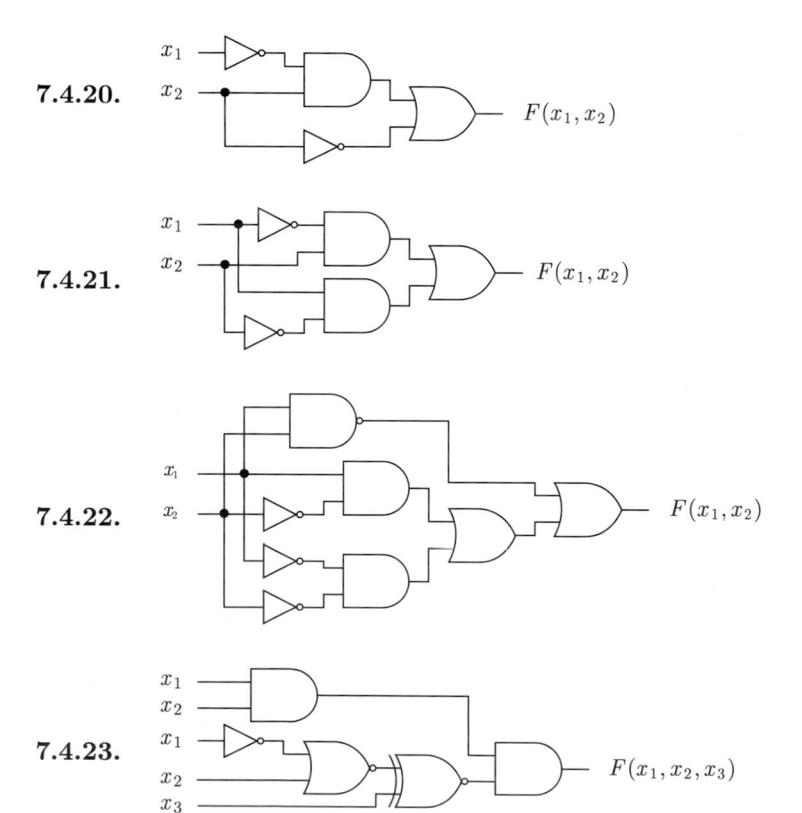

7.4.20.

$F(x_1, x_2)$

7.4.21.

$F(x_1, x_2)$

7.4.22.

$F(x_1, x_2)$

7.4.23.

$F(x_1, x_2, x_3)$

Exercises 24–27: Logic Circuits for Boolean Expressions

For each Boolean expression given below, do the following:

a. *Draw a logic-gate circuit diagram to realize the given Boolean expression.*

b. *Simplify the Boolean expression as much as possible, using the laws of Boolean Algebra.*

c. *Draw the logic-gate circuit corresponding to your simplified expression.*

7.4.24. $\overline{x}y + \overline{(x+y)}$ **7.4.26.** $(xy + \overline{y})(x + \overline{x}\,\overline{y})$

7.4.25. $xy + x\overline{y} + \overline{x}y$ **7.4.27.** $xyz + x\,\overline{y}\,z + \overline{x}yz + \overline{xyz}$

Exercises 28–29: Logic Circuits for Boolean Functions

For each Boolean function given below, do the following:

a. *Determine a Boolean expression to represent the given function.*

b. *Draw a logic-gate circuit that will implement the function.*

7.4.28. $f(x, y, z) = 1$ if and only if $y = 1$ and $x \neq z$.

7.4.29. $f(x, y, z)$ is given by the following table:

x	y	z	$f(x, y, z)$
0	0	0	0
0	0	1	1
0	1	0	1
0	1	1	0
1	0	0	1
1	0	1	0
1	1	0	0
1	1	1	1

Exercises 30–36: Logic-Gate Equivalents

Show how the following logic gates can be simulated using only NOT, AND, and OR gates, i.e., find a logic-gate circuit using only these gates to represent the same Boolean function as the given one.

7.4.30. NAND   **7.4.32.** XOR

7.4.31. NOR   **7.4.33.** XNOR

7.4.34. Show how AND can be simulated using only NOT and OR gates.

7.4.35. Show how OR can be simulated using only NOT and AND gates.

7.4.36. Can NOT be simulated using only AND and OR gates? Explain.

Exercises 37–40: Logic Gates Having One or Two Inputs

The text gives seven basic logic gates/Boolean functions, one with one variable and six with two variables. These are not the only ones possible. The following problems explore logic circuits for other connectives.

7.4.37. Is there a basic logic gate corresponding to the logical connective $\rightarrow$? corresponding to the logical connective $\leftrightarrow$? Explain.

7.4.38. How many distinct Boolean functions of a single variable are there? Write down function tables for those not already included among the basic logic gates and give them a name of your choice.

7.4.39. How many distinct Boolean functions of two variables are there? Write down function tables for those not already included among the basic logic gates and give them a name of your choice.

7.4.40. NAND-*Gate Simulations*: *a* NAND *Gate is Universal*
   a. Show how NOT can be simulated using only NAND gates.
   b. Show how AND can be simulated using only NAND gates.
   c. Show how OR can be simulated using only NAND gates.

Exercises 41–45: Ordinary Computation Done By Logic Circuits
The following problems explore integer arithmetic done via logic circuits.

7.4.41. *Need for Full Adders*
Explain why a half adder is inadequate for adding two binary numbers and why adding two binary numbers in each place value may require three inputs but only two outputs.

7.4.42. *Half Adder Diagrams*
   a. Provide an alternative circuit diagram for a half adder, using only NOT, AND, and OR logic gates.
   b. Design a half adder circuit that uses only NAND logic gates.
   c. Design a half adder circuit that uses only NOR logic gates.

7.4.43. *Full Adders from Half Adders*
   a. By knowing what is needed to add three bits (see Example 10), show that a full adder can be built out of three half adders that each add two bits at a time. For simplicity, you may represent each half adder as a rectangle with two inputs and two outputs, labeled appropriately. Explain how your circuit works on the inputs $(1, 0, 1)$ and $(1, 1, 1)$.
   b. By analyzing the possibilities that can occur in adding three bits, show that the final half adder in part *a* can be replaced with a simple logic gate and identify what that gate is.
   c. Using the logic-circuit diagram in Example 9 or your work in Exercise 42, draw a logic circuit for a full adder that uses only basic logic gates.

7.4.44. *Addition of Three-Bit Numbers*
   a. Show how one half adder and two full adders can add two three-bit numbers, $x_2 x_1 x_0 + y_2 y_1 y_0$, yielding the sum $z_3 z_2 z_1 z_0$. To simplify your diagram, represent each type of adder as a rectangle with two or three inputs and two outputs, labeling each input and output appropriately.
   b. Explain how your circuit from part *a* adds the binary numbers $101 + 11$ to yield 1000.

7.4.45. *Ordinary Multiplication using Logic Circuits*
   a. Create a function table for ordinary integer multiplication of two one-bit numbers ($z_0 = x_0 \cdot y_0$) and then draw a logic circuit to implement it.
   b. Design a logic circuit to multiply a two-bit number $x_1 x_0$ by a one-bit number y_0 to output a two-bit number $z_1 z_0$.
   c. Design a logic circuit to multiply two two-bit numbers $x_1 x_0 \cdot y_1 y_0$, yielding a four-bit number $z_3 z_2 z_1 z_0$.

7.4.46. *Ordering Numbers in Binary Notation*
To simplify terminology for this problem, consider x to be *smaller than* y if and only if $x \leq y$.   *Recall*: Boolean algebras are also Boolean lattices.
  a. Design a simple logic circuit to compare two one-bit numbers x_0 and y_0 and then output the smaller number.
  b. Design a logic circuit to compare two genuine two-bit numbers $x_1 x_0$ and $y_1 y_0$ and then output the smaller number.
  c. Design a logic circuit to compare two-bit numbers $x_1 x_0$ and $y_1 y_0$ and output the smaller number. This time let the lead bit be either 0 or 1.

7.5 Representing Boolean Functions

We're now ready to explore some ways to represent Boolean functions and expressions by generating them from elementary components, called *minterms* and *maxterms*. And, since Boolean functions are used to model truth-functional connectives, our findings will give us *normal form representations* for PL formulas. We'll also establish the pleasantly surprising fact that certain small sets of logical connectives are *expressively complete*, that is, they suffice to represent all possible connectives of any degree of complexity. Finally, we'll learn an interesting implication of our work for developing the theories of *Boolean Algebra* and *Set Theory*.

7.5.1 *Boolean Functions and Boolean Expressions*

Boolean functions are simpler than functions encountered in algebra and calculus, because their inputs and outputs only involve 0 and 1. However, since Boolean computations differ from ordinary arithmetic, the way functions and expressions are related is also different. For real-valued functions of a real variable, distinct polynomial expressions represent distinct functions. With Boolean polynomials, there is a great deal of duplication.

✦**Example 7.5.1**
  Compare the functions $f(x) = x$ and $g(x) = x^2$, both as functions from $\mathbb{R}$ to $\mathbb{R}$ and as functions from $\mathcal{B}$ to $\mathcal{B}$.

  Solution
  As functions from $\mathbb{R}$ to $\mathbb{R}$, $f \neq g$: they act differently on inputs and have very different graphs—f's graph is a straight line, while g's is a parabola. However, $f(x)$ and $g(x)$ agree on inputs 0 and 1, so as functions from $\mathcal{B}$ into $\mathcal{B}$, $f = g$. In fact, in *Boolean Algebra*, $x^2 = x$ (an *Idempotence Law*). Thus, different Boolean polynomials may define the same function.

  We can make the same point even more strongly with another example.

✦**Example 7.5.2**
  Determine all functions $f : \mathcal{B} \to \mathcal{B}$.

Solution
- Since $\mathcal{B} = \{0, 1\}$, there are exactly four distinct functions from $\mathcal{B}$ to $\mathcal{B}$. We can even identify them using standard Boolean formulas: they're $f_{00}(x) = 0$, $f_{01}(x) = x$, $f_{10}(x) = \overline{x}$, and $f_{11}(x) = 1$ (see Exercise 1).
- Thus, while there are infinitely many distinct polynomial formulas of a single variable, they all reduce to one of four basic expressions.

We can represent Boolean functions with *Boolean expressions* involving addition, multiplication, and complementation.

Definition 7.5.1: *Boolean Expressions*
A Boolean expression in the n variables $x_1, x_2, \ldots, x_n$ is any expression $E(x_1, x_2, \ldots, x_n)$ formed from these variables and the constants 0 and 1 using the operations of Boolean Algebra. More officially:
a) *Base case: constants 0 and 1 are Boolean expressions in $x_1, x_2, \ldots, x_n$, as are the variables $x_1, x_2, \ldots, x_n$.*
b) *Recursion step: if E and F are Boolean expressions in $x_1, x_2, \ldots, x_n$, then so are $E + F$, $E \cdot F$, and $\overline{E}$.*
c) *Closure clause: all Boolean expressions in $x_1, x_2, \ldots, x_n$ are generated in finitely many steps by recursively applying Boolean operations to Boolean constants and variables.*

✤ Example 7.5.3
Explain why each of the following is a Boolean expression in x and y, and then simplify the expression.

a) $x^2 + xy + 1$ b) $(x^3 + \overline{x}y)^2$

Solution
a) $x^2 + xy + 1$ is generated by multiplying x with x and x with y, then adding these products to 1. Thus, $x^2 + xy + 1$ is a Boolean expression in x and y.
- Since 1 added to anything is 1 in *Boolean Algebra*, $x^2 + xy + 1 = 1$.
b) The formula $(x^3 + \overline{x}y)^2$ is formed from the variables x and y by first multiplying three x factors together and multiplying the complement of x by y, then adding these products together, and finally multiplying the result by itself. Thus, $(x^3 + \overline{x}y)^2$ is a Boolean expression in x and y.
- Since squaring an input yields that input, $(x^3 + \overline{x}y)^2 = x^3 + \overline{x}y$. For the same reason, $x^n = x$, so $x^3 + \overline{x}y = x + \overline{x}y$. This last expression can be simplified by a *Redundancy Law* to $x + y$. Thus, $(x^3 + \overline{x}y)^2 = x + y$.

If Boolean expressions $E(x_1, x_2, \ldots, x_n)$ and $F(x_1, x_2, \ldots, x_n)$ satisfy the identity $E(x_1, x_2, \ldots, x_n) = F(x_1, x_2, \ldots, x_n)$ in *Boolean Algebra*, this must hold for all Boolean algebras, including $\mathcal{B}$. Thus, functions from $\mathcal{B}^n$ to $\mathcal{B}$ defined by these expressions are also equal since the identity holds for any assignment of 0's or 1's to the variables. At this point we don't know that the converse is true—that two expressions which agree for all assignments of 0's and 1's are a *Boolean* identity, but this is a corollary to *Theorem 2*.

7.5.2 *Minterm Expansion Representations*

Boolean expressions define Boolean functions, but can we always find a Boolean expression to represent any Boolean function? The answer here is a satisfying "yes." In fact, this can be done in more than one way. Our focus in this section is on minterm representations; maxterm representations will be left as a parallel exploration in the Exercise Set (see Exercises 18–19).

✤**Example 7.5.4**

Determine a Boolean expression for the following function $f : \mathcal{B}^2 \to \mathcal{B}$.

x_1	x_2	$f(x_1, x_2)$
0	0	1
0	1	0
1	0	0
1	1	1

Solution

· One way to solve this uses *Propositional Logic*. We did this in Example 7.4.6, where we recognized that this represented $x_1 \leftrightarrow x_2$, which is equivalent to $(x_1 \wedge x_2) \vee (\neg x_1 \wedge \neg x_2)$, yielding Boolean expression $x_1 x_2 + \overline{x}_1 \overline{x}_2$.

· A more systematic way uses what we know about addition, multiplication, and complementation in the target Boolean algebra $\mathcal{B}$. We'll concentrate on when the output is 1, i.e., on which input strings the function accepts. We'll work our way up from the bottom of the table.

$$f(x_1, x_2) = 1 \leftrightarrow x_1 = 1 \ \& \ x_2 = 1 \text{ (row 4), or } x_1 = 0 \ \& \ x_2 = 0 \text{ (row 1);}$$
$$\leftrightarrow x_1 = 1 \ \& \ x_2 = 1, \text{ or } \overline{x}_1 = 1 \ \& \ \overline{x}_2 = 1;$$
$$\leftrightarrow x_1 \cdot x_2 = 1 \text{ or } \overline{x}_1 \cdot \overline{x}_2 = 1;$$
$$\leftrightarrow x_1 \cdot x_2 + \overline{x}_1 \cdot \overline{x}_2 = 1.$$

· Thus, $f(x_1, x_2) = x_1 x_2 + \overline{x}_1 \overline{x}_2$ is the desired Boolean representation.

We can generalize the reasoning in this example and make it more explicit, based upon the following proposition. Each part extends to the case where several variables are involved (see Exercise 11ce).

Proposition 7.5.1: *Acceptance Conditions for Boolean Operations*

a) *Complements*: $\overline{x} = 1$ *if and only if* $x = 0$.

b) *Products*: $xy = 1$ *if and only if* $x = 1$ *and* $y = 1$.

c) *For x and y in $\mathcal{B} = \{0, 1\}$, $x + y = 1$ if and only if $x = 1$ or $y = 1$.*

Proof:

a) See Exercise 11a.

b) See Exercise 11b.

c) See Exercise 11d.

The first two parts of this proposition are simple theorems of *Boolean Algebra* and so hold for any Boolean algebra. The third part, however, restricts the variables to 0 and 1 because the forward direction of the biconditional isn't true for Boolean algebras in general (see Exercise 11f).

Proposition 1 focuses on the notion of input *acceptance*, i.e., inputs yielding output 1. This approach helps us generate a Boolean expression for a Boolean function based on its table. We'll formalize the procedure of the last example after first stating and illustrating some relevant definitions.

Definition 7.5.2: *Literal, Minterm, Minterm Expansion*
a) A **literal** is a Boolean variable or its complement.
b) A **minterm** in the Boolean variables $x_1, x_2, \ldots, x_n$ is an n-fold product $m_1 m_2 \cdots m_n$, where for each i either $m_i = x_i$ or $m_i = \overline{x}_i$.
c) A **minterm expansion** in $x_1, x_2, \ldots, x_n$ is a sum of distinct minterms.

The product of n literals is called a *minterm* because, thinking about this in lattice terms, a product is a meet, which lies below the factors involved, being their greatest lower bound.

Proposition 7.5.2: *Minterm Expansions' Acceptance & Uniqueness*
Boolean functions defined by minterm expansions accept as many strings as they have minterms. Furthermore, Boolean functions defined by distinct minterm expansions (ignoring minterm order) are distinct.

Proof:
· Let $m_1 m_2 \cdots m_n$ be a minterm in the variables $x_1, x_2, \ldots, x_n$. Then, by *Proposition* 1ab, the function defined by this minterm accepts exactly one input string $(s_1, s_2, \ldots, s_n)$, the one in which $s_i = 1$ if $m_i = x_i$ and $s_i = 0$ if $m_i = \overline{x}_i$. It's clear from this association that functions defined by different minterms accept different input strings.
· By *Proposition* 1c, functions defined by an expansion with k minterms will accept k input strings, the ones associated with each of its minterms.
· Thus, distinct minterm expansions define distinct Boolean functions.   ∎

Proposition 2 provides the necessary groundwork for proving the following central theorem for Boolean functions.

Theorem 7.5.1: *Minterm Representation of Boolean Functions*
Every non-zero Boolean function is represented by the unique minterm expansion that encodes the acceptance information in its function table.

Proof:
· Let $f(x_1, x_2, \ldots, x_n)$ be a non-zero Boolean function.
· For each input row having output 1, form the minterm that accepts that input string.
· Sum these minterms to form the minterm expansion $E(x_1, x_2, \ldots, x_n)$.
· By *Proposition* 2, the function defined by this minterm expansion accepts exactly the same input strings as f, so $f(x_1, x_2, \ldots, x_n) = E(x_1, x_2, \ldots, x_n)$.
· Also by *Proposition* 2, this minterm expansion representation is unique up to minterm order.   ∎

✣ Example 7.5.5
Determine the minterm expansion for the ternary majority function f:

x_1	x_2	x_3	$f(x_1, x_2, x_3)$
0	0	0	0
0	0	1	0
0	1	0	0
0	1	1	1
1	0	0	0
1	0	1	1
1	1	0	1
1	1	1	1

Solution

The minterms associated with the strings accepted by f are (working from the bottom up) $x_1 x_2 x_3$, $x_1 x_2 \overline{x}_3$, $x_1 \overline{x}_2 x_3$, and $\overline{x}_1 x_2 x_3$.

Adding these gives the minterm expansion for the ternary majority function:
$f(x_1, x_2, x_3) = x_1 x_2 x_3 + x_1 x_2 \overline{x}_3 + x_1 \overline{x}_2 x_3 + \overline{x}_1 x_2 x_3$.

7.5.3 *Minterm Expansions for Boolean Expressions*

We now have a uniform way to generate a minterm expansion for a Boolean function—write out its Boolean function table and use it to determine which minterms to add together.

While this process works, it may not be optimal, particularly if the function is already given in terms of a Boolean expression. In that case there's another method we can use to generate the function's minterm expansion. We'll demonstrate this for the ternary majority function just considered.

✦Example 7.5.6

Determine the minterm expansion for the ternary majority function.

Solution

· We know from Example 7.4.4 that a formula for the ternary majority function is $f(x_1, x_2) = x_1 x_2 + x_1 x_3 + x_2 x_3$ since its output is 1 when any two inputs are 1. This isn't yet a minterm expansion, though, since each term lacks one literal factor.

· We can correct this by multiplying each term by 1 in the form $x_i + \overline{x}_i$ for the missing literal and then dropping duplicate minterms.

$$x_1 x_2 + x_1 x_3 + x_2 x_3 = x_1 x_2 (x_3 + \overline{x}_3) + x_1 x_3 (x_2 + \overline{x}_2) + x_2 x_3 (x_1 + \overline{x}_1)$$
$$= x_1 x_2 x_3 + x_1 x_2 \overline{x}_3 + x_1 x_2 x_3 + x_1 \overline{x}_2 x_3 + x_1 x_2 x_3$$
$$+ \overline{x}_1 x_2 x_3$$
$$= x_1 x_2 x_3 + x_1 x_2 \overline{x}_3 + x_1 \overline{x}_2 x_3 + \overline{x}_1 x_2 x_3$$

· Checking, this is the same result we found in Example 5.

The procedure used in Example 6 gives us a way to find minterm expansions for Boolean expressions in general, but before we prove this, let's think about how this ties in with *Theorem* 1. Since any Boolean expression defines a Boolean function, and since a Boolean function can be uniquely represented by a minterm expansion, we'd like to conclude that this expansion represents (is identical to) the original expression. At the moment, though, we only know they agree when the variables are evaluated as 0 and 1, not that they must agree for all values in any Boolean algebra. The next theorem will extend this agreement, giving us a genuine Boolean identity. To prove this, we'll give a lemma worth stating in its own right, first illustrating it with an example.

❖ **Example 7.5.7**

Determine a sum-of-products decomposition for $(x + \overline{y}z)^2 + y(\overline{yz}) + z$.

Solution

$$
\begin{aligned}
(x + \overline{y}z)^2 + y(\overline{yz}) + z &= x + \overline{y}z + y(\overline{y} + \overline{z}) + z &&\text{Idem, DeM} \\
&= x + \overline{y}z + y\overline{y} + y\overline{z} + z &&\text{Distrib} \\
&= x + \overline{y}z + y\overline{z} + z &&\text{Compl, Iden} \\
&= x + y\overline{z} + z &&\text{Absorp}
\end{aligned}
$$

Both of the last two lines give a decomposition solution.

Lemma 7.5.1: *Sum-of-Products Decomposition*

Every non-zero Boolean expression in the variables $x_1, x_2, \ldots, x_n$ is identical to a sum of products of distinct literals in these variables.

Proof:

· We'll argue this informally via an induction argument, using procedures and results legitimate for any Boolean algebra.
· First, note that if an expression contains a 1, it can be replaced by $x_i + \overline{x}_i$ (any i). Furthermore, 0's can be eliminated by doing the indicated operation on them and simplifying. So we only need to consider expressions without constants, expressions with only literals in the given variables.
· *Base case*: note that each variable x_i is already in the required form: it's the trivial sum of a trivial product.
· *Induction step*: suppose that expressions E and F are non-zero sums of products of distinct literals.
  ○ Then a non-zero $E + F$ is as well: convert each part separately and add them together. Eliminate duplicate terms by using the *Idempotence Law*.
  ○ A non-zero $E \cdot F$ can be expanded using the *Distributive Law*, as needed. The resulting sum may contain product terms with duplicate literals or the product of a literal with its complement. These can be simplified using the *Idempotence Law*, the *Complementation Law*, the *Annihilation Law*, or other laws of *Boolean Algebra*. Duplicate summands and 0 terms are eliminated as described above.
  ○ If $\overline{E}$ is the complement of a sum of products of distinct literals, this equals the product of sums of literals via *De Morgan's Laws*. By what was said about sums and products, this reduces to the form needed.
· The final result, which need not be unique, will be a non-zero sum of products of distinct literals in the variables $x_1, x_2, \ldots, x_n$. ∎

Theorem 7.5.2: *Minterm Representation for Boolean Expressions*
Every non-zero Boolean expression in the variables $x_1, x_2, \ldots, x_n$ *is uniquely represented by a minterm expansion in these variables.*

Proof:
- From *Lemma* 1, every non-zero Boolean expression is a sum of products of distinct literals. As in Example 6, multiply each of these product terms by $x_i + \overline{x}_i$ if neither x_i nor $\overline{x}_i$ is present. This result equals the original one since we're merely multiplying each product by 1.
- The resulting expansion is a sum of minterms. Dropping duplicate minterms, we end up with a minterm expansion equal to the original Boolean expression.
- Furthermore, since minterm expansions uniquely define Boolean functions, this expansion uniquely represents the original Boolean expression. ∎

7.5.4 *Minterm Expansions and Boolean Algebra*

The *Minterm Representation Theorem for Boolean Expressions* justifies a practice silently adopted by many discrete mathematics textbooks and writers on *Boolean Algebra*: *truth tables suffice to prove the identities of Boolean Algebra*. In other words, showing that two expressions agree when the variables are 0 or 1 is enough to show that these expressions are identical for any values of the variables. The converse (*identical expressions share the same truth tables*) is obvious, but this new result is somewhat unexpected and may seem questionable. A Boolean algebra typically has many more elements than just 0 and 1. Why should the agreement of two Boolean expressions on all $0\,/\,1$ input combinations warrant their universal agreement? The *Minterm Representation Theorem for Boolean Expressions* is the key to this result.

Corollary 7.5.2.1: *Truth-Table Verification of Boolean Identities*
If $E(x_1, x_2, \ldots, x_n)$ *and* $F(x_1, x_2, \ldots, x_n)$ *are Boolean expressions in variables* $x_1, \ldots, x_n$, $E(x_1, x_2, \ldots, x_n) = F(x_1, x_2, \ldots, x_n)$ *if and only if their truth tables are identical.*

Proof:
- Suppose first that $E(x_1, x_2, \ldots, x_n) = F(x_1, x_2, \ldots, x_n)$ for all values of x_i. Then, since every Boolean algebra contains extreme elements 0 and 1, these values can replace the x_i in any input combination and yield the same $0\,/\,1$ output. Thus, these expressions' truth tables will be identical. ✓
- Conversely, suppose $E(x_1, x_2, \ldots, x_n)$ and $F(x_1, x_2, \ldots, x_n)$ have identical truth tables. Then the minterm expansions associated with these tables are likewise equal. By *Theorem* 2, these minterm expansions represent the given expressions, so the expressions themselves are also equal: $E(x_1, x_2, \ldots, x_n) = F(x_1, x_2, \ldots, x_n)$. ∎

❖**Example 7.5.8**

Use a truth table to prove the *Consensus Law* (*Proposition* 7.3.8) of *Boolean Algebra*: $xy + \overline{x}z + yz = xy + \overline{x}z$.

Solution

The following truth table proves the claim. The last two columns agree, so by the above corollary, the two expressions are equal for all x, y, and z. In other words, $xy + \overline{x}z + yz = xy + \overline{x}z$ is a theorem of *Boolean Algebra*.

x	y	z	$xy + \overline{x}z + yz$	$xy + \overline{x}z$
0	0	0	0	0
0	0	1	1	1
0	1	0	0	0
0	1	1	1	1
1	0	0	0	0
1	0	1	0	0
1	1	0	1	1
1	1	1	1	1

Checking Boolean identities by hand via truth tables gets tedious for three or more variables. It is even time-consuming for a computer if too many variables are involved. But occasionally a shortcut is possible. Note here that the two expressions differ only in the term yz. If $yz = 0$, then the two expressions are the same. On the other hand, if $yz = 1$, then both y and z must be 1. This makes the two expressions reduce to $x + \overline{x} + 1$ and $x + \overline{x}$, both of which are 1, so they also agree for these values.

By the *Corollary* to *Theorem* 2, truth tables provide a mechanical procedure for determining whether an equation is a theorem of *Boolean Algebra*, making its set of identities *decidable*. One need not creatively combine the axioms of *Boolean Algebra* in a complex argument to deduce an identity's validity. This means that the elementary theory of *Boolean Algebra* regarding identities is rather uninteresting from a proof-theoretic standpoint. Nevertheless, we did draw upon some significant mathematical results about Boolean expressions to justify using truth-tables for developing *Boolean Algebra*.

Moreover, given that portions of *Propositional Logic* and *Set Theory* can be interpreted in terms of *Boolean Algebra*, this result has implications for those theories. For *Propositional Logic*, it means that logical equivalences can be established via truth tables, which we already knew. For *Set Theory*, it means that set equality for sets involving intersections, unions, and complements can be demonstrated using Boolean truth tables. This is something we didn't do earlier. Let's explore what this means with an example.

❖**Example 7.5.9**

Use a Boolean truth table to demonstrate *De Morgan's Law* $\overline{S \cup T} = \overline{S} \cap \overline{T}$.

Solution

We'll first give the table and then explain how we can interpret it.

S	T	$\overline{S}$	$\overline{T}$	$S \cup T$	$\overline{S \cup T}$	$\overline{S} \cap \overline{T}$
0	0	1	1	0	1	1
0	1	1	0	1	0	0
1	0	0	1	1	0	0
1	1	0	0	1	0	0

One way to think about this table (as some texts do) is as follows. We can let the column headings stand for belonging to the sets mentioned and the table entries below them for whether an element is in (1) or not in (0) those sets. Full agreement between the final two columns then shows that an element is in $\overline{S \cup T}$ if and only if it is in $\overline{S} \cap \overline{T}$, so $\overline{S \cup T} = \overline{S} \cap \overline{T}$.

But the *Corollary* to *Theorem* 2 justifies a more surprising interpretation. Let S and T be subsets of some universal set U so that *De Morgan's Law* is a statement for the Boolean algebra $\mathcal{P}(U)$. Then $0 = \emptyset$ and $1 = U$, the entries in the first two columns represent all combinations of $\emptyset$ and U, and those in the later columns are the set-theoretic operation results on those sets. Since the final columns' entries agree, we again know that the law holds, but this time on the basis of an all-or-nothing verification—we've only checked whether the identity holds when S and T are either $\emptyset$ or U.

People who work with computerized reasoning are interested in *Boolean Algebra* because of the ease of verifying its identities. Besides using the system we've set up, researchers have explored using operations such as NAND and NOR, developing axiom systems for *Boolean Algebra* based upon them. They've found relatively simple systems containing two axioms or even one, using automated theorem proving.[5]

7.5.5 Normal Form Representation for Propositions

The *Minterm Representation Theorem for Boolean Expressions* yields an important consequence for *Propositional Logic*. There minterm expansions correspond to what are called *disjunctive normal forms*.

Definition 7.5.3: Disjunctive Normal Form
A PL formula in the variables $P_1, P_2, \ldots, P_n$ is in disjunctive normal form if and only if it is a disjunction of distinct conjuncts $M_1 \wedge M_2 \wedge \cdots \wedge M_n$, where M_i is either P_i or $\neg P_i$.

Corollary 7.5.2.2: Disjunctive Normal Form Representation
Every formula of Propositional Logic, regardless of the truth-functional connectives it uses, is logically equivalent to one in disjunctive normal form.

---

[5] For example, Stephen Wolfram proposed a single axiom involving six NANDs (or NORs) in **A New Kind of Science** (2002), pp. 808–811.

Proof:

This is a direct application of the *Minterm Representation Theorem for Boolean Expressions* to the case of *Propositional Logic*. ∎

✤ Example 7.5.10

Find a disjunctive normal form for $P \veebar Q$.

Solution

- Given XOR's meaning, $P \veebar Q \models\!\mid (P \vee Q) \wedge \neg(P \wedge Q)$. Although this isn't what's needed here, it shows how XOR can be expressed using the connectives $\neg$, $\wedge$, and $\vee$.
- A second equivalent—$(P \wedge \neg Q) \vee (\neg P \wedge Q)$—is in disjunctive normal form. If this latter form were overlooked, it could be generated from the first form by means of a series of PL equivalences (see Exercise 31a).

A disjunctive normal form for a *Propositional Logic* formula can be generated in a uniform way from the formula's truth table, much as we obtained a minterm expression for a Boolean function. Alternatively, if we have an expression that already involves these basic connectives, we can use a procedure similar to that used for generating a minterm expansion from a Boolean expression.

That every formula of *Propositional Logic* has a disjunctive normal form means that *every possible logical connective*, whether unary, binary, or *n*-ary, can be represented by a logically equivalent formula in the same variables that involves only negation, conjunction, and disjunction. Thus, the connectives $\neg$, $\wedge$, and $\vee$ form an *expressively complete* set of connectives.

The expressive completeness of $\{\neg, \wedge, \vee\}$ may not seem surprising to you at this point, but it can seem quite amazing to students learning elementary logic (see Section 2.1). Knowing what we do about logic (or *Boolean Algebra*), though, we can do even better—we can find a smaller expressively complete set of connectives. *De Morgan's Laws* warrant dropping either $\wedge$ or $\vee$, for $P \wedge Q \models\!\mid \neg(\neg P \vee \neg Q)$ and $P \vee Q \models\!\mid \neg(\neg P \wedge \neg Q)$. So two truth-functional connectives suffice: both $\{\neg, \wedge\}$ and $\{\neg, \vee\}$ are expressively complete.

Reducing the number of required connectives from three to two prompts us to ask whether any single connective suffices. The nineteenth-century American logician and mathematician C. S. Peirce answered this in the affirmative. The first published result of this fact, however, was by Henry Sheffer in 1913. Neither $\wedge$ nor $\vee$ suffice because they fail at inversion, but combining either of these with $\neg$ does work. The Sheffer stroke, known now as the NAND connective, suffices (see Exercise 1.3.28), as does the NOR connective (see Exercise 1.3.29), but no other unary or binary connective does. Restricting ourselves to one connective, or even two or three, complicates matters such as expressing and proving propositions, but it can be beneficial for other purposes. The expressive completeness (universality) of the NAND connective, for instance, is what makes it possible to build logic circuits solely from NAND gates (see Exercise 7.4.40). It's also what allows people to use it as the sole operator in certain axiomatizations of *Boolean Algebra*.

EXERCISE SET 7.5

Exercises 1–3: Boolean Functions and Boolean Expressions

The following problems explore the existence of particular Boolean functions.

7.5.1. *Example 2*
a. Explain why there are four functions from $\mathcal{B}$ to $\mathcal{B}$, and verify that those given in Example 2 are distinct.
b. Which of the four functions in Example 2 are one-to-one functions? onto functions? bijections?

7.5.2. *Boolean Functions of Several Variables*
a. Calculate the number of distinct functions from $\mathcal{B}^2$ to $\mathcal{B}$.
b. How many of these functions are one-to-one? onto? one-to-one-and-onto?
c. Determine Boolean expressions $F(x, y) = $ _____ for all the distinct Boolean functions $f : \mathcal{B}^2 \to \mathcal{B}$.
d. Calculate the number of distinct functions from $\mathcal{B}^n$ to $\mathcal{B}$.
e. How many of the functions in part d are one-to-one? onto? one-to-one-and-onto? Assume that $n \geq 2$.
f. Explain how to write Boolean expressions $F(x_1, x_2, \ldots, x_n) = $ _____ for all the distinct Boolean functions $f : \mathcal{B}^n \to \mathcal{B}$.

7.5.3. *Sum-of-Products Boolean Expressions*
For the following, find an equivalent sum of products of distinct literals.
a. $x^4 + x^3\,\overline{x} + \overline{x}\,^2$
b. $x^{12} + \overline{x}\,^6(x^4\,\overline{x} + x)^5(x^2 + \overline{x}\,^3)^8$
c. $(x^2 + xy^3 + \overline{x}\,y^2)\,\overline{x}\,^5\,\overline{y}\,^3$
d. $(x^3y^2 + \overline{x}\,y^2)\,(x^2 + x\,\overline{y}\,^3)$

Exercises 4–7: Algebra of Boolean Functions

The following problems consider the algebra of Boolean functions.
Definition: *Let f and g be functions from $\mathcal{B}^n$ to $\mathcal{B}$. Then define*
a. $(f + g)(x_1, x_2, \ldots, x_n) = f(x_1, x_2, \ldots, x_n) + g(x_1, x_2, \ldots, x_n)$
b. $(f \cdot g)(x_1, x_2, \ldots, x_n) = f(x_1, x_2, \ldots, x_n) \cdot g(x_1, x_2, \ldots, x_n)$
c. $\overline{f}(x_1, x_2, \ldots, x_n) = \overline{f(x_1, x_2, \ldots, x_n)}$

7.5.4. Explain why $f + g$, $f \cdot g$, and $\overline{f}$ are Boolean functions from $\mathcal{B}^n$ to $\mathcal{B}$.

7.5.5. Prove that if $f(x_1, x_2, \ldots, x_n) = E(x_1, x_2, \ldots, x_n)$ for Boolean expression E, then $\overline{f}(x_1, x_2, \ldots, x_n) = \overline{E(x_1, x_2, \ldots, x_n)}$.

7.5.6. Prove or disprove the following conditions, where 0 and 1 denote the constant functions with those values.
a. $f + \overline{f} = 1$
b. $f \cdot \overline{f} = 0$
c. $\overline{\overline{f}} = f$
d. $\overline{f}(\overline{x_1}, \ldots, \overline{x_n}) = f(x_1, \ldots, x_n)$

7.5.7. Does the set of all Boolean functions $f : \mathcal{B}^n \to \mathcal{B}$ form a Boolean algebra under the pointwise operations defined above? Check whether all of the *Boolean Algebra* axioms are satisfied. Explain your result.

Exercises 8–10: True or False

Are the following statements true or false? Explain your answer.

7.5.8. A sum of Boolean expressions is 1 if and only if some summand is 1.

7.5.9. Truth tables can be used to verify set-theoretic identities.

7.5.10. Any result that is true of the Boolean algebra $\mathcal{B} = \{0, 1\}$ is also true of all Boolean algebras. *Hint*: not all *Boolean Algebra* statements are identities. Does your answer contradict *Theorem 2*'s *Corollary*? Explain.

Exercises 11–12: Simple Propositions for Boolean Functions
Prove the following propositions, using Mathematical Induction where needed.

7.5.11. *Proposition* 1: *Acceptance for Boolean Operations*
  a. *Complements*: $\overline{x} = 1$ if and only if $x = 0$.
  b. *Products*: $xy = 1$ if and only if $x = 1$ and $y = 1$.
  c. *Generalized products*: generalize part b to the n-variable case.
  d. *Sums*: for w and z in $\mathcal{B} = \{0, 1\}$, $w + z = 1$ if and only if $w = 1$ or $z = 1$.
  e. *Generalized sums*: generalize part d to the n-variable case.
  f. *Counterexample*: explain where and why the biconditional of part d fails for Boolean algebras in general. *Hint*: consider $\mathcal{B}^2$.

7.5.12. *Rejection for Boolean Operations*
  a. *Complements*: $\overline{x} = 0$ if and only if $x = 1$.
  b. *Sums*: $x + y = 0$ if and only if $x = 0$ and $y = 0$.
  c. *Generalized sums*: generalize part b to the n-variable case.
  d. *Products*: for w and z in $\mathcal{B} = \{0, 1\}$, $wz = 0$ if and only if $w = 0$ or $z = 0$.
  e. *Generalized products*: generalize part d to the n-variable case.
  f. *Counterexample*: explain where and why the biconditional of part d fails for Boolean algebras in general. *Hint*: consider $\mathcal{B}^2$.

Exercises 13–17: Minterm Expansions
The following problems deal with minterm expansions.

7.5.13. Develop minterm expansions for the following Boolean expressions:
  a. $x + y$ d. $x + \overline{y} + z$
  b. xy e. $xy + \overline{x}z + y\overline{z}$
  c. $x + \overline{x}y$ f. $x(z + \overline{y}\,\overline{z}) + \overline{x}z$
  g. Does 0 have a minterm expansion? Explain.

7.5.14. Find minterm expansions that accept the following input strings.
  a. $00, 10$ c. $010, 011, 101$
  b. $01, 10, 11$ d. $000, 001, 100, 101$

7.5.15. *Counting Minterm Expansions*
  a. How many distinct minterms in $x_1, x_2, \ldots, x_n$ are there? Explain.
  b. Let $F(x_1, x_2, \ldots, x_n)$ denote the minterm expansion for the constantly 1 function. How many minterms does F contain? Explain.
  c. How many distinct minterm expansions in $x_1, x_2, \ldots, x_n$ are there?

7.5.16. *Minterms and Boolean Functions*
  a. How many distinct input strings are accepted by a minterm? Explain.
  b. How many distinct input strings are accepted by a minterm expansion? Explain.
  c. Explain why distinct minterm expansions represent distinct Boolean functions.

7.5.17. *Minterm Expansions for Complements of Functions*
Find a minterm expansion for each $\overline{f}$, where f is given by the following minterm expansion. (The definition for $\overline{f}$ is stated prior to Exercise 4.)
a. $xy + \overline{x}\,y + \overline{x}\,\overline{y}$ b. $xyz + x\,\overline{y}\,z + \overline{x}\,y\,\overline{z} + \overline{x}\,\overline{y}\,\overline{z}$
c. Explain how to find the minterm expansion for the complement $\overline{f}$ of a function f whose minterm expansion is given.

Exercises 18–19: Maxterm Expansions
The following problems explore the concept of maxterm expansions.

7.5.18. *Maxterm Expansions for Boolean Functions and Expressions*
Boolean functions and expressions can be represented by *maxterm expansions*, i.e., products of distinct n-fold sums of distinct literals. This is based on *rejection conditions* (see Exercise 12). Minterm expansion results have duals for maxterm expansions. Find maxterm expansions for the following:
a. $x + y$ c. $x + \overline{x}y$
b. xy d. $xy + xz + yz$

7.5.19. *Conjunctive Normal Forms for PL Formulas*
Maxterm representations give *conjunctive normal forms* for PL, i.e., conjunctions of distinct disjuncts. Find conjunctive normal forms for the following:
a. $P \wedge \neg(P \wedge Q)$ c. P NAND Q
b. $P \leftrightarrow (Q \leftrightarrow P)$ d. P NOR Q

Exercises 20–26: Proving *Boolean Algebra* Theorems
Prove the following Boolean Algebra identities using the method of truth-table verification.

7.5.20. *Proposition 7.3.2: Complements of Elements Laws*
a. $\overline{0} = 1$ c. $\overline{\overline{x}} = x$
b. $\overline{1} = 0$

7.5.21. *Proposition 7.3.3: Annihilation and Absorption Laws*
a. $x \cdot 0 = 0$ c. $x(x + y) = x$
b. $x + 1 = 1$ d. $x + xy = x$

7.5.22. *Proposition 7.3.4: Idempotence Laws*
a. $x \cdot x = x$ b. $x + x = x$

7.5.23. *Proposition 7.3.5: De Morgan's Laws*
a. $\overline{xy} = \overline{x} + \overline{y}$ b. $\overline{x + y} = \overline{x}\,\overline{y}$

7.5.24. *Proposition 7.3.7: Redundancy Laws*
a. $x(\overline{x} + y) = xy$ b. $x + \overline{x}y = x + y$

7.5.25. *Proposition 7.3.8: Consensus Laws*
$(x + y)(\overline{x} + z)(y + z) = (x + y)(\overline{x} + z)$ (see also Example 8)

7.5.26. Are the following equations Boolean identities? Explain.
a. $x + x\overline{y} + \overline{x}y = x + \overline{y}$ c. $x + xy + \overline{x}z = xy + x\overline{y} + z$
b. $x + yz = (x + y)z$ d. $xy + x\,\overline{y}\,z = xy + xz$

Exercises 27–30: Proving *Set Theory* Theorems
Prove the following identities using the method of truth-table verification. Explain what the entries in your truth tables stand for.

7.5.27. *Proposition 4.1.7: Associative Laws*
   a. $R \cap (S \cap T) = (R \cap S) \cap T$ b. $R \cup (S \cup T) = (R \cup S) \cup T$

7.5.28. *Proposition 4.1.8: Distributive Laws*
   a. $R \cap (S \cup T) = (R \cap S) \cup (R \cap T)$ b. $R \cup (S \cap T) = (R \cup S) \cap (R \cup T)$

7.5.29. *Proposition 4.1.10: De Morgan's Laws*
$\overline{S \cap T} = \overline{S} \cup \overline{T}$ (see also Example 9)

7.5.30. *Proposition 4.1.9: Absorption Laws*
Explain how to modify the method of truth-table verification to demonstrate a subset relationship instead of a set identity. Then show the following:
   a. $S \cap T \subseteq S$; $S \cap T \subseteq T$ b. $S \subseteq S \cup T$; $T \subseteq S \cup T$

Exercises 31–33: Disjunctive Normal Forms
The following problems explore disjunctive normal forms for sentences of PL.

7.5.31. *Example* 10
   a. Beginning with $P \veebar Q \vDash (P \vee Q) \wedge \neg (P \wedge Q)$, transform this via various *Replacement Rules* and *Contraction Rules* (see Exercises 1.9.16–1.9.19) to show that $P \veebar Q \vDash (P \wedge \neg Q) \vee (\neg P \wedge Q)$, which is a disjunctive normal form for the exclusive-or connective.
   b. Write down the truth table for $P \veebar Q$ and use the method of finding a minterm expansion for Boolean functions to determine a disjunctive normal form for $P \veebar Q$.

7.5.32. Determine disjunctive normal forms for the following formulas.
   a. $P \wedge \neg (P \wedge Q)$ c. $P \rightarrow (Q \wedge \neg R)$
   b. $P \leftrightarrow (Q \leftrightarrow P)$ d. $P \vee (\neg Q \rightarrow P \wedge R)$

7.5.33. Determine disjunctive normal forms for the following:
   a. P NAND Q c. P XNOR Q
   b. P NOR Q d. $\neg P$

7.6 Simplifying Boolean Functions

As mentioned earlier, computer scientists and engineers are interested in Boolean functions because of their connection to logic-circuit design. We've seen how circuits physically realize Boolean functions, and how *Boolean Algebra* can help determine equivalent Boolean expressions for a given function. We also know that all Boolean functions/logic circuits can be constructed using basic Boolean operators/logic gates, and that there are two ways to do this (minterm expansions, maxterm expansions). However, these forms are usually not the simplest ones possible, a prime concern for those who use logic circuits. In this section we'll explore how to simplify Boolean functions.

❖**Example 7.6.1**
   Show that the ternary majority function formula $f(x, y, z) = xy + xz + yz$ is simpler than its associated minterm and maxterm expansions.

Solution

The minterm expansion for this function is $xyz + xy\overline{z} + x\overline{y}z + \overline{x}yz$ (see Example 7.5.6), while $(x + y + z)(x + y + \overline{z})(x + \overline{y} + z)(\overline{x} + y + z)$ is its maxterm expansion (see Exercise 7.5.18d). Neither of these is as simple as the given expression—both contain more components and more operations.

While the phrase *simplest equivalent form* isn't completely unambiguous, simplifying an expression should lead to fewer terms or fewer operators or both (see Exercises 1–6). Engineers would want to take this topic further and focus on ways that reduce the total cost or increase the speed of a circuit.

7.6.1 *Algebraic Ways to Simplify Boolean Expressions*

Boolean Algebra provides us with one way to simplify a Boolean expression—use its identities to reduce the number of terms or operations.

✦**Example 7.6.2**

Simplify the minterm expansion of the ternary majority function, given by $f(x, y, z) = xyz + xy\overline{z} + x\overline{y}z + \overline{x}yz$.

Solution

This comes at the function of Example 1 from the other direction.

$$xyz + xy\overline{z} + x\overline{y}z + \overline{x}yz = xyz + xy\overline{z} + xyz + x\overline{y}z + xyz + \overline{x}yz$$
$$= xy(z + \overline{z}) + xz(y + \overline{y}) + yz(x + \overline{x})$$
$$= xy + xz + yz$$
$$= x(y + z) + yz\,.$$

Here we duplicated xyz twice (*Idempotence Law*) so it could be combined with each of the other terms, and then we simplified pairs of terms via a *Distributive Law*, a *Complementation Law*, and an *Identity Law*. The two final lines are both simplified equivalents.

While *Boolean Algebra* provides the basic tools for simplifying an expression, it's not always clear which law to apply when. Trial and error may be needed to decide what path to take and when to stop, because *Boolean Algebra* does not spell out how we should proceed.

People have found a few mechanical but insightful ways to simplify expressions. We'll look at two standard tools for this task: *Karnaugh Maps* and the *Quine-McCluskey method*.

7.6.2 *K-Maps for Two-Variable Boolean Functions*

Maurice Karnaugh introduced his *map method* in 1953 while working as an engineer at Bell Labs. *K-maps*, as they are now called, provide a visual device for charting and simplifying Boolean functions with a small number of variables. We'll first look at their use for two-variable Boolean functions and then show how to extend the technique to functions with three or four variables.

❖ **Example 7.6.3**

Exhibit the K-map for the Boolean function $f(x,y) = x + \overline{x}y$.

Solution

Treating f as a Boolean operator, we get the following table. We put x-inputs on the side, y-inputs on the top, and the calculated outputs $f(x,y)$ in the table's cells. This is simply a condensed version of a function table.

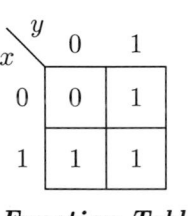

Function Table

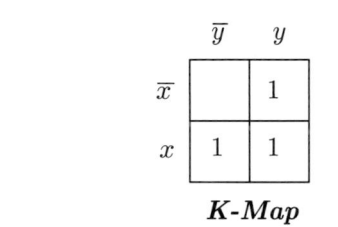

K-Map

The exhibited K-map is a slight variant of the operation table. It treats output cells as product minterms (see Proposition 7.5.1). We place a 1 in a cell (let's call the result a 1-*cell*) if the associated string is accepted by the function and otherwise leave it blank: 1's are in the x-row since x is a term in the formula for $f(x,y)$, and a 1 is also present in the $\overline{x}y$ cell. We'll often use 0's and 1's in column and row labels, as in the function table, but using letters indicates which literal is being represented where.

7.6.3 K-Maps for Three-Variable Boolean Functions

To represent Boolean functions with three variables in a two-dimensional chart, we'll place the first variable along the side and the other two along the top. So that adjacent cells will represent minterms with shared factors, we'll change the value of just one variable as we move from a cell to its neighbor. The third column thus represents yz (11) instead of the numerically next value $y\overline{z}$ (10). In this way, the middle two columns represent z. The last two columns represent y, the first two represent $\overline{y}$, and the last and first columns (which we'll consider adjacent since we can wrap the right edge around to join the left edge) represent $\overline{z}$. The following diagrams show the resulting template, both with letters and binary labels.

Karnaugh Maps for Three-Variable Boolean Functions

❖ Example 7.6.4

Determine the minterm Boolean expansions represented by the following K-maps and then simplify them, comparing your final result to the K-map:

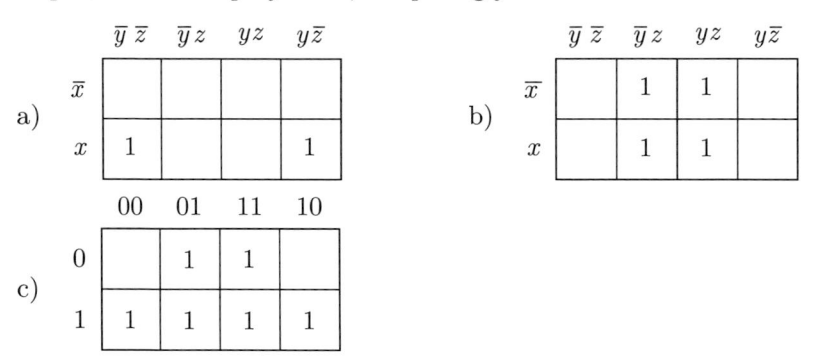

Solution

a) The minterm expansion represented by this K-map is $x\overline{y}\,\overline{z} + xy\overline{z}$, which simplifies, after factoring, to $x\overline{z}$. This can be read off the K-map, for the block of 1-cells is the intersection of the x row and the $\overline{z}$ columns.

b) The minterm expansion is $\overline{x}\,\overline{y}z + \overline{x}yz + x\overline{y}z + xyz$. This simplifies to $\overline{y}z + yz$ (corresponding to adding the second and third columns), which further simplifies to z. This fits the K-map we're given; the z columns are completely filled with 1's.

c) This K-map is the union of the last two. The minterm expansion, therefore, is given by $\overline{x}\,\overline{y}z + \overline{x}yz + x\overline{y}\,\overline{z} + x\overline{y}z + xyz + xy\overline{z}$. The sum of the simplified expressions is $x\overline{z} + z$. However, this can be further simplified to $x + z$ using a *Redundancy Law*.

Alternatively, the complement of this function is $\overline{x}\,\overline{z}$, found by inspection of the K-map. The original function is thus its complement, $\overline{\overline{x}\,\overline{z}} = x + z$. The simplified form $x+z$ can also be seen in the K-map—the set of 1-cells is the union of the x row and the z columns.

It should be clear that blocks of 1, 2, or 4 adjacent 1-cells represent products of three, two, or one literals, respectively (see Exercise 37). To determine a simplified expression for a Boolean function, then, we can decompose the patterned set of 1-cells into a small union of large blocks, possibly overlapping. This will generate a sum of products of literals to represent the function. (Note that products correspond to intersections and sums to unions.) To talk about this more precisely using standard terminology, we'll define some terms.

Definition 7.6.1: *Adjacent Cells, Blocks, Implicants, and Coverings*

a) *Two K-map cells are **adjacent** if and only if they share a common edge or are the outer cells in a row or column (adjacent in a wrapped sense).*

b) *A **block** for a Boolean function is a rectangular set of 2^k adjacent cells in the function's K-map.*

c) *An **implicant** for a Boolean function is a product of literals corresponding to a K-map block of 1-cells for that function.*

d) A **prime implicant** *for a Boolean function is an implicant associated with a maximal block of adjacent 1-cells for that function.*

e) An **essential prime implicant** *is a prime implicant whose block contains a 1-cell not in any other prime implicant's block.*

f) A **covering** *for a Boolean function is a collection of blocks whose union is the set of 1-cells for that function's K-map.*

❖ Example 7.6.5

Find the implicants, prime implicants, and essential prime implicants for the function of Example 4c. Then relate a covering associated with prime implicants to a simplified expression for that function.

Solution

· For the sake of convenience, let's repeat the function's K-map.

	00	01	11	10
0		1	1	
1	1	1	1	1

· Each of the six minterms $\overline{x}\,\overline{y}z$, $\overline{x}yz$, $x\overline{y}\,\overline{z}$, $x\overline{y}z$, xyz, and $xy\overline{z}$ correspond to one of the K-map's 1-cells, so they're all implicants of the function.

· Similarly, the products $\overline{x}z$, $x\overline{y}$, xz, xy, $x\overline{z}$, $\overline{y}z$, and yz are implicants—they correspond to 1-cell blocks of size 2.

· Finally, the trivial "products" x and z are implicants, corresponding to 1-cell blocks of size 4. These are the only prime implicants. Furthermore, both are essential—each is needed to cover some 1-cell.

· The covering associated with these prime implicants consists of the bottom row of four 1-cells and the middle square block of four 1-cells. From these blocks, we can generate the simplified representation $x + z$ for the function.

❖ Example 7.6.6

Determine the prime implicants for the K-map of the ternary majority function to find a simplified expression for the function.

Solution

· The ternary majority function's K-map is as follows (see Example 2).

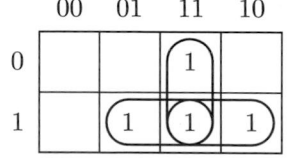

· There are no blocks of size 4 here, but there are three blocks of size 2 (encircled in the K-map). They correspond to the following prime implicants: xz, xy, and yz, each of which is essential. Adding these three together gives a simplified expression for the ternary majority function: $f(x, y, z) = xy + xz + yz$.

Inspecting a function's K-map, we can decompose its pattern of 1-cells into a minimal union of maximal blocks, find the products associated with these blocks, and add them together to generate a simplified representation. While we need to be able to recognize block patterns, particularly when they stretch around from one side to the other, the process isn't too complicated.

7.6.4 K-Maps for Boolean Functions of Four Variables

What if a Boolean function has more than three variables? A slight variation of the K-map for three-variable functions works for functions of four variables. We can use the top of a K-map to represent two variables and the side for two more. The next example illustrates the process.

✤**Example 7.6.7**

For the Boolean function $f(x_1, x_2, x_3, x_4) = x_1x_2\overline{x}_3\overline{x}_4 + x_1\overline{x}_2\overline{x}_3 + x_1x_3\overline{x}_4 + \overline{x}_1x_2x_4 + \overline{x}_1\overline{x}_2\overline{x}_3 + \overline{x}_1x_3x_4 + \overline{x}_1\overline{x}_3x_4 + \overline{x}_2x_3\overline{x}_4$, exhibit a K-map and then use it to determine a simplified equivalent for this function.

Solution

· Following are two ways to label and draw the K-map for this function (check that this is so by mapping each term of the formula). Covering blocks have been drawn in the second K-map for the function's prime implicants.

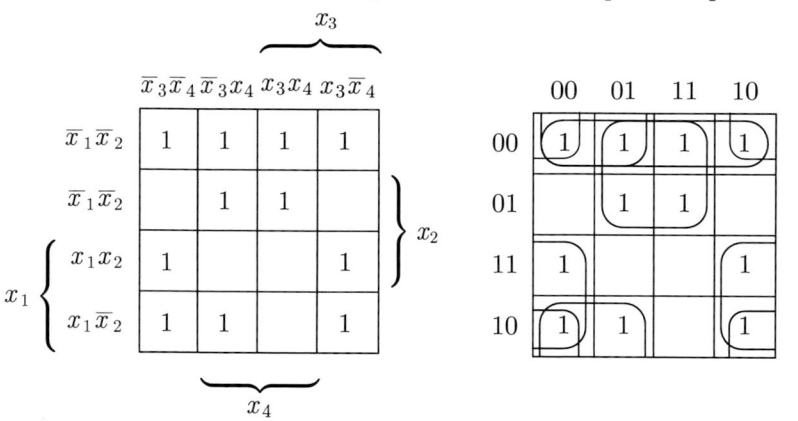

· The essential prime implicants for this function are $\overline{x}_1x_4$ (middle top four-square), $x_1\overline{x}_4$ (bottom left-right four-square), and $\overline{x}_2\overline{x}_3$ (left top-bottom four-square). Taking these essential prime implicants, we're still missing cell 0010 (top right corner). Other prime implicants are $\overline{x}_1\overline{x}_2$ (top row) and $\overline{x}_2\overline{x}_4$ (the "adjacent" four corners). Both of these cover the missing 1-cell. Thus, we have two possible minimal equivalents for our function: $f(x_1, x_2, x_3, x_4) = \overline{x}_1x_4 + x_1\overline{x}_4 + \overline{x}_2\overline{x}_3 + \overline{x}_1\overline{x}_2$, and $f(x_1, x_2, x_3, x_4) = \overline{x}_1x_4 + x_1\overline{x}_4 + \overline{x}_2\overline{x}_3 + \overline{x}_2\overline{x}_4$. No further simplifications are possible.

K-maps for Boolean functions of four variables will have blocks of size 1, $2 = 1 \times 2 = 2 \times 1$, $4 = 1 \times 4 = 4 \times 1 = 2 \times 2$, $8 = 2 \times 4 = 4 \times 2$, and $16 = 4 \times 4$.

Each of these (except for the last, which represents the constant expression 1) corresponds to a product of literals. We can always cover a K-map's 1-cells by maximal blocks of these binary sizes, and these will be associated with the function's prime implicants. All essential prime implicants must be used for a covering, but as in the last example, we may need additional prime implicants to complete it and determine a simplified formula, which may not be unique.

7.6.5 *The Quine-McCluskey Method*

K-maps are still used to represent Boolean functions of five or six variables. A six-variable K-map can be constructed either by representing eight three-literal minterms in some order along both the top and side of the K-map, or by stacking four four-variable K-maps on top of each other. This gets more complicated than what we've done so far, especially since blocks can wrap around and be overlooked, so we won't pursue this further here.

Instead we'll explore the *Quine-McCluskey method*, proposed independently in the mid-1950s by the philosopher Willard van Orman Quine for logic and by the electrical engineer Edward McCluskey for Boolean functions. The procedure goes as follows:

Quine-McCluskey Method
   0) *Minterm Expansion*
      To get started, we need a minterm representation for the function. This isn't really part of the method, so we'll label it stage 0.
   1) *Reduced Sum-of-Products Expansion*
      Rewrite the function's minterm expansion as a sum involving fewer shorter products (implicants with fewer factors):
      a) Replace *each pair* of added minterms by a single shorter product, if possible. This is accomplished by factoring and simplifying, as in $xy + x\overline{y} = x(y + \overline{y}) = x$.
      b) Apply the process of part a to the resulting sums of products, if possible.
      c) Repeat the process on each new sum of shorter products until no further combining and simplifying can occur.
   2) *Miminized Sum-of-Products Expansion*
      Stage 1 generates a set of prime implicants whose sum represents the function, but not all implicants may be either essential or needed to cover the function. Conclude by further reducing the expression to a minimal sum of essential and other prime implicants.

We'll illustrate this method with two simple three-variable examples and then work one for a more complex function of four variables.

✤**Example 7.6.8**
   Apply the Quine-McCluskey method to the ternary majority function.

Solution

This is essentially how we worked Example 2, but we'll work through it again to illustrate the steps of the Quine-McCluskey method.

0) The minterm expansion for this function is $f(x, y, z) = xyz + xy\overline{z} + x\overline{y}z + \overline{x}yz$.

1) We can combine and replace $xyz + xy\overline{z}$ by xy, $xyz + x\overline{y}z$ by xz, and $xyz + \overline{x}yz$ by yz; no other pairs combine. We thus have $f(x, y, z) = xy + xz + yz$.

   · These three new terms cannot be further combined in the required way, so all the prime implicants of the function have been found.

2) In this case, we can't further reduce our sum. If any term is dropped, the resulting sum will no longer generate the function—some of the function's original implicants will no longer be covered. So the minimal expansion for the ternary majority function is given by $f(x, y, z) = xy + xz + yz$.

♣ **Example 7.6.9**

Apply the Quine-McCluskey method to the function $f(x, y, z) = xyz + xy\overline{z} + x\overline{y}\,\overline{z} + \overline{x}yz + \overline{x}\,\overline{y}z$ of Examples 4c and 5.

Solution

0) The minterm expansion for this function is already given.

1) We can combine and replace $xyz + xy\overline{z}$ by xy, $xyz + x\overline{y}z$ by xz, $xyz + \overline{x}yz$ by yz; $xy\overline{z} + x\overline{y}\,\overline{z}$ by $x\overline{z}$; $x\overline{y}z + x\overline{y}\,\overline{z}$ by $x\overline{y}$, $x\overline{y}z + \overline{x}\,\overline{y}z$ by $\overline{y}z$; and $\overline{x}yz + \overline{x}\,\overline{y}z$ by $\overline{x}z$.

   · All possible pairs of terms have been considered in this process, so the new representation is $f(x, y, z) = xy + xz + yz + x\overline{y} + x\overline{z} + \overline{x}z + \overline{y}z$.

   · The terms in this expansion can be further combined. We can replace $xy + x\overline{y}$ by x; $xz + x\overline{z}$ by x, $xz + \overline{x}z$ by z; and $yz + \overline{y}z$ by z. All terms have now been incorporated into a shorter product, so the new representation is $f(x, y, z) = x + z$, omitting duplicates. No further combining can be done, because we have the prime implicants of our function (see Example 4c).

2) Since both terms are needed (neither one alone suffices), our minimal expansion is as stated: $f(x, y, z) = x + z$.

In both of the above examples, the Quine-McCluskey procedure was rather straightforward, so we could illustrate the essentials of the process without getting bogged down by complications. This is not typical, however. To illustrate this, we'll rework the more complex Example 7, which will allow us to check our work there.

♣ **Example 7.6.10**

Use the Quine-McCluskey method to find the prime implicants and a minimal expansion for Example 7's function: $f(x_1, x_2, x_3, x_4) = x_1 x_2 \overline{x}_3 \overline{x}_4 + x_1 \overline{x}_2 \overline{x}_3 + x_1 x_3 \overline{x}_4 + \overline{x}_1 x_2 x_4 + \overline{x}_1 \overline{x}_2 \overline{x}_3 + \overline{x}_1 x_3 x_4 + \overline{x}_1 \overline{x}_3 x_4 + \overline{x}_2 x_3 \overline{x}_4$.

Solution

0) The minterm expansion here is given by $f(x_1, x_2, x_3, x_4) = \overline{x}_1\overline{x}_2\overline{x}_3\overline{x}_4 + \overline{x}_1\overline{x}_2\overline{x}_3x_4 + \overline{x}_1\overline{x}_2x_3\overline{x}_4 + \overline{x}_1\overline{x}_2x_3x_4 + \overline{x}_1x_2\overline{x}_3x_4 + \overline{x}_1x_2x_3x_4 + x_1\overline{x}_2\overline{x}_3\overline{x}_4 + x_1\overline{x}_2\overline{x}_3x_4 + x_1\overline{x}_2x_3\overline{x}_4 + x_1x_2\overline{x}_3\overline{x}_4 + x_1x_2x_3\overline{x}_4$.

1) To facilitate our work, we'll use a table, replacing minterms with their bit strings labeled by the associated binary number they represent and grouping them in increasing order according to how many 1's appear in the representation. These appear in the first column of Table 7.2. Strings in each group may be combined with ones from the next group, since they differ in a place where one has the variable (1) and the other has its complement (0). This corresponds to adjacent 1-cells in a K-map. Combining these leads to new terms having one fewer factor; these are placed in the second column. For example, $0011 + 0111$ simplifies to $0\text{-}11$ (a dash indicates a missing variable), because $\overline{x}_1\overline{x}_2x_3x_4 + \overline{x}_1x_2x_3x_4 = \overline{x}_1x_3x_4$. Used minterms are marked with a $*$ and are omitted from our final sum-of-products representation; thus, both minterm 3 and minterm 7 have an asterisk after them.

Table 7.2 Combined product terms for Example 7.6.10

Minterms			Two-Cell Combinations			Four-Cell Combinations	
0	0000	$*$	$(0,1)$	000-	$*$	$(0,1;2,3)$	00--
1	0001	$*$	$(0,2)$	00-0	$*$	$(0,1;8,9)$	-00-
2	0010	$*$	$(0,8)$	-000	$*$	$(0,2;1,3)$	00--
8	1000	$*$	$(1,3)$	00-1	$*$	$(0,2;8,10)$	-0-0
3	0011	$*$	$(1,5)$	0-01	$*$	$(0,8;1,9)$	-00-
5	0101	$*$	$(1,9)$	-001	$*$	$(0,8;2,10)$	-0-0
9	1001	$*$	$(2,3)$	001-	$*$	$(1,3;5,7)$	0--1
10	1010	$*$	$(2,10)$	-010	$*$	$(1,5;3,7)$	0--1
12	1100	$*$	$(5,7)$	01-1	$*$	$(8,10;12,14)$	1--0
7	0111	$*$	$(8,9)$	100-	$*$	$(8,12;10,14)$	1--0
14	1110	$*$	$(8,10)$	10-0	$*$		
			$(8,12)$	1-00	$*$		
			$(3,7)$	0-11	$*$		
			$(10,14)$	1-10	$*$		
			$(12,14)$	11-0	$*$		

- The new combined product terms are labeled by which minterms gave rise to them. For example, the result $0\text{-}11$ is labeled $(3,7)$ because it resulted from combining minterm 3 with minterm 7.
- Once the entire second column of two-cell combined terms is generated, the same process is repeated with these strings. Two-cell blocks that are adjacent are combined into a four-cell block and put in the third column. Additional columns are added as necessary, but in this case, no new combinations can be made with the four-cell blocks.
- In the end, all strings *not* marked as contributing to a further combination will correspond to maximal blocks and so will yield prime implicants

for the function. Here these appear in the last column, but this may not always happen. The associated products are then used to form a sum-of-products representation for the function, duplicates being omitted.

· The final result is $f(x_1, x_2, x_3, x_4) = \overline{x}_1\overline{x}_2 + \overline{x}_2\overline{x}_3 + \overline{x}_2\overline{x}_4 + \overline{x}_1x_4 + x_1\overline{x}_4$.

2) Comparing the end result of stage one with what we obtained in Example 7 by using the function's K-map, we see that we have not yet achieved a minimal expansion. To determine which prime implicants are essential, we'll determine which minterms are covered by which implicants. We'll do this using the following table:

Prime	Minterms (Binary Cell Label)										
Implicants	0	1	2	3	5	7	8	9	10	12	14
$(0,1,2,3)$	X	X	X	X							
$(0,1,8,9)$	X	X					X	X			
$(0,2,8,10)$	X		X				X		X		
$(1,5,3,7)$		X		X	X	X					
$(8,10,12,14)$							X		X	X	X

· We must cover all the minterms in the function's minterm expansion. From the chart above we can see that 5 and 7 are only covered by the prime implicant $(1, 5, 3, 7)$; 9 is only covered by the prime implicant $(0, 1, 8, 9)$; and 12 and 14 are only covered by the prime implicant $(8, 10, 12, 14)$. We've drawn horizontal lines through the Xs covered by these essential prime implicants. Together these three essential prime implicants cover minterms 0, 1, 3, 5, 7, 8, 9, 10, 12, and 14—all but 2. We've drawn vertical lines to indicate this coverage. The remaining minterm can be covered either by implicant $(0, 1, 2, 3)$ or implicant $(0, 2, 8, 10)$.

· Translating this into a Boolean expression, we get the following two sum-of-products representations (see Table 7.2 for the products associated with the four-cell combinations): $f(x_1, x_2, x_3, x_4) = \overline{x}_1\overline{x}_2 + \overline{x}_2\overline{x}_3 + \overline{x}_1x_4 + x_1\overline{x}_4$, and $f(x_1, x_2, x_3, x_4) = \overline{x}_2\overline{x}_3 + \overline{x}_2\overline{x}_4 + \overline{x}_1x_4 + x_1\overline{x}_4$. These agree with the expansions we obtained in Example 7.

7.6.6 Boolean Algebra: Theory and Practice

We've now explored *Boolean Algebra* both theoretically and practically, both abstractly and concretely. To sum up, we began by looking at the theory of posets and lattices, eventually proving the key result that all finite Boolean lattices are essentially power set lattices ordered by the subset relation. Using 0's and 1's to represent elements in subsets gave us a concrete way to represent these structures as $\mathcal{B}^n$. We next saw that Boolean lattices can be converted into Boolean algebras, and conversely, which gave us another way to connect *Boolean Algebra* and *Set Theory*.

    Boolean Algebra can be considered an abstract generalization of *Propositional Logic* as well as *Set Theory*. *Boolean Algebra* had its roots in Boole's ideas about logic, a connection that was developed further in the twentieth century by Shannon, who investigated relations between Boolean functions and logic circuits. We saw that all Boolean functions/expressions had certain normal representations, such as minterm expansions, which means that Boolean logic circuits can be constructed using only the most basic logic gates AND, OR, and NOT. Theoretically, this means that the equational part of *Boolean Algebra* can be decided by truth tables, showing important connections between *Boolean Algebra*, *Propositional Logic*, and *Set Theory*.

Besides using *Boolean Algebra* to symbolize logic circuits, Shannon showed that we can construct logic circuits to do ordinary computations, a significant development underlying twentieth-century computer calculations.

Finally, we looked at some systematic ways to simplify Boolean expressions. Karnaugh maps provide a visual method for handling simple functions, while the Quine-McCluskey method gives a powerful tabular technique that can be developed into an efficient algorithm implemented on a computer. Simplification of Boolean expressions is obviously important to anyone designing complex logic circuits, both in engineering generally and computer science in particular.

Thus, a topic (*Boolean Algebra*) that is fairly abstract and theoretical is also very concrete and practical, forming the theoretical basis of some of the most important technological developments of the last century. This field demonstrates the interplay between these different aspects of mathematics, attracting people with varied interests and attitudes toward theoretical and practical mathematics. Given the concrete applications it has, *Boolean Algebra* makes an excellent first introduction to advanced abstract mathematics. And, given its links to *Propositional Logic* and *Set Theory*, it also helps unify different aspects of discrete mathematics in a foundational manner, providing reinforcement and a theoretical context for those fields.

EXERCISE SET 7.6

Exercises 1–6: Simplifying Boolean Functions via Boolean Algebra
Simplify the following functions using Boolean Algebra. Explain, by comparing the number of literals and number of operations involved, why your final answer is simpler than the given formula.

7.6.1. $f(x, y) = x\overline{y} + \overline{x}\,\overline{y}$

7.6.2. $f(x, y) = xy + x\overline{y} + \overline{x}\,\overline{y}$

7.6.3. $f(x, y, z) = xy + x\overline{y}\,\overline{z} + \overline{x}yz$

7.6.4. $f(x, y, z) = xy + x\overline{y}z + x\overline{z} + \overline{x}\,\overline{y}z + yz$

7.6.5. $f(x, y, z) = xy + xyz + xy\overline{z} + \overline{x}yz + \overline{x}y\overline{z}$

7.6.6. $f(w, x, y, z) = wz + \overline{w}y\overline{z} + x\overline{y}\,\overline{z} + \overline{x}\,\overline{y}\,\overline{z}$

Exercises 7–10: True or False
Are the following statements true or false? Explain your answer.

7.6.7. The minterm expansion for a Boolean function yields a simplified expression for the function.

7.6.8. An implicant for a Boolean function is one of the function's minterms.

7.6.9. An essential prime implicant for a Boolean function is a term that must appear in a simplified sum-of-products representation of the function.

7.6.10. The only terms that appear in a simplified sum-of-products representation of a Boolean function are essential prime implicants.

Exercises 11–14: Determining Implicants for K-Maps
For the following K-maps, identify the following, using alphabetic order $\{x, y\}$, $\{x, y, z\}$, and $\{w, x, y, z\}$ for two, three, and four variables:
  a. *all the implicants of the associated Boolean function;*
  b. *all the function's prime implicants; and*
  c. *all the function's essential prime implicants.*

7.6.11.

	0	1
0	1	1
1	1	

7.6.12.

	00	01	11	10
0	1		1	1
1	1			1

7.6.13.

	00	01	11	10
0	1			1
1	1	1	1	1

7.6.14.

	00	01	11	10
00	1	1	1	1
01		1		
11		1	1	
10		1	1	1

Exercises 15–18: Simplifying Boolean Functions from K-maps
For each of the following K-maps, determine:
  a. *the minterm expansion for the associated Boolean function, and*
  b. *a simplified sum-of-products-of-literals representation for the minterm expansion.*

7.6.15. The K-map of Exercise 11   **7.6.17.** The K-map of Exercise 13
7.6.16. The K-map of Exercise 12   **7.6.18.** The K-map of Exercise 14

Exercises 19–28: Simplifying Boolean Functions Using K-maps
Draw K-maps for the following Boolean functions. Then use these K-maps to determine simplified sums of products for the given functions.

7.6.19. $f(x, y) = xy + \overline{x}$

7.6.20. $f(x, y) = xy + x\overline{y} + \overline{x}\,\overline{y}$

7.6.21. $f(x, y, z) = xyz + xy\overline{z} + \overline{x}\,\overline{y}z + \overline{x}\,\overline{y}z$

7.6.22. $f(x, y, z) = xy + x(\overline{y + z}) + \overline{x}\,\overline{y}z + \overline{y}\,\overline{z}$

7.6.23. $f(x, y, z) = xyz + \overline{x}\,\overline{y}z + \overline{x}\,\overline{z} + yz + \overline{y}\,\overline{z}$

7.6.24. $f(x, y, z) = x\overline{y}\,\overline{z} + \overline{x}yz + \overline{x}\,\overline{y}z + \overline{x}\,\overline{y}\,\overline{z}$

7.6.25. $f(x, y, z) = xyz + xy\overline{z} + x\overline{y}z + x\overline{y}z + \overline{x}\,\overline{y}z + \overline{x}\,\overline{y}\,\overline{z}$

7.6.26. $f(w, x, y, z) = wxz + w\overline{x}\,\overline{y}\,\overline{z} + xz + y\overline{z}$

7.6.27. $f(w, x, y, z) = w\overline{x}y\overline{z} + w\overline{x}\,\overline{y}\,\overline{z} + \overline{w}xyz + \overline{w}\,\overline{x}y\overline{z} + \overline{w}\,\overline{x}\,\overline{y}\,\overline{z}$

7.6.28. $f(w, x, y, z) = wxy + x\overline{y}z + wy\overline{z} + wxyz + \overline{w}x\overline{y}\,\overline{z}$

Exercises 29–36: Quine-McCluskey Method
Use the Quine-McCluskey method to determine simplified representations for the following Boolean functions. For three-variable functions, use the procedure of Examples 8 and 9; for four-variable functions use Example 10's tabular method.

7.6.29. The function of Exercise 21   **7.6.33.** The function of Exercise 25
7.6.30. The function of Exercise 22   **7.6.34.** The function of Exercise 26
7.6.31. The function of Exercise 23   **7.6.35.** The function of Exercise 27
7.6.32. The function of Exercise 24   **7.6.36.** The function of Exercise 28

Exercises 37–39: Blocks in K-Maps
The following problems explore blocks of cells in K-maps.

7.6.37. *Boolean Expressions for Blocks in a Three-Variable K-map.*
  a. Explain why in a K-map for three-variable functions a rectangular block of two adjacent 1-cells represents a product of two literals.
  b. Explain why in a K-map for three-variable functions a rectangular block of four adjacent 1-cells represents a single literal.

7.6.38. *Boolean Expressions for Blocks in a Four-Variable K-map.*
  a. Explain why in a K-map for four-variable functions a rectangular block of two adjacent 1-cells represents a product of three literals.

b. Explain why in a K-map for four-variable functions a rectangular block of four adjacent 1-cells represents a product of two literals.

c. Explain why in a K-map for four-variable functions a rectangular block of eight adjacent 1-cells represents a single literal.

7.6.39. *Boolean Expressions for Blocks in an N-Variable K-map.*

a. Generalize the results of Exercises 37 and 38: in a K-Map for an n-variable Boolean function, a rectangular block of 2^k adjacent 1-cells represents a product of _____ literals, for $k = 0, 1, \ldots, n-1$. What function does the full block of 2^n 1-cells represent?

b. Explain why your formula in part a is correct.

Exercises 40–42: Implicants for Boolean Functions

The following problems explore implicants for Boolean Functions.

7.6.40. *Implicants*

Claim: if a function's implicant equals 1, then the function as a whole equals 1.

a. Show from the definition for an implicant that xy is an implicant for $f(x, y, z) = x + z$. Then use this to illustrate the given claim.

b. Argue the given claim in general.

c. Explain why the term *implicant* is an appropriate one.

d. Is the converse of the above claim also true? What if the implicant is a prime implicant? an essential prime implicant?

7.6.41. *Maximal Blocks and Implicants*

a. Explain why maximal blocks in a K-map correspond to prime implicants. Must maximal blocks also be essential prime implicants?

b. Prove or disprove: the sum of a Boolean function's essential prime implicants is a Boolean expression that represents the function.

7.6.42. *The Quine-McCluskey Method and Prime Implicants*

Explain why all products that haven't been checked off in the tabular approach to the Quine-McCluskey method are prime implicants of the function.

Chapter 8
Topics in Graph Theory

8.1 Eulerian Trails

For most people, the term *graph* brings to mind pictures of straight lines, circles, parabolas, and other figures drawn within a rectangular coordinate system. That's what it means in elementary algebra, calculus, and physical science, but discrete mathematics uses this term in a different sense. Here a graph is a set of vertices connected by edges. We've seen graphs like this when we drew production graphs for well-formed formulas in Chapters 1 and 3 and when we used Hasse diagrams for binary relations in Chapter 7. Now we'll explore graphs more systematically. *Graph Theory* is a well-developed branch of mathematics, though, so we'll only introduce some basics in this chapter.

8.1.1 *Graphs: Leisurely Pastime or Serious Business?*

Graph Theory can be traced to a 1736 paper of Euler, submitted as his solution to what has become known as the *Königsberg Bridge Problem*. Other well-known recreational problems connected to *Graph Theory* were investigated in the mid-nineteenth century. These include a game invented by Hamilton for traversing the edges of a dodecahedron to visit all its vertices exactly once and a puzzle posed to De Morgan about how many colors are needed to color a map when neighboring counties have distinct colors. Of the three, only Hamilton's game initially had an explicit connection to a graph, but all of them lead to important ideas in *Graph Theory*. We'll use these problems to introduce the topics explored in this chapter.

As a branch of mathematics, *Graph Theory* is mostly a twentieth- and twenty-first-century phenomenon, though discrete graphs began to appear in areas like chemistry and electricity during the last half of the nineteenth century. To exhibit the chemical structure of molecules, for instance, chemists used bonding diagrams such as the ones shown for nitric acid and methane gas. Scientists also began to explore the flow of electricity through a collection of circuits and proved fundamental laws about such networks.

Over the last hundred years or so, networks have found many more applications—in natural science, social science, transportation, communication, computer science, and engineering. *Graph Theory* began in earnest in the early-to-mid-twentieth century to provide a basis for recreational and other interests. Our focus in this chapter will be on the underlying ideas, taking a more geometric approach than we've done so far in the book.

© Springer Nature Switzerland AG 2019
C. Jongsma, *Introduction to Discrete Mathematics via Logic and Proof*,
Undergraduate Texts in Mathematics,
https://doi.org/10.1007/978-3-030-25358-5_8

8.1.2 *Euler's Bridges of Königsberg Solution*

Fig. 8.1 Euler

Leonhard Euler (Figure 8.1) was the most prolific mathematician who ever lived, known for his many methods, formulas, notations, and theorems. He continued producing first-rate mathematics even during the final dozen years of his life after losing his eyesight, aided by a photographic memory and an amazing ability to perform complex mental calculations. He contributed to every area of mathematics, organizing ideas in ways that influenced textbooks from then on. His 1736 paper on the *Königsberg Bridge Problem*, an expository gem, was offered as a contribution to what was then called Leibniz's geometry of position, but today it is seen as the earliest example of graph-theoretic reasoning.

The problem posed to Euler was this: "[If] an island in the city of Königsberg [is] surrounded by a river spanned by seven bridges, ... [can one] traverse the separate bridges in a connected walk in such a way that each bridge is crossed only once."[1]

Euler schematized the problem with a diagram of the Pregel river's two branches flowing around island A on its way toward the Baltic Sea, labeling its seven bridges by $a, b, c, d, e, f, g,$ and other parts of the city by B, C, D in Figure 8.2.

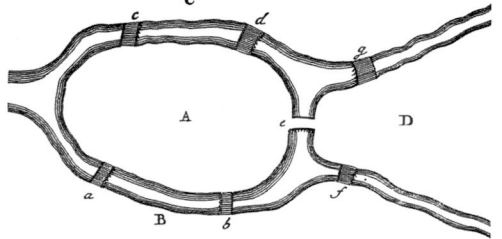

Fig. 8.2 Euler's schematic of the Pregel river

Euler solved the *Königsberg Bridge Problem* in nine paragraphs and then generalized it to any number of river branches, regions, and bridges. A sequence of capital letters such as $ABDC$ can represent a walk starting at A, crossing bridges b, f, and g, and ending at C. As there are seven bridges, a complete tour would be indicated by a sequence of eight letters. Euler rejected systematically enumerating all the possibilities to look for one that works, choosing instead to determine whether such a tour is even possible and, if so, what conditions it must satisfy. For example, A would have to be listed next to B twice since they are connected by two bridges.

To decide whether a tour is possible, Euler evaluated how often each area must be visited/listed. If region X has one bridge to it, a tour sequence must

---

[1] See *The Truth about Königsberg* by Brian Hopkins and Robin J. Wilson in the May 2004 issue, of **The College Mathematics Journal** for a nice discussion of Euler's argument. Euler's paper (and lots more) is in **Graph Theory: 1736–1936** (Oxford University Press, 1976) by Norman L. Biggs, E. Keith Lloyd, and Robin J. Wilson.

visit X once; if it has three bridges connecting it, it must be listed twice (regardless of how one goes in and out of X); and if five bridges connect to it, it must be listed three times. In general, if X has $2n + 1$ bridges, it must be listed $n + 1$ times (see Exercise 2a). Because all regions in the problem have an odd number of bridges connecting them, the associated eight-letter word must contain A three times and each of B, C, and D twice—an impossibility, since an eight-letter word can't contain nine letters.

Having solved the particular problem posed to him, Euler moved on to show how to solve any similar problem. A region with an odd number of bridges connected to it (an *odd region*) can use the method just outlined. For a region with an even number of connecting bridges (an *even region*), how often it must be listed depends on whether or not one starts/ends there. An intermediate region with $2n$ connecting bridges must be visited n times; a terminal region must be listed one more time than this.

Based on his analysis, Euler presented a method for solving any bridge-tour problem.

❖ Example 8.1.1

We'll exhibit Euler's tabular method for deciding whether a bridge-crossing tour exists for the later nineteenth-century arrangement of the Königsberg bridges, in which regions B and C were directly connected downriver by an additional bridge that we'll call h (add it onto Figure 8.2).

Solution

- Let A, B, C, D represent the various regions, and suppose that the eight bridges are positioned as described. Then a complete tour is represented by a nine-letter word. To determine whether such a tour is possible, Euler created a chart like the one shown.

	Bridges	Visits
A	5	3
B*	4	2
C*	4	2
D	3	2

- The first column represents the regions, the second column gives the numbers of their connecting bridges, and the third column shows how many times the regions should be visited/listed in a tour sequence.
- B and C have an asterisk because they are even regions. If one of these is chosen as a starting point, we must add 1 to the last entry in its row.
- Since the final column adds up to 9, as needed, there is a possible tour, but we can't start at either B or C. One tour sequence from A to D, with bridges inserted to make the walk definite, is $AaBbAcCdAeDfBhCgD$.

Euler claimed that his method always works for determining whether a complete tour of bridges is possible, but then he offered a simpler criterion based on some elementary observations. The first of these is often called the *Handshake Lemma* (see Exercise 5a).

Proposition 8.1.1: *The Handshake Lemma*

The sum of the number of bridge connections to all regions is twice the total number of bridges.

Proof:
See Exercise 5b.

Euler observed, as a corollary, that there can't be an odd number of odd regions (see Exercise 5c), but he also noted that there can't be more than two odd regions if there's a legitimate tour of bridges. For, if n denotes the number of bridge connections to a region, a tour sequence must visit an odd region $(n + 1)/2$ times and an even region $n/2$ times. Two odd regions will give a visit total of one more than the number of bridges, which still satisfies the condition for creating a tour sequence, but if there are four or more odd regions, we'd have to visit these regions too many times to satisfy the condition for a bona fide tour sequence (see Exercise 2c). Also, according to Euler, if there are no odd regions, a tour can be created starting anywhere.

Euler conclusively argued that if a tour is possible, then the setup must satisfy this condition about the total number of odd regions (equivalently, if this condition is violated, then a tour is impossible). He also asserted that this necessary condition is sufficient for undertaking a bridge tour, but he did not give a proof—he only suggested that deleting all [redundant] pairs of bridges joining the same regions would make it easier to determine such a tour. This advice doesn't provide a method for finding a tour, however.

There is an algorithm that completes Euler's analysis and establishes his claim, but we'll introduce some standard *Graph Theory* terminology first. As a lead-in, let's revisit the *Königsberg Bridge Problem*, once again using Euler's reasoning but now with a different diagram.

✤**Example 8.1.2**
Analyze the *Königsberg Bridge Problem* using a graph-theoretic diagram.

Solution
· We'll represent Königsberg landmasses $A - D$ as vertices and bridges $a - g$ as solid-line connecting edges. This sort of diagram only came into use toward the end of the nineteenth century (and is *not* found in Euler, contrary to what some say).

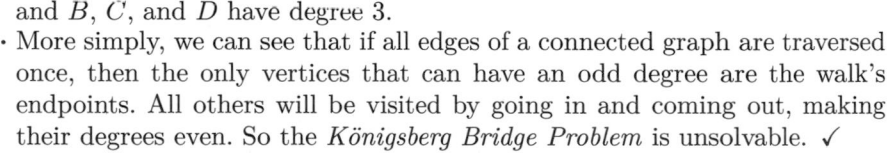

· Euler's reasoning tells us that the edges cannot be traversed exactly once because four vertices have odd degrees—A has degree 5 (five edges attached to it), and B, C, and D have degree 3.
· More simply, we can see that if all edges of a connected graph are traversed once, then the only vertices that can have an odd degree are the walk's endpoints. All others will be visited by going in and coming out, making their degrees even. So the *Königsberg Bridge Problem* is unsolvable. ✓
· If we were to modify the solid-line Königsberg graph by adding the dashed edge h between B and C, we'd obtain a graph whose edges can be traversed in the required way, as shown in Example 1: $AaBbAcCdAeDfBhCgD$.

8.1.3 Basic Definitions of Graph Theory

Let's now sharpen the meaning of terms we've been using informally.

Definition 8.1.1: Graphs, Vertices, and Edges
a) A **graph** G of **order** n consists of a finite non-empty set V of **vertices** with $|V| = n$ and a set E of **edges** (two-element subsets of V). An edge $e = uv$ **connects** distinct vertices u and v as **endpoints**. A **subgraph** is a subset of a graph; a **supergraph** is a superset of a graph.
b) Two vertices are **adjacent** if and only if they are connected by an edge.
c) A vertex is **isolated** if and only if it isn't connected to another vertex.
d) The **degree** of a vertex v, denoted $\deg(v)$, is the number of edges connected to it. An **even/odd vertex** is one of even/odd degree.

Unfortunately, graph-theoretic terms (of which there are many) haven't been completely standardized. Graphs, as defined above, have at most one edge for any two vertices. Thus, our diagram in Example 2 isn't a graph because, for example, A and B are connected by two edges—such structures are called *multigraphs*. And edge endpoints must be distinct—if the two vertices are the same, the "edge" is called a *loop*, and graphs that allow both multiple edges and loops are called *pseudographs*. We also associate edges with pairs of vertices rather than ordered pairs, so $uv = vu$. If order is important, the edges are drawn using arrows, giving what's called a *directed graph*.

Definition 8.1.2: Walks, Trails, Paths, Connected Graphs
a) A **walk** of length n with **endpoints** v_0 and v_n is a sequence of n edges $e_i = v_{i-1}v_i$ connecting a sequence of vertices $v_0, v_1, \ldots, v_n$.
b) A walk is **closed** (**open**) if and only if $v_n = v_0$ ($v_n \neq v_0$).
c) A **trail** is a walk whose edges are distinct; a **circuit** is a closed trail.
d) A **path** is a walk whose vertices are distinct; a **cycle** is a closed path.
e) An **Eulerian trail** (**circuit**) is a trail (circuit) that includes all edges of the graph once.
f) An **Eulerian graph** is a graph containing an Eulerian circuit.
g) A graph is **connected** if and only if every pair of vertices is connected by a walk.

Once again, a warning: not all graph theorists use the same terminology. *Walks, trails, circuits, paths, cycles*—all of these refer to related concepts, but they're not used in the same way by everyone. Always refer back to the definition in force if you need clarity on how a graph-theoretic term is being used in a discussion. We'll try to minimize the number of technical terms to avoid adding to the confusion.

✤ Example 8.1.3

Can a simple *barbell graph* be traced without lifting your pencil off the page? Does it have an Eulerian trail? An Eulerian circuit?

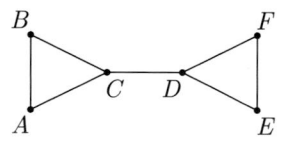

Solution
· *CABCDEFD* is an Eulerian trail that you can use to trace the graph.
Such a trail must start and end at either *C* or *D*. Note that while a trail
never repeats an edge, vertices may be visited more than once.
· There is no Eulerian circuit, however; both *C* and *D* are odd vertices.

8.1.4 *Eulerian Trails: A Necessary Condition*

Let's now prove Euler's necessary condition, for connected graphs.

Theorem 8.1.1: *Necessary Condition for Eulerian Circuits & Trails*
    a) *If a connected graph has an Eulerian circuit, then all vertices are even.*
    b) *If a connected graph has an Eulerian trail that is not a circuit, then
      exactly two vertices are odd, the trail's initial and terminal endpoints.*

Proof:
· The reasoning used toward the end of Example 2 works here, too.
  Note that in a connected graph, an Eulerian trail must contain every vertex.
· If a vertex is odd, at some point a trail will enter it and not leave or will
  leave it and not return. An Eulerian trail can contain only two of these—the
  initial vertex and the terminal vertex. All other vertices must be even.
· If an Eulerian trail ends where it started (i.e., is an Eulerian circuit), all
  vertices are even.   ∎

❖**Example 8.1.4**
Do the following graphs have Eulerian trails or circuits?
  a)   b)   c)

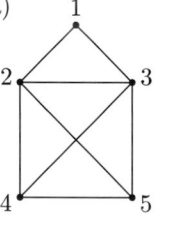
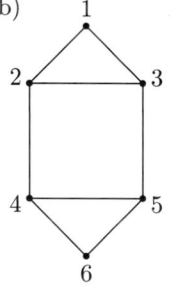
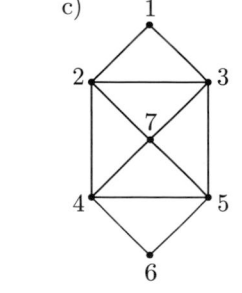

Solution
  a) Two vertices are odd in this graph, so there is no Eulerian circuit. But it
    has an Eulerian trail, starting and ending at an odd vertex: 421352345.
    · Note that while the diagram has intersecting diagonals, their point of
      intersection is *not* a vertex of the graph.
  b) No Eulerian trail can be found in this graph since it has four odd vertices.
  c) Adding two diagonals with their intersection point to the second graph
    gives this final graph an Eulerian circuit: 1237427546531. Such a circuit
    can start anywhere.
    · If edge 23 in this graph were dropped, it wouldn't have an Eulerian
      circuit, but an Eulerian trail would still exist: 213564572473.

8.1.5 *Eulerian Trails: A Sufficient Condition*

We'll now show that Euler's necessary condition is also sufficient.

❖**Example 8.1.5**

Systematically construct an Eulerian circuit through the graph shown, which has
only even vertices.

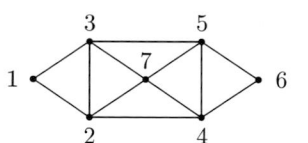

Solution

- It's easy to find an Eulerian circuit in a graph this small, as we showed in
  Example 4c, but here we will outline a uniform *least-first procedure* that
  can be used for any (labeled) graph. At each vertex we'll choose the open
  edge going to the vertex with the least label until a circuit is completed.
- For this graph, stage 1 gives the circuit 1231.
- Vertex 2 has the first unused attached edge, so we'll circle around from
  there and break out at the end, creating the next circuit: 2312 45372.
- This new circuit has an unused edge at 4, so we'll again reorder the circuit
  and continue from there, generating 453723124 6574—an Eulerian circuit.
- We can schematize this process, reordering the circuit and breaking out to
  expand it, as follows:

  1231
  2312 45372
    453723124 6574

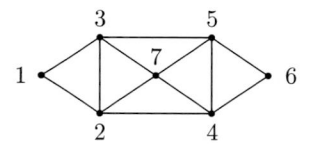

- If this graph had been missing edge 37, making 3 and 7 odd vertices, we
  would temporarily add it in, develop the above circuit, and then rearrange it
  to start at the first 7 and end at 3, yielding the Eulerian trail 723124657453.

We can generalize the procedure just used to prove that Euler's necessary
condition is also sufficient. This was first proved (much in the way we will)
about a century and a half after Euler first asserted it.

Theorem 8.1.2: *Sufficient Condition for Eulerian Circuits & Trails*
   a) *If all vertices in a connected graph are even, then the graph has an Eule-
      rian circuit, starting at any vertex.*
   b) *If all vertices except two in a connected graph are even, then the graph
      has an Eulerian trail starting and ending at the odd vertices.*

Proof:
- We'll prove the first part using a breakout-expansion procedure that will
  produce an Eulerian circuit in any graph satisfying the condition.
- Let G be a connected graph with even vertices. Choose any vertex v and
  move along a trail of unused edges until there is no exit from a vertex. This
  must be v, because every other vertex entered along the trail can also be
  departed, having an even degree. This trail is thus a circuit, which can be
  circumnavigated from any of its vertices.

· If a vertex u in the circuit has an unused edge, circle around the circuit from u and then break out along this new edge. Continue on to create an expanded circuit with edges not used earlier, starting and ending at u.

· Repeat this process as long as the expanding circuit has a protruding unused edge. Eventually (since G is finite), the circuit won't be further expandable. This final circuit C will be an Eulerian circuit.

· To see this, let e be any edge in G and let w be one of its endpoints. Because G is connected, there is a trail $v_0 v_1 \cdots v_n$ from $v = v_0$ to $v_n = w$ (see Exercise 19). The first edge $v_0 v_1$ of this trail must be in C because C has no protruding edges at v_0. Similarly, every successive edge out to w along this trail must be in C, and so C passes through w. Since C has no protruding edges, e must be in C. ✓

· To handle the case where there are two odd vertices, first note that the argument just made holds for multigraphs as well as graphs.

· So, connect the two odd vertices by a temporary edge, which may create a multigraph. Then an Eulerian circuit exists starting at either of these vertices. Since this temporary edge is in the circuit, we can use it as the final edge in the circuit. Removing this temporary edge, our Eulerian circuit becomes an Eulerian trail through the original graph. ∎

EXERCISE SET 8.1

Exercises 1–3: Königsberg Bridge Problem
The following relate to Euler's solution of the Königsberg Bridge Problem.

8.1.1. *Counting Possible Tours*
Let landmasses be denoted by A, B, C, D, and connecting bridges by $a, b, c, d,$ e, f, g, as in Euler's solution of the *Königsberg Bridge Problem*.

  a. A 7-bridge tour can be represented by a 15-letter word, 8 capital letters alternating with 7 lowercase letters. How many such words are there?

  b. A 7-bridge tour can also be denoted by an 8-letter word with all capital letters. How many such words are there?

8.1.2. *Bridge Tours with Even and Odd Regions*
  a. Prove Euler's claim that if a region X has $2n + 1$ connecting bridges that are crossed exactly once, then a tour sequence of capital letters representing regions contains $n + 1$ occurrences of X.

  b. Prove Euler's claim that if a region X has $2n$ connecting bridges that are crossed exactly once, then a tour sequence of capital letters representing regions contains n occurrences of X, unless it is a starting region, in which case it is $n + 1$.

  c. Explain why a bridge tour must visit one more region than the number of bridges, and then tell why having more than two odd regions means that no bridge tour is possible.

8.1.3. *Königsberg Bridge Problem Extension*

a. How many new bridges must be added to the Königsberg setting to make
a complete bridge-crossing circuit possible? Where should the new bridges
be placed? Illustrate your solution with a multigraph like in Example 2.

b. Euler also considered the bridge setup in Figure 8.3 in his 1736 paper.

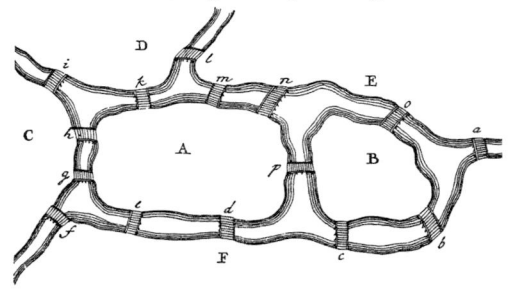

Fig. 8.3 Euler's new bridge setup

Can this set of 15 bridges $\{a, b, c, d, e, f, g, h, i, k, l, m, n, o, p\}$ connecting
regions A, B, C, D, E, F be traversed without repetition? Develop a multigraph chart and diagram like in Examples 1 and 2 to support your answer.

c. To verify your answer in part *b*, modify the algorithm presented in Example 5 to develop a trail/circuit over the edges of your multigraph.

8.1.4. *Shared Degrees in a Graph*

a. Prove that every graph with two or more vertices has at least two vertices
that share the same degree.

b. Apply part *a* to numbers of handshakes made by guests at a party.

8.1.5. *The Handshake Lemma*

a. Reformulate the *Handshake Lemma* (*Proposition* 1) for graphs with vertices and edges instead of for landmasses and bridges. How do you think
this lemma got its name? (See Exercise 4b.)

b. Prove your graph-theoretic formulation of the *Handshake Lemma*.

c. Explain why the *Handshake Lemma* implies that there are an even number of odd vertices in a graph.

Exercises 6–8: True or False

Are the following statements true or false? Explain your answer.

8.1.6. Euler's method of determining whether a tour of bridges is possible or
not (see Example 1) is conclusive.

8.1.7. All cycles are circuits.

8.1.8. A maximal trail through a connected graph is an Eulerian trail.

Exercises 9–12: Graph-Theoretic Models

The following problems explore using graphs to model real-life situations.

8.1.9. *Traversing City Streets*

Snowville has had a winter storm and sent its snowplow out at midnight to
clear both sides of each street in the town. Discuss what following an efficient

plowing route is in terms of graph theory, identifying what the vertices and edges are as well as the potential routes.

8.1.10. *Round-Robin Letters*
A group of six friends keeps in touch by each sending news of themselves to the next person, removing any letter they previously sent and adding a new one. Model this phenomenon with a graph, explaining what the vertices and edges represent. How many essentially different round-robin setups are possible?

8.1.11. *Running the Bases*
A baseball field is outlined as shown. How should the team mascots run the bases and along the foul lines and warning track in one pass, without retracing their steps? Can they finish at home plate?

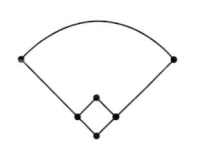

8.1.12. *Chalking a Tennis Court*
The outline of a tennis court is shown with a dashed line where the net will be placed. Can the outline be chalked without stopping and starting at different points? Explain how you would chalk the court to have the fewest stops and starts.

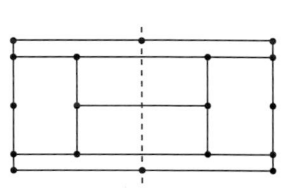

Exercises 13–19: Edges, Trails, Paths, and Cycles
The following explore relations between edges, trails, and cycles in graphs.

8.1.13. *Constructing a Graph from Given Conditions*
Draw a graph G satisfying the following conditions, if possible. If none exist, explain why not.
  a. 4 vertices with degrees 2, 2, 1, and 1.
  b. 4 vertices with degrees 3, 2, 2, and 1.
  c. 5 vertices and 8 edges.
  d. 5 vertices and 11 edges.
  e. 5 vertices with degrees 4, 3, 2, 2, and 1.
  f. 5 vertices with degrees 4, 3, 3, 2, and 1.
  g. 5 vertices with degrees 4, 3, 2, 1, and 0.

8.1.14. *Trails and Paths*
  a. Prove that every path is a trail.
  b. Prove that every walk between two vertices can be converted into a path.
  c. Prove that any two vertices in a connected graph with n vertices can be connected by a path of length less than n.

8.1.15. *Trees*
A graph is a *tree* if and only if there is exactly one path joining each pair of vertices. A *spanning tree* of a graph G is a tree subgraph of G containing all of G's vertices and some of its edges.
  a. Prove that a connected graph is a tree if and only if it contains no cycles.
  b. Prove that a tree has at least two vertices of degree one.
  c. Prove that a connected graph always contains a spanning tree. Illustrate this with Exercise 12's graph.

8.1.16. *Cycles in an Eulerian Graph*
Prove that a connected graph is Eulerian if and only if it can be partitioned into cycles.

8.1.17. *Number of Cycles in a Connected Order-4 Graph*
Consider cycles identical if and only if they produce the same loop, in either direction. Construct connected graphs G on 4 vertices satisfying the following:
a. G has exactly one cycle.
b. G has exactly three cycles.
c. G has no cycles.
d. How many distinct cycles does G have if all pairs of vertices are adjacent, i.e., is a *complete* graph?

8.1.18. *Number of Cycles in a Connected Order-5 Graph*
Let G be a connected graph with 5 vertices.
a. Show that G must have at least 4 edges.
b. Show that G has no cycles if it has exactly 4 edges.
c. Show that G has at least one cycle if it has more than 4 edges.

8.1.19. *Number of Cycles in a Connected Graph*
Let G be a connected graph with n vertices and m edges. Prove the following:
a. $m \geq n - 1$.
b. $m < n$ (i.e., $m = n - 1$) if and only if G contains no cycle, i.e., is *acyclic*.
c. If $m = n$ then G contains exactly one cycle.
d. If $m > n$ then G contains more than one cycle.
e. Explain why the converses of c and d are also true.

Exercises 20–22: Trails and Circuits
The following problems explore trails and circuits in graphs.

8.1.20. *Maximal Trails and Circuits*
List a maximal trail for each of the following graphs. Is your trail Eulerian? Is it a circuit?

a. 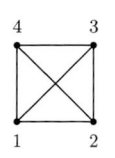 b. 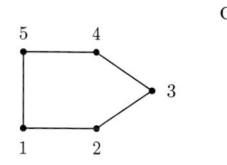 c. 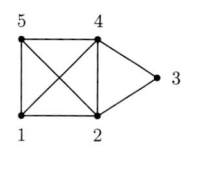 d.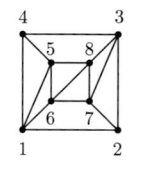

8.1.21. *Eulerian Trails and Circuits*
Decide for each of the following graphs whether it contains an Eulerian circuit or an Eulerian trail. If it does, use the algorithm of Example 5 to determine it. If it doesn't, tell why not.

a. 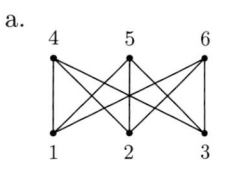 b. 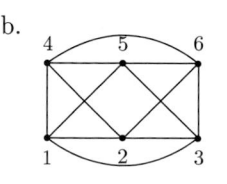 c.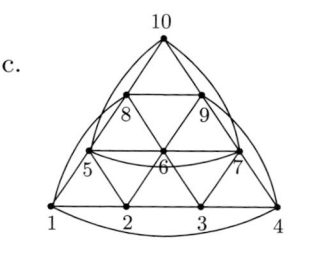

8.1.22. *Fleury's Algorithm* produces an Eulerian circuit/trail as follows: *Begin at any vertex (at an odd vertex if two such are present) and proceed to a connected vertex along an existing edge, removing it as you proceed. Choose an edge so that the next vertex in the trail has an outgoing edge, until this is not possible.* Show that this procedure produces an Eulerian circuit/trail for the following graphs. Compare your trail with what was found earlier.

   a. The graph of Example 5.

   b. The graph of Exercise 21.b.

   c. The graph of Exercise 21.c.

8.2 Hamiltonian Paths

Graphs with Eulerian trails can be completely traced in one continuous motion. Such trails include every edge of the graph without duplication. They also touch all of the graph's vertices. But what if this is all we want, a path that visits every vertex, whether or not all edges are traversed? This is the sort of subgraph one would want for traveling to every place in a graph without taking every road leading to those places.

    Hamilton's *icosian game* was like this. It leads us into a second area of *Graph Theory*.

8.2.1 *Hamilton's Icosian Game*

William Rowan Hamilton, pictured in Figure 8.4, is Ireland's most important mathematician. His brilliance was recognized when he was appointed both Professor of Astronomy and Royal Astronomer of Ireland while still an undergraduate at Trinity College, Dublin. He made valuable contributions to physics and mathematics, introducing quaternions in 1843 as the noncommutative system of algebra needed for three-dimensional geometry. Ten years earlier he had proposed an algebraic approach for making various number extensions more rigorous (see Section 6.4).

Fig. 8.4 Hamilton

    In 1857 Hamilton invented a puzzle that he called his *icosian game*, which involved taking a trip to 20 places around the globe along various edges. This is pictured in two dimensions by Hamilton's diagram in the next example.

✦**Example 8.2.1**

    Determine a cycle through all the vertices of the dodecahedron graph at the right, if $BCDFG$ is the beginning of the path. (This is the puzzle Hamilton supplied when registering the game.)

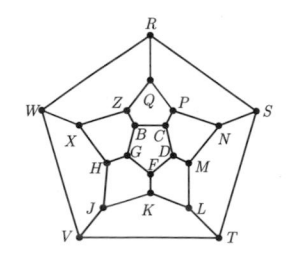

Solution
- Hamilton noted that we can simply continue with the consonants in alphabetical order, obtaining cycle $BCDFGHJKLMNPQRSTVWXZB$.
- However, he also pointed out that one could continue from $BCDFGH$ by moving to X, creating a different cycle.

Hamilton's game was an outgrowth of research into what he called his icosian calculus, an algebra associated with cycles passing through the 20 vertices of a dodecahedron. The game was a commercial failure, but the idea of traversing graphs in this way gave rise to the term *Hamiltonian cycle*, even though others before him had explored traversing graphs in a similar way.

8.2.2 Hamiltonian Paths and Cycles

Definition 8.2.1: Hamiltonian Paths and Cycles
  a) A **Hamiltonian path** in a graph is a path that includes every vertex exactly once.
  b) A **Hamiltonian cycle** in a graph is a cycle that is a Hamiltonian path.
  c) A **Hamiltonian graph** is a graph containing a Hamiltonian cycle.

The graph for the icosian game contains a variety of Hamiltonian paths and cycles (see Exercise 1). Here are a few examples for other graphs.

✤**Example 8.2.2**
Do the following graphs have a Hamiltonian path or cycle? Identify them if they exist.

a) 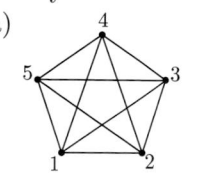 b) 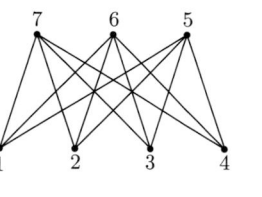 c)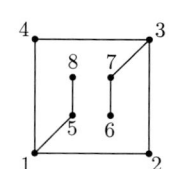

Solution
  a) A *complete graph* is one that includes every possible edge. This is the complete graph on 5 points, denoted by K_5. Having lots of edges makes it easy to produce a Hamiltonian cycle: 12345 is one such, but there are others as well (see Exercise 15b).
  b) This is the *complete bipartite graph* $K_{3,4}$. It contains the Hamiltonian path 1726354 (and others), but no Hamiltonian cycle (see Exercise 17b).
  c) This graph has no name, but if edges 56, 78, 26, and 48 were added, it would be the planar graph for a cube (see Section 8.3). This graph has no Hamiltonian path. We'll come back to this graph in later examples.

As the size of a graph increases, it gets more difficult to tell whether it has a Hamiltonian path or cycle. A simple necessary and sufficient condition like there was for the existence of Eulerian trails and circuits would help. Unfortunately, we don't know of any such condition beyond the definition. We can say some things separately, however, about when graphs do or do not have Hamiltonian paths and cycles.

8.2.3 *Sufficient Conditions for Hamiltonian Paths*

One thing that makes it likely for a graph to be Hamiltonian is its having lots of edges, so that there are many ways to get from vertex to vertex. We already saw in Example 2a that complete graphs are Hamiltonian. But even if all the diagonals were deleted from K_n, making each vertex of degree 2 instead of $n-1$, the resulting polygon would still have a Hamiltonian cycle, so Hamiltonian graphs don't need high-degree vertices. Can we say in general, though, how many edges *suffice* to *guarantee* a Hamiltonian cycle? *Dirac's Theorem* gives us a first criterion: if each vertex is connected to at least half of all vertices, then it has a Hamiltonian cycle.

Theorem 8.2.1: *Dirac's Theorem for Hamiltonian Cycles* (1952)
  In a connected graph G on $n \geq 3$ vertices, if the degree of each vertex is at least $n/2$, then G has a Hamiltonian cycle.

Proof:
  Let P be a path $v_1, v_2, \ldots, v_m$ of maximum length m in G. We can schematize P as follows:

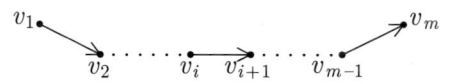

  Claim 1: Vertices v_1 and v_m are each adjacent to at least $n/2$ vertices in P.
  · Because P is of maximum length, neither v_1 nor v_m is adjacent to any vertex outside P—otherwise P could be enlarged by such a connection.
  · Thus, both v_1 and v_m are connected to at least $n/2$ vertices in P. ✓
  Claim 2: Some v_i is adjacent to v_m with v_{i+1} adjacent to v_1, $1 \leq i < m$.
  · If no pair of vertices v_i, v_{i+1} satisfies this condition, then there are at least $n/2$ path vertices v_{i+1} adjacent to v_1 such that v_i is not adjacent to v_m. But v_m is not adjacent to itself, either, so more than $n/2$ path vertices are not adjacent to v_m, which contradicts *Claim 1*. ✓
  Claim 3: P can be converted into a cycle P'.
  · Let v_i, v_{i+1} be a pair of vertices satisfying the adjacency condition of *Claim 2*. We'll use them to convert P into a cycle: proceed from v_1 along P to v_i, then go to v_m, continue backward along P to v_{i+1} and then go to v_1. This gives cycle $P' = v_1, v_2, \ldots, v_i, v_m, v_{m-1}, v_{m-2}, \ldots, v_{i+1}, v_1$. ✓

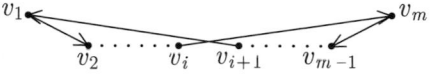

  · *Remark*: if $i = m - 1$, no genuine conversion will occur—P is the cycle.
  Claim 4: P' contains all vertices of G, making it a Hamiltonian cycle.
  · Suppose P' misses some vertex of G. Because G is connected, some missed vertex v must be adjacent to a vertex v_j in P'.
  · We can then enlarge P' by starting with v and circling around P', starting at v_j. This contradicts P's being a path of maximum length.
  · So P' must include all of G's vertices: P' is a Hamiltonian cycle. ∎

A natural question to ask is whether the degree specified in *Dirac's Theorem* is optimal. Would it suffice, for instance, to have each vertex connected to half of the *remaining* vertices, to have $\deg(v) \geq (n-1)/2$? The next example answers this in the negative.

✤ Example 8.2.3

Show that a graph G on n vertices may fail to have a Hamiltonian cycle if not all vertices have degree at least $n/2$.

Solution

$K_{3,4}$, which has no Hamiltonian cycle, shows this. It has 7 vertices, but some have degree $3 < 7/2$. ✓

Addendum: by connecting two bottom vertices, a Hamiltonian cycle becomes possible. For example, adding edge 23, a Hamiltonian cycle is 27164532.

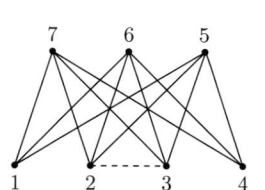

We can weaken the condition of *Dirac's Theorem* in another way. Instead of requiring that *each vertex* have degree at least $n/2$, we might require that *pairs of vertices* have a degree sum of at least n. Adjacent vertices, being directly connected by an edge, need not satisfy such a condition, but nonadjacent ones probably should, to make sure there are enough edges available to connect them. This is the intuition behind *Ore's Theorem*.

Theorem 8.2.2: *Ore's Theorem for Hamiltonian Cycles* (1960)

In a connected graph G on $n \geq 3$ vertices, if each pair of nonadjacent vertices has a degree sum of at least n, then G has a Hamiltonian cycle.

Proof:
- We'll prove the logically equivalent partial contrapositive of this result.
- Suppose G is a connected non-Hamiltonian graph on $n \geq 3$ vertices (e.g., the solid-line graph shown below).

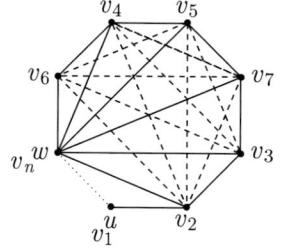

- Expand G to a maximal non-Hamiltonian graph H by successively adding edges (the dashed edges) between nonadjacent vertices.
- Let u and w be two vertices still nonadjacent in H. Since H is a maximal non-Hamiltonian graph, adding edge uw will create a Hamiltonian cycle C.

- Deleting uw from C gives a Hamiltonian path $v_1, \ldots, v_i, \ldots, v_n$ in H, with $u = v_1$ and $w = v_n$.
- For each of the $n-1$ pairs (v_i, v_{i+1}), at most one of $\{v_{i+1}v_1, v_iv_n\}$ can be in H; otherwise, as in the proof for *Dirac's Theorem*, H will contain the Hamiltonian cycle $v_1, v_2, \ldots, v_i, v_n, v_{n-1}, \ldots, v_{i+1}, v_1$.
- But then the degree sum in H of $v_1 = u$ and $v_n = w$ is at most $n-1$.
So not all nonadjacent vertices in G have a degree sum of at least n. ∎

❖**Example 8.2.4**

What does *Ore's Theorem* say about the graphs below, from Example 2?

a) 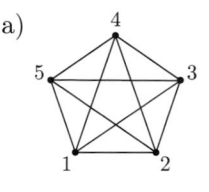   b) 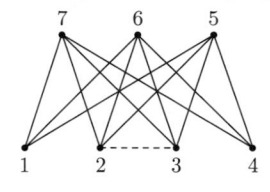   c)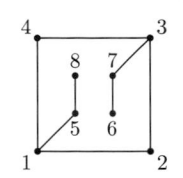

Solution

a) Complete graphs like K_5 have all possible edges. The degree sum of each pair here is 10. In general, for K_n ($n \geq 3$), the degree sum is $2(n-1) > n$. Thus, *Ore's Theorem* implies these graphs are Hamiltonian.

b) $K_{3,4}$ doesn't satisfy the condition of *Ore's Theorem*. But then neither does its expansion with the added dashed edge ($\deg(1) + \deg(4) = 3 + 3 = 6 < 7$), and it *does* have a Hamiltonian cycle: for example, 27164532. This illustrates the fact that *Ore's Theorem*, like *Dirac's Theorem*, gives a sufficient but not a necessary condition for being Hamiltonian.

c) This graph also doesn't satisfy the condition of *Ore's Theorem*, but again, that doesn't warrant us to conclude that it lacks a Hamiltonian cycle, though that's the case here. For more on this, see Example 6.

We saw in Example 3 that weakening the condition in *Dirac's Theorem* to vertices having degree $(n-1)/2$ no longer guarantees the existence of a Hamiltonian cycle. A similar thing is true for *Ore's Theorem*—the same graph $K_{3,4}$ shows that if the degree sum for a nonadjacent pair falls below n, the graph need not be Hamiltonian. However, this graph does have a Hamiltonian path, as noted in Example 2: 1726354.

This leads us to ask whether these weakened versions suffice for a graph to have a Hamiltonian *path*. This time, the answer is yes. We'll state these without proofs (see Exercise 20).

Theorem 8.2.3: *Degree-Sum Condition for Hamiltonian Paths*
If each pair of nonadjacent vertices of a connected graph has a degree sum of at least $n - 1$, then the graph contains a Hamiltonian path.

Corollary 8.2.3.1: *Degree Condition for Hamiltonian Paths*
If each vertex of a connected graph has degree at least $(n - 1)/2$, then the graph has a Hamiltonian path.

❖**Example 8.2.5**

In-Depth TV News schedules a special report each night of the week during election season. If no reporter is featured two nights in a row, how many different weekly schedules for the *reports* are possible? Could the same *reporter-schedule* be run the following week if desired?

Solution
- Let seven vertices represent the various special reports for the week, and connect them with an edge if they can be presented on successive nights. Nonadjacent vertices will represent reports done by the same reporter.
- Since each reporter has at most four reports in a week, the degree of each vertex is at least 3. Thus, there is a Hamiltonian path through the graph. This path will represent a viable schedule for the week. ✓
- There are at least $144 = 4 \cdot 3 \cdot 3 \cdot 2 \cdot 2 \cdot 1 \cdot 1$ schedules (two reporters) and at most $5040 = 7!$ schedules (for seven reporters) possible for the week's reports. ✓
- If one reporter for the organization is always responsible for four reports, the *reporter-schedule* cannot be repeated a second week without having that reporter featured on successive nights. Otherwise, the schedule can be repeated weekly. ✓

8.2.4 Necessary Conditions for Hamiltonian Paths

It's often very difficult to determine whether a graph contains a Hamiltonian path or cycle. In fact, this belongs to a class of problems whose known algorithmic solution takes much longer to solve as the number of vertices increases. The related *Traveling Salesman Problem*—find the shortest Hamiltonian cycle in a graph whose edges have given lengths—has been seriously investigated since the 1930s and is similarly difficult. Our description of the problem's difficulty is vague, but a more precise formulation would take too long to outline, so we'll leave it to be explored independently.

This doesn't mean that there are no necessary conditions for determining whether a graph is Hamiltonian, only that these don't make the job of deciding this much easier in general.

One simple set of joint criteria that must be satisfied by a Hamiltonian graph is that provided by the relevant definitions. If a graph G is Hamiltonian, then it has a Hamiltonian cycle, which means that G must have a spanning subgraph H with the following properties:
1. H has the same vertices as G;
2. H is connected;
3. H has the same number of edges as vertices; and
4. every vertex of H has degree 2.

So, if G is Hamiltonian, it must be possible to drop edges from G to get a connected subgraph H with the right number of vertices and vertex degrees.

Theorem 8.2.4: *Necessary Condition for a Hamiltonian Cycle*
If G is a Hamiltonian graph, then it contains a connected spanning subgraph H having the same number of edges as vertices, all of degree 2.

Proof:
See Exercise 21a.

What we're really interested in here is the contrapositive of this theorem. If satisfying some of these four criteria makes another one impossible, the graph will be non-Hamiltonian. On the other hand, if a graph jointly satisfies all of these conditions, it must be Hamiltonian, i.e., this necessary condition is also a sufficient condition (see Exercise 21bc). This fact is not an advance on the definition, however—it amounts to showing that a graph is Hamiltonian because it contains a Hamiltonian cycle.

✤ **Example 8.2.6**

Do the following graphs have a Hamiltonian cycle?

a) 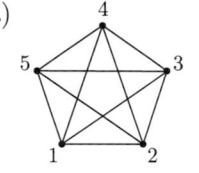   b) 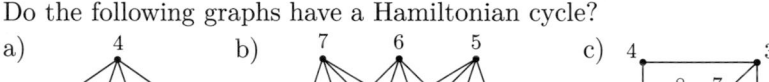 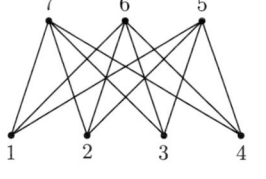   c)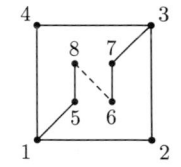

Solution

These are the same graphs we looked at in Example 2.

a) K_5 satisfies the conditions of *Theorem* 4: if we delete all the diagonals, a Hamiltonian cycle around the outside remains. ✓

b) $K_{3,4}$ has no Hamiltonian cycle. There are 12 edges in this graph, and four must be deleted from the bottom vertices so they'll each have degree 2. This leaves eight edges for the top three vertices, so two more edges must be dropped, which will reduce the degree of two bottom vertices below what's needed, so the criteria of *Theorem* 4 can't be satisfied. ✓

c) This graph can't be reduced to a Hamiltonian subgraph, either, for several reasons. Most simply, vertices 6 and 8 both have degree 1. ✓
However, even if we added the dashed edge 68 to remedy this, the graph would still not be Hamiltonian. Reducing the degrees of vertices 1 and 3 to 2 means two of $\{2, 4, 5, 7\}$ would still have degree 1. ✓
Adding edges 26 and 48, though, would give a Hamiltonian graph. ✓

Our method of solving parts *b* and *c* in the last example is a strategy that can be used for showing that some graphs are not Hamiltonian. To determine whether a graph is Hamiltonian, compare the vertices' degrees and the total number of edges in the graph to what's required in the cycle's subgraph. Then remove edges/reduce degrees while leaving the graph connected, with enough edges and sufficient degrees. Continue until you find a Hamiltonian cycle or until you can see that the procedure must violate connectivity or the degree requirement. Note that this process may not be very efficient when there are many ways to delete edges. But for small graphs done by hand, it's worth trying. Along the way, it also simplifies the graph needing a Hamiltonian path.

✤ **Example 8.2.7**

Determine whether the following graphs have Hamiltonian paths.

a)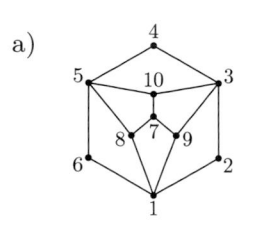

b)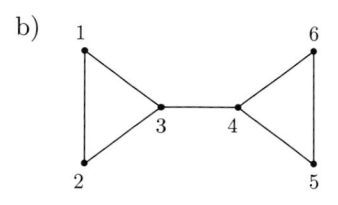

Solution

a) This hexagonal graph doesn't satisfy the sufficiency condition of *Theorem 3*, because the degree sum of all pairs of vertices is less than 9, but that still doesn't tell us that it has no Hamiltonian path. However, this graph is a version of the complete bipartite graph $K_{4,6}$, which has no Hamiltonian path (see Exercise 17d).

· One way to see this is to assign two colors to the vertices so that adjacent vertices have different colors. Vertices $\{1, 3, 5, 7\}$ can be colored red and vertices $\{2, 4, 6, 8, 9, 10\}$ colored blue. A Hamiltonian cycle or path through this graph must then have alternating colors, either with equal numbers of each color or one more of one color than the other. As there are 6 blues and only 4 reds, no Hamiltonian path is possible.

b) Coloring vertices doesn't help here to decide whether this barbell graph has a Hamiltonian path: three colors are needed if adjacent vertices are different colors.

· But bridge edge 34 must be crossed twice in any attempt to cycle around the graph, so no Hamiltonian cycle is possible.

· A Hamiltonian path is possible, though: for instance, 123456.

· Alternatively, notice that 3 and 4 are each vertices whose removal (along with the attached edges) would make the remaining subgraph disconnected. This makes a Hamiltonian cycle impossible (see *Theorem 5*).

Definition 8.2.2: *Graph Component, Cut Point, Bridge*

a) A ***component*** of a graph is a maximal connected subgraph.

b) A ***cut point*** in a connected graph is a vertex whose removal creates a disconnected graph/a graph with more than one component.

c) A ***bridge*** in a connected graph is an edge whose removal creates a disconnected graph/a graph with more than one component.

Theorem 8.2.5: *Hamiltonian Paths, Cut Points, and Bridges*

If a graph is Hamiltonian, then it has no cut point and no bridge.

Proof:

See Exercise 21d.

This is a rather specialized necessary condition, but it gives another way to show that some graphs have no Hamiltonian cycle. It is not sufficient, however—the graph of Example 7a is not Hamiltonian, though it has no cut point or bridge. Moreover, graphs with cut points and bridges can still have Hamiltonian paths, as Example 7b shows.

We'll state a generalization of *Theorem 5* (where $k = 1$) without proof.

Theorem 8.2.6: *Component Condition for Hamiltonian Cycles*
If removing k vertices along with their edges from a graph G creates a subgraph H with more than k components, then G is not Hamiltonian.

✤**Example 8.2.8**
Show that the graph below is not Hamiltonian.

Solution

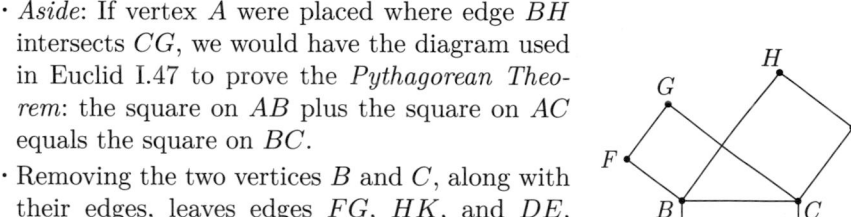

· *Aside*: If vertex A were placed where edge BH intersects CG, we would have the diagram used in Euclid I.47 to prove the *Pythagorean Theorem*: the square on AB plus the square on AC equals the square on BC.
· Removing the two vertices B and C, along with their edges, leaves edges FG, HK, and DE, three components of the resulting graph.
· Thus, by *Theorem 6*, this graph has no Hamiltonian cycle. It does have an easy-to-find Hamiltonian path, however (again, though *Theorem 3*'s condition isn't satisfied): $GFBDECKH$. ✓
· Adding vertex A, though, where suggested, would permit a Hamiltonian cycle: CKHAGFBDEC.

EXERCISE SET 8.2

Exercises 1–3: Hamiltonian Games
The following problems relate to Hamilton's icosian game and other games.

8.2.1. *The Icosian Game*
Find a Hamiltonian path through the icosian game's graph (see Example 1) given each of the following initial segments. (These are examples Hamilton gave for potential games.)

a. Cycle from $BCPNM$
b. Cycle from $JVTSR$
c. Cycle from $LTSRQ$

d. Path from BCD, ending at T
e. Path from BCD, ending at W
f. Path from BCD, ending at J

g. Path from BCD, missing P and ending at F
h. Path from BCD, missing L and ending at a point nonadjacent to L

8.2.2. *Traversing a Grid Graph*
A *grid graph* is an $m \times n$ rectangular array of lattice points with line segments connecting all adjacent horizontal and vertical points.

a. Describe traversing a grid graph to visit each lattice point exactly once in graph-theoretic terms.
b. What is the least number of colors needed to color a grid graph using different colors for adjacent points?

c. Does a 2×3 grid graph have a Hamiltonian path? A Hamiltonian cycle? What about a $2 \times n$ graph for positive integers n?

d. Does a 3×3 grid graph have a Hamiltonian path? A Hamiltonian cycle? Which $3 \times n$ grid graphs have a Hamiltonian cycle?

e. Which $m \times n$ grid graphs have a Hamiltonian cycle? Prove your conjecture.

8.2.3. *The Knight's Tour Problem*

A *knight's tour* is a sequence of chessboard moves in which a knight visits every square once, moving in the usual L-jump manner (two squares one way, then one sideways). A *closed knight's tour* ends where it began.

Note: Euler explored this in the late 1750s, a century before Hamilton's game.

a. Explain how to model a sequence of knight's moves with a graph. What do the vertices and edges represent? In terms of such a graph, what is a knight's tour? a closed knight's tour?

b. For a chessboard of size 3×3, can a knight's tour be made, starting from any square on the board? Can a closed tour be made? Explain.

c. For a chessboard of size 4×4, can a knight's tour be made, starting from any square on the board? Can a closed tour be made? Explain.

d. Trace out a closed knight's tour on a normal 8×8 chessboard.

e. Look up online to see which $m \times n$ chessboards permit a knight's tour.

Exercises 4–7: Graph-Theoretic Models

The following problems use graphs to model real-life situations.

8.2.4. *Solving a Rubik's Cube*

a. Solving a scrambled Rubik's Cube is done by going through a sequence of quarter turns until all faces have a solid color. A Rubik's Cube circuit is a sequence of all the positions a cube can have, without repeats, ending with the initial face arrangement. Describe this in graph-theoretic terms.

b. Look up whether such a circuit exists and how many positions it contains. What is the maximum number of moves needed to solve a Rubik's Cube?

8.2.5. *The Itinerant Student*

Taking a gap year before attending college, Anneke wants to travel to a number of locations without backtracking to any place already visited, until she returns home. What is she looking to do in graph-theoretic terms?

8.2.6. *Security Sites*

A night watchman is responsible for checking a number of buildings once each night. Describe this job in graph-theoretic terms.

8.2.7. *Making Deliveries*

A grocery store delivery truck goes out to deliver a number of orders. Describe the route taken by the truck in graph-theoretic terms. What else about the route might the store be interested in?

Exercises 8–10: True or False

Are the following statements true or false? Explain your answer.

8.2.8. If a graph with n vertices has a degree sum for all pairs of nonadjacent vertices of at least $n - 1$, then it has a Hamiltonian cycle.

8.2.9. If every vertex of a connected graph has degree 2 except for two vertices of degree 1, then it has a Hamiltonian path.

8.2.10. Connected graphs exist that are ...
a. both Eulerian and Hamiltonian.   c. Hamiltonian but not Eulerian.
b. not Eulerian nor Hamiltonian.   d. Eulerian but not Hamiltonian.

Exercises 11–14: Hamiltonian Paths and Cycles

Find Hamiltonian cycles or paths in the following graphs. If none is possible, explain why not.

8.2.11. *Graphs with Squares*

a. 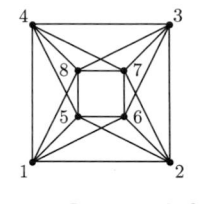   b. 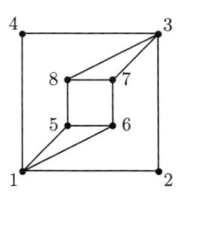   c.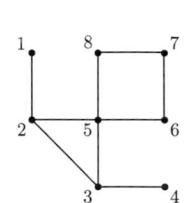

8.2.12. *Diamond Graphs*

a. 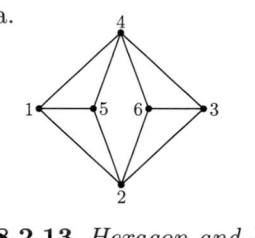   b. 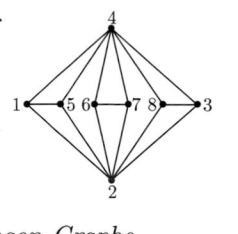   c.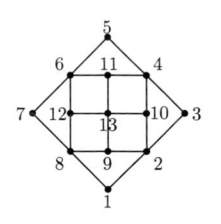

8.2.13. *Hexagon and Pentagon Graphs*

a. 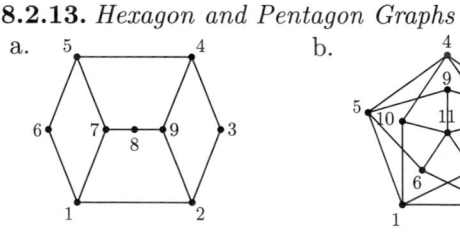   b. 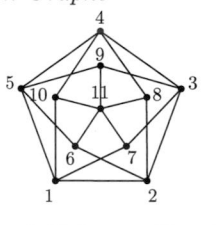   c.

8.2.14. *Hexagon, Heptagon, and Octagon Graphs*

a. 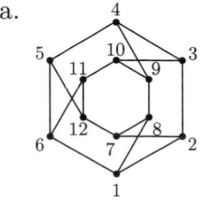   b. 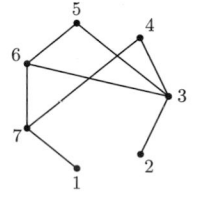   c.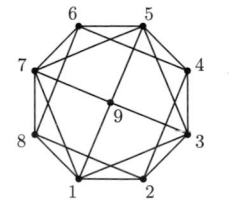

Exercises 15–17: Complete Graphs and Bipartite Graphs

*A **complete graph** K_n on n vertices is a graph having n vertices and all possible edges between vertices.*

*A **bipartite graph** is a graph whose vertices lie in two disjoint sets (parts), with no vertices in either set adjacent to ones in the same set.*

*A **complete bipartite graph** $K_{m,n}$ is a bipartite graph whose m vertices in the one part are connected to all n vertices in the other part.*

8.2.15. *Disjoint Cycles in Complete Graphs*
Consider two cycles disjoint if and only if they share no undirected edges.
  a. How many pairwise disjoint Hamiltonian cycles does K_3 have? K_4?
  b. How many pairwise disjoint Hamiltonian cycles does K_5 have? K_6? K_7?
  c. Based on your answers to parts *a* and *b*, formulate a conjecture about the number of pairwise disjoint Hamiltonian cycles in K_n for some category of positive integers n. Check your conjecture for another value of n. Can you prove your conjecture?

8.2.16. Draw the following complete bipartite graphs.
  a. $K_{2,2}$ c. $K_{2,4}$ e. $K_{3,3}$ g. $K_{3,5}$
  b. $K_{2,3}$ d. $K_{2,5}$ f. $K_{3,4}$ h. $K_{4,4}$

8.2.17. *Hamiltonian Paths and Cycles in Complete Bipartite Graphs*
For $n \geq 2$, prove the following:
  a. $K_{n,n}$ is Hamiltonian. How many distinct Hamiltonian cycles does it have? Check your formula for $K_{2,2}$ and $K_{3,3}$, and tell how many cycles $K_{4,4}$ has.
  b. $K_{n,n+1}$ has a Hamiltonian path but no Hamiltonian cycle. How many distinct Hamiltonian paths does it have? Check your formula on $K_{2,3}$. How many paths does $K_{3,4}$ have? Relate these to cycles in $K_{3,3}$ and $K_{4,4}$.
  c. $K_{n,n+m}$ has no Hamiltonian path for $m \geq 2$. Illustrate this with $K_{2,4}$.
  d. Redraw Example 7a's graph to show that it is a bipartite graph. What does this imply about its having a Hamiltonian path?
  e. $K_{n,n+m}$ has a Hamiltonian path if and only if $m \leq 1$, and it has a Hamiltonian cycle if and only if $m = 0$.

Exercises 18–21: Conditions for Hamiltonian Graphs
The following problems explore necessary and sufficient conditions for Hamiltonian paths and cycles.

8.2.18. *Dirac's Theorem*
Explain why *Dirac's Theorem* is a special case of *Ore's Theorem*, i.e., why it follows as an immediate corollary of the latter.

8.2.19. *Ore's Theorem*
  a. Show that *if all nonadjacent vertices of a graph have degree sums of at least $n-1$ (or if every vertex has degree at least $(n-1)/2$), then the graph is connected.* Consequently, the antecedent conditions in *Theorems 1–3* can be "weakened" by dropping connectivity.
  b. Explain why successively adding edges to a connected non-Hamiltonian graph with at least three vertices will eventually create a maximal non-Hamiltonian graph, i.e., one such that adding another edge creates a Hamiltonian graph.

8.2.20. *Sufficient Conditions for Hamiltonian Paths*
  a. Prove *Theorem 3* by constructing a Hamiltonian path whose degree sum for pairs of nonadjacent vertices is at least $n - 1$.

b. Prove the *Corollary* to *Theorem* 3: *If each vertex of a connected graph has degree at least $(n-1)/2$, then the graph contains a Hamiltonian path.*

8.2.21. *Necessary Conditions for Hamiltonian Graphs*

a. Prove *Theorem* 4, which states joint necessary conditions for having a Hamiltonian cycle.

b. Prove that the necessary conditions given in *Theorem* 4 are also sufficient: *if a subgraph H of G has the same number of vertices as G, is connected, has the same number of edges as vertices, and has only vertices of degree 2, then G contains a Hamiltonian cycle.*

c. Based on your argument in part *b*, explain why the converse of *Theorem* 4 isn't very useful.

d. Prove *Theorem* 5: *Hamiltonian graphs contain no cut points or bridges.*

e. Show that the graph of Example 7a contains neither a cut point nor a bridge. What, if anything, does this say about *Theorem* 5?

8.2.22. *Cut Points, Bridges, Cycles, and Paths*

a. Prove that *an edge of a connected graph is a bridge if and only if it lies on no cycle in the graph.* Consequently, *Hamiltonian graphs have no bridges.*

b. Prove that *Eulerian graphs have no bridges.* Can they have cut points?

c. Prove that *a vertex w in a connected graph is a cut point if and only if there exist vertices u and v distinct from w such that every path connecting u and v passes through w.*

8.3 Planar Graphs

Mathematicians, scientists, and philosophers have been interested in regular planar and solid figures since ancient times. Early astronomers used circles and spheres in plotting the paths of the stars and planets because of their uniform curved shapes. In his mathematically based cosmology, Plato used regular polyhedra—convex solids whose faces are congruent regular polygons—as the shapes of the fundamental elements of earth, fire, air, and water, with a fifth one representing the universe itself. Greek mathematicians before him had discovered that there were exactly five such solids—the cube, tetrahedron, octahedron, icosahedron, and dodecahedron. Euclid included this result as *Proposition* XIII.18 in his **Elements**.

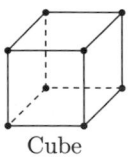
Cube
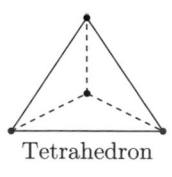
Tetrahedron
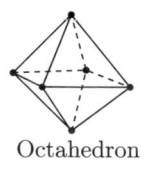
Octahedron

Icosahedron

Dodecahedron

A thousand years later, Kepler also succumbed to the magic of Platonic solids in his quasi-mystical defense of Copernican astronomy. He proposed that these figures could be nested between inscribing and circumscribing spheres in which the six known planets orbited, thus revealing God's elegant geometric blueprint for the universe.

Kepler also investigated semiregular polyhedra—convex figures whose faces are different regular polygons, similarly joined at each vertex. These had been studied earlier by Archimedes, who, like Kepler, identified all 13, though none of his extant works discuss them. We can create many of these polyhedra by truncating Platonic solids, cutting corners off in regular way. A truncated cube, for instance, has eight triangular faces (the sliced corners) and six octagonal faces (the trimmed faces). Kepler investigated these solids along with prisms and star polyhedra as he explored geometrical figures.

8.3.1 *Euler's Polyhedral Formula*

The following chart lists the number of vertices, edges, and faces for the Platonic solids and two semiregular truncated solids. What conjecture can we make based on this and similar data for other polyhedra, such as prisms? It took some time before this question was asked and answered.

Solid	Vertices	Edges	Faces
Cube	8	12	6
Tetrahedron	4	6	4
Octahedron	6	12	8
Icosahedron	12	30	20
Dodecahedron	20	30	12
Truncated Cube	24	36	14
Truncated Tetrahedron	12	18	8

Over the centuries, various properties of polyhedra had been discovered by mathematicians, including Descartes, who, in an unpublished manuscript, asserted a result about the relation between the number of faces and the number of plane angles of a polyhedron. It wasn't until 1750, however, that Euler discovered a very simple relationship between the numbers of vertices V, edges E, and faces F. That nobody had noticed it earlier surprised Euler. The formula, which can be verified in the above chart, is $V - E + F = 2$. This result, known as *Euler's Polyhedral Formula*, is considered by some as one of the most beautiful and important results in mathematics.[2]

Euler published a proof of the formula a year later, but one part of his argument was flawed. Legendre first published a correct proof toward the end of the century, but in 1813 Cauchy ingeniously reformulated Euler's result in terms of planar graphs. Thinking of a polyhedron as a surface instead of a solid, he flattened it so that its vertices, edges, and faces became a two-dimensional graph with no intersecting edges. His proof then showed that such planar graphs satisfy *Euler's Formula*.

---

[2] See Joseph Malkevitch's two 2005 AMS Feature Columns on *Euler's Polyhedral Formula* at http://www.ams.org/samplings/feature-column/fcarc-eulers-formula

The five Platonic solids are represented by the following flattened graphs. Each of these has an infinite outer face bounded by the outside edges.

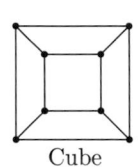 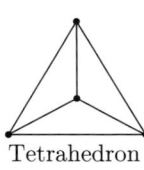  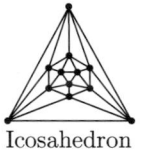 

   Cube   Tetrahedron   Octahedron   Icosahedron   Dodecahedron

8.3.2 Euler's Formula for Planar Graphs

Definition 8.3.1: *Planar Graph*
  a) *A **planar graph** is a graph that can be embedded (drawn) in a plane so that its edges intersect only at a vertex.*
  b) *A **face** of an embedded planar graph is a connected component of the plane minus the graph's boundary edges and vertices.*

Graphs can be drawn in different ways while keeping fixed the numbers of vertices and edges and the incidence relations. We'll accept without proof the fact that the number of planar faces is independent of the embedding.

✤**Example 8.3.1**
Show that the following graphs are planar and that they satisfy *Euler's Formula*. Recall that the region outside the graph's border is also a face.

a)   b)   c)

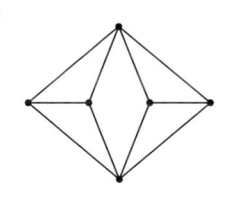 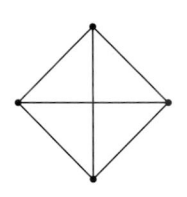 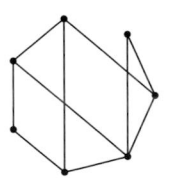

Solution
  a) This graph is already drawn in a planar fashion—all edges intersect at vertices. It has 6 vertices, 10 edges, and 6 faces, so $V - E + F = 2$.
  b) The diagonals of this graph, K_4, intersect at a non-vertex point. The graph can be redrawn, however, by looping one diagonal around the outside. Or, it can be reshaped into the tetrahedron's graph (see below). These satisfy *Euler's Formula* because $4 - 6 + 4 = 2$.
  c) This graph is also not drawn in a planar fashion. However, two of the intersecting diagonals can be redrawn around the outside, as shown below. *Euler's Formula* is satisfied for this graph because $7 - 10 + 5 = 2$.

   b)   c)

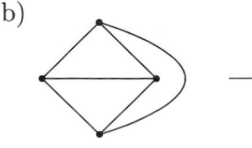  

Theorem 8.3.1: *Euler's Formula for Planar Graphs*
*If G is a connected planar graph, then $V - E + F = 2$, where V, E, and F
are the numbers of vertices, edges, and faces of its embedded graph.*

Proof:
- We'll prove this by mathematical induction on the number of edges E.
- For 0 edges, there is 1 vertex and 1 face, so *Euler's Formula* holds for this
  trivial base case: $1 - 0 + 1 = 2$. ✓
- Now suppose that *Euler's Formula* holds for any connected planar graph
  with k edges, and consider such a graph G having $E = k + 1$ edges.
- If there is a vertex of degree 1 in G, remove it, along
  with its edge, as indicated.
- This gives a connected planar subgraph G' having $V' =$
  $V - 1$ vertices, $E' = E - 1 = k$ edges, and $F' = F$ faces.
- By the induction hypothesis, $V' - E' + F' = 2$, so (adding
  1 to V' and E') $V - E + F = 2$, too. ✓

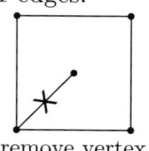
remove vertex

- If all vertices have degree greater than 1, then G has a
  bounded region surrounded by edges (see Exercise 17).
- Removing one of these edges, as shown, creates a subgraph G' with $V' = V$ vertices, $E' = E - 1$ edges, and
  $F' = F - 1$ faces.
- Again, by the induction hypothesis, $V' - E' + F' = 2$,
  so (adding 1 to E' and F') $V - E + F = 2$, too. ✓

remove one edge

- Thus, by induction, *Euler's Formula* holds for connected planar graphs of
  all sizes. ∎

This theorem gives a necessary condition for being planar, but it can't
show that a graph is not planar, if for no other reason than that we can't
identify faces for such graphs. We can revert to the definition, however.

✤**Example 8.3.2**
Show that the graphs K_5 and $K_{3,3}$ are not planar.

a)
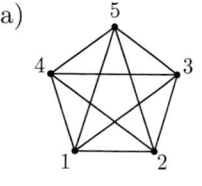

b)
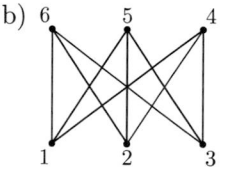

Solution
a) While K_4 was planar, K_5 is not, because it has too many intersecting
   diagonals. We won't be able to reroute these diagonals to create a planar
   graph, because an outer arc will always enclose a vertex whose diagonal
   edge still needs rerouting. The diagram below shows one unsuccessful
   attempt: edges 24 and 25 are moved, but 15 can't be drawn inside or
   outside the pentagon without crossing another edge. Arguing that all
   such attempts will fail is tedious—plus, we may overlook some possibility.
   Clearly, another criterion for nonplanarity would be helpful.

b) $K_{3,3}$ also has too many intersecting diagonals. We can redraw the graph as a hexagon with diagonals between opposite vertices, and then we can reroute one diagonal around the outside, but not both. The diagram below moves diagonal 36, but diagonal 24 can't be drawn inside or outside the hexagon without intersecting diagonal 15 or arc 36.

a) 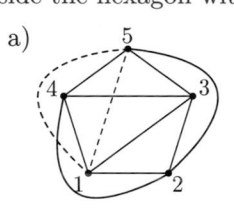   b)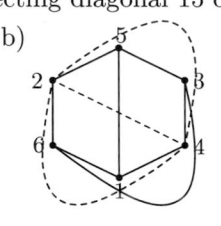

8.3.3 The Edge-Vertex Inequality for Planar Graphs

As noted above, *Euler's Formula* can't be used directly to show that graphs are not planar because the notion of a face isn't meaningful for them. But perhaps we can transform *Euler's Formula* into one involving only edges and vertices. If so, we may be able to use that to identify graphs as nonplanar.

Let's begin by considering very simple planar graphs—a cycle surrounding a face. This takes at least 3 edges, so we might initially conclude that $E \geq 3F$. But this doesn't take into consideration that cycles have both inner and outer faces; it seems that the best inequality we can get is $2E \geq 3F$, with equality holding for triangles. In fact, this *Edge-Face Inequality* holds for all planar graphs (see Exercise 14).

Tripling *Euler's Formula* gives $3V - 3E + 3F = 6$. Since $2E \geq 3F$, by replacing $3F$ with $2E$, we're increasing the left-hand side of the formula, yielding $3V - 3E + 2E \geq 6$, i.e., $3V - E \geq 6$, or $E \leq 3(V - 2)$. This gives us an *Edge-Vertex Inequality* that we can test on planar and nonplanar graphs to see whether it gives us a useful condition for deciding between them.

✦**Example 8.3.3**

Test the *Edge-Vertex Inequality* on the following graphs.

a) 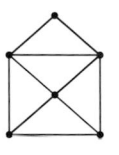   b) 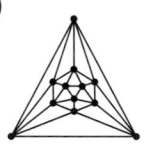   c) 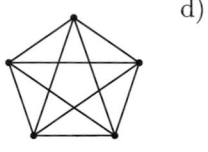   d)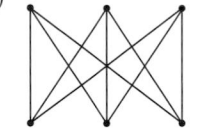

Solution

a) This graph has $E = 10$ and $V = 6$. It certainly satisfies $E \leq 3(V - 2)$, because $10 \leq 12$. This fits with the graph being planar.

b) The icosahedron's planar graph has $V = 12$ and $E = 30$, which makes it satisfy the *Edge-Vertex Inequality*: $30 \leq 3(12 - 2)$. Notice that the equality here corresponds to all 20 faces being bordered by triangles.

c) K_5 doesn't satisfy the *Edge-Vertex Inequality*: $10 \not\leq 3(5 - 2)$. Failing this inequality seems to disqualify it from being planar.

d) As we noted above, $K_{3,3}$ is also nonplanar. This graph, though, still satisfies the *Edge-Vertex Inequality*: $9 \leq 3(6-2)$. This makes the test inconclusive—the inequality seems to be a necessary condition for being planar (and thus a sufficient condition for being nonplanar), but evidently not all nonplanar graphs fail the inequality. We'll see in Example 4, however, that $K_{3,3}$ does fail a sharpened inequality.

Theorem 8.3.2: *The Edge-Vertex Inequality*
If G is a connected planar graph with V vertices and $E \geq 2$ edges, then $E \leq 3(V-2)$.

Proof:
· First note that this inequality is satisfied for connected planar graphs with 2 edges, for such a graph has 3 vertices, and $2 \leq 3(3-2)$. ✓
· So, suppose G has at least 3 edges, and assume first that it has no edges with an endpoint of degree 1.
· Then every face is bordered by a cycle of at least 3 edges, and each edge borders 2 faces. Thus, as argued above, $3F \leq 2E$, which yields the desired inequality $E \leq 3(V-2)$. ✓
· In the case where G has protruding edges with an endpoint of degree 1, these add equal numbers of new vertices and edges to be considered, which only sharpens the inequality. ✓
· Thus, $E \leq 3(V-2)$ holds for all connected planar graphs with more than one edge. ■

The proof of the *Edge-Vertex Inequality* generalizes to show the next theorem as well. We'll first define a term needed to formulate that result.

Definition 8.3.2: *Girth of a Graph*
*The **girth** of a graph G is the length of its shortest cycle, if it has one.*

The girth of any graph containing a cycle is at least 3. The graphs of Example 3abc all had girth 3, and these graphs interacted with the *Edge-Vertex Inequality* in the way we had hoped—the two planar graphs satisfied the inequality, and the nonplanar K_5 failed to satisfy it. The girth of $K_{3,3}$ in Example 3d is 4, and that graph didn't meet our expectations. An improved *edge-vertex inequality* that takes girth into consideration, however, will show that $K_{3,3}$ is nonplanar (see Example 4).

Theorem 8.3.3: *The Generalized Edge-Vertex Inequality*
If G is a connected planar graph with V vertices, E edges, and girth $g \geq 3$, then $E \leq \dfrac{g}{g-2}(V-2)$.

Proof:
See Exercise 16a.

❖ **Example 8.3.4**
Demonstrate that the following graphs are nonplanar.

a) 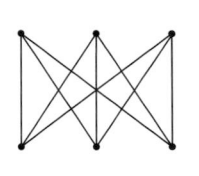   b) 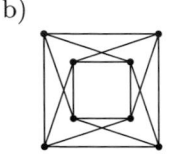   c)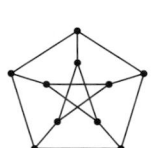

Solution
We'll use the *Generalized Edge-Vertex Inequality* to test these graphs. As we'll see, a little theory now makes definitive conclusions easy to deduce.

a) $K_{3,3}$ has $V = 6$, $E = 9$, and $g = 4$. Since $9 \not\leq (4/2)(6 - 2) = 8$, $K_{3,3}$ is not planar. ✓

b) This graph doesn't look planar because of all the crossing diagonals. In fact, it's $K_{4,4}$ in disguise (see Exercise 21a). Since $K_{3,3}$ isn't planar, its supergraph $K_{4,4}$ isn't, either. Alternatively, $K_{4,4}$ has $V = 8$, $E = 16$, and $g = 4$. Because $16 \not\leq (4/2)(8 - 2) = 12$, the graph is not planar. ✓

c) This graph is known as the *Petersen graph*, first introduced around 1900 by Julius Petersen, an early researcher in *Graph Theory*. For this graph, $V = 10$, $E = 15$, and $g = 5$. Since $15 \not\leq (5/3)(10 - 2) = 40/3$, the Petersen graph is not planar. ✓

8.3.4 *Kuratowski's Theorem for Planar Graphs*

We don't know yet whether the *Generalized Edge-Vertex Inequality* is a sufficient condition as well as a necessary one for a connected graph to be planar. A slight modification of the cube's planar graph, though, shows that it's not.

❖ **Example 8.3.5**
Show that the twisted-cube graph satisfies the *Generalized Edge-Vertex Inequality*, but that it is nevertheless nonplanar.

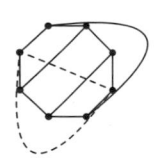

Solution
· This graph has $V = 8$, $E = 12$, and $g = 4$. The *Generalized Edge-Vertex Inequality* is therefore satisfied because $12 \leq (4/2)(8 - 2) = 12$. ✓

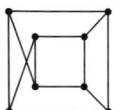

· However, it seems that this graph is nonplanar. By redrawing it as an octagon with diagonals, we've attempted to present it as planar. We then rerouted one diagonal around the outside, but another (dashed) can't be kept in place or moved outside without crossing an edge. But will all attempts fail? We'll say more on this below.

The counterexample just given indicates that the *Edge-Vertex Inequality*, even in its sharpest form, may not help us identify all nonplanar graphs.

Evidently, satisfying the inequality doesn't prove that a graph is planar. Is there any other simple property we can use to guarantee planarity? Or must we redraw the graph to show that it's planar or argue that it isn't?

Perhaps we can characterize nonplanar graphs instead. As graphs expand by adding more edges, even planar ones tend to become nonplanar. But this approach seems unproductive. We can't conclude, for instance, that graphs with a relatively large number of edges are the nonplanar ones, because some nonplanar graphs have smaller edge-to-vertex ratios than planar graphs (see Exercise 16e).

On the other hand, supergraphs of nonplanar graphs do stay nonplanar— no crossings get erased by adding in new vertices and edges. Could it be that complex nonplanar graphs fail to be planar because they contain simpler ones like K_5 and $K_{3,3}$ inside? Amazingly enough, something very close to this is true, as Kazimierz Kuratowski discovered in 1930. To formulate his result more precisely, we must first define some new relations between graphs.

Definition 8.3.3: *Subdivisions and Fusions of Graphs*
  a) *G' is a **subdivision** of G if and only if G' arises from G by adding degree-two vertices along any edges of G and replacing those edges with the new segments.*
  b) *G is a **fusion** of G' if and only if G' is a subdivision of G.*

The relation *is-a-fusion-of* is the converse of *is-a-subdivision-of*. Subdivisions are segmented expansions of a graph constructed by replacing edges with paths created by inserting degree-two vertices (let's call them *elbows*) along those edges. Conversely, fusions are constructed by fusing edges at elbows and then dropping those degree-two vertices.

The next proposition looks at how these ideas relate to planarity.

Proposition 8.3.1: *Necessary Conditions for Planar Graphs*
  a) *All subgraphs of a planar graph are planar.*
  b) *All subdivisions of a planar graph are planar.*
  c) *All fusions of a planar graph are planar.*

Proof:
  a) The subgraph of a planar graph is also planar, because deleting edges or vertices won't generate any illegal crossings.
  b) A planar graph doesn't gain any unwanted crossings by adding degree-two vertices along edges, so any subdivision of a planar graph is planar.
  c) Fusing edges at elbows will also not create any illegal crossings. (No fusing occurs where more than two edges meet; it's only done at degree-two vertices.) Thus, any fusion of a planar graph is planar.

Each part of *Proposition 1* gives a necessary condition for being planar. Their contrapositives thus provide sufficient conditions for being nonplanar. If a subgraph, subdivision, or fusion of a graph is nonplanar, then so is the graph itself. But we're most interested in considering subgraphs and fusions, since those give simpler graphs for us to examine. *Kuratowski's Theorem* combines these two results and sharpens them further by claiming that all we need to look for in such simplified graphs is K_5 or $K_{3,3}$.

Theorem 8.3.4: *Kuratowski's Theorem for Planar Graphs*
G is planar if and only if no subgraph of G is a subdivision of K_5 or $K_{3,3}$. Equivalently, G is nonplanar if and only if G has a subgraph that's a subdivision of K_5 or $K_{3,3}$. Such subgraphs are called **Kuratowski subgraphs**.

Proof:
- We'll prove the easier backward direction of the contrapositive formulation. The other direction is beyond the scope of this text.
- Suppose G is a graph containing a Kuratowski subgraph H, i.e., a graph such that K_5 or $K_{3,3}$ is a fusion of H.
- Since these are not planar, neither is H by *Proposition* 1c.
- But then neither is G by *Proposition* 1a. ∎

   We now have a new strategy for showing that a graph is nonplanar. *If we can drop edges or fuse them at degree-two vertices in some way that ends up with either K_5 or $K_{3,3}$, then the graph is nonplanar; otherwise it is planar.*

   Kuratowski's Theorem doesn't yield a procedure for redrawing a graph to show that it is planar, but it gives something to check: either a graph has a Kuratowski subgraph, in which case it is nonplanar, or no such subgraph exists, in which case it is planar. This gives a satisfying sufficient condition for being a planar graph, even if its practicality is somewhat limited.

✤ Example 8.3.6
   Show that the following graphs are nonplanar using *Kuratowski's Theorem*.

a)
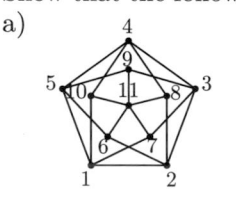

b)
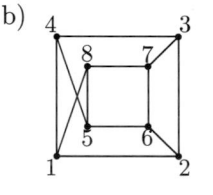

c)
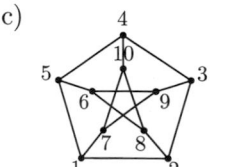

Solution
a) This is known as the Grötzsch graph. It has an outer pentagon shape with degree-four vertices. To see that it contains K_5, drop central vertex 11 along with its edges (dotted). This leaves a Kuratowski subgraph: K_5 results if we now fuse the resulting degree-two vertices 6, 7, 8, 9, and 10.

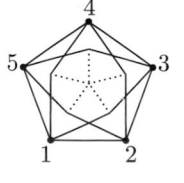

b) This graph's vertices are all degree 3, so it won't contain K_5. If we drop edge 58 (dotted), we'll have a Kuratowski subgraph: fusing elbows 5 and 8 to create new edges 46 and 17 yields $K_{3,3}$, with $\{1, 3, 6\}$ being one part and $\{2, 4, 7\}$ the other.

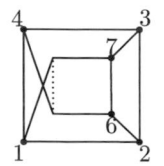

c) The Petersen graph has a pentagon outer boundary, but as all vertices in the graph have degree 3 and fusing edges doesn't increase vertex degrees, it obviously can't contain K_5. So we'll need to look for $K_{3,3}$ somewhere inside. This requires four fewer vertices and six fewer edges.
  · Deleting one edge will give two vertices to fuse, which will further reduce the edges by two more. Doing this twice will give six fewer edges and four fewer vertices, as needed.
  · In order to create a bipartite graph, we'll need to remove edges judiciously; not every pair of edges will produce what's needed. But deleting edges 12 and 4 10 (dotted) yields a Kuratowski subgraph— fusing vertices 1, 2, 4, and 10 creates edges 35, 38, 57, and 78, which yields $K_{3,3}$. The two disjoint parts are $\{3,6,7\}$ and $\{5,8,9\}$.

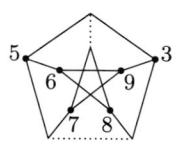

EXERCISE SET 8.3

Exercises 1–7: True or False
Are the following statements true or false? Explain your answer.

8.3.1. Platonic solids are named after Plato, because he discovered them.

8.3.2. Every planar graph must contain a vertex of degree 4 or less.

8.3.3. The subgraph of a planar graph is planar.

8.3.4. A graph that satisfies the *Generalized Edge-Vertex Inequality* is planar.

8.3.5. K_5 has only planar proper subgraphs.

8.3.6. If a subdivision of a graph is planar, then the graph is planar.

8.3.7. Planar graphs are Eulerian.

Exercises 8–10: Regular and Semiregular Polyhedra
The following problems explore Platonic and Archimedean solids.

8.3.8. *Determining Platonic Solids*
Explain why there are exactly five Platonic solids by arguing as follows:
  a. In order for the faces of a Platonic solid to be flattened around a vertex there must be a positive *angle defect*, the difference between the angle of a full circle and the angle sum of the faces around that vertex.
  b. At least three faces are needed at each vertex for a solid's surface. As no more than five triangles, three squares, or three pentagons yield a positive angle defect, these are the only combinations found in a Platonic solid.
  c. The different possible arrangements for regular polygon faces of a solid are all realized in the five Platonic solids.

8.3.9. *Planar Graphs for Platonic Solids*
Verify that the planar graphs for the Platonic solids have the correct number of vertices, edges, and faces and that they satisfy *Euler's Polyhedral Formula*.

8.3.10. *Archimedean Solids*

Tell how many faces of what shapes the following figures have, and verify that *Euler's Polyhedral Formula* holds for them. Use online diagrams to help you visualize these figures, where necessary. Explain any numerical answers.

  a. Truncated tetrahedron

  b. Truncated octahedron

  c. Truncated icosahedron

  d. Truncated dodecahedron

  e. Snub cube: 24 vertices, 60 edges. How many faces must it have?

  f. Truncated cuboctahedron: 48 vertices, 72 edges, 12 square faces, and 6 octagon faces. How many hexagon faces (the remaining faces) are there?

  g. Snub dodecahedron: 92 faces—80 triangles and 12 pentagons. How many edges and vertices does it have?

Exercises 11–13: Euler's Formula

The following problems explore Euler's Formula for various polyhedra.

8.3.11. *Euler's Polyhedral Formula for Prisms*

A *prism* is a polyhedron whose end faces are parallel congruent polygons and whose sides are parallelograms. Show the following:

  a. A triangular prism satisfies *Euler's Polyhedral Formula.*

  b. A rectangular prism satisfies *Euler's Polyhedral Formula.*

  c. All polygonal prisms satisfy *Euler's Polyhedral Formula.*

  d. Describe the planar graph of a prism. Show that such a graph satisfies *Euler's Formula* for planar figures.

8.3.12. *Euler's Polyhedral Formula for Pyramids*

A *pyramid* is a solid whose base is a polygon and whose sides are triangles meeting at a common point. Show the following:

  a. A pentagonal pyramid satisfies *Euler's Polyhedral Formula.*

  b. A hexagonal pyramid satisfies *Euler's Polyhedral Formula.*

  c. All polygonal pyramids satisfy *Euler's Polyhedral Formula.*

  d. Describe the planar graph of a pyramid. Show that such a graph satisfies *Euler's Formula* for planar figures.

8.3.13. *Possible Polyhedra*

Do any polyhedra have the following features? Give an example or argue that no such polyhedron exists.

  a. 10 edges and 6 faces   c. 7 edges and 5 faces

  b. 24 vertices and 14 faces   d. 7 edges

Exercises 14–19: Vertex-Edge-Face Relations and Planar Graphs

The following problems explore vertex-edge-face relations and planar graphs.

8.3.14. *Edge-Face Inequalities*

  a. Use the fact that edges are counted at most twice in adding up the *sides* of a planar graph's faces to prove the *Edge-Face Inequality* $2E \geq 3F$.

  b. Prove the *Generalized Edge-Face Inequality* $2E \geq gF$ for planar graphs, where g denotes a graph's girth.

8.3.15. *Vertex-Edge-Face Relations in Maximal Planar Graphs*
A *maximal planar graph* is a planar graph in which any additional edge drawn between the graph's vertices causes the graph to become nonplanar. Such a graph is *nontrivial* if and only if it is on three or more vertices.
  a. Draw a maximal planar graph on six vertices.
  b. Explain why a nontrivial maximal planar graph has all of its faces (including the outer one) bounded by triangles.
  c. Prove that $2E = 3F$ for nontrivial maximal planar graphs. Verify that this holds for the planar graph on six vertices.
  d. Using *Euler's Formula*, prove that $E = 3(V - 2)$ for nontrivial maximal planar graphs. Verify that this holds for the graph on six vertices.

8.3.16. *Edge-Vertex Relations*
  a. Prove the *Generalized Edge-Vertex Inequality* using the *Generalized Edge-Face Inequality* (see Exercise 14b).
  b. Prove the *Edge-Vertex Inequality* as a corollary of the *Generalized Edge-Vertex Inequality*.
  c. Prove that if the girth g of a planar graph is at least 4, then the graph satisfies the inequality $E \leq 2(V - 2)$.
  d. Discuss why the *Generalized Edge-Vertex Inequality* is an improvement on the *Edge-Vertex Inequality*.
  e. Show with examples that some nonplanar graphs have relatively fewer edges than some planar graphs, i.e., compare their *edge : vertex* ratios. What can you conclude from this about using these ratios?

8.3.17. *Cycles in Planar Graphs*
Explain why a planar graph whose vertices all have degree greater than 1 has a cycle surrounding a bounded region. (A rigorous proof of this result requires the *Jordan Curve Theorem*, which says that any continuous simple closed curve divides the plane into disjoint inner and outer regions.)

8.3.18. *Five Neighbors Theorem*
  a. Prove the *Five Neighbors Theorem*: *every planar graph contains a vertex of degree $d \leq 5$.*
  b. Can the *Five Neighbors Theorem* be improved by making the vertex-degree inequality $d \leq 4$? Explain.

8.3.19. *Euler's Formula for Planar Graphs*
  a. Show by direct calculation that *Euler's Formula for Planar Graphs* holds if a graph G consists of two vertices connected by an edge; if G is K_3.
  b. Does *Euler's Formula* hold if a graph G is a cycle? If G is a path? Explain.
  c. Does *Euler's Formula* hold if G is a graph with two vertices and no edges? If G is a graph with three vertices and one edge? If G is a graph consisting of two disjoint edges? Does any of this violate *Theorem 1*?
  d. If G is graph containing more than one component, does *Euler's Formula for Planar Graphs* hold? Conjecture a generalization for this situation and then prove it.

Exercises 20–23: Planar and Nonplanar Graphs

The following problems explore planar and nonplanar graphs.

8.3.20. *The Three-Utilities Puzzle*

a. A well-known puzzle asks: *can three houses in a plane be connected to water, gas, and electricity utilities in the same plane without having their lines intersect or pass through another house or location?* Interpret this puzzle in terms of graph theory.

b. Solve the three-utilities puzzle using what you know about its graph.

8.3.21. *Complete Bipartite Graphs*

a. Show that the nonplanar graph in Example 4b is $K_{4,4}$ in disguise.

b. For which m, n is $K_{m,n}$ planar? Prove your conjecture.

8.3.22. *Planar or Nonplanar?*

Determine whether each of the following graphs is planar or nonplanar.

   i) If the graph is planar, draw a representation in which no edges cross.

  ii) If the graph is nonplanar, use the *Generalized Edge-Vertex Inequality* to test it.

 iii) If the graph is nonplanar but satisfies the *Generalized Edge-Vertex Inequality*, prove that it is nonplanar by locating K_5 or $K_{3,3}$ inside it.

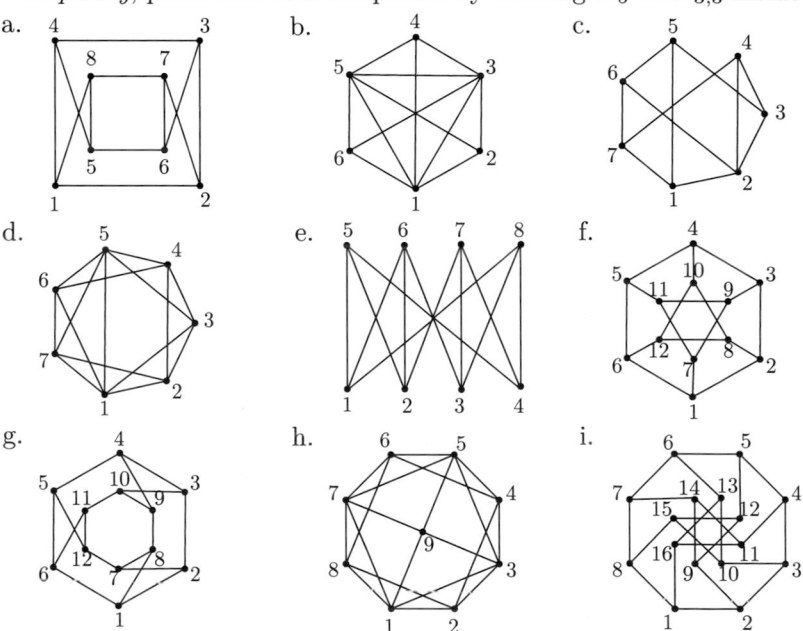

8.3.23. *Kuratowski's Theorem*

a. Explain why a graph is nonplanar if it can be transformed into K_5 or $K_{3,3}$ by dropping edges and fusing elbows in any order.

b. Give counterexamples to show that nonplanar graphs need not contain either K_5 or $K_{3,3}$ as an ordinary subgraph. Hence, edges may need to be fused as well as dropped to conclude that a graph is nonplanar.

8.4 Coloring Graphs

We've now looked at two different ways to do walks through a graph, and we've explored when graphs can be drawn without crossings. For our final section, we'll look at the topic of coloring a graph, which was initially linked to planarity. Once again, this has an interesting history.

8.4.1 *Origin of the Four-Color Problem*

In 1852, one of Augustus De Morgan's students asked him whether one can always color a map using only four colors, if regions sharing a border receive different colors. His brother, the student said, had drawn this conclusion while coloring the map of England (see Figure 8.5).

De Morgan was intrigued by the question. Failing to answer it, he tried to enlist Hamilton's assistance, noting that "it is tricky work, and I am not sure of the convolutions." Hamilton replied by return mail: "I am not likely to attempt your "quaternion" of colours very soon."

Fig. 8.5 Map of England

De Morgan continued to work on the problem and mentioned the result in a book review he wrote in 1860 for a literary magazine. It took nearly 20 years, though, before it received broader attention and interest from the mathematical community. Arthur Cayley commented on it at the 1878 London Mathematical Society meeting. The sticking point in solving it, he later pointed out, is that one can't always extend a four-coloring for n regions of a map to the next region—the new region might enclose all n areas, so one might need to backtrack and alter the earlier coloring.

✦**Example 8.4.1**

Show that the exhibited simple map requires 4 colors, and illustrate Cayley's point about what could happen if it were part of a larger map.

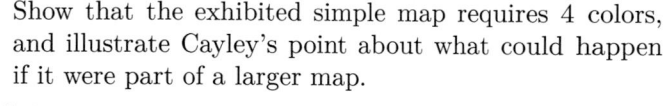

Solution

If we color the central region red, the three surrounding sectors will need colors like blue, green, and yellow, since they each share a boundary with the other three.

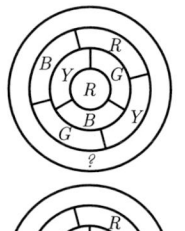

If more rings are placed around this map, one might start coloring them as indicated, but that runs into trouble with the outer ring, if only four colors can be used.

Backtracking, this impasse can be avoided by more judiciously choosing colors for the second ring, as shown. Now four colors are sufficient.

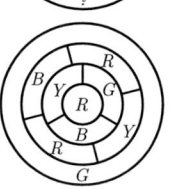

One year after Cayley publicized the problem, Alfred Kempe, who had been his student earlier at Cambridge, published a proof in the **American Journal of Mathematics** that four colors did, in fact, suffice.[3] His argument purported to show that if regions with four or fewer colors surrounded a region yet to be colored, then these colors could in all instances be reduced to three by appropriately shifting reds and greens or blues and yellows along chains of neighboring regions, freeing up a fourth color for the inner region.

8.4.2 *Kempe Chains and the Four-Color Problem*

Near the end of his paper, Kempe noted that one can conceptualize maps as *linkages*, now called graphs. Points would represent regions, and points of neighboring regions would be connected by an edge (see Exercise 8). Properly coloring a map, then, would translate into coloring/labeling the vertices of a planar graph so that adjacent vertices have different colors/labels. Using this graph-theoretic interpretation, we'll sketch Kempe's argument for the case when a central region is surrounded by four other regions.

❖**Example 8.4.2**

Prove that if a map is four-colorable except for a region surrounded by four other regions, then the entire map can be colored with four colors.

Proof:
· Let $\mathcal{G}$ be a planar graph representing such a map, and let u be a vertex that is adjacent to four other vertices, as pictured.
If the surrounding vertices are colored with fewer than four colors, the interior vertex u can be colored with one of the remaining colors, making this configuration—and thus the entire graph—four-colorable. ✓

· Now suppose the surrounding vertices are colored red, blue, green, and yellow, in counterclockwise order. Label these vertices R, B, G, Y. We'll consider two possibilities for the four-colorable subgraph $\mathcal{G}'$ that omits u.

· First, suppose $\mathcal{G}'$ contains no walk of alternating red-green vertices connecting R and G. We can then interchange colors for all red and green vertices along such walks connected to R, freeing up the color red for u and making $\mathcal{G}$ four-colorable. ✓

· Otherwise there will be a walk of alternating red-green vertices in $\mathcal{G}'$ (dotted) from R to G.

· Since $\mathcal{G}$ is planar, no alternating blue-yellow walk in $\mathcal{G}'$ will connect B to Y, being blocked by the red-green walk.

· Thus, we can interchange blue and yellow in any such walks connected to B and then color u blue, again making the given configuration and the entire graph $\mathcal{G}$ four-colorable. ∎

---

[3] Robin Wilson's **Four Colors Suffice** (Princeton University Press, 2013) gives a fascinating and very readable account of the entire history of the four-color problem.

8.4.3 *Demise of Kempe's Four-Color Theorem*

Kempe's overall proof strategy for map coloring was to select a region adjacent to at most five regions—guaranteed by the *Five Neighbors Theorem* (see Exercise 8.3.18)—and first color the rest of the map in a valid way. To finish, then, he had only to show how this chosen region and its neighbors could be colored with at most four colors, even if it was adjacent to five regions. Using an alternating-color chain argument, he showed that this configuration, too, could be recolored so that it and therefore the entire graph would be four-colored. Or so he and everyone else thought at the time.

Kempe's proof of his *Four-Color Theorem* stood for a decade. Then in 1890, Percy Heawood published an article giving a counterexample of 28 regions that invalidated Kempe's argument. Later, mathematicians discovered simpler counterexamples to Kempe's method of proof.

✤**Example 8.4.3**

Show that Kempe's argument fails for a graph with fewer than 28 vertices.

Solution

- A colored nine-point planar graph that can't be recolored using Kempe's method of color exchanges along a chain of alternating colors is given below.[4] Here the uncolored vertex u is surrounded by a pentagon of five points colored red, blue, green, yellow, and blue in counterclockwise order.
- This graph has an alternating red-green walk connecting the pentagon's red and green vertices, but interchanging these colors won't make either of them available for u.
- And, unlike Kempe's argument for the case in which a central vertex is surrounded by four vertices, the existence of this red-green walk doesn't show that no blue-yellow walk connects vertices on the pentagon. Interchanging blue and yellow won't free up a color for u.

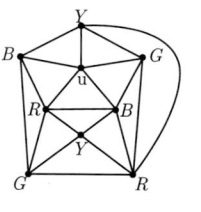

- The same conclusion follows if we begin with red and yellow: neither blue nor green is freed up for u (see Exercise 9ab).
- Thus, the vertex u will in all cases remain attached to vertices having four colors.
- Hence, Kempe's argument using color interchanges along an alternating two-color chain fails for this graph.

Having a counterexample to Kempe's argument doesn't mean, of course, that this graph can't be four-colored, only that a different color assignment is needed if it can (see Exercise 9c). So the *Four-Color Theorem* itself might still be true. As a modest step toward that result, we'll show that five colors suffice. Kempe's chain method still works here, as Heawood showed.

---

[4] This is Alexander Soifer's minimal counterexample. See ***The Mathematical Coloring Book*** (Springer, 2009), p. 182. The *Fritsch Graph* provides another counterexample on nine vertices.

Theorem 8.4.1: *The Five-Color Theorem*
Every planar graph can be colored with at most five colors.

Proof:
- We'll first prove this using induction on the number of vertices.
- Obviously, any graph with five or fewer vertices can be colored with five or fewer colors. ✓
- Now suppose that all planar graphs containing k vertices can be colored with at most five colors, and let $\mathcal{G}$ be a planar graph with $k + 1$ vertices. Since $\mathcal{G}$ is planar, it has a vertex u with $\deg u \leq 5$.
- Let $\mathcal{G}'$ denote the subgraph of $\mathcal{G}$ with u and its edges deleted. By our induction hypothesis, $\mathcal{G}'$ can be colored with five or fewer colors.
- If less than five colors are used for the vertices adjacent to u, color u with an unused color, and $\mathcal{G}$ will be colored with at most five colors. ✓
- If five colors are used, suppose the five vertices R, B, G, Y, W surrounding u counterclockwise have the colors red, blue, green, yellow, and white.
- As in Example 2, we have two cases to consider.
- First, if there is no walk of alternating red-green vertices in $\mathcal{G}'$ connecting R and G, interchange all red and green vertices connected by such walks to R and then color u red. ✓
- Else there is such a walk (shown as dotted) of alternating red-green vertices in $\mathcal{G}'$ from R to G.
- In this case, since $\mathcal{G}$ is planar, no alternating blue-yellow walk in $\mathcal{G}'$ connects B to Y, so we can interchange blue and yellow in any such chains connected to B and then color u blue. ✓
- Thus, any planar graph can be colored with at most five colors. ∎

- This theorem can also be proved by contradiction. Later proofs follow this basic approach, so we'll sketch it out.
- If a planar graph exists that requires six colors, then by the *Well-Ordering Principle* there must be minimal-order graph like this, i.e., a graph such that all graphs with fewer vertices are five-colorable.
- Choosing a vertex in this graph of degree less than or equal to five, it and its neighbors can in all cases be colored with five colors, and in such a way that the original graph can be as well—a contradiction.
- Fleshing out this argument is left as an exercise (see Exercise 11). ∎

8.4.4 *Progress and Success on the Four-Color Theorem*

After 1890 the four-color problem was once again open. Is the *Four-Color Theorem* true, or might there be a counterexample to the result itself?

The first decades of the twentieth century saw renewed interest in the problem, and several American mathematicians made progress by using new ideas. Although we won't go into these developments, a brief description will give a general idea of how the problem was tackled and eventually resolved.

The overall approach taken was indirect, as we just described. Suppose there were a counterexample of minimal size. If such a map must contain one of a set of *unavoidable configurations*, one which at the same time permits the original map to be recolored with four colors (as in Example 2), called a *reducible configuration*, then no such minimal counterexample can exist.

Determining an *unavoidable set* of *reducible configurations* proved to be a long and arduous task, requiring many maps to be checked. How could one shrink a promising set of unavoidable configurations down to a manageable size and show they were all reducible? Or, how could one take a large set of reducible configurations and select some to constitute an unavoidable set? People tried both approaches, inventing new ideas for generating and evaluating such configurations.

It would be interesting to know how many fruitless hours (even on a honeymoon!) mathematicians spent trying to prove or disprove the *Four-Color Theorem*— drawing, coloring, and analyzing graphs. Many undoubtedly shared Hamilton's view that the result wasn't tempting enough to spend much time on it. But some were addicted, and the longer the problem went unsolved, the more challenging it became and the more fame it promised whoever would solve it. And, once computers became available in the 1960s and '70s, mathematicians had a powerful resource never before available.

In the April 1975 issue of the **Scientific American**, the eminent mathematical puzzler Martin Gardner announced a breakthrough on the four-color problem by exhibiting a map of 110 regions that he claimed required five colors. This created a brief buzz of excitement and thousands of hours of coloring by mathematics aficionados, until people realized they had been pranked by an April Fool's jokester and that the map could be colored with only four colors.[5]

About a year later, Wolfgang Haken and Kenneth Appel announced that they had a computer-assisted proof of the *Four-Color Theorem*. By the early 1970s, a number of mathematicians had already begun to suspect that success might be imminent. Haken, who had been working on the problem for almost a decade, decided to first construct promising unavoidable sets of configurations and test them for reducibility later. At one point he was ready to postpone further work on the problem because it looked like the job would be beyond a computer's capability at the time, but then Appel offered his expert assistance with the project's computer programming. Together they and a few assistants eventually came up with an unavoidable set of about 2000 configurations, which were then checked for reducibility. Using 1200 hours of computer time and hours of human checking for components of the work, on July 22, 1976 they finally announced to their colleagues and the world that the problem had been solved—as the University of Illinois' postage meter proudly proclaimed: *four colors suffice!*

---

[5] For more information on Gardner's map and its four-coloring, see the Wolfram Math-World posting on the Four-Color Theorem at http://mathworld.wolfram.com/Four-ColorTheorem.html.

The immediate reception was mixed. Several mathematicians, close to solving the problem themselves, confirmed substantial parts of the proof. A few minor problems were found, but these were soon rectified. Many mathematicians were excited about the solution to a famous 125-year-old problem. But others were lukewarm or even antagonistic. The issue was whether a theorem could be considered proved if a computer was an essential collaborator. How could one know for sure that there were no bugs in the computer code and that all possible cases had been properly considered? Should an argument be accepted as a proof if it couldn't be surveyed or verified by human experts?

In 1996, four mathematicians presented a simplified proof of the theorem, involving fewer configurations and a smaller set of rules for proving unavoidability. This time, the computer was used for all aspects of the argument, because it was now judged to be more reliable than checking them by hand.

About a decade later, a proof for the *Four-Color Theorem* was formalized using *Coq*, a reputable interactive computer proof checker. Gradually, most mathematicians have come to accept the legitimacy of some computer assistance in their work. To date, though, no humanly surveyable proof of the *Four-Color Theorem* has been found. If such exists, it will likely require completely new ideas and developments.

8.4.5 *Graph Coloring Equivalents*

Nothing about coloring a graph requires it to be planar—that connection arose because the original focus was on planar maps. We can ask the same question of any graph: how many colors suffice? That number is the *chromatic number* of the graph.

Definition 8.4.1: *Chromatic Number of a Graph*
The chromatic number $X(G)$ of a graph G is the least number of colors required to color G so that adjacent vertices have different colors.

Some natural questions that arise in this connection are: Given graphs having such and such properties, what can we say about their chromatic number? Or, given a numerical value, how can we characterize graphs having that chromatic number? Let's first look at an example and then generalize.

✤**Example 8.4.4**
Find the chromatic number of $K_{3,4}$.

Solution
- $K_{3,4}$ is a bipartite graph.
- Since the vertices in each disjoint part are not adjacent to one another, they form a *color class*—they can all be colored with the same color.
- Thus $X(K_{3,4}) = 2$, as indicated.

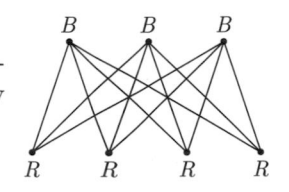

The same result holds for any bipartite graph: two colors suffice. In fact, the converse is also true. And we can also say something about the sorts of cycles in such graphs. First, a definition to lay down some terminology.

Definition 8.4.2: *Colorable Graphs*
  a) *A graph G is k-colorable if and only if $X(G) \leq k$.*
  b) *A graph G is k-chromatic if and only if $X(G) = k$.*

Theorem 8.4.2: *2-Chromatic Graphs*
  The following are logically equivalent for nontrivial connected graphs G:
  a) *G is 2-chromatic.*
  b) *G is bipartite.*
  c) *G has no cycle of odd length.*

Proof:
· *Note*: for nontrivial connected graphs (ones with at least two adjacent vertices), being 2-chromatic and being 2-colorable are the same thing: each pair of points connected by an edge requires two different colors.
· The following proof links two equivalence arguments.
· First suppose G is 2-chromatic.
  Then we can separate G into two parts—the non-empty color classes.
  As adjacent vertices have different colors, the only edges joining vertices are those connecting a vertex in one color class to a vertex in the other.
  Thus, G is bipartite. ✓
· Furthermore, G can't have an odd cycle, for such cycles require 3 colors (see Exercise 16a). ✓
· Now suppose G is bipartite.
  Then we can color all the vertices in one part with one color and those in the other part with another.
  Thus G is 2-chromatic. ✓
· Finally, suppose G has only even cycles, as in the solid-lined diagram below. We'll show that G is 2-chromatic by outlining a coloring procedure.
  Stage 1: Choose some vertex [1 in the diagram] and color it red.
  Stage 2: Color all vertices adjacent to the initial vertex blue [2 and 6]. None of these are adjacent, else G would have a cycle of length three created by two of these and the first vertex [2–6–1].
  Stage 3: Color red all uncolored vertices, if any, adjacent to those colored blue in stage two.

These will not be adjacent to those colored in stage 1 [such as 3 to 1] or to one another [such as 3 to 5] because that would create an odd cycle of length three [3–1–2] or five [3–5–6–1–2].
We can continue in this way, alternating blue and red, until all vertices have been properly colored (see Exercise 14a).
Thus, G is 2-chromatic. ■

❖ **Example 8.4.5**

Find the chromatic number of the exhibited Petersen graph.

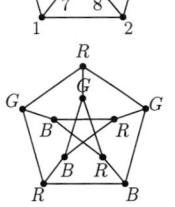

Solution

· The Petersen graph has an odd cycle (1–2–3–4–5), so it's not 2-chromatic.
· It is 3-chromatic, though: we can color $\{1, 4, 8, 9\}$ red, $\{2, 6, 7\}$ blue, and $\{3, 5, 10\}$ green.

We've characterized connected graphs that are 2-colorable; can we do the same thing for graphs that are 3-colorable or k-colorable? The answer to this seems to be "no"—it's difficult to decide which graphs can be k-colored for $k \geq 3$. Maybe the best we can do is to say something about an upper bound on the chromatic number of a graph.

8.4.6 Calculating Chromatic Numbers for Graphs

A rather obvious upper bound on $X(G)$ is the number of vertices in G. And for complete graphs, $X(K_n) = n$ (see Exercise 13bc). Typically, though, a graph's chromatic number is quite a bit less than the number of its vertices. Its value depends in some way on how many vertices are adjacent to one another, something that's captured by the vertices' degrees. For a complete graph, $X(G)$ is one more than its maximum degree. The following theorem says this is an upper bound for $X(G)$ in general.

Definition 8.4.3: Maximum Degree of a Graph
 The **maximum degree** of a graph G is $\Delta(G) = \max_{v \in G} (\deg(v))$, taken over all vertices v in G.

Theorem 8.4.3: Maximum Degree Bound for Chromatic Number
 $X(G) \leq \Delta(G) + 1$

Proof:
· We'll use a *Greedy Coloring Algorithm* to prove this result.
· List the vertices in reverse order of degree, a largest-degree vertex first, and so on, choosing any order among vertices having the same degree.
· Assign color 1 to the first vertex and to every later vertex that can be so colored after earlier vertices have been colored or bypassed.
· Assign color 2 in the same way to those vertices that remain uncolored.
· Continue coloring vertices like this until all vertices are colored.
· When final color m is used to color a vertex v, there will have been at least $m - 1$ vertices adjacent to v colored with other colors, so $\deg(v) \geq m - 1$, i.e., $m \leq \deg(v) + 1$.
· But $X(G) \leq m$ and $\deg(v) \leq \Delta(G)$.
· Combining these inequalities, we have $X(G) \leq \Delta(G) + 1$. ∎

Because vertices may be listed in a number of different orders when the *Greedy Coloring Algorithm* is used, the number of colors m produced by this procedure is not invariant, and it may not even give a good estimate for $X(G)$. We know that $X(G) \leq m \leq \Delta(G) + 1$, but we can't say anything more definitive than this. The next example illustrates both extremes for m.

✤Example 8.4.6
Use the *Greedy Coloring Algorithm* to calculate a color-bound m for $X(G)$ for the graph shown.

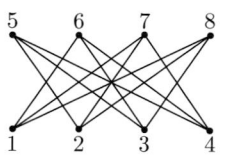

Solution
- As this is a bipartite graph with $\Delta(G) = 3$, we know from the last theorem that the *Greedy Coloring Algorithm* gives $2 = X(G) \leq m \leq \Delta(G) + 1 = 4$.
- There are lots of order choices here—all vertices have degree 3.
- Ordering vertices numerically, we get the bipartite 2-coloring—we can color the bottom vertices red, and the top blue. Here $m = X(G) = 2$.
- Ordering them as $1, 5; 2, 6; 3, 7; 4, 8$ (bottom to top; left to right) yields a 4-coloring: color vertices 1 and 5 red; 2 and 6 blue; 3 and 7 green; 4 and 8 yellow. In this case, $m = \Delta(G) + 1 = 4$.

A slight improvement on this bound for $X(G)$ holds for most graphs. *Brooks' Theorem* (1941) says that *if a connected graph is neither an odd cycle nor a complete graph, then $X(G) \leq \Delta(G)$*. Its proof is beyond this text.

8.4.7 *Coloring Graphs: a Productive Diversion*

We've seen that some key topics in *Graph Theory* emerged from recreational interests—solving a puzzle, playing a game, coloring a map. Thought-provoking problems attract mathematical effort, regardless of origin. Solutions to problems beget more difficult problems, and the cycle continues. As mathematicians searched for solutions to coloring a graph, they gradually developed a network of new ideas and results in *Graph Theory* whose ramifications extended far beyond the original problem (which seems in any case not to have been of genuine interest to cartographers). Ideas arising in recreational contexts often turn out to have important implications for real-world problems. We'll close this chapter by illustrating this sort of connection.

✤Example 8.4.7
Explain how graph coloring can model and solve scheduling problems.

Solution
- Suppose one wants to schedule a number of events (track and field events, academic classes, business tasks) so that people can participate in as many of them as they need or want to attend over some time period.
- Events can be considered vertices joined by edges (somewhat counterintuitively) if they need to be offered at different times, i.e., if they involve the same participants.

· Coloring such a graph will then partition the events into color classes of vertices that are not adjacent to one another, and so events in each class can be scheduled at the same time. The chromatic number of the graph gives the minimum number of time periods needed for the events.

EXERCISE SET 8.4

Exercises 1–6: True or False
Are the following statements true or false? Explain your answer.

8.4.1. If four regions of a planar map all border each other, one of them must be enclosed by the other three.

8.4.2. No planar map has five countries mutually adjacent to one another.

8.4.3. The outer bounded regions of a planar map are three-colorable.

8.4.4. An Eulerian graph is 2-colorable.

8.4.5. Every 2-colorable graph is Hamiltonian.

8.4.6. Every 2-colorable graph with more than 5 vertices is nonplanar.

Exercises 7–12: Coloring Planar Maps
The following problems focus on coloring planar maps and their graphs.

8.4.7. *Möbius' Five Kingdoms Puzzle*
  a. A ruler wants to pass his kingdom on to his five heirs so that each (simply connected) region borders all the others. Can this be done? Explain.
  b. Would a positive solution to this puzzle prove that a planar map requires five colors? Explain.
  c. Would a negative solution to this puzzle show that a planar map can always be colored with four colors, as some have thought? Explain.

8.4.8. *Map Coloring and Planar Graphs*
  a. Carefully explain how planar maps can be transformed into planar graphs.
  b. Transform the final map of Example 1 into a planar graph, and color it as that map eventually was.

8.4.9. *Kempe's Alternating-Color Chain Argument*
  a. Show for Example 3 that if one begins by focusing on the vertices R and Y, a similar conclusion can be drawn that Kempe's chain method fails.
  b. Show that any other color swap in Example 3 fails to free up a color for the central vertex u.
  c. Assign colors to the graph in Example 3 to show that it's four-colored.

8.4.10. *The Six-Color Theorem*
Without citing the *Four-* or *Five-Color Theorems*, prove the *Six-Color Theorem* (*any planar map can be colored with at most six colors*) as follows:
  a. Using mathematical induction, as in the proof of the *Five-Color Theorem*.
  b. Using proof by contradiction, as suggested in the remarks following the proof of the *Five-Color Theorem*.

8.4.11. *The Five-Color Theorem*
Prove the *Five-Color Theorem* using proof by contradiction, as suggested in
the remarks following its proof in the text.

8.4.12. *Three-Dimensional Maps*
A *three-dimensional map* is a set of connected solid regions without holes.
Adjacent regions are ones that share a connected two-dimensional surface.
De Morgan's student Frederick Guthrie asked the following: Does some mini-
mum number of colors suffice to color all three-dimensional maps? Show that
three-dimensional maps exist that require any given number of colors.

Exercises 13–17: Chromatic Number
The following problems explore the chromatic number of a graph.

8.4.13. *Prove or disprove*:
 a. If H is a subgraph of G, then $\chi(H) \leq \chi(G)$.
 b. If $\chi(G) = n$ for a graph G on n vertices, then $G = K_n$.
 c. If G contains a complete subgraph on m of its vertices, then $\chi(G) \geq m$.

8.4.14. *Theorem 2*
 a. Explain why the procedure outlined in the final part of *Theorem 2*'s proof
    will never lead to two adjacent red or blue vertices.
 b. Schematically indicate with *PL* symbolism how the proof of *Theorem 2*
    demonstrates that all three of its claims are logically equivalent.

8.4.15. *Brooks' Theorem*
Verify *Brooks' Theorem* for the following graphs:
 a. The flattened graph of the octahedron (see Section 8.3.1).
 b. The flattened graph of the icosahedron (see Section 8.3.1).

8.4.16. *Coloring Cycles*
 a. Prove that a cycle is 3-colorable. When is it 2-chromatic and when is it
    3-chromatic?
 b. Explain why the conclusion of *Brooks' Theorem*—$\chi(G) \leq \Delta(G)$—fails
    when G is an odd cycle.

8.4.17. *Coloring Cubic Graphs*
A cubic graph is one in which each vertex has degree 3.
 a. Tell why a cubic graph can be colored with at most four colors.
 b. Characterize cubic graphs that are 3-chromatic. Are any 2-chromatic?

8.4.18. *Graph Coloring for Computers*
In compiling a program, variables are assigned to available registers. Variables
can be assigned to a shared register if they are not in use at the same time.
Given a limited number of registers, an optimal allocation will assign as many
variables to the same register as possible. Interpret this task as a graph-
theoretic coloring procedure.

Exercises 19–21: Graph Coloring
Color or determine the chromatic number of the following graphs.

8.4.19. *Graph Coloring*
Color the following graphs with at most $\Delta(G)$ colors. Explain whether the
number of colors you used is $\chi(G)$.

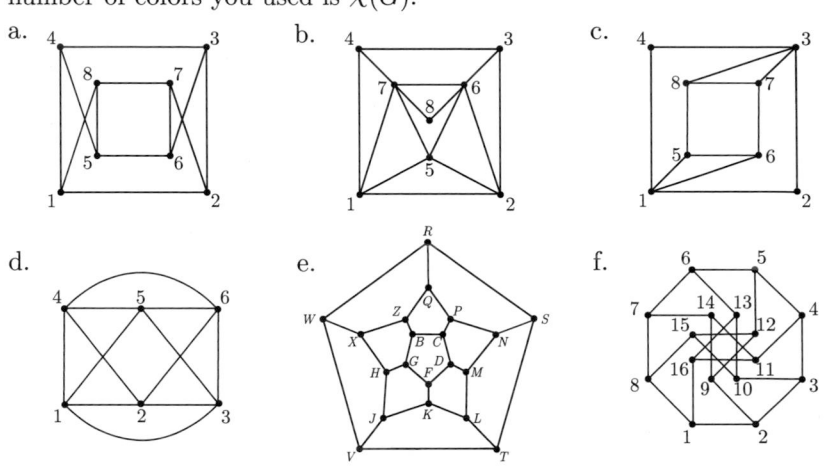

8.4.20. *Bipartite Graphs*
Are the following graphs bipartite? Explain. Find the chromatic number of
each graph.

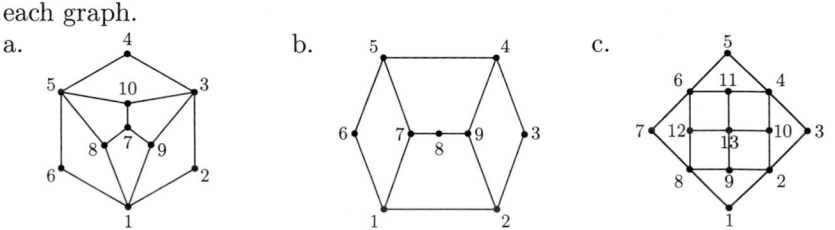

8.4.21. *The Greedy Coloring Algorithm and Chromatic Number*
Use the *Greedy Coloring Algorithm* to determine a bound on $\chi(G)$ for the
following graphs. Did your procedure produce $\chi(G)$? Explain.

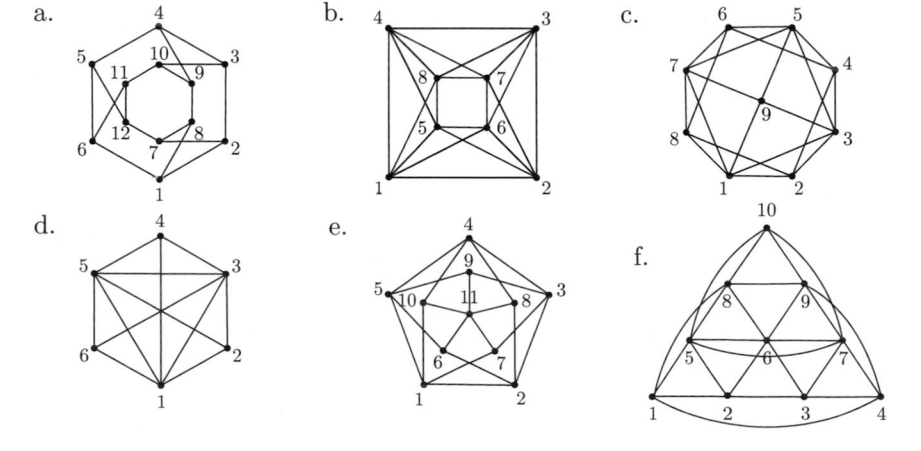

Image Credits

Page 2: Bust of Aristotle, Ludovisi Collection, photograph by Marie-Lan Nguyen, accessed at
https://commons.wikimedia.org/wiki/File:Aristotle_Altemps_Inv8575.jpg/
Public domain

Page 3: Portrait of George Boole, *The Illustrated London News*, 21 January 1865, accessed at
https://commons.wikimedia.org/w/index.php?curid=39762976/Public domain

Page 3: Image of Claude Shannon posing in front of bookshelf, accessed at
https://s3.amazonaws.com/bell-labs-microsite-news/news/2001/february/26/shannon2_lg.jpeg/Reused with permission of Nokia Corporation

Page 10: Alfred Tarski, 1968. Author: George M. Bergman. Source: Archives of the Mathematisches Forschungsinstitut Oberwolfach. Reproduced and modified with permission

Page 184: Giuseppe Peano, School of Mathematics and Statistics, University of St Andrews, Scotland, accessed at
https://commons.wikimedia.org/wiki/File:Giuseppe_Peano.jpg/Public domain

Page 255: Georg Cantor, accessed at
https://commons.wikimedia.org/wiki/File:Georg_Cantor2.jpg/Public domain

Page 266: Richard Dedekind, accessed at
https://commons.wikimedia.org/wiki/File:Dedekind.jpeg/Public domain

Page 278: Kurt Gödel, accessed at
https://en.wikipedia.org/wiki/File:Kurt_g%C3%B6del.jpg/Public domain

Page 282: Ernst Zermelo, accessed at
https://commons.wikimedia.org/wiki/File:Ernst_Zermelo_1900s.jpg/
Public domain

Page 283: Bertrand Russell, accessed at
https://commons.wikimedia.org/wiki/File:Russell1907-2.jpg/Public domain

Page 291: Passport photo of Alan Turing at age 16, accessed at
https://commons.wikimedia.org/w/index.php?curid=22828488/Public domain

© Springer Nature Switzerland AG 2019
C. Jongsma, *Introduction to Discrete Mathematics via Logic and Proof*,
Undergraduate Texts in Mathematics,
https://doi.org/10.1007/978-3-030-25358-5

Page 420: *Portrait of Leonhard Euler*, painted by Emanuel Handmann, 1753, photograph by Kuntsmuseum Basel, accessed at https://commons.wikimedia.org/wiki/File:Leonhard_Euler.jpg/Public domain

Page 420: Euler's diagram of Königsberg bridge, accessed at https://www.maa.org/sites/default/files/pdf/upload_library/22/Polya/hopkins.pdf/Public domain

Page 430: William Rowan Hamilton, accessed at https://commons.wikimedia.org/wiki/File:William_Rowan_Hamilton_portrait_oval_combined.png/Public domain

Page 455: Map showing the ceremonial counties of England excluding the City of London in 2010, created by Nilfanion from Ordnance Survey Open-Data, accessed at https://commons.wikimedia.org/wiki/File:English_ceremonial_counties_2010.svg/Reproduced with permission

All other images are the author's own.

Appendix A
Inference Rules for PL and FOL

INFERENCE RULES FOR PROPOSITIONAL LOGIC

Elementary Rules of Inference

Premises (Prem)

A premise may be put down at any line in a deduction.

Reiteration (Reit)

A sentence may be reiterated at any line in a subproof.

Elimination and Introduction Rules for $\wedge$

Simplification (Simp)

$$\frac{P \wedge Q}{P} \qquad \frac{P \wedge Q}{Q}$$

Conjunction (Conj)

$$\frac{P \quad Q}{P \wedge Q}$$

Elimination and Introduction Rules for $\rightarrow$

Modus Ponens (MP)

$$\frac{P \rightarrow Q \quad P}{Q}$$

Modus Tollens (MT)

$$\frac{P \rightarrow Q \quad \neg Q}{\neg P}$$

Conditional Proof (CP)

$$\frac{\begin{array}{|l} P \\ \hline Q \end{array}}{P \rightarrow Q}$$

Hypothetical Syllogism (HS)

$$\frac{P \rightarrow Q \quad Q \rightarrow R}{P \rightarrow R}$$

Elimination and Introduction Rules for $\leftrightarrow$

BiCndnl Elim (BE)

$$\frac{P \leftrightarrow Q \quad P}{Q} \qquad \frac{P \leftrightarrow Q \quad Q}{P}$$

BiTrans (BT)

$$\frac{P \leftrightarrow Q \quad Q \leftrightarrow R}{P \leftrightarrow R}$$

Neg BiCndnl Elim (NBE)

$$\frac{P \leftrightarrow Q \quad \neg P}{\neg Q} \qquad \frac{P \leftrightarrow Q \quad \neg Q}{\neg P}$$

BiCndnl Int (BI)

$$\frac{\begin{array}{|l} P \\ \hline Q \end{array} \quad \begin{array}{|l} Q \\ \hline P \end{array}}{P \leftrightarrow Q}$$

(BICon)

$$\frac{\begin{array}{|l} P \\ \hline Q \end{array} \quad \begin{array}{|l} \neg P \\ \hline \neg Q \end{array}}{P \leftrightarrow Q}$$

Cyclic BiCndnl Int (CycBI)

$$\frac{P_1 \rightarrow P_2 \quad P_2 \rightarrow P_3 \quad P_3 \rightarrow P_1}{P_i \leftrightarrow P_j} \qquad \text{any } i, j$$

© Springer Nature Switzerland AG 2019
C. Jongsma, *Introduction to Discrete Mathematics via Logic and Proof*,
Undergraduate Texts in Mathematics,
https://doi.org/10.1007/978-3-030-25358-5_A

Elimination and Introduction Rules for ¬

Negation Introduction (NI)

| P > > >
| Q
| ¬Q
|=====
| ¬P

Negation Elimination (NE)

| ¬P > > >
| Q
| ¬Q
|=====
| P

Elimination and Introduction Rules for ∨

Disjunctive Syllgsm (DS)

| P ∨ Q | P∨Q
¬P	¬Q
Q	P

| ¬P ∨ Q | P ∨ ¬Q
P	Q
Q	P

(LEM)

|=====
| P ∨ ¬P

Constructive Dilemma (Cases)

| P ∨ Q
| | P
| |---
| | R
|
| | Q
| |---
| | R
|------
| R

Addition (Add)

| P | Q
|===========|====
| P ∨ Q | P ∨ Q

Either-Or (EO)

| | ¬P | | ¬Q
	Q		P
P ∨ Q	P ∨ Q

Replacement Rules

De Morgan's Rules (DeM)

¬(P ∧ Q) :: ¬P ∨ ¬Q
¬(P ∨ Q) :: ¬P ∧ ¬Q

Double Negation (DN)

¬¬P :: P

Negative Conditional (Neg Cndnl)

¬(P → Q) :: P ∧ ¬Q

Contraposition (Conpsn)

P → Q :: ¬Q → ¬P

Negative Bicndnl (Neg BiCndnl)

¬(P ↔ Q) :: (P ∧ ¬Q) ∨ (¬P ∧ Q)

Biconditional Equiv (Bicndnl)

P ↔ Q :: (P → Q) ∧ (Q → P)
      :: (P → Q) ∧ (¬P → ¬Q)
      :: (P ∧ Q) ∨ (¬P ∧ ¬Q)

Conditional Equiv (Cndnl)

P → Q :: ¬(P ∧ ¬Q)
      :: ¬P ∨ Q

Distribution (Dist)

P ∧ (Q ∨ R) :: (P ∧ Q) ∨ (P ∧ R)
(P ∨ Q) ∧ R :: (P ∧ R) ∨ (Q ∧ R)
P ∨ (Q ∧ R) :: (P ∨ Q) ∧ (P ∨ R)
(P ∧ Q) ∨ R :: (P ∨ R) ∧ (Q ∨ R)

Exportation (Exp)

P → (Q → R) :: (P ∧ Q) → R
P → (Q ∧ R) :: (P → Q) ∧ (P → R)

Idempotence (Idem)

P ∧ P :: P
P ∨ P :: P

Commutation (Comm)

P ∧ Q :: Q ∧ P
P ∨ Q :: Q ∨ P

Association (Assoc)

P∧(Q∧R) :: (P∧Q)∧R
P∨(Q∨R) :: (P∨Q)∨R

INFERENCE RULES FOR FIRST-ORDER LOGIC

Propositional Rules of Inference
See *Inference Rules for Propositional Logic*
> Take all letters as standing for well-formed formulas of *First-Order Logic*

Introduction and Elimination Rules for =

Substitution of Equals (Sub)

$$\begin{array}{|l} \mathbf{P}(\cdot \mathbf{t_1} \cdot) \\ \mathbf{t_1} = \mathbf{t_2} \\ \hline \mathbf{P}(\cdot \mathbf{t_2} \cdot) \end{array}$$

Transitivity of = (Trans)

$$\begin{array}{|l} \mathbf{t_1} = \mathbf{t_2} \\ \mathbf{t_2} = \mathbf{t_3} \\ \hline \mathbf{t_1} = \mathbf{t_3} \end{array}$$

Law of Identity (Iden)

$$\begin{array}{|l} \\ \hline \mathbf{t} = \mathbf{t} \end{array} \qquad [\text{t any term}]$$

Symmetry of = (Sym)

$$\begin{array}{|l} \mathbf{t_1} = \mathbf{t_2} \\ \hline \mathbf{t_2} = \mathbf{t_1} \end{array}$$

Introduction and Elimination Rules for ∀

Universal Instantiation (UI)

$$\begin{array}{|l} \forall \mathbf{x} \mathbf{P}(\mathbf{x}) \\ \hline \mathbf{P}(\mathbf{t}) \end{array} \qquad [\text{t any term}]$$

Universal Generalization (UG)

$$\begin{array}{|l} \mathbf{P}(\mathbf{a}) \\ \hline \forall \mathbf{x} \mathbf{P}(\mathbf{x}) \end{array} \qquad [\text{a an arbitrary constant}]$$

Introduction and Elimination Rules for ∃

Existential Generalization (EG)

$$\begin{array}{|l} \mathbf{P}(\mathbf{t}) \qquad [\text{t any term}] \\ \hline \exists \mathbf{x} \mathbf{P}(\mathbf{x}) \end{array}$$

Existential Instantiation (EI)

$$\begin{array}{|l} \exists \mathbf{x} \mathbf{P}(\mathbf{x}) \\ \quad \begin{array}{|l} \mathbf{P}(\mathbf{a}) \qquad [\text{a an arbitrary constant}] \\ \mathbf{Q} \qquad [\text{a not in } \mathbf{Q}] \\ \hline \end{array} \\ \hline \mathbf{Q} \end{array}$$

Replacement Rule for ∃!

Unique Existence (Uniq Exis)
$$\exists! \mathbf{x} \mathbf{P}(\mathbf{x}) :: \exists \mathbf{x} \mathbf{P}(\mathbf{x}) \wedge \forall \mathbf{x} \forall \mathbf{y} (\mathbf{P}(\mathbf{x}) \wedge \mathbf{P}(\mathbf{y}) \rightarrow \mathbf{x} = \mathbf{y})$$

Replacement Rules for Negations ¬∀ and ¬∃

Universal Negation (UN)
$$\neg \forall \mathbf{x} \mathbf{P}(\mathbf{x}) :: \exists \mathbf{x} (\neg \mathbf{P}(\mathbf{x}))$$

Existential Negation (EN)
$$\neg \exists \mathbf{x} \mathbf{P}(\mathbf{x}) :: \forall \mathbf{x} (\neg \mathbf{P}(\mathbf{x}))$$

Index

Symbols

1-cell, *see* cell

A

absorption law, 28, 211, 371, 403, 404
abstract mathematics, vi–ix, 10, 48, 58,
   114, 119, 120, 216, 344, 346, 352,
   355, 368–370, 374, 413, 414
acceptance, 384, 393, 394, 402
acyclic graph, 428, 429
addition, 185, 186, 188–193, 335,
   338–340, 342
addition rule, 92, 93, 135, 470
additive counting principle, 246, 247
adjacent cells, *see* cell
adjacent vertices, *see* vertex
affirming the consequent, 52, 59
algebra, 2, 3, 17, 18, 28, 32, 33, 102,
   125, 126, 133, 136, 153, 184, 327,
   333, 335, 336, 343, 366, 368, 369,
   372, 376, 378, 381, 391, 430, 431
algebra and logic, 3, 4, 8, 103, 368, 372,
   373, 375, 376, 381, 385, 414
algebraic number, 265, 269, 273
algebraic solution method, 316
annihilation law, 371, 396, 403
antecedent, 24, 30, 64
antisymmetric relation, 208, 343, 344,
   346
Appel, Kenneth, 459
Archimedean solid, 451, 452
Archimedes, 152, 443
Aristotelian logic, 122, 133, 137, 146,
   372, 376
Aristotle, vii, 2, 5, 14, 75, 258
arithmetic sequence, 166
arithmetic series, 167
association rule, 45, 46, 94, 470
associative law, 369, 378, 404
asymmetric relation, 346
atom, 366, 367

axiom, 2, 25–27, 125, 186–188, 191,
   194, 205, 207, 246, 255, 258, 278,
   281–285, 287, 289, 290, 295, 366,
   368–370, 378
axiom of choice, 265, 267, 270, 282,
   289, 290
axiom of extensionality, 283, 284, 292,
   295
axiom of foundation, 288, 290
axiom of infinity, 289, 295
axiom of mathematical induction,
   186–188, 190
axiom of replacement, 290
axiom of separation, 283–286, 290, 292,
   295
axiomatic theory, viii, 2, 26, 27, 29, 35,
   47, 281–283, 289, 290, 292

B

backward-forward method, 46–48, 63,
   71, 82, 83, 93, 135, 136, 143, 156,
   209, 220
Banach-Tarski paradox, 265, 270
base case, 151, 152, 159, 160, 163, 173
Bernoulli, Jean, 206
Bernstein, Felix, 257
biconditional elimination, 54, 58, 469
biconditional equivalence rule, 57, 67,
   470
biconditional introduction, 67, 94, 469
   biconditional introduction, contrapositive form, 69, 70, 469
biconditional sentence, 37, 38, 54
biconditional transitivity, 56, 469
bijection, *see* one-to-one-and-onto
   function
binary relation, 322–324, 329, 330, 332
binomial coefficients, 234, 235, 245
bipartite graph, 431, 440, 451, 460, 461,
   463, 466
block, 407–410, 412, 416, 417

© Springer Nature Switzerland AG 2019
C. Jongsma, *Introduction to Discrete Mathematics via Logic and Proof*,
Undergraduate Texts in Mathematics,
https://doi.org/10.1007/978-3-030-25358-5